AIM Airman's Information Manual '93

Edited and Updated by
Aviation Supplies & Academics, Inc.

Department
of Transportation

- **U.S. Customs Guide for Private Flyers**
- **Pilot/Controller Glossary**
- **Cross-Referenced for Private, Commercial, Instrument, CFI, CFII, and ATP Written Exams**
- **Basic Flight Information and ATC Procedures**
- **Recent Changes Clearly Identified**

Aviation Supplies & Academics, Inc.
7005 132nd Place SE
Renton, WA 98059-3153

AIM (Airman's Information Manual)

This publication contains current
regulations as of October 15, 1992.

ISBN 1-56027-145-0
ASA-93-AIM

Published by
Aviation Supplies & Academics, Inc.
7005 132nd Place SE
Renton, WA 98059-3153

Printed in the United States of America

Contents

Airman's Information Manual (AIM)

Exam Cross-Reference Index

Applicants preparing for written exams should have full knowledge of the following applicable AIM paragraphs:

Recreational / Private / Commercial / CFI

AIM ¶	AIM ¶	AIM ¶	AIM ¶	AIM ¶	AIM ¶	AIM ¶
1-2	3-1	3-32	4-11	4-54	5-13	7-43
1-3	3-2	3-41	4-12	4-55	6-1	7-44
1-4	3-11	3-42	4-13	4-56	6-2	7-45
1-5	3-12	3-43	4-15	4-63	6-11	7-46
1-6	3-14	3-44	4-18	4-64	6-12	7-48
1-7	3-15	3-45	4-31	4-66	6-13	7-49
1-14	3-21	3-46	4-32	4-68	6-15	8-1
1-19	3-22	3-47	4-33	4-70	6-32	8-2
1-31	3-23	3-61	4-34	4-74	7-6	8-3
1-32	3-24	3-62	4-36	4-81	7-14	8-4
2-2	3-25	4-1	4-42	4-85	7-15	8-5
2-6	3-26	4-2	4-43	4-89	7-16	8-6
2-7	3-27	4-3	4-44	5-1	7-26	8-8
2-11	3-28	4-7	4-51	5-3	7-32	
2-13	3-30	4-8	4-52	5-4	7-41	
2-21	3-31	4-9	4-53	5-11	7-42	

135 / Airline Transport Pilot

AIM ¶	AIM ¶	AIM ¶	AIM ¶	AIM ¶	AIM ¶	AIM ¶
1-1	2-2	3-44	4-15	4-93	5-37	5-60
1-2	2-4	3-45	4-16	5-1	5-41	5-91
1-4	2-5	3-46	4-18	thru	5-43	5-92
1-8	2-21	3-61	4-56	5-13	5-44	6-13
1-9	3-14	3-62	4-60	5-21	5-46	6-21
1-10	3-23	3-64	4-72	5-23	5-47	thru
1-11	3-24	4-2	4-83	5-25	5-50	6-25
1-14	3-25	4-3	4-84	5-26	5-52	6-31
1-19	3-26	4-7	4-85	5-31	5-55	7-1
1-31	3-27	4-9	4-87	5-32	5-56	thru
1-33	3-28	4-11	4-90	5-34	5-57	7-83
1-34	3-42	4-12	4-91	5-35	5-58	
2-1	3-43	4-13	4-92	5-36	5-59	

Instrument-Airplane and Flight Instructor-Instrument on next page.

Exam Cross-Reference *Continued*

Instrument-Airplane / Flight Instructor-Instrument

AIM ¶	AIM ¶	AIM ¶	AIM ¶	AIM ¶	AIM ¶	AIM ¶	AIM ¶
1-1	2-12	3-45	4-53	5-2	5-45	5-80	7-24
1-2	2-13	3-46	4-54	5-3	5-46	5-81	7-25
1-3	2-21	3-47	4-62	5-6	5-47	5-82	7-26
1-4	3-1	3-61	4-64	5-7	5-48	5-83	7-31
1-5	3-2	3-62	4-68	5-9	5-49	5-84	7-32
1-6	3-11	4-1	4-69	5-10	5-50	6-1	7-33
1-7	3-13	4-2	4-70	5-12	5-51	6-2	7-41
1-8	3-14	4-3	4-71	5-13	5-52	6-12	7-42
1-9	3-15	4-5	4-72	5-21	5-53	6-21	7-43
1-10	3-21	4-8	4-73	5-23	5-55	6-22	7-44
1-11	3-22	4-9	4-81	5-24	5-56	6-24	7-45
1-13	3-23	4-11	4-82	5-25	5-57	6-31	7-46
1-31	3-24	4-12	4-83	5-26	5-58	6-32	7-47
1-32	3-25	4-13	4-84	5-31	5-59	6-33	7-49
1-33	3-26	4-14	4-85	5-32	5-60	7-1	8-1
1-34	3-27	4-18	4-86	5-33	5-71	7-2	8-2
2-1	3-28	4-31	4-87	5-34	5-72	7-12	8-3
2-2	3-29	4-32	4-89	5-35	5-73	7-13	8-4
2-3	3-31	4-33	4-90	5-36	5-74	7-15	8-5
2-4	3-32	4-34	4-91	5-37	5-75	7-16	8-6
2-5	3-41	4-36	4-92	5-41	5-76	7-17	8-8
2-6	3-42	4-43	4-93	5-42	5-77	7-18	
2-7	3-43	4-51	4-94	5-43	5-78	7-19	
2-11	3-44	4-52	5-1	5-44	5-79	7-20	

U.S. Customs Guide for Private Flyers
(General Aviation Pilots)

CONTENTS

Introduction

Customs Guide for Private Flyers is for you—the private, corporate, and charter pilot—on business or pleasure flights to and from foreign countries. It sets forth the basic Customs requirements, provides a list of airports at which Customs clearance may be obtained and explains overtime charges.

You can facilitate your air travel if you know Customs regulations and follow them. Additional valuable information on regulations concerning international flights is available in the International Flight Information Manual published by the Federal Aviation Administration (FAA) and sold by the Superintendent of Documents, U.S. Government Printing Office, Washington, D.C. 20402.

If you are in doubt, don't guess. Ask your local Customs officer or the Customs officer at the airport of your intended return to the United States.

Happy Landing!

The Narcotics Smuggling Threat

The United States Customs Service is requesting the assistance of Fixed Base Operators, Fuel Service Facilities, Aircraft Brokers, Aircraft Charter Services, Airport Security, pilots, crew, airport personnel, and others associated with general aviation activities.

Over the past several years, Puerto Rico, Florida, Louisiana, Texas, Arizona, New Mexico, and California have been inundated by persons and/or organizations that are responsible for smuggling large amounts of narcotics into the United States by air. In an attempt to identify these individuals and/or groups, we are asking for your assistance. Through the use of the following indicators, it is believed that you can assist us in deterring the smuggling of narcotics into the United States.

Remember these smugglers endanger you as the general aviation community and your family as the public community. They tarnish the highly thought of general aviation community name.

The basic indicators that you should be cognizant of are as follows (It should be noted these are only guidelines and may vary with individuals. The presence of one or more of these indicators does not necessarily mean that you have encountered an air smuggler):

1. Passenger seats removed from the aircraft.

2. Carrying gas cans inside the aircraft or persons without an aircraft buying large amounts of aviation gas in containers.

3. Carrying numerous cardboard boxes, duffle bags, plastic bags, etc., inside the aircraft, or seeds, green vegetable matter, fragments of various colored butcher or cellophane paper indicating possible marijuana debris visible inside the aircraft; tape markings or residue around aircraft tail number.

4. Maps or other evidence of flights to Mexico, the Caribbean, Central, or South America present in aircraft when the pilot avoids reference to such flights, or a pilot requesting maps or information pertaining to areas in Mexico, the Caribbean, Central, or South America when it appears he is not going to follow official procedures for such trips.

5. Strong odors from the aircraft (perfumes and deodorizers are often used to disguise the odor of marijuana and cocaine).

6. FAA registration numbers on the aircraft which appear to be incomplete, crooked, or to have otherwise been altered or concealed.

7. Muddy wheels, dirty or dusty aircraft, beat-up props, pitted undercarriage or other evidence of having landed on unpaved strips, fields, or in sand, etc.

8. Vans, panel trucks or campers meeting the aircraft at an isolated location on the field.

9. Pilot or passengers reluctant to leave the immediate area of the aircraft or to allow others close to the aircraft during refueling or servicing.

10. Payment of cash for fuel or services or display of large amounts of cash by the pilot or passengers.

11. Persons who on aircraft rental applications list themselves as being self employed operating out of their residences.

12. Persons who rent hangars for one month or similarly short term basis, particularly when paying cash in advance and giving minimal information.

13. Pilots who own or operate expensive aircraft with no visible means of support, or any other factors that might lead you to believe the individual might be trafficking in narcotics.

14. Pilots reluctant to discuss destination or point of origin, or reluctant to discuss any of the above listed conditions.

Any of the aforementioned points, when coupled with other suspicious behavior by the aircraft operator or occupants, may possibly indicate the aircraft is being used in an illegal activity.

When the above conditions are observed and you feel there is a possibility these individuals might be engaged in smuggling, please note any information regarding the identity of the pilot(s), other occupants, aircraft, description and license numbers of vehicles. Under no circumstances should you take any direct action on your own. Immediately, or as soon as it can be done safely, notify U.S. Customs. Use the national 1–800–BE–ALERT number or contact the nearest Customs Office of Enforcement. All information received will be held in confidence.

If, as a result of your information, the United States Customs Service seizes and forfeits currency or monetary instruments, a reward of up to $150,000 can be paid. In the case of vessel, aircraft, vehicle or other property, a reward of up to $250,000 can be paid for information.

The Customs Service realizes that the vast majority of private aircraft operators and crewmembers are honest and have no intention of violating our laws. Nevertheless, our officers must sometimes ask a private aircraft operator to have his aircraft undergo a thorough Customs search, possibly including the removal of panels and the opening of compartments.

Ordinarily, the Customs officers who question an aircraft operator who has arrived from foreign or who appears to have arrived from foreign will be in full Customs uniform and will be wearing a Customs badge. This is always the case with Customs inspectors who are stationed at airports for the purpose of inspecting and clearing reported arrivals from and departures to foreign countries.

Customs also has enforcement functions and conducts special enforcement operations during which the wearing of a Customs uniform would be self-defeating; the officers may be observing suspicious activities, for example. Should any Customs officer approach a citizen for questioning, inspection or search, the officer must identify himself or herself, including providing the citizen with his or her badge number on request. This is especially crucial for Customs officers who are not in uniform. Customs officers, if they believe that they may be harassed by the citizen because of the questioning, do not have to provide their names. In any case, a Customs officer's badge number is sufficient for our Service to determine his or her identity should an investigation be warranted.

In any instance when a person claiming to be a Customs officer does not clearly and fully identify himself (by being in Customs uniform, by providing his badge number upon request, and by showing the citizen his badge when not in uniform), the citizen should ask to speak with the officer's supervisor. If a supervisor is not available, the citizen should contact the nearest Customs office or the District Director of Customs having jurisdiction over the location where the incident occurred. This procedure also is appropriate whenever a Customs officer has properly identified himself, but the citizen believes he was treated with less than courtesy, professionalism, or fairness.

Should the citizen believe that the person who is questioning him is impersonating a Customs officer, he should cooperate to a reasonable extent for purposes of safety and then contact the nearest Customs office or other law enforcement agency.

Please be assured that the Customs Service is fully aware that the effective enforcement of our laws depends as much upon the cooperation of honest citizens as it does upon our own enforcement efforts.

Scope and Definitions

1. The information in this pamphlet applies to private aircraft of both United States and foreign registry.

2. Customs requirements vary according to whether an aircraft is operating as a commercial or private flight. It is the nature of each particular flight that determines whether an aircraft is operating in a private or commercial capacity. The owners, aircraft type, or predominant usage of the aircraft has little bearing on this determination. In fact, **many corporate and business aircraft typically operate as "private aircraft" for Customs purposes.**

   Aircraft not qualifying as private aircraft must comply with the applicable entry and clearance requirements for commercial aircraft as specified in Part 6, Customs Regulation (19 CFR Part 6). For additional information on these requirements, please contact the nearest U.S. Customs office.

3. For Customs purposes, a "private" aircraft is any civilian aircraft not being used to transport persons or property for compensation or hire. A *"commercial"* aircraft is any civilian being used in the transportation of persons or property for compensation or hire.

   The term "person transported for compensation or hire" means a person who would not be transported unless there was some payment or other consideration, including monetary or services rendered, by or for the person and who is not connected with the operation of the aircraft or its navigation, ownership or business. The major criterion for determining whether an aircraft is private or commercial will be the use of the aircraft on a particular flight, and this determination would be the same if the owner or lessee is a corporation, partnership, sole proprietorship or an individual.

   An aircraft will be presumed not to be carrying persons or merchandise for hire, and thus will be a private aircraft for Customs purposes, when the aircraft is transporting only the aircraft owner's employees, invited guests, or the aircraft owner's own property. This presumption may be overcome by evidence that the employees, "guests", or property are being transported for compensation or other consideration. If an aircraft is used by a group of individuals, one of whom is the pilot making the flight for his own convenience, and all persons aboard the aircraft including the pilot contribute equally toward payment of the expense of operating the aircraft owned or rented by them, the aircraft would be considered private.

Usually those aircraft arriving in the United States which have raised the question as to whether they are private or commercial are operated by the owner or lessee. For purposes of determining if an aircraft is private or commercial, the lessee of an aircraft will be considered as its owner if the pilot is a regular employee of the lessee and the lessee has complete control of the aircraft and its itinerary. Accordingly, aircraft operated by the owner or by a lessee as described in the preceding sentence will be classified as private or commercial depending on the use of the aircraft on a particular flight, that is, whether the aircraft is being used in the transportation of persons or property for compensation or hire. On the other hand, a leased aircraft will be considered commercial for Customs purposes if the pilot and/or crew are part of the leasing arrangement of the aircraft.

An amendment to Part 6.14 of the Customs Regulations, effective December 31, 1984, redefined certain air taxi and air charter flights as "private"-*for Customs reporting purpose only* Air charter or air taxi flights are now subject to Customs reporting requirements for private aircraft if they have a passenger seating capacity of 30 or less and they have a payload of 7,500 pounds or less. Therefore, the affected aircraft now must land at one of Customs 26 designated southern border airports when arriving in the United States by crossing our southern borders. For additional information on this subject see "Special Reporting Requirements" in the "Inward Flights" section of this booklet.

It must be emphasized that although considered "private" for Customs reporting purposes, these smaller air taxi and air charter flights are still commercial for other than reporting purposes. This means that these aircraft may be subject to Customs processing fees for commercial flights rather than the fees for private aircraft ($25 limit). The air taxi and air charter flights must also provide Customs, upon arrival into the United States, with the normal commercial entry documents (e.g., the General Manifest). Further, Federal Aviation Administration designations, as part 91 (private) or part 135 (commercial) aircraft, need not necessarily coincide with Customs designations. Thus a given flight might be considered private under FAA regulations, commercial for Customs entrance and clearance purposes, and private for Customs reporting purposes.

4. The term "international airport" means any airport designated by the Secretary of the Treasury or the Commissioner of Customs as a port of entry for civil aircraft arriving in the United States from any place outside thereof and for cargo carried on such aircraft. It is also designated by the Attorney General as a port of entry for certain classes of aliens arriving on such aircraft, and by the Secretary of Health and Human Services as a place for quarantine inspection. This pamphlet lists all such airports that have been officially designated to date.

   (Note: Frequently the word "International" is included in the name of an airport for other than Customs purposes, in which case it has no special Customs meaning.)

5. The term "landing rights airport" means an airport at which permission to land may be granted by the appropriate Customs officer with acknowledgement of the Immigration and Naturalization Service, the Public Health Service, and the Animal and Plant Health Inspection Service of the Department of Agriculture. Such landing

rights are required before an aircraft may land at an airport which has not been designated for Customs purposes as an international airport.

6. The term "United States" for Customs purposes includes all 50 states, the District of Columbia, and Puerto Rico. In some circumstances aircraft from Puerto Rico will have to provide Customs with an advance notice of penetration of U.S. airspace. See "Special Reporting Requirements" in the "Inward Flight" section of this booklet for further details. For information regarding U.S. possessions refer to the FAA International Flight Information Manual.

Outward Flights

advance notice of arrival in Canada and Mexico

Many countries, including Canada and Mexico, require advance notice of the intent of pilots to arrive in those countries.

Under agreements between the United States, Canada, and Mexico, operators of private planes may, in most cases, include this advance notice in a flight plan to be filed prior to departure from the United States at the nearest FAA Flight Service Station. That station will then cause the message to be transmitted to the proper authorities in the country of destination. Contact the nearest FAA Flight Service Station regarding this procedure.

Aircraft carrying passengers or cargo for hire or compensation on either the outbound or inbound portions of flights are not considered to be "private aircraft" for Customs purposes and must comply with the clearance requirements specified for commercial aircraft as set forth in Part 6, Customs Regulations. Aircraft leaving the U.S. for a foreign destination may be subjected to a search by Customs Officers.

Inward Flights

providing notification to Customs

In order to have an officer present to provide Customs service for you and your aircraft, Customs must be notified of your intention to land and time of arrival

This notification may be provided to Customs by telephone, radio, or other means, or may be furnished by means of an ADCUS (advise Customs) message in the flight plan through the Federal Aviation Administration to Customs. The term ADCUS should be clearly identifiable in the "Remarks" section of your flight plan if you wish to have Customs notified. This procedure, however, entails the relaying of information and is not as timely or reliable as direct communication (telephone, radio, or other means). It is recommended that if possible, pilots attempt to communicate directly with Customs by telephone or other means to insure that an officer will be available at the time requested. **It is the ultimate responsibility of the pilot to insure Customs is properly notified, and the failure to do so may subject the pilot to penalty action (see "Penalties for Violations").** The last section of this guide lists telephone numbers to facilitate this notification process.

Special reporting requirements—(Southern border, Pacific, Gulf of Mexico and Atlantic Coastlines)

The U.S. Customs Service has identified general aviation aircraft as the highest-risk vehicles for narcotics smuggling. This is especially true for aircraft flying from areas south of the U.S. Therefore, the following special reporting procedures have been instituted.

All private aircraft arriving in the United States via the United States/Mexican border or the Pacific Coast from a foreign place in the Western Hemisphere south of 33 degrees north latitude shall furnish a notice of intended arrival to the Customs Service at the *nearest* designated airport to the point of first border crossing listed below. All private aircraft arriving in the United States via the Gulf of Mexico, and Atlantic coast from a foreign place in the Western Hemisphere south of 30 degrees north latitude, from any place in Mexico, or from the U.S. Virgin Islands, shall furnish a notice of intended arrival to the Customs Service at the nearest designated airport to the point of crossing listed below. They then must land at this airport for inspection, unless they have an overflight exemption. Landing rights must be obtained from Customs to land at designated airports that are *not* also approved as international airports. See the individual airport listings in this booklet for further information on this subject. The requirement to furnish a notice of intended arrival shall *not* apply to private aircraft departing from Puerto Rico and conducting their flights under instrument flight rules (IFR) until crossing the United States coastline or proceeding north of 30 degrees north latitude. The notice must be furnished at least one hour before crossing the United States coastline or border. The notice may be furnished directly to Customs by telephone, radio, or other means, or may be furnished by means of an ADCUS message in the flight plan through the Federal Aviation Administration to Customs. This notice will be valid as long as actual arrival is within 15 minutes of the original ETA. If you cannot land within 15 minutes of the original ETA, you must give Customs a new notice. Customs and the FAA will accept these notices up to 23 hours in advance.

A penetration notice (but not landing) requirement also exists for private aircraft arriving in the continental United States from Puerto Rico which are *not* conducting their flight in accord with Instrument Flight Rules, and those private aircraft which have flown beyond the inner boundary of the Air Defense Identification Zone south of 30 degrees north latitude on the Atlantic Coast, beyond the inner boundary of the Gulf Coast ADIZ, south of the United States/Mexican border, or beyond the inner boundary of the Pacific Coast ADIZ south of 33 degrees north latitude and *have not landed in a foreign place.* This notice requirement may be satisfied by either filing a flight plan with the FAA or contacting Customs directly at least one hour prior to the inbound crossing of the U.S. border or coastline. Customs may be contacted by telephone at (305) 536–6591 in Miami, Florida, and at (714) 351–6674 in Riverside, California. If you cannot cross within 15 minutes of the original estimated time, you must give Customs a new notice. Customs and the FAA will accept these notices up to 23 hours in advance.

The notice to Customs required by this section shall include the following:

(1) Aircraft registration number;
(2) Name of aircraft commander;
(3) Number of United States citizen passengers;
(4) Number of alien passengers;
(5) Place of last departure;
(6) Estimated time and location of crossing United States border/coastline;
(7) Name of United States airport of first landing;
(8) Estimated time of arrival.

Private aircraft arriving from a foreign place are required to furnish a notice of intended arrival in compliance with these special reporting requirements and must land for Customs processing at the *nearest* designated airport to the border or coastline crossing point as listed below, unless exempted (see below) from this requirement. In addition to the requirements of this paragraph, private aircraft commanders must comply with all other landing and notice of arrival requirements. This landing requirement shall *not* apply to private aircraft which have *not landed in a foreign place* or are arriving directly from Puerto Rico or if inspected (precleared) by Customs Officers in the U.S. Virgin Islands.

Location	Name
Beaumont, TX	Jefferson County Airport
Brownsville, TX	*Brownsville International Airport
Calexico, CA	*Calexico International Airport
Corpus Christi, TX	Corpus Christi Int'l Airport
Del Rio, TX	*Del Rio International Airport
Douglas, AZ	*Bisbee-Douglas Int'l Airport
Eagle Pass, TX	*Eagle Pass Municipal Airport
El Paso, TX	*El Paso International Airport
Fort Lauderdale, FL	Fort Lauderdale Executive Airport
Fort Lauderdale, FL	Fort Lauderdale-Hollywood Airport
Fort Pierce, FL	St. Lucie County Airport
Houston, TX	William P. Hobby Airport
Key West, FL	Key West International Airport
Laredo, TX	*Laredo International Airport
McAllen, TX	*Miller International Airport
Miami, FL	Miami International Airport
Miami, FL	Opa-Locka Airport
New Orleans, LA	New Orleans Int'l AP (Moisant Field)
New Orleans, LA	New Orleans Lakefront Airport
Nogales, AZ	*Nogales International Airport
Presidio, TX	*Presidio-Lely Int'l Airport
San Diego, CA	*Brown Field
Tampa, FL	Tampa International Airport
Tucson, AZ	Tucson International Airport
West Palm Beach, FL	Palm Beach International Airport
Yuma, AZ	Yuma International Airport

The asterisk (*) denotes that 24-hour free service is provided Monday through Saturday.

exemption from special landing requirements

Applicants for overflight exemption request the privilege of being exempt from one specific provision of the U.S. Customs Regulations. If approved, the applicant is bound to comply with all other requirements, including advance notice of penetration to U.S. Customs at least one hour in advance of crossing the border or coastline, furnishing advance notice of arrival at the intended airport of landing, etc.

The U.S. Customs Service district offices have been instructed to review carefully all requests for overflight exemption. The two main concerns are completeness and accuracy of the specific required information and the demonstration by the applicant of genuine need for this privilege. Control of aircraft arrivals in the United States is of paramount importance if the Customs Service is to successfully combat the flow of illegal drugs and contraband. In order for an exemption to be granted to an individual or corporation, the deciding official must be reasonably assured that the party requesting exemption will not compromise the mission of the U.S. Customs Service.

The information contained in your application is reviewed and background checks are performed in conjunction with law enforcement agencies. Some elements of the application are required by law, others are strictly voluntary. Processing of background checks may be a lengthy and prolonged procedure. By providing information which is voluntary, e.g. social security numbers, you will substantially reduce the time required to act on your application. A limited number of "identifiers" for people or corporations can not only create a cloud of uncertainty but delay background reports forwarded to the deciding official.

Your application should be addressed to the District Director of U.S. Customs having jurisdiction over the airport you will utilize most frequently when arriving from points south of the U.S. You should request exemption for either a single specific flight or term (one year) approval. Applications for a single overflight exemption must be received at least fifteen days in advance; for term exemption at least thirty days in advance.

Air charters or taxi service cannot be granted an unqualified term exemption since they cannot reasonably comply with the requirements of a term application, namely, comprehensive details of the passengers they will transport in the course of one year. By submitting all other details, air charters/taxis will accrue the benefit of "conditional" approval. This approval is "conditional" because the operator must receive the concurrence of the District Director prior to each trip. Concurrence will be based upon factors such as the foreign point of departure to the U.S. and the passengers being transported. The benefit realized by the charter/taxi operator is that the time constraints listed above for timely submission of single overflight exemptions can be drastically reduced. Local Customs districts will establish minimum time frames in accordance with their own requirements.

Required elements of any overflight exemption include the following:

—Aircraft registration number and serial number.

—Identify of the aircraft (make, model, color scheme, and type, such as turboprop, etc.)

—A statement that the aircraft is equipped with a functioning mode C (altitude reporting) transponder which will be in use during the overflight.

—A statement that the aircraft is capable of flying above 12,500 feet and that it will be operated at such an altitude when utilizing the overflight exemption unless ordered to fly at a lower altitude by FAA air traffic controllers.

—Names, home addresses, social security numbers (optional), and dates of birth of owners of the aircraft (if the aircraft is being operated under a lease, the name and address of the lessee, in addition to that of the owner).

—Names, home addresses, social security numbers (optional), dates of birth, and any FAA certificate numbers of all crewmembers that the applicant wishes to have approved.

Individual applications from each crewmember must also be attached and should take the form of a signed letter from the crewmember in question.

The applicant must verify the accuracy of the information provided by the crewmember to the best of his/her ability. The application must contain a statement to this effect.

—Names, home addresses, social security numbers (optional), and dates of birth of usual and potential passengers to the greatest extent possible. **An approved passenger must be on board to utilize the overflight exemption.**

—Description of usual or anticipated cargo or baggage.

—Description of the company's usual business activity, if the aircraft is company owned.

—Name of intended airport(s) of first landing in the U.S. (The overflight exemption will only be valid to fly to airports pre-approved by Customs).

—Foreign place(s) from which the flight(s) will originate.

—Reason for the request of overflight exemption.

Your information should be as complete and accurate as possible, and should be specific rather than generalized. The following points will assist you in preparing an acceptable application:

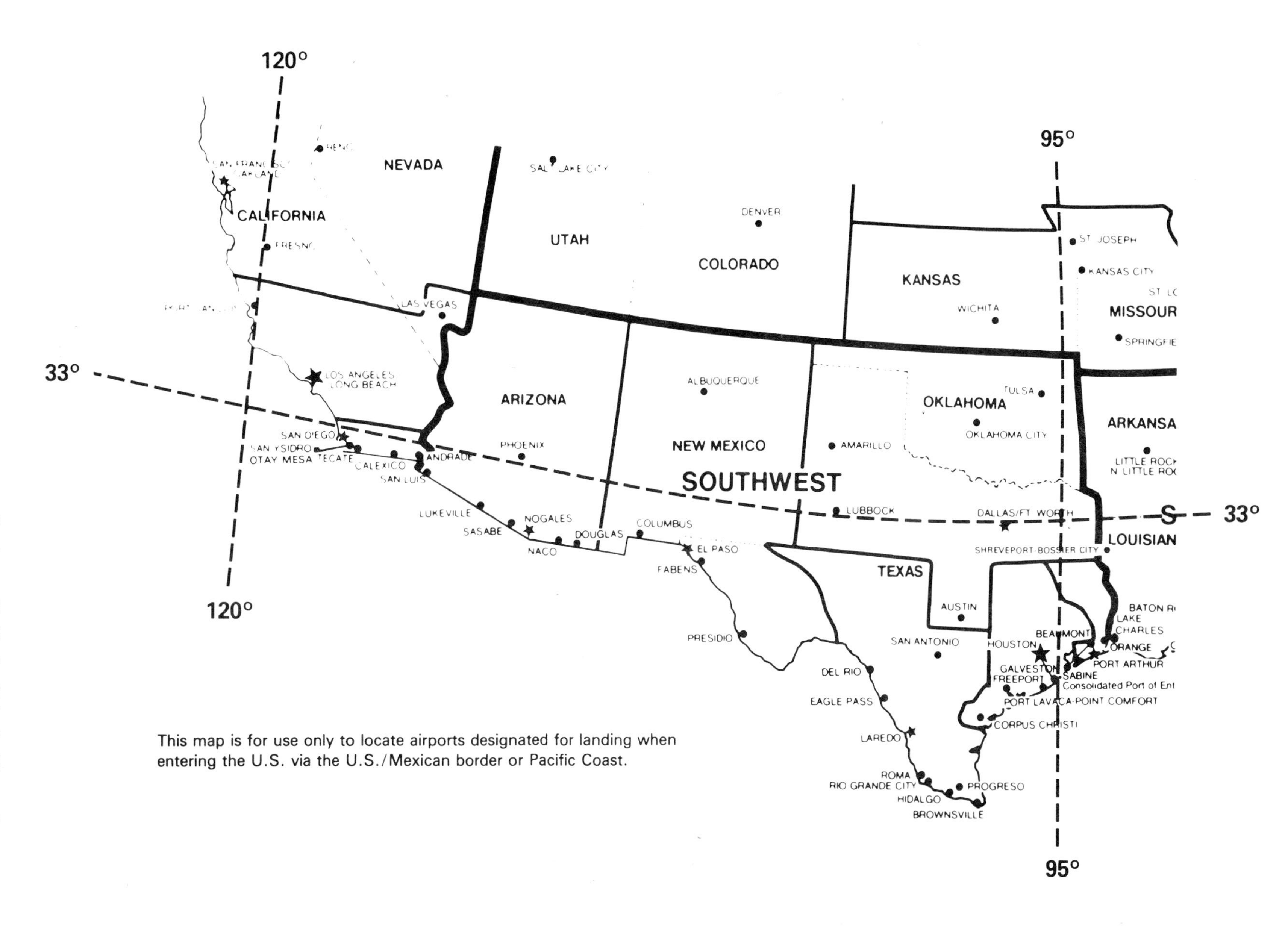

This map is for use only to locate airports designated for landing when entering the U.S. via the U.S./Mexican border or Pacific Coast.

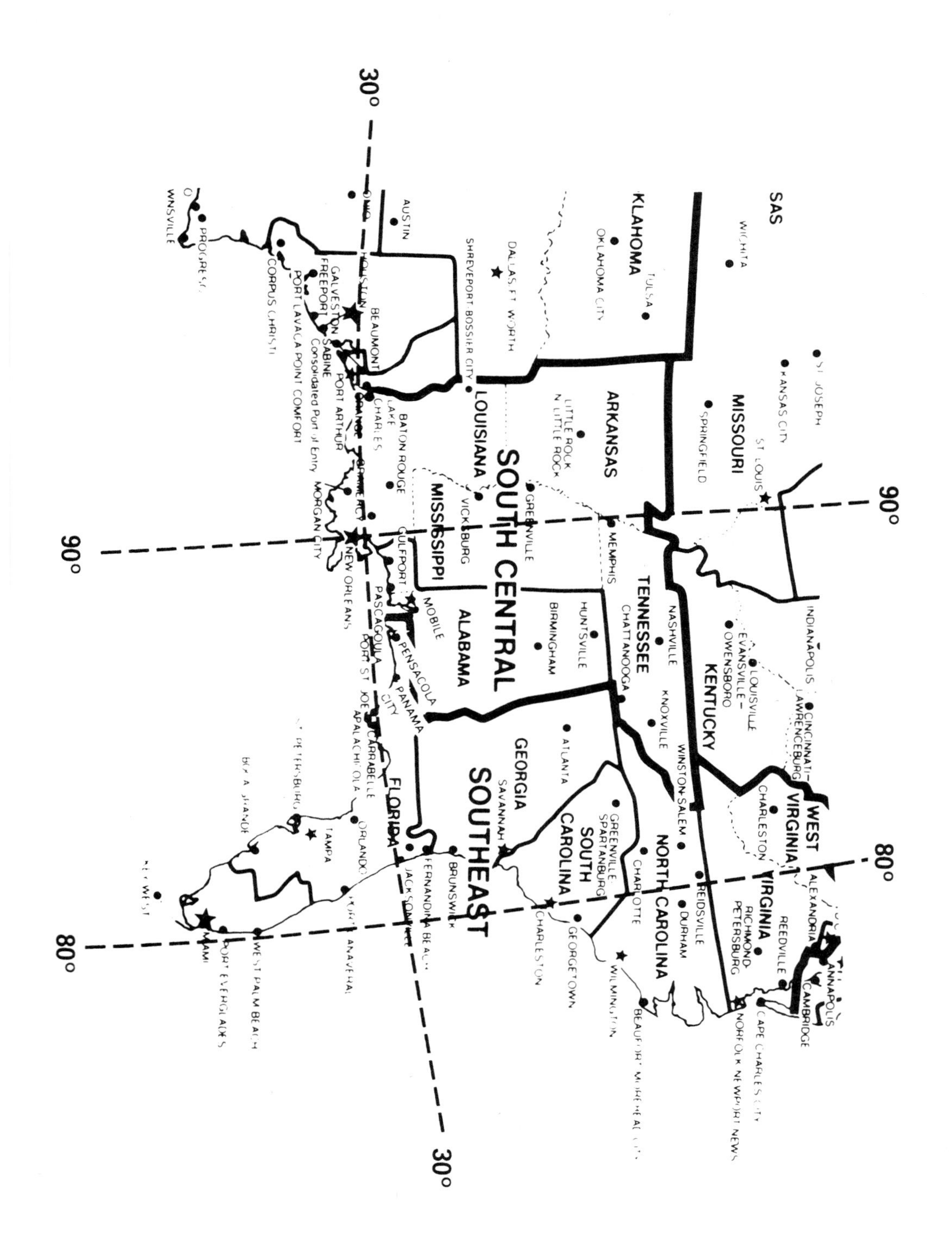

This map is for use only to locate airports designated for landing when entering the U.S. via the Gulf of Mexico and Atlantic Coast, from any place in Mexico or the U.S.V.I.

—Include all potential crewmembers which might be present on the aircraft during the term of the desired exemption. **In order for overflight exemptions to remain valid, at least one crewmember listed on your application must be present on every flight.** It is to your advantage to make your listing as comprehensive as possible.

—Provide as many "identifiers" as possible for all crew members and passengers. Social security numbers, passport numbers, aircraft pilot license numbers, etc., will contribute greatly to expediting background investigations.

—Describe the type of business your corporation is engaged in. If the corporation which owns the aircraft is merely an air transportation service for the benefit of an affiliated company, please provide details.

—List the foreign cities and countries your aircraft will visit. It is to your advantage to describe the nature of your business in each, or to indicate that certain destinations are vacation/entertainment locations.

—The reason for overflight exemption requests should be as tangible and concretely stated as possible. Estimate the costs incurred by making an extra landing at a "designated airport" (fuel, wear on aircraft components, landing fees, additional time/distance). Your flight crew or aviation references can advise you of the above and others, e.g. safety factors.

—Provide an estimate of the number of nautical flying miles which will be saved on an annual basis if the exemption is granted. Determine the distance from the last foreign point(s) of departure to the inland airport where the aircraft will be inspected. Then calculate the flying distance from the same point of departure to the same destination with an intermediate stop at the most convenient (i.e., closest to a straight line route) designated airport.

Subtracting the mileage for the direct flight from the mileage for the flight that stops at the designated airport will yield the mileage saved on a given flight.

—Companies involved in air ambulance type operations may be granted a single overflight exemption when emergency situations arise, as well as in the case of non-emergency transport for individuals seeking medical treatment. Both U.S. and foreign registered aircraft will be eligible for the special exemption. The applicant must provide all the necessary information normally required for an overflight exemption. Customs should be notified at least 24 hours prior to departure. If this cannot be accomplished, Customs will allow receipt of the overflight exemption application up to departure time, as well as in flight through a Federal Aviation Administration Flight Service Station.

Applicants should be aware that the processing of term applications requires time for all background reports to be prepared for the deciding official. Incomplete applications will not be processed and the applicant will be notified of additional information which must be supplied.

Should your application for overflight be denied at the district level, an appeal process is available. Letters of denial will include the name and address of the Regional Commissioner of Customs responsible for the district in which your application was denied. You may petition the Regional Commissioner for reconsideration of your request.

PRECLEARANCE FROM THE VIRGIN ISLANDS

Private aircraft departing the U.S. Virgin Islands destined for the mainland or Puerto Rico may request preclearance service at St. Croix and St. Thomas. This service will be provided 0800 to 1700 hours, Monday through Sunday, workload permitting. Preclearance outside of these hours may be authorized at the discretion of the District Director on a case-by-case basis. All private aircraft pilots desiring preclearance to the U.S. are required to make advance arrangements for clearance directly with Customs or through the Federal Aviation Administration by radio or other means. FAA will notify Customs of all preclearance requests.

At time of preclearance, you will complete a General Declaration (CF-7507) which will be authenticated by the inspecting officer as evidence of your having been precleared.

If for any reason you find that you cannot proceed directly to the U.S. (must stop for refueling or any other reason), your preclearance becomes invalid and you must report to Customs on your arrival in the U.S. or Puerto Rico.

Use of the General Declaration, although not required for general aviation aircraft unless carrying persons or cargo for hire, will serve as evidence of your having been precleared if challenged upon your arrival in the Customs territory. However, this does not preclude reinspection by Customs, although such reinspection would be rare.

entry and clearance—Cuba

Aircraft arriving from or departing for Cuba must land at or depart from Miami International Airport as provided by section 6.3a, Customs Regulations.

Upon arrival the pilot will present a manifest of all passengers on board to an officer of the United States Immigration Service or to a Customs officer acting as an Immigration Officer. No passenger arriving from Cuba by aircraft will be released by Customs, nor will the aircraft be cleared or permitted to depart, before the passenger is released by an Immigration officer or a Customs officer acting on behalf of that agency.

Aircraft proceeding to Cuba are required to have a validated license issued by the Department of Commerce or a license issued by the Department of State.

These special requirements do not apply to aircraft arriving from or departing to the U.S. Naval Base at Guantanamo Bay. Aircraft from this base must meet the same requirements as aircraft arriving from other Caribbean nations.

international airports

It is unnecessary for aircraft arriving at these airports to request permission to land or "landing rights" from Customs (see next page). However, an advance notice of the estimated time of arrival (in local time) is required to be transmitted to U.S. Customs for each flight. In general a 1-hour advance notice of arrival is sufficient, although you will note in the "Special Arrangements or Restrictions" column that a longer time is required at certain airports. Aircraft operators must bear in mind that this advance notice is predicated on the time that the Customs officer receives the notification, and not on the time that the flight plan or message is filed.

Except in the case of "Short Flights" (see below), requests to transmit such notices may be included in flight plans to be filed at certain airports in Canada and Mexico under agreements with those countries. Information concerning the availability of this service at United States airports has been included in the column "FAA Flight Plan Notification Available." If "Advise Customs" or "ADCUS" is not included on the flight plan, FAA will not advise Customs of intended arrival. **The pilot is ultimately responsible for insuring that Customs is properly notified.**

At those airports where the flight plan notification service is not available, notices of arrival must be transmitted directly to U.S. Customs.

For private aircraft arriving in the United States from certain foreign areas south of the United States, refer to "Special Reporting Requirements" under "Inward Flights".

International Airports

Location	Name
Akron, Ohio	Akron Municipal Airport
Albany, New York	Albany County Airport
Baudette, Minnesota	Baudette International Airport
Bellingham, Washington	Bellingham Int'l Airport
Brownsville, Texas	Brownsville, Texas
Burlington, Vermont	Burlington Int'l Airport
Calexico, California	Calexico International Airport
Caribou, Maine	Caribou International Airport
Chicago, Illinois	Midway Airport
Cleveland, Ohio	Cleveland Hopkins Int'l Airport
Cut Bank, Montana	Cut Bank Airport
Del Rio, Texas	Del Rio International Airport
Detroit, Michigan	Detroit City Airport
Do	Detroit Metropolitan Wayne County Airport
Douglas, Arizona	Bisbee Douglas Int'l Airport
Duluth, Minnesota	Duluth International Airport
Do	Sky Harbor Airport
Eagle Pass, Texas	Eagle Pass Int'l Airport
El Paso, Texas	El Paso International Airport
Fort Lauderdale, Florida	Fort Lauderdale-Hollywood International Airport
Friday Harbor, Washington	Friday Harbor Seaplane Base
Grand Forks, North Dakota	Grand Forks Int'l Airport
Great Falls, Montana	Great Falls Int'l Airport
Havre, Montana	Havre City-County Airport
Houlton, Maine	Houlton International Airport
Int'l Falls, Minnesota	Falls International Airport
Juneau, Alaska	Juneau Municipal Airport
Do	Juneau Harbor Seaplane Base
Ketchikan, Alaska	Ketchikan Harbor Seaplane Base
Key West, Florida	Key West International Airport
Laredo, Texas	Laredo International Airport
Massena, New York	Richards Field
McAllen, Texas	Miller International Airport
Miami, Florida	Chalk Seaplane Base
Do	Miami International Airport
Minot, North Dakota	Minot International Airport
Nogales, Arizona	Nogales International Airport
Ogdensburg, New York	Ogdensburg Harbor
Do	Ogdensburg International Airport
Oroville, Washington	Dorothy Scott Airport
Do	Dorothy Scott Seaplane Base
Pembina, North Dakota	Pembina International Airport
Portal, North Dakota	Portal Municipal Airport
Port Huron, Michigan	St. Clair County Int'l Airport
Port Townsend, Wash.	Jefferson County Int'l Airport
Ranier, Minnesota	Ranier Int'l Seaplane Base
Rochester, New York	Rochester-Monroe County Airport
Rouses Point, New York	Rouses Point Seaplane Base
San Diego, California	San Diego Int'l Airport

Sandusky, Ohio	Griffing-Sandusky Airport
Sault Ste. Marie, Mich.	Sault Ste. Marie City-County Airport
Seattle, Washington	King County Int'l Airport
Do	Lake Union Air Service (Seaplanes)
Tampa, Florida	Tampa International Airport
Tucson, Arizona	Tucson International Airport
Watertown, New York	Watertown New York International Airport
West Palm Beach, Florida	Palm Beach International Airport
Williston, North Dakota	Sloulin Field International Airport
Wrangell, Alaska	Wrangell Seaplane Base
Yuma, Arizona	Yuma International Airport

landing rights airports

In addition to advance notice of arrival as described under "International Airports," specific permission to land at a "Landing Rights" airport must be obtained in advance of contemplated use. Except in the case of "Short Flights" advance notice of arrival may be transmitted to Customs in flight plans where flight notification service is rendered. Such notices will be treated as application for permission to land although the pilot is still ultimately responsible for insuring that Customs is properly notified, and pilots should be aware that "landing rights" can be denied if inspection service cannot be provided. If the notification service is not available, pilots must submit applications for landing rights and provide arrival notices directly to U.S. Customs.

Customs officers may at their discretion grant blanket "Landing Rights" to persons to land at certain airports for a specified period of time, in which event advance notices of arrival will be the only requirement.

to other airports

Permission to land at airports other than those listed in this booklet may be obtained in some limited cases; however, advance arrangements (preferably in writing) must be made with the Customs office nearest the airport of intended arrival. Advance notice of arrival is required as usual. Pilots should be aware that mileage and per diem costs may be accrued in addition to any overtime charges if applicable.

what to report

Except at airports where the flight plan notification service is rendered, applications for landing rights and arrival notices shall specify:

Type of aircraft
Registration marks
Name of commander
Place of last departure
Other countries visited
Airport of arrival and FAA airport code designation
Number of alien passengers
Number of citizen passengers
Estimated time of arrival

The above requirements do not apply to private aircraft arriving in the United States from certain foreign areas south of the United States. (See "Special Reporting Requirements" under "Inward Flights")

short flights

If flying time from the foreign airport to the U.S. airport is less than one hour, the pilot should request his application for landing rights, when required, and transmit his arrival notice directly to the U.S. Customs office before departure from the foreign airport (unless prior arrangements were made). This is necessary to allow inspectional personnel to be assigned and at the airport prior to arrival of the aircraft. FAA cannot guarantee delivery of a message (under the flight plan notification arrangement) in sufficient time on such short notice. **It is still the pilot's responsibility to give timely notice even though a flight plan has been filed.**

For private aircraft arriving in the United States from certain foreign areas south of the United States, refer to "Special Reporting Requirements" under Inward Flights.

changing destination en route

Pilots may find it necessary or convenient to change the intended airport of destination en route; however, definite confirmation *from Customs* permitting a change must be received. Failure to obtain permission in advance to alter the airport of destination may result in Customs initiated penalty action.

documentation and examination on arrival

Private aircraft are required to report directly to Customs for inspection immediately upon arrival. Normally a Customs office (or an officer from the Immigration and Naturalization Service or Department of Agriculture) will be present if a pilot has given proper advance notice of arrival. Should no inspecting officer be present, the pilot should report his arrival to Customs by telephone or most convenient means. He should keep the aircraft, passengers, crewmembers, baggage, food and cargo intact in a segregated place until the officer arrives or until he has received special instructions from a Customs Officer.

The pilot should provide the necessary information to assist the inspecting officer in the preparation of the required documentation unless the pilot has prepared the necessary documents in advance. Customs Form 178 must be completed for all private aircraft arrivals. This form is obtainable from Customs.

The pilot may also be requested to produce for inspection a valid airman's certificate, medical certificate (14 CFR 61.3) and the aircraft registration certificate.

Crew and passenger baggage will be examined in the same manner as that of other international travelers. A verbal declaration of articles acquired abroad will suffice, except that a written declaration, Customs Form 6059–B (or appropriate substitute), shall be presented when duty is to be collected or when the inspecting officer deems a written declaration necessary. Noncommercial cargo and unaccompanied baggage carried on board private aircraft shall be accounted for on a baggage declaration (CF 6059–B) prepared by the pilot in command, and appropriate entry for same shall be required. Customs officers will furnish the necessary forms. In addition the inspecting officer may require that baggage and cargo be removed from the aircraft for inspection, and he may physically inspect the aircraft. It is the responsibility of the pilot to assist in opening baggage and compartments. Commercial aircraft operators may have to employ the services of a certified mechanic in the event of intensive examination.

personal exemptions

Persons engaged in the operation of a private aircraft are not considered crewmembers for tariff purposes, and they are to be treated as returning residents for exemption purposes (19 CFR 148.61). Charter aircraft crews may be considered crewmembers for exemptions and the importation of typical tourist items, refer to the Customs pamphlets *Know Before You Go,* and *Customs Hints for Visitors (Nonresident).*

Immigration Requirements for Private Aircraft

Persons arriving from a foreign place (including Canada) must report for immigration inspection immediately upon arrival in the United States.

With few exceptions, all persons who are not U.S. citizens or permanent resident aliens are required to present a visa upon entering the U.S. Visas can be obtained only at a U.S. Consulate or Embassy abroad. Canadian citizens are exempt from visa requirements in most cases.

U.S. and Canadian citizens are advised to carry proof of citizenship which may include a passport, birth certificate, naturalization papers, or other documents that support your claimed citizenship. A driver's license alone is not proof of citizenship. Most aliens are required to present a valid unexpired passport. If you are in doubt as to what documents are required for entry, call the nearest U.S. Consulate, Embassy, or U.S. Immigration Office.

Most alien passengers must execute and present Form I–94, Arrival/Departure Record. INS has revised Form I–94, effective March 1, 1986, and prior editions may not be used. Form I–94 must be completed by all persons except U.S. citizens, returning resident aliens, aliens with immigrant visas, and Canadians visiting or in transit. Mexican nationals in possession of Immigration Form I–86 or Form I–586, are exempted from Form I–94 reporting requirements when their itinerary is limited to California, Arizona, New Mexico or Texas, and will not exceed 72 hours in duration. This exemption does not apply when travel will exceed 25 miles from the international border between Mexico and the U.S. Travel to Nevada by Mexican nationals is exempted for periods of the less than 30 days. Mexican nationals proceeding to destinations more than 25 miles from the border in these states will have to obtain a visitors permit I–444 when arriving in the U.S. Mexican nationals presenting official or diplomatic passports and destined to the U.S. for purposes other than permanent assignment are exempted from Form I–94 reporting requirements.

The form is available for sale by the Superintendent of Documents, Government Printing Office, Washington, D.C. 20402. Forms and printing specifications are also available through the INS Form I–94 Support Desk, Appalachian Computer Services, Inc., Highway 25 South, London, Kentucky 40241 (606–878–7900). The forms are available in English and in 13 foreign languages.

Individuals entering the U.S. by private aircraft should indicate PRIVATE in block #7—Airline and Flight Number. Individuals entering by private aircraft do not need to complete block #9—City Where You Boarded. All other items on the form are self-explanatory and should be completed prior to actual arrival in the U.S. The completed forms must be presented to either a U.S. Immigration or U.S. Customs inspector upon arrival.

Aircraft owners are responsible for the proper completion and submission of Form I–94 for all crew and passengers affected by the reporting requirement. Departure documents should be annotated on the reverse of the document to indicate Port of Departure and Date of Departure. Following Carrier, print the word PRIVATE. In the space provided for Flight Number/Ship Name, print the tail number of the aircraft. Departure documents should be submitted to a U.S. Immigration or U.S. Customs Inspector, at the time of departure from the U.S., or mailed to the Appalachian Computer Service address in London, Kentucky. Aircraft owners are responsible for the submission of all I–94 Departure Records upon departure to a foreign destination.

plant and animal quarantines of the U.S. Department of Agriculture

To prevent the entry of dangerous agriculture pest the following are restricted: fruits, vegetables, plants, plant products, soil, meats, meat products, birds, snails, and other live animals or animal products. Failure to declare all such items to a Customs/ Agriculture Officer can result in fines or other penalties. Agricultural items, whether in baggage, stores, lunches or garbage must be called to the attention of the inspecting officer. Prohibited items on board will be confiscated. Agricultural items should not be brought to the United States unless you are informed in advance by Agriculture inspectors of the Animal and Plant Health Inspection Service (APHIS) or Customs inspectors that such items are admissible. An APHIS officer must be present when you return from an area where there has been an outbreak of destructive pests or diseases. Customs will advise APHIS of flights arriving from, or stopping in such an area. For further information on this subject, refer to the FAA International Flight Information Manual or contact your local USDA APHIS Office.

Endangered species of plant and animal wildlife and products thereof may be prohibited or require permits or certification by the U.S. Fish and Wildlife Service.

repairs to private aircraft

Aircraft belonging to a resident of the U.S. and taken abroad for noncommercial purposes and returned by the resident shall be admitted free of duty upon being satisfactorily identified. Repairs made abroad to such aircraft, must be reported to Customs but are not subject to duty. The reimportation of U.S. made parts may require entry. It is advisable that the nearest U.S. Customs office be contacted for clarification of this matter in each particular instance.

in case of emergency

If an emergency landing is made in the United States, the pilot should report as promptly as possible by telephone or most convenient means to the nearest Customs office. He should keep all merchandise or baggage in a segregated place and should not permit any passenger or crewmember to depart the place of arrival or come in contact with the public without official permission, unless other action is necessary for preservation of life, health, or property.

Companies involved in air ambulance type operations may be granted a single overflight exemption when emergency situations arise, as well as in the case of non-emergency transport for individuals seeking medical treatment. **Both U.S. and foreign registered aircraft will be eligible for the special exemption.** The applicant must provide all the necessary information normally required for an overflight exemption. Customs should be notified at least 24 hours prior to departure. If this cannot be accomplished, Customs will allow receipt of the overflight exemption application up to departure time, as well as in flight through a Federal Aviation Administration Flight Service Station.

hours of service

Generally speaking, free service is provided at airports during regular business hours (usually 8 AM to 5 PM), Monday through Saturday, and from 8 AM to 5 PM on Sundays and national holidays. However, tours of duty at airports are based on the need for services and are altered at some ports to coincide with schedule changes and peak workloads. The normal hours of service are listed in this booklet, however to be positive that processing fees and/or overtime charges will not accrue, private aircraft operators may contact, prior to departure, the Customs officer in charge at the U.S. airport of intended arrival in order to ascertain those hours during which free service can normally be expected. Phone numbers which may be used for this purpose are listed herein.

A private aircraft will be charged a processing fee of $25 once every calender year. This will be charged the first time the aircraft arrives from a foreign place in the calender

DEPARTMENT OF THE TREASURY
UNITED STATES CUSTOMS SERVICE

PRIVATE AIRCRAFT INSPECTION REPORT

6.2, 6.14, C.R.

PARTIAL LIST OF COUNTRY CODES

AC	ANTIGUA	DR	DOMINICAN	MX	MEXICO	ST	ST LUCIA
BB	BARBADOS		REPUBLIC	NA	NETHER	TD	TRINIDAD &
BD	BERMUDA	EL	EL SALVADOR		LANDS		TOBAGO
BF	BAHAMAS	GJ	GRENADA		ANTILLES	TK	TURKS &
BH	BRITISH	GP	GUADELOUPE	NU	NICARAGUA		CAICOS IS
	HONDURAS	GT	GUATEMALA	PN	PANAMA	VC	ST VINCENT
CA	CANADA	GY	GUYANA	PQ	CANAL ZONE	VE	VENEZUELA
CJ	CAYMAN IS	HA	HAITI	SC	ST CHRISTO	VI	BRITISH VI
CK	COCOS IS	HO	HONDURAS		PHER NEVIS	VQ	VI (U.S.)
CO	COLOMBIA	JM	JAMAICA		ANGUILLA		
CS	COSTA RICA	MB	MARTINIQUE	SQ	SWAN IS		

1. AIRCRAFT NUMBER 31: •

2. SYSDINO 20: A •

3. PILOT NAME 38: Last Name / First Name / M.I.

4. FOREIGN DEPARTURE Last Foreign City / Country Code

5. ARRIVAL IN U.S. U.S. Airport Code / Time of Arrival / Arrival Date (MMDDYY) •

6. AIRCRAFT 61: Make / Model

Color(s) •

7. PILOT DATA 66: Pilot's name as in Item no. 3 above / Date of Birth (MMDDYY) / License No. / Nationality •

8. PILOT ADDRESS

9. OWNER NAME AND ADDRESS

10. FOREIGN ITINERARY

11. DATE/TIME OF DEPARTURE FOR U.S.

12. U.S.-BASED AIRCRAFT ONLY ▶ a. U.S. Airport of Departure

b. Date/Time of U.S. Departure

13. PASSENGERS

a. Last Name	b. First Name	c. M.I.	d. Nationality	Date of Birth

INSPECTOR SIGNATURE & BADGE NO.

☐ INS ☐ USDA ☐ OTHER: ____________

Agriculture Data

☐ Fruits ☐ Plants
☐ Meats ☐ Veget.

Country of Origin

PAIRS Entry

Sta. Code:

Date:

Time:

Inspection Data

Travel Time:

Waiting Time:

Inspection Time:

(PAIRS ENTRY—New Record: Items 1, 3, 4, 5, 6, & 7, Modify: Items 1, 2, 3, 4, 5, & 7 if new.)

Customs Form 178 (10-30-81)

GPO 946-690

year or may be paid in advance. This fee is charged to the aircraft, not the pilot, and the receipt should be kept with the aircraft.

Commercial aircraft operators will be charged a processing fee of $5 per paying passenger for each arrival from foreign to the U.S. This fee will not be charged for passengers arriving from Canada, Mexico, and certain nearby Caribbean countries.

overtime charges

Private aircraft will be processed free of any overtime charges during regular hours of duty (usually 8 AM to 5 PM on Sundays and holidays. Overtime charges will accrue after regular hours of duty on weekdays, and before 8 AM and after 5 PM on Sundays and holidays. However, at the specially designated airports along the southwest border (see "Special Reporting Requirements"), Customs overtime charges will accrue only before 8 AM and after 5 PM on Sundays and holidays. If an officer from an inspecting agency other than Customs is providing the service, overtime charges after regular working hours during weekdays may be incurred.

A maximum charge of $25 has been established as a private aircraft pilot's liability for the cost of all overtime services performed by Federal inspection service employees in connection with each arrival or departure. Overtime charges are prorated if services are performed for more than one operator or owner by the same inspector during an overtime assignment. Depending on the overtime charge and the number of inspections performed, the cost can be less than the $25 maximum. However, private aircraft operators should expect to pay $25 for overtime services. Refunds, if applicable, will usually be made at a later date by Customs National Finance Center.

For information about requirements of other federal inspectional agencies and the manner in which overtime services are performed by them, operators should contact the respective services listed below:

APHIS Port Operations Staff
U.S. Department of Agriculture
Federal Building
Hyattsville, Maryland 20782

Chief, Quarantine Branch
Epidemiology Program
Center for Disease Control
Atlanta, Georgia 30333

Associate Commissioner, Management
U.S. Immigration and Naturalization Service
425 Eye Street, N.W.
Washington, D.C.

A Customs bond on Form 301 could be of particular advantage to aircraft owners and operators who frequently travel abroad. It is recommended that a copy of the bond be carried aboard the aircraft to simplify verification. For further information, contact the nearest Customs office.

penalties for violations

The Anti-Drug Abuse Act of 1986 has substantially increased civil penalties for violations of Customs Regulations. Specifically, penalties have increased from $500 to $5,000 for the first offense and from $1,000 to $10,000 for the second offense. Seizure of aircraft may occur at any time depending upon the circumstances behind the violation.

Since the law provides for substantial penalties for violations of the Customs Regulations, aircraft operators and pilots should make every effort to comply with them. Examples of the more common violations and resulting penalties include:

- Failure to report arrival [19 CFR 6.2 (a)—$5,000].
- Failure to obtain landing rights [19 CFR 6.2(b)—$5,000]
- Failure to provide advance notice of arrival [19 CFR 6.2(b)—$5,000].
- Failure to provide penetration report on southern border [19 CFR 6.14(a)—$5,000].
- Departing without permission or discharging passengers or cargo without permission [19 CFR 6.2(c)—$5,000].
- Importation of contraband, including agriculture materials, or undeclared merchandise can result in penalty action and seizure of aircraft which varies according to the nature of the violation and pertinent provision of law. In some instances Customs utilizes a notice of violation or warning instead of a penalty. Repeat violators, however, can normally expect to be penalized.

If a penalty is incurred, application may be made to the Customs officer in charge for reduction in amount or cancellation, giving the grounds upon which relief is believed to be justified. If the operator or pilot desires to further petition for relief of the penalty, he may appeal to the appropriate District Director of Customs. If still further review of the penalty is desired, written appeal may be made to the proper Regional Commissioner of Customs and, in some cases, to Customs Headquarters.

Narcotics Smuggling

Aircraft used to smuggle narcotics are subject to seizure. Title 19. Sec. 1595(a) of the U.S. code provides that such an aircraft may be forfeited. In addition, aircraft which are illegally modified in order to facilitate the importation of contraband are also subject to forfeiture under Section 1590. It makes no difference whether the narcotics are being smuggled by a crewmember or a passenger. These individuals are subject to criminal prosecution.

customs districts

Address all correspondence to the District Director of Customs at the following locations (Designations preceding addresses indicate the region in which the district is located):

Pacific	Anchorage, Alaska 995–1/620 E. Tenth Ave.
Northeast	Baltimore, Maryland, 21202/405 S. Gay St.
Northeast	Boston, Massachusetts 02222 10 Causeway St.
Northeast	Buffalo, New York 14202/111 W. Huron St.
Southeast	Charleston. South Carolina 29402/200 E. Bay St.
Southeast	Charlotte Amalie, St Thomas-Virgin Islands 00801/Main P.O. Sugar Estate
North Central	Chicago, Illinois 66607/610 S. Canal St.
North Central	Cleveland, Ohio 44114/55 Erieview Plaza
Southwest	Dallas/Fort Worth, Texas 75261/700 Parkway Plaza, P.O. Box 619050
North Central	Detroit, Michigan 48226/477 Michigan Ave.
North Central	Duluth, Minnesota 55802/515 W. First St., 209 Fed. Bldg.
Southwest	El Paso, Texas 79985/Bldg. B, Room 134 Bridge of the Americas (P.O. Box 9516)

North Central	Great Falls, Montana 59401/600 Central Plaza, Suite 200.
Pacific	Honolulu, Hawaii 96813/335 Merchant St., P.O. Box 1641
Southwest	Houston/Galveston, Texas 77052/701 San Jacinto St., P.O. Box 52790
Southwest	Laredo, Texas 78040/Mann Rd. & Santa Maria P.O. Box 3130
Pacific	Los Angeles/Long Beach, California 90731/300 S. Ferry St. Terminal Island
Southeast	Miami, Florida 33131/77 S.E. 5th St.
North Central	Milwaukee, Wisconsin 53202/517 E. Wisconsin Ave.
North Central	Minneapolis, Minnesota 55401/110 S. Fourth St.
South Central	Mobile, Alabama 36602/250 N. Water St., P.O. Box 2748
South Central	New Orleans, Louisiana 70130/423 Canal St.
New York	New York, New York New York Seaport Area, New York, New York 10048 Customhouse, 6 World Trade Center Kennedy Airport Area, Jamaica, New York 11430 Cargo Bldg. 80, Room 2E Newark Area, Newark, New Jersey 07114 Airport International Plaza
Southwest	Nogales, Arizona 85621/International & Terrace Sts., P.O. Box 670
Southeast	Norfolk, Virginia 23510/101 E. Main St.
Northeast	Ogdensburg, New York, 13669/127 N. Water St.
North Central	Pembina, North Dakota 58271/Post Office Bldg.
Northeast	Philadelphia, Pennsylvania 19106/2nd & Chestnut Sts. Room 102
Southwest	Port Arthur, Texas 77640/4550 75th Street
Northeast	Portland, Maine 04111/312 Fore St., P.O. Box 4688
Pacific	Portland, Oregon 97209/511 N.W. Broadway
Northeast	Providence, Phode Island 02903/24 Weybosset St.
Northeast	St. Albans, Vermont 05478/Main & Stebbins St., P.O. Box 111
North Central	St. Louis, Missouri 63105/7911 Forsyth Bldg., Suite 625.
Southeast	St. Thomas Virgin Islands 00801/P.O. Box 510
Pacific	San Diego, California 92188/880 Front St., Suite 559
Pacific	San Francisco, California 94126/555 Battery St., P.O. Box 2450
Southeast	San Juan, Puerto Rico 00903/P.O. Box 2112
Southeast	Savannah, Georgia 31401/1 East Bay St.
Pacific	Seattle, Washington 98174/909 First Ave.
Southeast	Tampa, Florida 33602/301 S. Ashley Dr.
Southeast	Washington, D.C. 20041/P.O. Box 17423 Gateway 1 Bldg., Dulles Int'l. Aprt. Chantilly, Va. 22021
Southeast	Wilmington, North Carolina 28401/One Virginia Ave.

(Note: New York has Area Directors instead of District Director)

customs regions

Address all correspondence to the Regional Commissioner of Customs at the following locations:

Northeast	Boston, Massachusetts 02222/10 Causeway St.
New York	New York, New York 10048/6 World Trade Center
Southeast	Miami, Florida 33131/909 S.E. 1st Ave.
South Central	New Orleans, Louisiana 70112/423 Canal St.
Southwest	Houston, Texas 77057/5850 San Felipe St.
Pacific	Los Angeles, California 90053/300 N. Los Angeles St.
North Central	Chicago, Illinois 60603/55 E. Monroe St.

customs headquarters

Address all correspondence to the Commissioner of Customs, 1301 Constitution Avenue N.W., Washington, D.C. 20229.

(Operations questions) Attn: Office of Inspection & Control

(Legal questions) Attn: Office of Regulations & Rulings

Airports at Which Customs Service is Normally Available

Information in this section is provided in the following order:

1st line:

Location/Name of Airport/Code/FAA Flight Plan Notification/Customs (USCS) Phone Number, (2nd line used if more than one phone number).

Remaining lines:

Special Arrangements or Restrictions.

- indicates USCS 24-hour numbers; FSS indicates FAA Flight Service Station number.

The Term "advance arrangement must be made" means that inspection services cannot be provided on a timely basis if only advance notice of arrival is given. Arrangements in writing or by telephone must be made prior to the flight by the pilot.

The term "regular business hours" means 0800 to 1700 hours, Monday through Saturday, except where otherwise indicated and these are the hours that the Customs office is normally staffed to make arrangements for service.

The notation "MP/F: 10," or other number, indicates maximum number of passenger which may be cleared per flight.

ALABAMA

Birmingham/Birmingham Municipal/BHM/FAA: Yes/USCS 205-731-1464.
On-call basis. Two hours advance notice required. MP/F: 20

Huntsville/Huntsville Madison County Jetplex/HSV/FAA: Yes/USCS 205-772-3404.
On-call basis. Two hours advance notice required.

Mobile/Bates Field/FAA: Yes/USCS 205-290-2111.
On-call basis. Two hours advance notice required during regular business hours, 0800-1700, Monday through Saturday. Three hours advance notice after regular business hours.

Montgomery/Dannelly/MGM/FAA: Yes/USCS 205-731-1464 (Birmingham).
Not staffed by Customs. Advance arrangements must be made for inspection by Customs through Birmingham.

ALASKA

Note: to assure timely inspection in Alaska, Customs requests that all calls (notice of advance arrival), regardless of when service is required, be made during regular business hours, 0800–1700, Monday through Friday, unless other hours are indicated.

Anchorage/Anchorage Int'l ANC/FAA: Yes/USCS 907-243-4312.
One hour advance notice required.

Cold Bay/Cold Bay/CDB/FAA: Yes/USCS 907-532-2482, 907-274043.
Prior written permission required to land for refueling. MP/F: 15. Customs inspection cannot be effected at Cold Bay. The written request must be sent to the District Director in Anchorage. The aircraft will be issued a "permit to proceed" at Cold Bay, for presentation at the airport where Customs inspection is to be made. Landing rights must be requested if that airport is not designated an international airport. It is recommended that written confirmation from all Customs offices be obtained prior to initiation of flight.

Eagle/Eagle Municipal/EFA/FAA: No/USCS none. Phone 907-547-2211.
Report on Arrival. (Contact U.S. Postmaster, Eagle.) MP/F: 15.

Fairbanks/Fairbanks Int'l/FAI/FAA: Yes/USCS 907-474-0307.
One hour advance notice required. Regular business hours, 0800–2400, Monday through Sunday.

Juneau/•Juneau Municipal/JNU/FAA: Yes/USCS 907-586-7211.
On-call basis. Two hours advance notice required. Regular business hours 0800–1700, Monday through Saturday.

Juneau/•Juneau Harbor Seaplane Base/JSE/FAA: Yes/USCS 907-586-7211.
On-call basis. Two hours advance notice required. Regular business hours 0800–1700, Monday through Saturday. MP/F: 15.

Ketchikan/•Ketchikan Int'l/KTN/FAA: No/USCS 907-225-2254.
One hour advance notice required. On-call basis only during 0630–2130 Regular business hours, 0800–1700, Monday through Saturday.

Ketchikan/•Ketchikan Harbor Seaplane Base/ECH/FAA: Yes/USCS 907-225-2254.
On-call basis. Two hours advance notice required. Regular business hours, 0800–1700, Monday through Saturday.

Northway/Northway/ORT/FAA: Yes/USCS 907-778-6223.
One hour advance notice required. Regular business hours, 0800–1700, Monday through Friday, MP/F: 15. Service May–September; October through April; contact Customs at 907-774-2252 or UNICOM 122.8 MHZ for Information.

Sitka/Sitka/SIT/FAA: Yes/USCS 907-983-3374.
One hour advance notice required. Regular business hours, 0800–1700, Monday through Friday. MP/F: 15.

Skagway/Skagway Municipal/SGY/FAA: Yes/USCS 907-983-2325.
One hour advance notice required. Regular business hours, 0800–2400, Monday through Sunday. MP/F: 15.

Wrangell/•Wrangell Seaplane Base/WRG/FAA: No/USCS 907-874-3415.
On-call basis. Two hours advance notice required. Regular business hours, 0800–1700, Monday through Friday. MP/F: 15.

Yukon/Ft. Yukon/FYU/FAA: No/USCS 907–662–2366 (Fairbanks).
Advance arrangements must be made through Fairbanks. MP/F: 15.

Valdez/Valdez/VDZ/FAA: Yes/USCS 907–835–2355.
On call basis. One hour advance notice required.

ARIZONA

Douglas/•Bisbee Douglas Int'l/DUG/FAA: Yes/USCS 602–364–8486, FSS 602–364–8458.
Communicate direct via radio to FAA or by telephone to Customs of intention to land and advise intended point and time of border penetration no later than one hour prior to entering U.S. airspace.

Nogales/•Nogales Int'l/OLS/FAA: Yes/USCS 602–287–5679, FSS 602–287–5161.
Communicate direct via radio to FAA or by telephone to Customs of intention to land and advise intended point and time of border penetration no later than one hour prior to entering U.S. airspace.

Phoenix/Sky Harbor Int'l/PHX/FAA: Yes, if radio contact with tower/ USCS 602–261–3515, FSS 1–800–228–4160 (Landing Rights Airport).
Private aircraft entering U.S from south of the border not designated to land at Sky Harbor Int'l. Advance arrangements must be made. Regular business hours, 0830–1630, Monday through Friday. Saturday and Sunday requests for service must be made prior to 1630 hours on Friday. MP/F: 20 (including crew).

Tucson/•Tucson Int'l/TUS/FAA: Yes/USCS 602–629–6461, FSS 602–889–9689.
Communicate direct via radio to FAA or by telephone to Customs of intention to land and advise intended point and time of border penetration no later than one hour prior to entering U.S. airspace. During one 2-week period, service is free 0800–1700 hours, Monday through Sunday. The following 2-week period, service is free 0800–2000 hours, Monday through Saturday and 0800–1700 hours on Sundays and holidays. All other times on an overtime basis.

Yuma/•Yuma Int'l/YUM/FAA: Yes/USCS 602–627–8326, FSS 602–762–2550, 1000 to 1800 hours only.
Communicate directly via radio to FAA or by telephone to Customs of intention to land and advise intended point and time of border penetration no later than one hour prior to entering U.S. airspace. Free Customs service provided 0900–1700 hours, Monday through Sunday; overtime basis all other times.

ARKANSAS

Little Rock/Adams Field/LIT/FAA: Yes/USCS 501–378–5289.
On-call basis. Two hours advance notice required.

CALIFORNIA

Calexico/•Calexico Int'l/CXL/FAA: Yes/USCS 714–357–1195, 714–357–4841, FSS 619–352–8740.
Communicate directly via radio to FAA or telephone to Customs of intention to land and advise intended point and time of border penetration no later than one hour prior to entering U.S. airspace.

Eureka/Eureka/EKA/FAA: Yes/USCS 707–442–4822.
On-call basis. Two hours advance notice required.

Fresno/Fresno Municipal/FAT/FAA: Yes/USCS 209–487–5460.
On-call basis. Two hours advance notice required during regular business hours,

0800–1700, Monday through Friday. Advance notice by 1630 Friday for all flights requesting service on Saturdays and Sundays. MP/F: 10.

Los Angeles/Los Angeles Int'l/LAX/FAA: Yes/USCS 213–215–2418, 213–215–2240, 213–646–0356.
One hour advance notice required. Regular business hours, 0800–1700, Monday through Friday. Advance arrangements must be made during other times. Int'l Arrivals Bldg. usually not available for private aircraft. MP/F: 10.

Oakland/Metropolitan Oakland Int'l/Oak/FAA: Yes/USCS 415–273–7706, 415–273–7552. (San Francisco).
On-call basis. Four hours advance notice required. Regular business hours, 0800–1700, Monday through Friday. Advance arrangements must be made for service during other times, in writing or by telephone, prior to flight by pilot.

Sacramento/Sacramento Metropolitan/SMF/FAA: Yes/USCS 415–876–2812 (San Francisco).
On-call basis. Four hours advance notice required during regular business hours, 0800–1700 Monday through Friday. Pilot must make advance arrangements in writing or by telephone prior to flight.

San Diego/•San Diego Int'l (Lindbergh)/SAN/FAA: Yes/USCS 619–428–7201, FSS 619–291–6381 or 800–532–3815.
No longer a designated airport. Private aircraft arriving from areas south of the U.S. (as defined in Special Reporting Requirement in the Inwards Flight section of this book) cannot proceed directly to this airport unless they have an overflight exemption to do so. Communicate direct via radio to FAA or by telephone to Customs intention to land and advise (if applicable) intended point and time of border penetration no later than one hour prior to entering U.S. airspace.

San Diego/•Brown Field/SDM/FAA: Yes/USCS 619–428–7201, FSS 619–291–6381 or 800–532–3815.
Communicate direct via radio to FAA or by telephone to Customs of intention to land and advise intended point and time of border penetration no later than one hour prior to entering U.S. airspace.

San Francisco/San Francisco Int'l/SFO/FAA: Yes/USCS 415–876–2812.
Two hours advance notice required during regular business hours, 0800–1700, Monday through Saturday; 4 hours advance notice required between, 1700–2400 hours, Monday through Saturday. Advance arrangements must be made during other times, in writing or by telephone, prior to flight by pilot.

San Jose/San Jose Municipal/SJC/FAA: Yes/USCS 415–876–2812 (San Francisco).
Four hours advance notice required. MP/F: 6

COLORADO

Colorado Springs/Colorado Springs Municipal/COS/FAA: No/USCS 303–574–6607.
Two hours advance notice required during regular business hours. Regular business hours are 0830–1700 hours.

Denver/Stapleton Int'l/DEN/FAA: Yes/USCS 303–361–0716, 1–800–843–5619.
Two hours advance notice required during regular business hours. Regular business hours are 0830–1700 hours. At least three hours otherwise.

CONNECTICUT

Bridgeport/Igor I. Sikorsky Memorial/BDR/FAA: Yes/USCS 203–579–5606.
Not staffed by Customs. Advance arrangements must be made for Customs inspection prior to departure. Regular business hours, 0800–1700, Monday through Friday.

Requests for Saturday and Sunday service must be made prior to 1700 hours on Friday.

Groton/Groton-New London/GON/FAA: Yes/USCS 203–442–7123, 203–773–2040/2041.
Not staffed by Customs. Advance arrangements must be made for Customs inspection prior to departure. Regular business hours, 0800–1700, Monday through Friday. Requests for Saturday and Sunday service must be made prior to 1700 hours on Friday.

New Haven/Tweed-New Haven Municipal/HVN/FAA: Yes/USCS 203–773–2040/2041.
Not staffed by Customs. Advance arrangement must be made for Customs inspection prior to departure. Regular business hours, 0800–1700, Monday through Friday. Request for Saturday and Sunday service must be made prior to 1700 hours on Friday.

Windsor Locks/Bradley Int'l/BDL/FAA: Yes/USCS 203–240–4306,
Sector Communications 1–800–343–2480.
Two hours advance notice required during regular business hours, 0800–1700, Monday through Friday. Requests for Saturday and Sunday service must be made prior to 1700 hours on Friday.

DELAWARE

Wilmington/Greater Wilmington/ILG/FAA: No/USCS 302–573–6191.
Not staffed by Customs personnel. Advance arrangements must be made for Customs inspection by party of interest prior to departure. Regular business hours 0800–1700, Monday through Friday. Saturday and Sunday requests for service must be made prior to 1700 hours Friday. Requests for holiday service must be made prior to 1700 hours on the preceding work day.

DISTRICT OF COLUMBIA

Washington/Dulles Int'l/IAD/FAA: Yes/USCS 703–471–5885, 202–566–8213/14.
Two hours advance notice required after regular business hours, 0800–1700, Monday through Saturday. For flight with over 10 persons, including crew, and passengers, all baggage must be transported to inspection area for clearance.

FLORIDA

Fort Lauderdale/•Fort Lauderdale-Hollywood Int'l/FLL/FAA: Yes/USCS 305–527–7411.
Regular business hours 0900–1700, Monday through Saturday.

Fort Lauderdale/Fort Lauderdale Executive/FXE/FAA: Yes/USCS 305–527–7412.
Regular business hours, 1100–1900, Monday through Sunday. Tower will direct flyers to Customs ramp. MP/F: 20.

Fort Pierce/St. Lucie County/FPR/FAA: Yes/USCS 305–461–1200.
One hour advance notice required. Regular business hours 0800–1900, Monday through Saturday.

Jacksonville/Jacksonville Int'l/JAX/FAA: Yes/USCS 904–356–4731.
Not staffed by Customs. Advance arrangements must be made for Customs inspection by party of interest prior to departure. Regular business hours, 0800–1700, Monday through Friday. Saturday and Sunday requests for service must be made prior to 1700 hours Friday.

Key West/•Key West Int'l/EYW/FAA: Yes/USCS 305–296–5411.
One hour advance notice required at all times. Regular business hours 0800–1700, Monday through Saturday.

Miami/Miami Int'l/MIA/FAA: Yes/USCS 305-526-2875.
Twenty-four hour service provided. Landing rights granted after 2100 hours with concurrence of FAA.

Miami/Opa-Locka/OPF/FAA: Yes/USCS 305-688-0832.
One hour advance notice required. Regular business hours, 1100-1900, Sunday-Saturday.

Orlando/Orlando Int'l/MCO/FAA: Yes/USCS 305-648-6308.
Regular business hours, 0800-1700, Monday through Saturday. Tower will direct to Customs ramp. One hour advance notice required. After hours call 305-859-6244.

Panama City/Panama City-Bay County (Fannin Field)/PEN/FAA:
Yes/USCS 904-785-4688.
Advance arrangements must be made during regular business hours, 0800-1700, Monday through Friday. MP/F: 20.

Penscola/Penscola Municipal/PNS/FAA: Yes/USCS 904-432-6811.
On-call basis. Three hours advance notice required. MP/F: 20.

St. Petersburg/Clearwater/St. Petersburg-Clearwater/PIE/FAA:
Yes/USCS 813-536-7311.
One hour advance notice required. Regular business hours 0830-1700, Monday through Saturday.

Tampa/•Tampa Int'l/TPA/FAA: Yes/USCS 813-228-2395. Nights: 813-228-2385.
One hour advance notice required. Communicate direct via radio to FAA or by telephone to Customs of intention to land and advise of ETA no later than one hour prior to entering U.S. airspace. Regular business hours, 0830-1700, Monday through Saturday. MP/F: 18. With 19 or more passengers and crew, report to gate 31, Airside C, main passenger terminal.

West Palm Beach/•Palm Beach Int'l/PBI/FAA: Yes/USCS 305-683-1806.
Regular business hours, 0800-1900, Monday through Sunday. One hour advance notice required.

GEORGIA

Atlanta/Charlie Brown County/FTY/FAA: Yes/USCS 404-763-7125.
On-call basis. Two hours or more advance notice required.

Atlanta/DeKalb-Peachtree/PDK/FAA: Yes/USCS 404-763-7125.
On-call basis. Two hours or more advance notice required.

Atlanta/William B. Hartfield Int'l (Atlanta)/ATL/FAA: Yes/USCS 404-763-7125/7126.
On-call basis. Two hours or more advance notice required.

Brunswick/Glyno Jetport/BQK/FAA: Yes/USCS 912-267-2803.
Limited personnel available. Daylight hours only. Advance arrangements must be made for inspection by Customs. MP/F: 10.

Brunswick/Malcolm-McKinnon/SSI/FAA: Yes/USCS 912-267-2803.
Limited personnel available. Advance arrangements must be made for inspection by Customs. MP/F: 10.

Savannah/Savannah Municipal/SAV/FAA: Yes/USCS 912-232-7507.
On-call basis. Two hours or more advance notice required, during regular business hours, 0800-1700, Monday through Friday.

HAWAII

Hanapepe/Port Allen/PAK/FAA: No/USCS 808-822-5521.
Advance arrangements must be made during business hours, 0800-1700, Monday through Friday. On-call basis. MP/F: 15.

Hilo/General Lyman Field/ITO/FAA: No/USCS 808-935-6976.
On-call basis. Advance arrangements must be made during regular business hours, 0800-1700, Monday through Friday.

Honolulu/Honolulu Int'l/HNL/FAA: Limited basis, pilot must specify that Government inspection agencies must be given prior notice/USCS 808-836-3613.
Advance arrangements must be made for landing rights. One hour advance notice required during regular business hours, 0800-1600, Monday through Saturday; at least two hours all other hours.

Kahului/Kahului/OGG/FAA: Yes/USCS 808-877-6013.
On-call basis. Advance arrangements must be made during regular business hours, 0800-1700, Monday through Friday. MP/G: 15.

IDAHO

Porthill/Eckhart Int'l/1S1/FAA: No/USCS 208-267-5309.
Landing rights required. Grass airstrips. Float plane landing also. Radio communications. One hour advance notice required which must be made during regular business hours, 0800-1700, Monday through Friday. MP/F: 15.

Boise/Boise Air Terminal-Gowen Field/BOI/FAA: Yes/USCS 208-334-9062, FSS 208-343-2525.
One hour advance notice required. Regular business hours 0800-1700, Monday through Friday, for weekend service, prior arrangements must be made no later than Thursday. MP/F: 15.

ILLINOIS

Chicago/•Chicago Midway/MDW/FAA: Yes/USCS 312-458-1238.
One hour advance notice required. Regular business hours, 0830-1700, Monday through Friday. MP/F: 15. After hours call 312-686-2131/2133.

Chicago/Chicago-O'Hare Int'l/ORD/FAA: Yes/USCS 312-686-2133.
One hour advance notice required during regular business hours, 0800-2130, Monday through Friday; 1200-2130 on Saturdays. Two hours advance notice required at other times. After hours call 312-686-2131.

Chicago/Merill C. Meigs Field/CGX/FAA: Yes/USCS 312-458-1238.
Nights: 312-626-8266.
One hour advance notice required during regular business hours 0830-1700, Monday through Friday. MP/F: 10.

Peoria/Greater Peoria/PIA/FAA: Yes/USCS 309-671-7047. Nights: 309-673-0195.
On-call basis. Two hours advance notice required during regular business hours, 0830-1700, Monday through Friday, 3 hours advance notice required after regular business hours, Saturdays, Sundays and holidays. MP/F: 10.

INDIANA

Indianapolis/Indianapolis Int'l/IND/FAA: Yes/USCS 317–248–4060.
Advance arrangements for the granting of landing rights must be made. Regular business hours, 0830–1700, Monday through Friday. Saturday and Sunday requests for service must be made prior to 1700 hours on Friday.

IOWA

Des Moines/Des Moines Municipal/DSM/FAA: Yes/USCS 317–284–4403.
Nights: 515–285–4640.
Advance arrangements for the granting of landing rights must be made. Regular business hours, 0800–1630, Monday through Friday. Saturday and Sunday requests for service must be made prior to 1630 hours on Friday.

KANSAS

Kansas City/Fairfax/KCK/FAA: Yes/USCS 816–374–3958. Nights: 816–374–3068.
Advance arrangements for the granting of landing rights must be made. Regular business hours, 0800–1630, Monday through Friday. Saturday and Sunday requests for service must be made prior to 1630 hours on Friday.

Wichita/Mid-Continent/ICT/FAA: Yes/USCS 316–269–7040. Nights: 316–942–4131.
Advance arrangements for the granting of landing rights must be made. Regular business hours, 0830–1700, Monday through Friday. Saturday and Sunday requests for service must be made prior to 1700 hours on Friday. MP/F: 15.

KENTUCKY

Louisville/Bowman Field/LOU/FAA: Yes/USCS 502–582–5183.
Advance arrangements for the granting of landing rights must be made at least two hours in advance of arrival. Regular business hours 0800–1700, Monday through Friday. Saturday and Sunday requests for service must be made prior to 1700 hours on Friday.

Louisville/Standiford Field/SDF/FAA: Yes/USCS 502–582–5183.
Advance arrangements for the granting of landing rights must be made at least two hours in advance of arrival. Regular business hours 0800–1700, Monday through Friday. Saturday and Sunday requests for service must be made prior to 1700 hours on Friday.

Owensboro/Davies County Airport/OWB/FAA: No/USCS 502–683–5133.
Twenty-four hours notice of arrival required. Regular business hours 0800–1630 Monday through Friday. Saturday and Sunday request for service must be made prior to 1630 hours on Friday.

LOUISIANA

Baton Rouge/Ryan/BTR/FAA: Yes/USCS 504–389–0261.
On-call basis. Two hours advance notice required.

Lake Charles/Lake Charles Municipal/LCH/FAA: Yes/USCS 318–439–5512.
Two hours advance notice required.

New Orleans/New Orleans Int'l (Moisant Field)/MSY/FAA: Yes/USCS 504–467–4319.
Nights: 504–589–6804.
Two hours advance notice required. Regular business hours, 0800–1700 Monday through Saturday.

New Orleans/New Orleans Lakefront Airport/NEW/FAA: Yes/USCS 504–589–2510. Nights: 504–589–6809.
On-call basis. Two hours advance notice required. Regular business hours, 0800–1700, Monday through Saturday. MP/F: 20.

MAINE

Bangor/Bangor Int'l/BGR/FAA: Yes/USCS 207–947–7861.
One hour advance notice required. Regular business hours, 0800–2400, Monday through Saturday. Due to limited staffing, aircraft arrivals after 1700 may be subject to inspectional overtime charge.

Baring/St. Croix/FAA: No/USCS 207–454–3621 (Calais).
On-call basis. To insure timely service, pilots are requested to telephone Customs prior to takeoff from Canada. One hour advance notice required. Charge is made for mileage. Inspectional overtime charges may be applicable after 1700 depending upon manpower availability at the port of Fort Fairfield, Maine. MP/F: 10.

Eastport/Eastport Municipal/47B/FAA: No/USCS 207–853–4313.
On-call basis. To insure timely service, pilots are requested to telephone Customs prior to takeoff from Canada. One hour advance notice required. Regular business hours, 0800–1700 (daylight hours only), Monday through Saturday. MP/F: 10.

Fort Fairfield/Fairfield Municipal/1B4/FAA: No/USCS 207–473–7474.
On-call basis. To insure timely service, pilots are requested to telephone Customs prior to takeoff from Canada. One hour advance notice required. Regular business hours, 0800–1700 (daylight hours only), Monday through Saturday. MP/F: 10.

Frenchville/Northern Aroostook Regional Airport/FVE/FAA: Yes/USCS 207–728–4376 (Madawaska).
On-call basis. One hour advance notice during regular business hours, 0800–1700, Monday through Saturday. Two hours advance notice required after 1700, and on Sundays and holidays. Charge is made for mileage. Inspectional overtime charge may be applicable after 1700 depending upon the availability of manpower at the Port of Madawaska. MP/F: 10.

Houlton/•Houlton Int'l/HUL/FAA: Yes/USCS 207–532–2131, FSS 207–532–2475.
One hour advance notice required. Regular business hours: 24-hour service, Monday through Saturday, except holidays.

Jackman/Moose River Seaplane Base/60B/FAA: No/USCS 207–668–3711.
One hour advance notice required.

Jackman/Newton Field/59B/FAA: No/USCS 207–668–3711.
On-call basis to insure timely service, pilots are requested to telephone Customs prior to takeoff from Canada. One hour advance notice required. MP/F: 20.

Lubec/Lubec Municipal/65B/FAA: No/USCS 207–733–4331.
On-call basis. To insure timely service, pilots are requested to telephone Customs prior to takeoff from Canada. One hour advance notice required. Regular business hours, 0800–1700 (daylight hours only), Monday through Saturday. MP/F: 10.

Old Town/Old Town Municipal (DeWitt Field)/OLD/FAA: Yes/USCS 207–947–7861 (Bangor), 207–945–6320.
On-call basis. To insure timely service, pilots are requested to telephone Customs prior to takeoff from Canada. Charge is made for mileage. One hour advance notice required. MP/F: 20.

Portland/Portland Int'l Jetport/PWM/FAA: Yes/USCS 207–780–3328, 0800–1700 hours, FSS 207–833–0602, 0700–2300 hours.

To insure timely service, pilots are requested to telephone Customs prior to takeoff from Canada. One hour advance notice required. Due to limited staffing aircraft arrivals after 1700 may be subject to inspectional overtime charge.

Presque Isle/Presque Isle Municipal/PQI/FAA: Yes/USCS 207–473–7474, 207–473–7396 (Fort Fairfield).
On-call basis. To insure timely service, pilots are requested to telephone Customs prior to takeoff from Canada. One hour advance notice required for permission to land. Charge is made for mileage. Inspectional overtime charge may be applicable after 1700 depending upon manpower availability at the port of Fort Fairfield, Maine. MP/F: 10.

Princeton/Princeton Municipal/PNN/FAA: No/USCS 207–454–3621 (Calais).
On-call basis. To insure timely service, pilots are requested to telephone Customs prior to takeoff from Canada. One hour advance notice required. Charge is made for mileage. MP/F: 10.

Van Buren/St. John River Sea Plane Base/FAA: No/USCS 207–868–3391.
One hour advance notice required. MP/F: 10.

MARYLAND

Baltimore/Baltimore Washington Int'l/BAL/FAA: Yes/USCS 301–962–3170, 301–962–3843, 617–565–6180.
One hour advance notice required during regular business hours. Two hours advance notice required after 1700 hours and on Sundays and holidays. With 10 or more passengers report to International Arrivals Building. MP/F: 12.

MASSACHUSETTS

Bedford/Laurence G. Hanscom Field/BED/FAA: Yes/USCS 617–283–0425 (Gloucester) 617–565–6180.
Two hours advance notice required. Regular business hours, 0800–1700, Monday through Friday. Charge is made for mileage.

Beverly/Beverly Municipal/BVY/FAA: Yes/USCS 617–686–1363.
One hour advance notice required. Regular business hours, 0800–1700, Monday through Friday. After hours call 617–565–6180. MP/F: 15.

Boston/General Edward Lawrence Logan Int'l/BOS/FAA: Yes/USCS 617–565–6180, 617–565–4657.
Regular business hours, 0800–2400, Monday through Saturday.

Chicopee/Westover Air Park/CEF/FAA: Yes/USCS 413–785–0365.
One hour advance notice required. Regular business hours, 0800–1700, Monday through Friday.

Lawrence/Lawrence Municipal/LWM/FAA: Yes/USCS 617–283–0425, 565–6180.
One hour advance notice required. Regular business hours, 0800–1700, Monday through Friday. MP/F: 10.

New Bedford/New Bedford Municipal/EWB/FAA: Yes/USCS 617–994–5158.
Advance arrangements must be made for Customs inspection prior to departure. Regular business hours, 0800–1700, Monday through Friday.

Westfield/Barnes Municipal/BAF/FAA: Yes/USCS 413–785–0365.
One hour advance notice required. Regular business hours, 0800–1700, Monday through Friday.

Worchester/Worchester Municipal/ORH/FAA: Yes/USCS 617–793–0293.
On-call basis. To insure timely service, pilots are requested to telephone Customs

prior to takeoff from Canada. Two hours advance notice required for permission to land. Saturday and Sunday requests for service must be made prior to 1700 hours on Friday.

MICHIGAN

Alpena/Phelps Collins/APN/FAA: No/USCS 517-356-0910.
Telephone call from pilot requesting service before takeoff from Canada. Charge is made for mileage. MP/F: 5.

Battle Creek/Kellogg Regional/BTL/FAA: No/USCS 616-965-3349.
Nights: 616-964-8991, 616-657-5872.
Telephone call from pilot requesting service before takeoff from Canada. Regular business hours, 0800-1630, Monday through Friday. MP/F: 10.

Bay City/James Clements Municipal/3CM/FAA: No/USCS 517-892-4222,
Nights: 517-686-3720.
Telephone call from pilot requesting service before takeoff from Canada. One hour advance notice required. MP/F: 5.

Detroit/•Detroit City/DET/FAA: Yes/USCS 313-226-3140.
Telephone call from pilot requesting service before takeoff from Canada. One hour advance notice required. MP/F: 15.

Detroit/Detroit Metropolitan-Wayne County/DTW/FAA: Yes/USCS 313-226-3140.
Telephone call from pilot requesting service before takeoff from Canada during hours of 2400-0800. At other times flight service notification is sufficient.

Flint/Bishop Int'l/FNT/FAA: No/USCS 517-892-4222. Nights: 517-686-3720.
Telephone call from pilot requesting service before takeoff from Canada. One hour advance notice required. MP/F: 5.

Grand Rapids/Kent County Int'l/GRR/FAA: No/USCS 616-456-2515.
Nights: 616-949-6309.
Telephone call from pilot requesting service before takeoff from Canada. Regular business hours, 0800-1630, Monday through Friday. MP/F: 20.

Kalamazoo/Kalamazoo Municipal/AZO/FAA: No/USCS 616-965-3349.
Nights: 616-964-8991, 616-657-5872.
Telephone call from pilot requesting service before takeoff from Canada. Regular business hours, 0800-1630, Monday through Friday. MP/F: 10.

Marquette/Marquette County/MQT/FAA: No/USCS 906-475-4082.
Nights: 906-475-4082.
Telephone call from pilot requesting service before takeoff from Canada. Charge is made for mileage. MP/F: 5.

Port Huron/Baker's Field/90G/FAA: Yes/Direct radio contact not available./
USCS 313-985-9541.
One hour notice required. Seaplane use only. On-call from Blue Water Bridge. MP/F: 5.

Port Huron/•St. Clair County/PHN/FAA: Yes/USCS 313-985-9541.
Telephone call from pilot requesting service before takeoff from Canada with at least one hour advance notice. MP/F: 10.

Saginaw/Tri-City/MBS/FAA: No/USCS 517-892-4222.
One hour advance notice required. Telephone call from pilot requesting service before takeoff from Canada. MP/F: 10.

Sault Ste. Marie/Chippewa County Int'l/CIU/FAA: Yes/USCS 906-632-7222,
906-632-2631.

Telephone call from pilot requesting service before takeoff from Canada. One hour advance notice required. On-call from International Bridge. MP/F: 10.

MINNESOTA

Baudette/•Baudette Int'l/BDE/FAA: Yes/USCS 218–634–2661.
On-call from border station. One hour advance notice required. MP/F: 4.

Crane Lake/Scott's Seaplane Base/8Y8/FAA: No/USCS 218–993–2321.
One hour advance notice required.

Duluth/•Duluth Int'l/DHL/FAA: Yes/USCS 218–720–5203.
On-call basis. One hour advance notice required. Service furnished on request.

Duluth/•Sky Harbor/D36/FAA: Yes/USCS 218–720–5203.
On-call basis. One hour advance notice required. Service furnished on request.

Ely/Ely Municipal/ELO/FAA: No/USCS 218–365–3262.
Advance notice required. Communicate direct to airport base via radio or telephone to Customs. Open on a seasonal basis May 15 through October 14.

Ely Municipal/Shagawa Seaplane Base/MN41/FAA: No/USCS 218–365–3262.
Advance notice required. Communicate direct to seaplane base via radio or telephone to Customs. Open on a seasonal basis May 15 through October 14.

Grand Marais/Devil's Track/GRM/FAA: No/USCS 218–387–1750.
Nights: 218–475–2244.
Open on a seasonal basis from May 15 through October 14. Advance arrangements must be made.

International Falls/•Falls Int'l/INL/FAA: Yes/USCS 218–283–2541.
On-call basis. One hour advance notice required.

Minneapolis/Minneapolis-St. Paul Int'l/MSP/FAA: Yes/USCS 612–725–3689/90,
Sundays and holidays: 612–725–3690, FSS Day or Night 612–726–1130.
One hour advance notice required. Regular business hours, 0800–1700, Monday through Friday. Saturday and Sunday on-call basis.

Pinecreek/Pinecreek/48Y/FAA: No/USCS 218–463–1952.
Serviced from border station 0900–2200.

Ranier/•Int'l Seaplane Base/50Y/FAA: Yes/USCS 218–286–5211, 218–283–2541.
On-call basis. One hour advance notice required.

St. Paul/St. Paul Downtown (Holman Field) STP/FAA: Yes/USCS 612–725–3689,
FSS Day or Night: 612–725–3690.
One hour advance notice required. Regular business hours, 0800–1630, Monday through Friday. Saturday and Sunday on-call basis.

Warroad/Warroad Int'l Seaplane Base/75Y/FAA: No/USCS 218–386–1676.
On-call basis. One hour advance notice required. MP/F: 4.

Warroad/Sweet Carlson Field/D45/FAA: No/USCS 218–386–1676.
On-call basis. One hour advance notice required. MP/F: 4.

MISSISSIPPI

Gulfport/Gulfport Biloxi Regional/GPT/FAA: Yes/USCS 601–864–6794.
On-call basis. Two hours advance notice required during regular business hours, 0800–1700, Monday through Friday. MP/F: 10.

Pascagoula/Jackson County/PGL/FAA: No/USCS 610-762-7311.
On-call basis. Two hours advance notice required during regular business hours, 0800-1700, Monday through Friday.

Vicksburg/Vicksburg Municipal Airport/FAA: No/USCS 601-634-7186.
On-call basis. Two hours advance notice required during regular business hours, 0800-1700, Monday through Friday.

MISSOURI

Kansas City/Kansas City Int'l/MCI/FAA: Yes/USCS 816-374-3958.
Nights: 1-800-392-4220.
Advance arrangements for granting of landing rights must be made. Regular business hours, 0800-1630, Monday through Friday. Saturday and Sunday requests for service must be made prior to 1630 hours on Friday.

Kansas City/Downtown Airport/MKC/FAA: Yes/USCS 816-374-3958.
Nights: 1-800-392-4220.
Advance arrangements for the granting of landing rights must be made. Regular business hours 0800-1630, Monday through Friday. Saturday and Sunday requests for service must be made prior to 1630 hours on Friday.

Springfield/Springfield Regional/SGF/FAA: Yes/USCS 417-831-4035.
Nights: 417-865-9992.
Advance arrangements must be made. Regular business hours, 0830-1700, Monday through Friday. Saturday and Sunday requests for service must be made prior to 1700 on Friday. MP/F: 15.

St. Louis/Lambert-St. Louis Municipal/STL/FAA: Yes/USCS 314-428-8230, 314-425-7138. Nights and weekends: 314-532-1011.
On-call basis. Two hours advance notice required. Regular business hours, 0800-1700, Monday through Friday.

St. Louis/Spirit of St. Louis/SUS/FAA: Yes/USCS 314-428-8230.
Nights and weekends: 314-532-1011.
Advance arrangements must be made with Customs for granting of landing rights during regular business hours, 0830-1700, Monday through Friday. Nights and weekends, two hours advance notice through FAA.

MONTANA

NOTE: To assure timely inspection of locations in Montana, Customs requests that all calls (notice of advance arrival), regardless of when service is required, be made during regular business hours, unless other hours are specified.

Billings/Logan Int'l/BIL/FAA: Yes/USCS 406-457-8495, FSS 406-259-4545.
One hour advance notice required. Commercial cargo cannot be entered at this airport. On-call basis. MP/F: 15.

Buttle/Bert Mooney/BTM/FAA: Yes/USCS 406-494-3492, FSS 873-4154. One hour advance notice required. Regular business hours, 0800-1700, Monday through Friday.

Cut Bank/•Cut Bank/CTB/FAA: Yes/USCS 406-873-2711, FSS 406-873-4154.
One hour advance notice required. On-call basis. Commercial cargo cannot be entered at this airport. MP/F: 15.

Del Bonita/Del Bonita/FAA: No/USCS 406-336-2130.
Airstrip does not have landing lights. One hour advance not required during regular business hours, 0800-1700, Monday through Friday.

Glasgow/Glasgow Int'l/GGW/FAA: Yes/USCS 1-800-824-7706, 406-259-4545.
Two hours advance notice required. On-call basis. Commercial cargo cannot be entered at this airport. MP/F: 15.

Great Falls/•Great Falls Int'l/GTF/FAA: Yes/USCS 406-453-0861; FSS 406-761-7110.
One hour advance notice required. Regular business hours, 0800-1700, Monday through Friday.

Havre/Havre City-County/HVR/FAA: Yes/USCS 406-265-2512.
On-call basis. Commercial cargo cannot be entered at this airport. MP/F: 15.

Kalispell/Glacier Park Int'l/FCA/FAA: Yes/USCS 406-257-7034, FSS 406-761-7110.
On-call basis. Two hours advance notice required. Commercial cargo cannot be entered at this airport.

Missoula/Missoula County/MSO/FAA: No/USCS 406-721-2576.
One hour advance notice required. On-call basis. Commercial cargo cannot be entered. FAA 1-800-641-0078.

Morgan-Loring/Morgan/7U4/FAA: No/USCS 406-674-5248.
One hour advance notice required. Regular business hours, 0800-1700, Monday through Friday. MP/F: 15.

NEVADA

Las Vegas/McCurran Int'l/LAS/FAA: Yes/USCS 702-388-6480.
On-call basis. Two hours advance notice required during regular business hours, 0800-1700, Monday through Friday, for all flights MP/F: 20.

Reno/Reno Int'l/RNO/FAA: Yes/USCS 702-784-5585.
On-call basis. Two hours advance notice required. Arrangements should be made during regular business hours, 0800-1700, Monday through Friday, for all flights. MP/F: 25.

NEW JERSEY

Newark/Newark Int'l/EWR/FAA: Yes/USCS 201-645-3409, 201-645-2236, FSS 800-932-0835, Butler Aviation 201-961-2600.
Two hours advance notice required after regular business hours and on Sundays and holidays. MP/F: 20.

Teterboro/Teterboro/TEB/FAA: Yes/USCS 201-288-8799, 201-288-1740, Saturdays 201-645-3409.
On-call basis. One hour advance notice required during regular business hours. A minimum of four hours advance notice is required after 1700 hours and on Sundays and holidays. MP/F: 20.

NEW MEXICO

Albuquerque/Albuquerque Int'l/ABQ/FAA: Yes/USCS 505-766-2621.
Two hours advance notice required during regular business hours, 0800-1700, Monday through Friday. At least three hours otherwise.

NEW YORK

Albany/Albany County/ALB/FAA: Yes/USCS 518-472-3457.
Two hours advance notice required at all times, Monday-Friday, 0800-1700. After 1700 hours, Sundays and holidays, Burlington 802-863-5271.

Buffalo/Greater Buffalo Int'l/BUF/FAA: Yes/USCS 716–632–4727; if no answer, call 716–846–4316.
Advance arrangements for Customs inspection must be made at all times 2 hours prior to arrival. Regular business hours, 0800–1700, Monday through Saturday.

Massena/•Richards Field/MSS/FAA: Yes/USCS 315–769–3091.
On-call basis. One hour advance notice required.

New York/John F. Kennedy/JFK/FAA: Yes/USCS 718–917–1648 (24 hours), 718–917–1649.
Private aircraft handled only on an emergency basis.

New York/La Guardia/LGA/FAA: Yes/USCS 718–476–4378, 718–917–1648, (24 hours).
One hour advance notice required. MP/F: 15.

Newburgh/Stewart/SWF/FAA: Yes/USCS 914–564–7490, 800–522–5270 (N.Y.), 212–466–5472, 800–221–4265 (Outside N.Y.).
One hour advance notice required 0800–1700 hours Monday through Saturday. Advance notice must be made during regular business hours. All other hours, charge may be made for mileage. MP/F: 20.

Niagara Falls/Niagara Falls Int'l/IAG/FAA: Yes/USCS 846–4580/78 for service 0800–1700 hours Monday through Saturday. For service after 1700 hours and before 1800 hours weekdays, and all Sundays and U.S. holidays, call 716–846–4316.

Ogdensburg/•Ogdensburg Harbor/FAA: No/USCS 315–393–1390.
On-call basis. One hour advance notice required.

Ogdensburg/•Ogdensburg Int'l/OGS/FAA: Yes/USCS 315–393–1390.
On-call basis. One hour advance notice required.

Rochester/Mayers Marina/FAA: Yes/USCS 716–263–6295, 6294.
On-call basis. One hour advance notice required.

Rochester/•Rochester Monroe County/ROC/FAA: Yes/USCS 716–263–6295, 6294.
On-call basis. One hour advance notice required. After hour call 1–800–343–2840.

Rouses Point/•Rouses Point Seaplane Base/NY47/FAA: No/USCS 518–298–8321, 518–298–8341, 518–298–8345.
On-call basis. One hour advance notice required.

Schenectady/Schenectady County/SCH/FAA: Yes/USCS 518–472–3457 (Albany).
24 hours advance notice required. Prior permission to land must be obtained during regular business hours, Monday–Friday 0800–1700. Charge may be made for mileage.

Syracuse/Syracuse Hancock Int'l/SYR/FAA: Yes/USCS 315–445–2271.
Two hour advance notice required. Use Gate 6 (Flying Tigers). After hours 800–343–2840, 617–565–6180.

Watertown/Watertown New York Int'l/ART/FAA: Yes/USCS 315–482–2261.
On-call basis. Two hours advance notice required.

White Plains/Westchester County/HPN/FAA: Yes/USCS 914–428–7858, 800–522–5270 in New York, 800–221–4265 outside New York.
One hour advance notice required 0800–1700 Monday–Saturday. All other hours, advance notice must be made during regular business. MP/F: 20.

NORTH CAROLINA

Charlotte/Douglas Municipal/CLT/FAA: Yes/USCS 704–527–2167.
On-call basis. One hour advance notice required during regular business hours, 0830–1700, Monday–Friday. Three hour advance notice required after 1700 hours and on Saturdays, Sundays and holidays.

Greensboro/Greensboro-Highpoint-Winston Salem Regional/GSO/FAA: Yes/ USCS 919–761–3001 (Winston-Salem).
On-call basis. Two hour advance notice required during regular business hours, 0830–1700, Monday–Friday. Three hour notice required after 1700 hours and on Saturdays, Sundays and holidays. Charge is made for mileage.

Raleigh-Durham/Raleigh-Durham/RDU/FAA: Yes/USCS 919–541–5211.
Advance arrangements must be made for Customs inspection by party of interest prior to departure. Regular business hours, 0830–1700, Monday–Friday. Saturday and Sunday requests for service must be made prior to 1700 hours Friday. Requests for holiday service must be made prior to 1700 hours on the preceding work day. Charge is made for mileage.

Wilmington/New Hanover County/ILM/FAA: Yes/USCS 919–343–4616.
On-call basis. Two hour advance notice required during regular business hours, 0800–1700. Three hour advance notice required after 1700 hours and on Sundays and holidays.

Winston-Salem/Smith-Reynolds/INT/FAA: Yes/USCS 919–761–3001.
On-call basis. Two hour advance notice required during regular business hours, 0830–1700, Monday–Friday. Three hours notice required after 1700 hours and on Saturdays, Sundays, and holidays.

NORTH DAKOTA

Dunseith/Int'l Peace Garden/S28/FAA: No/USCS 701–263–4513, Nights: 701–772–7201 (Grand Forks).
Serviced from border station. Border station open 24 hours year-round. Radio communication is available. Airport is not lighted, hard surface. MP/F: 4.

Fargo/Hector Field/FAR/FAA: Yes/USCS 701–261–8124.
One hour advance notice required during regular business hours, 0800–1700. On-call basis during other hours and on Sundays and holidays.

Grand Forks/•Mark Andrews Int'l/GFK/FAA: Yes/USCS 701–772–3301, Nights: 701–772–7201 (Grand Forks).
One hour advance notice required during regular business hours, 0900–1700. On-call basis during other hours and on Sundays and holidays.

Minot/•Minot Int'l/MCT/FAA: Yes/USCS 701–838–6704, Nights: FSS 701–852–3696.
One hour advance notice required during regular business hours, 0800–1700. On-call basis all other times. MP/F: 20.

Pembina/•Pembina Municipal/PMB/FAA: Yes, through Grand Forks, ND/ USCS 701–825–6551 (24 hours).
One hour advance notice required. Contact airport manager to request lighted border.

Williston/•Sloulin Field Int'l/ISN/FAA: Yes/USCS 701–572–6552/2197.
On-call basis. One hour advance notice required. MP/F: 10.

OHIO

Akron/Akron-Canton/CAK/FAA: Yes/USCS 216–375–5496.
On-call basis. Two hour advance notice required, 0830–1700, Monday–Friday. Requests for service on Saturday, Sunday, and holidays must be made prior to 1700 hours on Friday. MP/F: 15.

Akron/•Akron Municipal/AKR/FAA: Yes/USCS 216–375–5496.
On-call basis. Two hours advance notice required, 0830–1700, Monday–Friday. Requests for service on Saturdays, Sundays, and holidays must be made prior to 1700 hours on Friday. MP/F: 15.

Cincinnati/Cincinnati Municipal Airport-Lunken Field/LUK/FAA: Yes/
USCS 513–684–3528.
Two hours advance arrangements for granting of landing rights must be made, 0830–1700, Monday–Friday. Saturday and Sunday requests for services must be made prior to 1700 hours on Friday. MP/F: 10.

Cincinnati (Covington, KY)/Greater Cincinnati/CVG/FAA: Yes/USCS 513–684–3528.
Two hours advance arrangements for granting of landing rights must be made, 0830–1700, Monday–Friday. Saturday and Sunday requests for service must be made prior to 1700 hours on Friday.

Cleveland/Cleveland Hopkins Int'l/CLE/FAA: Yes/USCS 216–522–7030.
One hour advance notice required. Regular business hours, 0800–2100, Monday–Friday; 0800–1700 Saturday.

Cleveland/Burke Lakefront/BKL/FAA: Yes/USCS 216–522–7030.
Advance arrangements for the granting of landing rights must be made. Service provided Monday–Friday, 0900–1500 only. Two hours advance notice required. MP/F: 8.

Columbus/Port Columbus Int'l/CMH/FAA: Yes/USCS 614–469–6670.
Advance arrangements for the granting of land rights must be made. Regular business hours, 0830–1700, Monday–Friday. Saturday and Sunday requests for service must be made prior to 1700 hours on Friday.

Dayton/Cox Int'l/DAY/FAA: Yes/USCS 513–225–2876/2877.
Advance arrangements for the granting of landing rights must be made. Regular business hours, 0830–1700, Monday–Friday. Saturday and Sunday requests for service must be made prior to 1700 hours on Friday.

Sandusky/•Friffing-Sandusky/SKY/FAA: Yes/USCS 419–625–2194.
One hour advance notice required. Regular business hours, 0830–1700, Monday–Friday.

Toledo/Toledo Express/TOL/FAA: No/USCS 419–259–6424.
Advance arrangements must be made during regular business hours. Regular business hours, 0830–1700, Monday–Friday. Saturday and Sunday requests for service must be made prior to 1700 hours on Friday. MP/F: 15.

OKLAHOMA

Oklahoma City/Will Rogers World/OKC/FAA: Yes/USCS 405–231–4347.
Advance arrangements must be made. Regular business hours, 0830–1700, Monday–Friday. Saturday and Sunday requests for service must be made prior to 1700 hours on Friday. MP/F: 8 (including crew).

Tulsa/Tulsa Int'l/TUL/FAA: Yes/USCS 918–835–7631.
Advance arrangements must be made. Regular business hours, 0830–1700,

Monday–Friday. Saturday and Sunday requests for service must be made prior to 1700 hours on Friday. MP/F: 8 (including crew).

OREGON

Portland/Portland Int'l/PDX/FAA: Yes/USCS 503–221–3515.
One hour advance notice required, 0800–1700, Monday–Saturday. Call 503–222–1699 for service between 1700–0800 hours and on Sundays and holidays.

PENNSYLVANIA

Allentown/Allentown-Bethelem-Easton Int'l/ABE/FAA: Yes/USCS 215–266–1042, 215–776–9226.
Two hours advance notice required. Regular business hours, 0800–1700, Monday–Friday. Saturday, Sunday, and holiday requests for service must be made prior to 1700 hours on the preceding workday. User fee charges are applicable. Notification can be made through FSS at 800–992–7433.

Erie/Erie Int'l/ERI/FAA: Yes/USCS 814–452–2905, ERI FSS 814–833–1345.
On-call basis. Two hours advance notice required during regular business hours, 0830–1700, Monday–Friday. Requests for service on Saturdays, Sundays, and holidays must be made before 1700 hours on the preceding workday.

Harrisburg/Harrisburg Int'l/MDT/FAA: Yes/USCS 717–782–4510.
Two hours advance notice required. Regular business hours, 0830–1700, Monday–Friday. Saturday, Sunday, and holiday requests for service must be made prior to 1700 hours on the preceding workday. Notification can be made through FSS, 800–992–7433.

Philadelphia/Philadelphia Int'l/PHL/FAA: Yes/USCS 215–596–1973.
Customs Officers are frequently available from 0800–2100 hours. Regular business hours, 0800–1700, Sunday–Saturday. Advance arrangements with Customs required during other than regular hours.

Pittsburgh/Allegheny County/AGC/FAA: No/USCS 412–644–3586/87/88.
On-call basis. Advance arrangements must be made for Customs inspection.

Pittsburgh/Greater Pittsburgh Int'l/PIT/FAA: No/USCS 412–644–3587/88.
Two hours advance notice required during regular business hours, 0800–1700, Monday–Friday. Requests for service outside these hours must be made directly with Customs during regular business hours.

Wilkes-Barre/Scranton/Wilkes-Barre/Scranton/AVP/FAA: Yes/USCS 717–654–3674.
Two hours advance notice required. Regular business hours, 0800–1700, Monday–Friday. FSS 800–992–7433.

PUERTO RICO

Culebra/Culebra/CPX/FAA: No/USCS 809–742–3531.
One hour advance notice required during regular business hours, 0900–1600, Monday–Saturday. Arrivals from U.S. Virgin Islands only.

Fajardo/Fajardo/FAA: No/USCS 809–863–0950, 0811, 0102.
One hour advance notice required during regular business hours, 0800–1700, Monday–Saturday. Arrivals from U.S. Virgin Islands only.

Isla de Vieques/Vieques/VQS/FAA: No/USCS 809–741–8366.
One hour advance notice required during regular business hours, 0800–1700, Monday–Saturday. Arrivals from U.S. Virgin Islands only.

Mayaguez/Mayaguez/MAZ/FAA: No/USCS 809-832-0042/0308.
On-call basis. One hour advance notice required.

Ponce/Mercedita/PSE/FAA: No/USCS 809-842-1030, 809-843-5998.
On-call basis. One hour advance notice required.

San Juan/Luis Munoz Marin Int'l/SJU/FAA: Yes/USCS 809-791-5245/0222/5927.
Regular business hours, 0800-2300, Monday-Saturday.

San Juan/Isla Grande/SIG/FAA: Yes/USCS 809-725-6911.
Regular business hours, 1000-1900, Monday-Saturday; Sunday 1000-1300 hours.

RHODE ISLAND

North Kingstown/Quonset State/OQU/FAA: Yes/USCS 401-528-5080, 617-565-6180, 1-800-343-2840.
Three hours advance notice required.

Pawtucket/North Central State/SFZ/FAA: Yes/USCS 401-528-5080, 617-565-6180.
Three hours advance notice required.

Providence/Theodore Francis Green State/PVD/FAA: Yes/USCS 401-528-5080, 617-565-6180, 1-800-343-2840.
One hour advance notice required during regular business hours. All other times, three hours advance notice is required.

SOUTH CAROLINA

Charleston/Charleston Int'l/CHS/FAA: Yes/USCS 803-767-7100, 803-723-1272, 803-724-4644.
One hour advance notice required.

Columbia/Columbia Metropolitan/CAE/FAA: No/USCS 803-794-0782.
One hour advance notice required during regular business hours, 0800-1700, Monday-Friday. Communication through FAA, 1700-0800.

Greer/Greenville-Spartanburg/GSP/FAA: No/USCS 803-877-8006.
One hour advance notice required during regular business hours, 0800-1700, Monday-Friday.

TENNESSEE

Chattanooga/Lovell Field/CHA/FAA: Yes/USCS 615-267-1327.
On-call basis. Two hours advance notice required.

Knoxville/McGhee Tyson/TYS/FAA: Yes/USCS 615-970-4158.
On-call basis. Two hours advance notice required.

Memphis/Memphis Int'l/MEM/FAA: Yes/USCS 901-521-3558.
On-call basis. Two hours advance notice required.

Nashville/Nashville Metropolitan/BNA/FAA: Yes/USCS 615-736-5861.
On-call basis. Two hours advance notice required.

TEXAS

Amarillo/Amarillo Int'l/AMA/FAA: Yes/USCS 806-376-2347.
Advance arrangements must be made. Regular business hours, 0830-1700, Monday-Friday. Saturday and Sunday requests for service must be made prior to 1700 hours on Friday. MP/F: 8 (including crew).

Austin/Robert Mueller Municipal/AUS/FAA: Yes/USCS 512–482–5309.
Advance arrangements must be made. Regular business hours, 0830–1700, Monday–Friday. Saturday and Sunday requests for service must be made prior to 1700 hours on Friday. MP/F: 10 (including crew).

Beaumont/Jefferson County/BPT/FAA: Yes/USCS 409–727–0285 (Port Arthur).
On-call basis. Two hours advance notice required. Regular business hours, 0800–1700, Monday–Friday.

Brownsville/•Brownsville Int'l/BRO/FAA: No/USCS 512–542–5661/8296.
Communicate direct via radio to FAA or by telephone to Customs of intention to land and advise intended point and time of border penetration no later than one hour prior to entering U.S. airspace.

Corpus Christi/Corpus Christi Int'l/CRP/FAA: Yes/USCS 512–888–3352.
Two hours advance notice required. Regular business hours, 0830–1700, Monday–Friday. All other times, 512–853–5859.

Dallas/DFW Regional/DFW/FAA: Yes/USCS 214–574–2130.
One hour advance notice required. Regular business hours, 0830–1700. Two hour advance notice required after 1700 hours. Contact FSS at 214–350–7340 after 1700 hours.

Dallas/Dallas Love Field/DAL/FAA: Yes/USCS 214–350–4170.
Two hours advance notice required. After 1700 hours contact FAA at 214–350–7340. MP/F: 10 (including crew).

Del Rio/•Del Rio Int'l/DRT/FAA: Yes/USCS 512–775–9321.
Communicate direct via radio to FAA or by telephone to Customs of intention to land and advise intended point and time of border penetration no later than one hour prior to entering U.S. airspace.

Eagle Pass/•Eagle Pass Municipal/EGP/FAA: No/USCS 512–773–5426/9468/9454.
Communicate direct via radio to FAA or by telephone to Customs of intention to land and advise intended point and time of border penetration no later than one hour prior to entering U.S. airspace.

El Paso/•El Paso Int'l/ELP/FAA: Yes/USCS 915–541–7123/24.
Communicate direct via radio to FAA or by telephone to Customs of intention to land and advise intended point and time of border penetration no later than one hour prior to entering U.S. airspace.

Ft. Worth/Meachem Field/FTW/FAA: Yes/USCS 214–574–2131/2130/2114/2175.
Two hours advance notice required. Contact FAA after 1700 hours. MP/F: 10 (including crew).

Galveston/Scholes Field/GLS/FAA: Yes/USCS 409–766–3787.
Landing rights will be granted in emergency situations with the concurrence of other interested agencies.

Houston/Houston Intercontinental/IAH/FAA: Yes/USCS 713–443–5926/5910.
One hour advance notice required during regular business hours; two hours otherwise. MP/F: 20 (including crew) with the concurrence of other interested agencies.

Houston/William P. Hobby/HOU/FAA: Yes/USCS 713–644–1174 (Monday–Friday); all other times 713–229–3478.
One hour advance notice required during regular business hours; two hours otherwise. MP/F: 20 (including crew) with the concurrence of other interested agencies.

Laredo/•Laredo Int'l/LRD/FAA: Yes/USCS 512–723–4411.
Communicate direct via radio to FAA or by telephone to Customs of intention to land and advise intended point and time of border penetration no later than one hour

prior to entering U.S airspace during regular business hours, 0900–1700, Monday–Saturday. After hours and Sundays call 512-726-2368.

Lubbock/Lubbock Int'l/LBB/FAA: Yes/USCS 806-743-7458.
Advance arrangements must be made. Regular business hours, 0830–1700, Monday–Friday. Saturday and Sunday requests for service must be made prior to 1630 hours on Friday. MP/F: 8 (including crew).

McAllen/•Miller Int'l/MFE/FAA: Yes/USCS 512-682-2331.
Communicate direct via radio to FAA or by telephone to Customs of intention to land and advise intended point and time of border penetration no later than one hour prior to entering U.S. airspace.

Presidio/•Presidio-Lely Int'l/TXO7/FAA: Yes/USCS 915-229-3349.
On-call basis. One hour advance notice required during regular business hours, 0830–1700, Monday–Friday. Two hours advance notice required on Saturday, Sunday, and holidays and after 1700 hours, Monday–Friday.

San Antonio/San Antonio Int'l/SAT/FAA: Yes/USCS 512-822-0471, 512-229-5137.
One hour advance notice required during regular business hours. All other times, two hours advance notice required.

UTAH

Salt Lake City/Salt Lake City Int'l/SLC/FAA: Yes/USCS 801-524-5093.
On-call basis. One hour advance notice required. MP/F: 40.

VERMONT

Burlington/•Burlington Int'l/BTV/FAA: Yes/USCS 802-864-5181, FSS 802-863-5074.
On-call basis. One hour advance notice required.

Highgate/Franklin County State/1B7/FAA: No/USCS 802-868-2778.
One hour advance notice required. MP/F: 12.

Newport/Newport State/EFK/FAA: No/USCS 802-873-3219
One hour advance notice required. MP/F: 8.

VIRGIN ISLANDS

Charlotte Amalie/Cyril E. King/STT/FAA: Yes/USCS 809-774-1719.
One hour advance notice required during regular business hours, 0800–1700 daily.

Christiansted/Alexander Hamilton/STX/FAA: Yes/USCS 809-778-0216.
One hour advance notice required during regular business hours, 0800–1700, Monday–Saturday. Two hour advance notice required on Sundays and holidays and after 1700 hours Monday–Saturday.

VIRGINIA

Chantilly/Dulles Int'l (See District of Columbia).

Newport News/Patrick Henry Int'l/PHF/FAA: Yes/USCS 804-245-6470 (24 hours).
Two hours advance notice required. MP/F: 10.

Norfolk/Norfolk Int'l/ORK/FAA: Yes/USCS 804-441-6778/6731/6741.
One hour advance notice required during regular business hours, 0800–1700, Monday–Saturday. Two hour advance notice required on Sundays and holidays and after 1700 hours, Monday–Saturday. Call 804-441-6741 on Sundays and holidays and after 1700 hours, Monday–Saturday.

Richmond/Richard E. Byrd Int'l/RIC/FAA: Yes/USCS 804-771-2552/2071.
Two hours advance notice required during regular business hours, 0800-1700, Monday-Friday, after hours call 804-746-8708. Requests for service on Saturdays and Sundays must be made prior to 1700 hours on the previous regular business day. MP/F: 10.

WASHINGTON

Note: To assure timely inspection at locations in Washington, Customs requests a phone call (notice of arrival) at all airports, regardless of when service is required, be made during regular business hours, unless other hours are indicated.

Anacortes/Anacortes/74S/FAA: No/USCS 206-293-2331.
On-call basis. One hour advance notice required. Regular business hours, 0800-1700, Monday-Friday. MP/F: 15.

Anacortes/Skyline Seaplane Base/WAo3/FAA: No/USCS 206-293-2331.
On-call basis. One hour advance notice required. Regular business hours, 0800-1700, Monday-Friday. MP/F: 15.

Bellingham/•Bellingham Int'l/BLI/FAA: Yes/USCS 206-734-5463.
On-call basis. One hour advance notice required. Regular business hours, 0800-1700, Monday-Friday. MP/F: 15.

Blaine/Blaine Municipal/WA-09/FAA: No/USCS 206-332-6318.
On-call basis. One hour advance notice required. Regular business hours, 0000-2400, Monday-Sunday. MP/F: 15.

Everette/Paine Field Snohomish County/PAE/FAA: No/USCS 206-259-0246.
On-call basis. Two hours advance notice required. Regular business hours, 0800-1700, Monday-Friday. Permission granted only if workload permits. MP/F: 15.

Friday Harbor/•Friday Harbor Seaplane Base/WA24/FAA: Yes/USCS 206-378-2080.
On-call basis. Requests for landing rights must be made at least one hour in advance during regular business hours, 0800-1700, Monday-Friday. Permission granted only if workload permits. MP/F: 15.

Friday Harbor/Friday Harbor/S19/FAA: No/USCS 206-378-2080.
On-call basis. One hour advance notice required. Regular business hours, 0800-1700, Monday-Friday. MP/F: 15.

Hoquaim/Bowerman/HQM/FAA: No/USCS 206-532-2030.
One hour advance notice required. Regular business hours, 0800-1700, Monday-Friday. Permission granted only if workload permits. MP/F: 15.

Kelso/Kelso-Longview/KLS/FAA: Yes/USCS 206-425-3710.
One hour advance notice required. Regular business hours, 0800-1700, Monday-Friday. Runway and facilities are not maintained. MP/F: 15.

Laurier/Avey Field State/69F/FAA: No/USCS 509-684-2100.
On-call basis. Request for landing rights must be made at least one hour in advance during regular business hours, 0800-2400, Monday-Sunday. Airstrip does not have landing lights.

Northport/James A. Lowry Municipal/WA42/FAA: No/USCS 509-732-4418/6215.
One hour advance notice required. Regular business hours, 0800-2400, Monday-Sunday. Permission granted only if workload permits. Airstrip does not have landing lights. MP/F: 15.

Olympia/Olympia Municipal/OLM/FAA: No/USCS 206-593-6338 (Tacoma).
Two hours advance notice required. Regular business hours, 0800–1700, Monday–Friday. Permission granted only if workload permits. MP/F: 15.

Oroville/•Dorothy Scott/OS7/FAA: No/USCS 206-593-6338 (Tacoma).
Two hours advance notice required. Regular business hours, 0800–1700, Monday–Friday. Permission granted only if workload permits. MP/F: 15.

Oroville/•Dorothy Scott Seaplane Base/006/FAA: Yes/USCS 206-593-6338 (Tacoma).
On-call basis. One hour advance notice required. Regular business hours, 0800–1700, Monday–Friday. Airstrip does not have landing lights. MP/F: 15.

Port Angeles/William R. Fairchild Int'l/CLM/FAA: No/USCS 206-457-4311.
One hour advance notice required. Regular business hours, 0800–1700, Monday–Friday. Permission granted only if workload permits. MP/F: 15.

Port Angeles/Port Angeles Marine Dock/NOW/FAA: No/USCS 206-457-4311.
One hour advance notice required. Regular business hours, 0800–1700, Monday–Friday. Permission granted only if workload permits. MP/F: 15.

Port Townsend/•Jefferson County Int'l/OS9/FA: No/USCS 206-385-3777.
On-call basis. One hour advance notice required. Regular business hours, 0800–1700, Monday–Saturday. Recommend advance arrangements for all flights be made during regular business hours. Permission granted only if workload permits. MP/F: 15.

Seattle/•Boeing Field-King County Int'l/BFI/FAA: Yes/USCS 206-442-1971.
One hour advance notice required. Regular business hours, 0800–1700, Monday–Saturday. After hours call 206-443-3830. MP/F:30.

Seattle/Kenmore Air Harbor/S60/FAA: Yes/USCS 206-442-1971.
Two hour advance notice required. Regular business hours, 0800–1700, Monday–Saturday. After hours call 206-443-3830. MP/F: 15.

Seattle/Kurtzer's Flying Service/WA56/FAA: Yes/USCS 206-442-1971.
Two hours advance notice required. Regular business hours, 0800–1700, Monday–Saturday. After hours call 206-443-3830. MP/F: 15.

Seattle/•Lake Union Air Service/WA57/FAA: Yes/USCS 206-442-1971.
Two hours advance notice required. Regular business hours, 0800–1700, Monday–Saturday. After hours call 206-443-3830. MP/F: 15.

Seattle/Seattle-Tacoma Int'l/SEA/FAA: Yes/USCS 206-442-1971.
Two hours advance notice required. Regular business hours, 0600–2200, Monday–Saturday. After hours call 206-443-3830.

Seattle/Renton Municipal/RNT/FAA: Yes/USCS 206-442-1971.
Two hours advance notice required. Regular business hours, 0800–1700, Monday–Saturday. After hours call 206-443-3830.

Spokane/Felts Field/SFF/FAA: Yes/USCS 509-456-4661.
On-call basis. Two hours advance notice required. Regular business hours, 0800–1700, Monday–Friday. MP/F: 15.

Spokane/Spokane Int'l/GEG/FAA: Yes/USCS 509-456-4661.
On-call basis. One hour advance notice required. Regular business hours, 0800–1700, Monday–Friday. MP/F: 15.

Tacoma/Tacoma Industrial/TIW/FAA: Yes/USCS 206-593-6338.
On-call basis. One hour advance notice required. Regular business hours, 0800–1700, Monday–Friday. MP/F: 15.

WEST VIRGINIA

Charleston/Yeager Field/CRW/FAA: Yes/ USCS 304–347–5204.
One hour advance notice required. Regular business hours, 0800–1700, Monday–Friday. Prior arrangements must be made for Saturday and Sunday arrivals.

Huntington/Tri-State/HTS/FAA: Yes/USCS 304–347–5204.
On-call basis. Advance arrangements must be made.

WISCONSIN

Green Bay/Austin Straubel/GRB/FAA: Yes/USCS 414–433–3923.
Advance arrangements for landing rights must be made at least two hours prior to arrival during regular business hours, 0800–1700, Monday–Friday. Arrangements for evening arrivals must be made prior to 1700. Saturday, Sunday, or holiday requests for service must be made prior to 1700 hours on the preceding Friday.

Milwaukee/General Mitchell Field/MKE/FAA: Yes/USCS 414–291–3932.
One hour advance notice required during regular business hours, 0800–1700, Monday–Friday. Nights and weekends, two hours advance notice is required through FAA.

Racine/Horlick-Racine/RAC/FAA: Yes/USCS 414–633–0286.
Advance arrangements for landing rights must be made at least two hours prior to arrival during regular business hours, 0800–1700 Monday–Friday. For evening hours, arrangements must be made before 1700. Saturday, Sunday, or holiday requests for service must be made prior to 1700 hours on the preceding Friday.

WYOMING

Casper/Natrona County Int'l/CPR/FAA: Yes/USCS 307–235–8513.
One hour advance notice required. Regular business hours, 0800–1700, Monday–Friday. FSS 307–235–1555. MP/F: 15.

Airman's Information Manual
(AIM)

AIRMEN'S INFORMATION MANUAL (AIM)
FEDERAL AVIATION REGULATIONS AND ADVISORY CIRCULARS

Federal Aviation Regulations—The FAA publishes the Federal Aviation Regulations (FAR's) to make readily available to the aviation community the regulatory requirements placed upon them. These *Regulations* are sold as individual *Parts* by the Superintendent of Documents.

The more frequently amended Parts are sold on subscription service with subscribers receiving Changes automatically as issued. Less active Parts are sold on a single–sale basis. Changes to single–sale Parts will be sold separately as issued. Information concerning these Changes will be furnished by the FAA through its *Status of Federal Aviation Regulations, AC 00–44.*

Advisory Circulars—The FAA issues advisory circulars (AC's) to inform the aviation public of nonregulatory material of interest. Unless incorporated into a regulation by reference, the contents of an advisory circular are not binding on the public. Advisory circulars are issued in a numbered subject system corresponding to the subject areas of the Federal Aviation Regulations (14 CFR Ch. 1).

The AC 00–2 Checklist contains advisory circulars that are for sale as well as those distributed free of charge by the Federal Aviation Administration.

NOTE—The above information relating to FAR's and AC's is extracted from AC 00–2, *Advisory Circular Checklist (and Status of Other FAA Publications).* Many of the FAR's and AC's listed in AC 00–2 are cross–referenced in the AIM. These regulatory and nonregulatory references cover a wide range of subjects and are a source of detailed information of value to airmen. AC 00–2 is issued annually and can be obtained free–of–charge from:

U.S. Department of Transportation
Distribution Requirements Section M–494.1
Washington, DC 20590

External References—All reference to Advisory Circulars and other FAA publications in the Airmen's Information Manual include the FAA Advisory Circular or Manual identification numbers (when available); however, due to varied publication dates, the basic publication letter is not included (Example: The Controller's Handbook 7110.65F is referenced as 7110.65). To insure that you are consulting the latest information when referencing one of these other documents, use the FAA Advisory Circular AC: 00–2, Advisory Circular Checklist (and Status of Other FAA Publications).

AIRMAN'S INFORMATION MANUAL

EXPLANATION OF MAJOR CHANGES

***** SPECIAL NOTICE *****

Sections in this publication dated other than **10/15/92** have no changes.

****

All references to FAR Part 91 Sections 1 through 311 have been redesignated to the new section numbers (1 through 905) effective August 18, 1990.

Para 2–9. TAXIWAY LIGHTS. Paragraph was modified to inform pilots that taxiway lights at most major airports have variable intensities.

Para 2–22. AIRPORT MARKING AIDS. Paragraph was modified to clarify Threshold, Displaced Threshold, and Paved Areas beyond the Runway End.

Para 4–58. BREAKING ACTION REPORTS AND ADVISORIES. Paragraph was updated and merged with Breaking Action Advisories.

Para 4–59. RUNWAY FRICTION REPORTS. A new paragraph was added.

Para 4–60. INTERSECTION DEPARTURES. An Example was added to show specific phraseology that pilots should use when requesting an intersection departure.

Para 4–68. TAXIING. Paragraph was modified to emphasize the requirement that air traffic controllers obtain a readback from pilots when runway hold short instructions have been issued.

Para 4–71. PRACTICE INSTRUMENT APPROACHES. A Note was added that clarifies the intent of a clearance to land.

Para 4–88. VFR/IFR FLIGHTS. Paragraph was modified to clarify Pilot/Controller responsibilities.

Para 5–5. FLIGHT PLAN – DEFENSE VFR (DVFR) FLIGHTS. Paragraph was modified to clarify that a pilot which files a Defense Visual Flight Rules (DVFR) flight plan is afforded the same search and rescue protection as a Visual Flight Rules (VFR) flight plan.

Para 5–72. AIR TRAFFIC CLEARANCE. Paragraph was modified to emphasize the requirement that air traffic controllers obtain a readback from pilots when runway hold short instructions have been issued, and a Note was added that clarifies the intent of a clearance to land.

Para 6–17. SEARCH AND RESCUE. Paragraph was modified as a result of the Department of Defense changing from the AUTOVON system to the Defense Switching Network (DSN), and to clarify that a pilot which files a Defense Visual Flight Rules (DVFR) flight plan is afforded the same search and rescue protection as a Visual Flight Rules (VFR) flight plan.

Para 7–29. INTERNATIONAL CIVIL AVIATION ORGANIZATION (ICAO) TERMINAL FORECAST (TAF). Paragraph was added.

Para 7–73. OBSTRUCTIONS TO FLIGHT. Paragraph was modified to more clearly defined objects or structures that could affect flight.

TABLE OF CONTENTS

(Bar indicates paragraphs updated for this edition)

Chapter 1. NAVIGATION AIDS

Section 1. AIR NAVIGATION RADIO AIDS

Section 2. RADAR SERVICES AND PROCEDURES

Chapter 2. AERONAUTICAL LIGHTING AND OTHER AIRPORT VISUAL AIDS

Section 1. AIRPORT LIGHTING AIDS

TABLE OF CONTENTS—*Continued*

Section 2. AIR NAVIGATION AND OBSTRUCTION LIGHTING

Section 3. AIRPORT MARKING AIDS AND SIGNS

Chapter 3. AIRSPACE

Section 1. GENERAL

Section 2. UNCONTROLLED AIRSPACE

Section 3. CONTROLLED AIRSPACE

Section 4. SPECIAL USE AIRSPACE

TABLE OF CONTENTS—*Continued*

Section 5. OTHER AIRSPACE AREAS

Chapter 4. AIR TRAFFIC CONTROL

Section 1. SERVICES AVAILABLE TO PILOTS

Section 2. RADIO COMMUNICATIONS PHRASEOLOGY AND TECHNIQUES

TABLE OF CONTENTS—*Continued*

Section 3. AIRPORT OPERATIONS

Section 4. ATC CLEARANCES/SEPARATIONS

TABLE OF CONTENTS—*Continued*

Chapter 5. AIR TRAFFIC PROCEDURES

Section 1. PREFLIGHT

Section 2. DEPARTURE PROCEDURES

Section 3. EN ROUTE PROCEDURES

Section 4. ARRIVAL PROCEDURES

TABLE OF CONTENTS—*Continued*

Section 5. PILOT/CONTROLLER ROLES AND RESPONSIBILITIES

Section 6. NATIONAL SECURITY AND INTERCEPTION PROCEDURES

Chapter 6. EMERGENCY PROCEDURES

Section 1. GENERAL

Section 2. EMERGENCY SERVICES AVAILABLE TO PILOTS

TABLE OF CONTENTS—*Continued*

Section 3. DISTRESS AND URGENCY PROCEDURES

Section 4. TWO-WAY RADIO COMMUNICATIONS FAILURE

Chapter 7. SAFETY OF FLIGHT

Section 1. METEOROLOGY

Section 2. ALTIMETER SETTING PROCEDURES

TABLE OF CONTENTS—*Continued*

Section 3. WAKE TURBULENCE

Section 4. BIRD HAZARDS AND FLIGHT OVER NATIONAL REFUGES, PARKS, AND FORESTS

Section 5. POTENTIAL FLIGHT HAZARDS

Section 6. SAFETY, ACCIDENT, AND HAZARD REPORTS

Chapter 8. MEDICAL FACTS FOR PILOTS

Section 1. FITNESS FOR FLIGHT

TABLE OF CONTENTS—*Continued*

Chapter 9. AERONAUTICAL CHARTS AND RELATED PUBLICATIONS

Section 1. TYPES OF CHARTS AVAILABLE

Appendix 1. AERONAUTICAL CHARTS

PILOT/CONTROLLER GLOSSARY

INDEX

Chapter 1. NAVIGATION AIDS

Section 1. AIR NAVIGATION RADIO AIDS

1–1. GENERAL

Various types of air navigation aids are in use today, each serving a special purpose. These aids have varied owners and operators, namely: the Federal Aviation Administration (FAA), the military services, private organizations, individual states and foreign governments. The FAA has the statutory authority to establish, operate, maintain air navigation facilities and to prescribe standards for the operation of any of these aids which are used for instrument flight in federally controlled airspace. These aids are tabulated in the Airport/Facility Directory.

1–2. NONDIRECTIONAL RADIO BEACON (NDB)

a. A low or medium frequency radio beacon transmits nondirectional signals whereby the pilot of an aircraft properly equipped can determine his bearing and "home" on the station. These facilities normally operate in the frequency band of 190 to 535 kHz and transmit a continuous carrier with either 400 or 1020 Hz modulation. All radio beacons except the compass locators transmit a continuous three-letter identification in code except during voice transmissions.

b. When a radio beacon is used in conjunction with the Instrument Landing System markers, it is called a Compass Locator.

c. Voice transmissions are made on radio beacons unless the letter "W" (without voice) is included in the class designator (HW).

d. Radio beacons are subject to disturbances that may result in erroneous bearing information. Such disturbances result from such factors as lightning, precipitation static, etc. At night radio beacons are vulnerable to interference from distant stations. Nearly all disturbances which affect the ADF bearing also affect the facility's identification. Noisy identification usually occurs when the ADF needle is erratic. Voice, music or erroneous identification may be heard when a steady false bearing is being displayed. Since ADF receivers do not have a "flag" to warn the pilot when erroneous bearing information is being displayed, the pilot should continuously monitor the NDB's identification.

1–3. VHF OMNI-DIRECTIONAL RANGE (VOR)

a. VORs operate within the 108.0 to 117.95 MHz frequency band and have a power output necessary to provide coverage within their assigned operational service volume. They are subject to line-of-sight restrictions, and the range varies proportionally to the altitude of the receiving equipment. The normal service ranges for the various classes of VORs are given in 1–8d *NAVAID SERVICE VOLUMES.*

b. Most VORs are equipped for voice transmission on the VOR frequency. VORs without voice capability are indicated by the letter "W" (without voice) included in the class designator (VORW).

c. The only positive method of identifying a VOR is by its Morse Code identification or by the recorded automatic voice identification which is always indicated by use of the word "VOR" following the range's name. Reliance on determining the identification of an omnirange should never be placed on listening to voice transmissions by the Flight Service Station (FSS) (or approach control facility) involved. Many FSSs remotely operate several omniranges with different names. In some cases, none of the VORs have the name of the "parent" FSS. During periods of maintenance, the facility may radiate a T–E–S–T code (– . ... –) or the code may be removed.

d. Voice identification has been added to numerous VORs. The transmission consists of a voice announcement, "AIRVILLE VOR" alternating with the usual Morse Code identification.

e. The effectiveness of the VOR depends upon proper use and adjustment of both ground and airborne equipment.

1. Accuracy: The accuracy of course alignment of the VOR is excellent, being generally plus or minus 1 degree.

2. Roughness: On some VORs, minor course roughness may be observed, evidenced by course needle or brief flag alarm activity (some receivers are more susceptible to these irregularities than others). At a few stations, usually in mountainous terrain, the pilot may occasionally observe a brief course needle oscillation, similar to the indication of "approaching station." Pilots flying over unfamiliar routes are cautioned to be on the alert for these vagaries, and in particular, to use the "to/from" indicator to determine positive station passage.

(a) Certain propeller RPM settings or helicopter rotor speeds can cause the VOR Course Deviation Indicator to fluctuate as much as plus or minus six degrees. Slight changes to the RPM setting will normally smooth out this roughness. Pilots are urged to check for this modulation phenomenon prior to reporting a VOR station or aircraft equipment for unsatisfactory operation.

1–4. VOR RECEIVER CHECK

a. The FAA VOR test facility (VOT) transmits a test signal which provides users a convenient means to determine the operational status and accuracy of a VOR receiver while on the ground where a VOT is located. The airborne use of VOT is permitted; however, its use is strictly limited to those areas/altitudes specifically authorized in the Airport/Facility Directory or appropriate supplement.

b. To use the VOT service, tune in the VOT frequency on your VOR receiver. With the Course Deviation Indicator (CDI) centered, the omni-bearing selector should read 0 degrees with the to/from indication showing "from" or the omni-bearing selector should read 180 degrees with the to/from indication showing "to". Should the VOR receiver operate an RMI (Radio Magnetic Indicator), it will indicate 180 degrees on any OBS setting. Two means of identification are used. One is a series of dots and the other is a continuous tone. Information concerning an individual test signal can be obtained from the local FSS.

c. Periodic VOR receiver calibration is most important. If a receiver's Automatic Gain Control or modulation circuit deteriorates, it is possible for it to display acceptable accuracy and sensitivity close into the VOR or VOT and display out-of-tolerance readings when located at greater distances where weaker signal areas exist. The likelihood of this deterioration varies between receivers, and is generally considered a function of time. The best assurance of having an accurate receiver is periodic calibration. Yearly intervals are recommended at which time an authorized repair facility should recalibrate the receiver to the manufacturer's specifications.

d. Federal Aviation Regulations (FAR 91.171) provides for certain VOR equipment accuracy checks prior to flight under instrument flight rules. To comply with this requirement and to ensure satisfactory operation of the airborne system, the FAA has provided pilots with the following means of checking VOR receiver accuracy:

1. VOT or a radiated test signal from an appropriately rated radio repair station.

2. Certified airborne check points.

3. Certified check points on the airport surface.

e. A radiated VOR test signal from an appropriately rated radio repair station serves the same purpose as an FAA VOR signal and the check is made in much the same manner as a VOT with the following differences:

1. The frequency normally approved by the FCC is 108.0 MHz.

2. Repair stations are not permitted to radiate the VOR test signal continuously; consequently, the owner or operator must make arrangements with the repair station to have the test signal transmitted. This service is not provided by all radio repair stations. The aircraft owner or operator must determine which repair station in his local area provides this service. A representative of the repair station must make an entry into the aircraft logbook or other permanent record certifying to the radial accuracy and the date of transmission. The owner, operator or representative of the repair station may accomplish the necessary checks in the aircraft and make a logbook entry stating the results. It is necessary to verify which test radial is being transmitted and whether you should get a "to" or "from" indication.

f. Airborne and ground check points consist of certified radials that should be received at specific points on the airport surface or over specific landmarks while airborne in the immediate vicinity of the airport.

1. Should an error in excess of plus or minus 4 degrees be indicated through use of a ground check, or plus or minus 6 degrees using the airborne check, IFR flight shall not be attempted without first correcting the source of the error.

CAUTION: No correction other than the correction card figures supplied by the manufacturer should be applied in making these VOR receiver checks.

2. Locations of airborne check points, ground check points and VOTs are published in the Airport/Facility Directory.

3. If a dual system VOR (units independent of each other except for the antenna) is installed in the aircraft one system may be checked against the other. Turn both systems to the same VOR ground facility and note the indicated bearing to that station. The maximum permissible variations between the two indicated bearings is 4 degrees.

1–5. TACTICAL AIR NAVIGATION (TACAN)

a. For reasons peculiar to military or naval operations (unusual siting conditions, the pitching and rolling of a naval vessel, etc.) the civil VOR/DME system of air navigation was considered unsuitable for military or naval use. A new navigational system, TACAN, was therefore developed by the military and naval forces to more readily lend itself to military and naval requirements. As a result, the FAA has been in the process of integrating TACAN facilities with the civil VOR/DME program. Although the theoretical, or technical principles of operation of TACAN equipment are quite different from those of VOR/DME facilities, the end result, as far as the navigating pilot is concerned, is the same. These integrated facilities are called VORTAC's.

b. TACAN ground equipment consists of either a fixed or mobile transmitting unit. The airborne unit in conjunction with the ground unit reduces the transmitted signal to a visual presentation of both azimuth and distance information. TACAN is a pulse system and operates in the UHF band of frequencies. Its use requires TACAN airborne equipment and does not operate through conventional VOR equipment.

1–6. VHF OMNI-DIRECTIONAL RANGE/ TACTICAL AIR NAVIGATION (VORTAC)

a. A VORTAC is a facility consisting of two components, VOR and TACAN, which provides three individual services: VOR azimuth, TACAN azimuth and TACAN distance (DME) at one site. Although consisting of more than one component, incorporating more than one operating frequency, and using more than one antenna system, a VORTAC is considered to be a unified navigational aid. Both components of a VORTAC are envisioned as operating simultaneously and providing the three services at all times.

b. Transmitted signals of VOR and TACAN are each identified by three-letter code transmission and are interlocked so that pilots using VOR azimuth with TACAN distance can be assured that both signals being received are definitely from the same ground station. The frequency channels of the VOR and the TACAN at each VORTAC facility are "paired" in accordance with a national plan to simplify airborne operation.

1–7. DISTANCE MEASURING EQUIPMENT (DME)

a. In the operation of DME, paired pulses at a specific spacing are sent out from the aircraft (this is the interrogation) and are received at the ground station. The ground station (transponder) then transmits paired pulses back to the aircraft at the same pulse spacing but on a different frequency. The time required for the round trip of this signal exchange is measured in the airborne DME unit and is translated into distance (Nautical Miles) from the aircraft to the ground station.

b. Operating on the line-of-sight principle, DME furnishes distance information with a very high degree of accuracy. Reliable signals may be received at distances up to 199 NM at line-of-sight altitude with an accuracy of better than ½ mile or 3 percent of the distance, whichever is greater. Distance information received from DME equipment is SLANT RANGE distance and not actual horizontal distance.

c. DME operates on frequencies in the UHF spectrum between 962 MHz and 1213 MHz. Aircraft equipped with TACAN equipment will receive distance information from a VORTAC automatically, while aircraft equipped with VOR must have a separate DME airborne unit.

d. VOR/DME, VORTAC, ILS/DME, and LOC/DME navigation facilities established by the FAA provide course and distance information from collocated components under a frequency pairing plan. Aircraft receiving equipment which provides for automatic DME selection assures reception of azimuth and distance information from a common source when designated VOR/DME, VORTAC, ILS/DME, and LOC/DME are selected.

e. Due to the limited number of available frequencies, assignment of paired frequencies is required for certain military noncollocated VOR and TACAN facilities which serve the same area but which may be separated by distances up to a few miles. The military is presently undergoing a program to collocate VOR and TACAN facilities or to assign nonpaired frequencies to those that cannot be collocated.

f. VOR/DME, VORTAC, ILS/DME, and LOC/DME facilities are identified by synchronized identifications which are transmitted on a time share basis. The VOR or localizer portion of the facility is identified by a coded tone modulated at 1020 Hz or a combination of code and voice. The TACAN or DME is identified by a coded tone modulated at 1350 Hz. The DME or TACAN coded identification is transmitted one time for each three or four times that the VOR or localizer coded identification is transmitted. When either the VOR or the DME is inoperative, it is important to recognize which identifier is retained for the operative facility. A single coded identification with a repetition interval of approximately 30 seconds indicates that the DME is operative.

g. Aircraft equipment which provides for automatic DME selection assures reception of azimuth and distance information from a common source when designated VOR/DME, VORTAC and ILS/DME navigation facilities are selected. Pilots are cautioned to disregard any distance displays from automatically selected DME equipment when VOR or ILS facilities, which do not have the DME feature installed, are being used for position determination.

1–8. NAVAID SERVICE VOLUMES

a. Most air navigation radio aids which provide positive course guidance have a designated standard service volume (SSV). The SSV defines the reception limits of unrestricted NAVAIDS which are usable for random/unpublished route navigation.

b. A NAVAID will be classified as restricted if it does not conform to flight inspection signal strength and course quality standards throughout the published SSV. However, the NAVAID should not be considered usable at altitudes below that which could be flown while operating under random route IFR conditions (FAR 91.177), even though these altitudes may lie within the designated SSV. Service volume restrictions are first published in the Notices to Airman (NOTAM) and then with the alphabetical listing of the NAVAID's in the Airport/Facility Directory.

c. Standard Service Volume limitations do not apply to published IFR routes or procedures.

Figure 1–8[1]

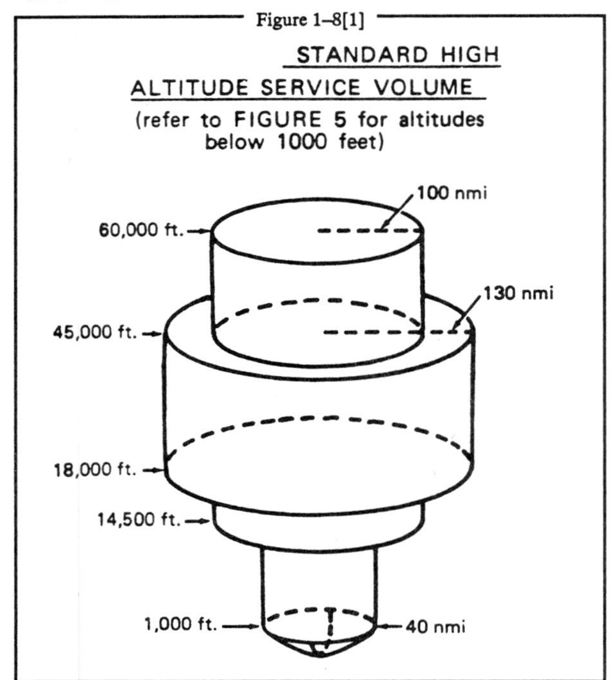

Figure 1–8[2]

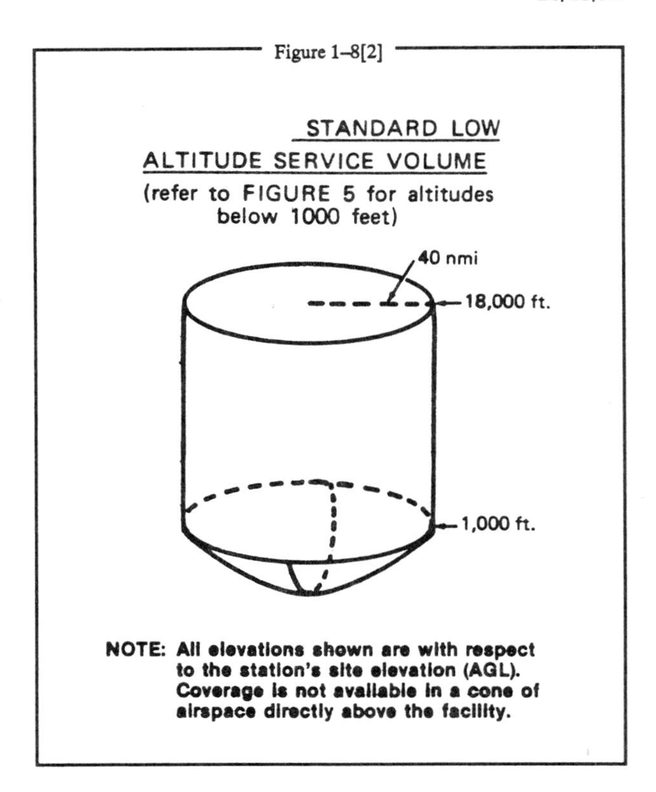

Figure 1–8[3]

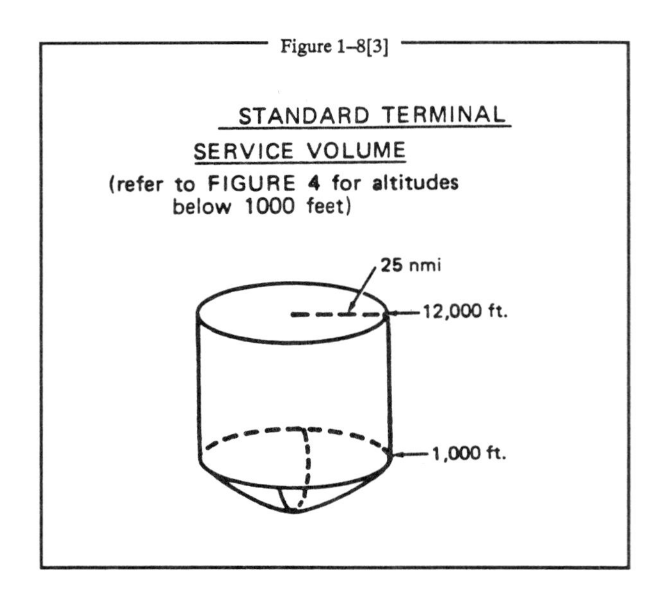

Figure 1–8[4]

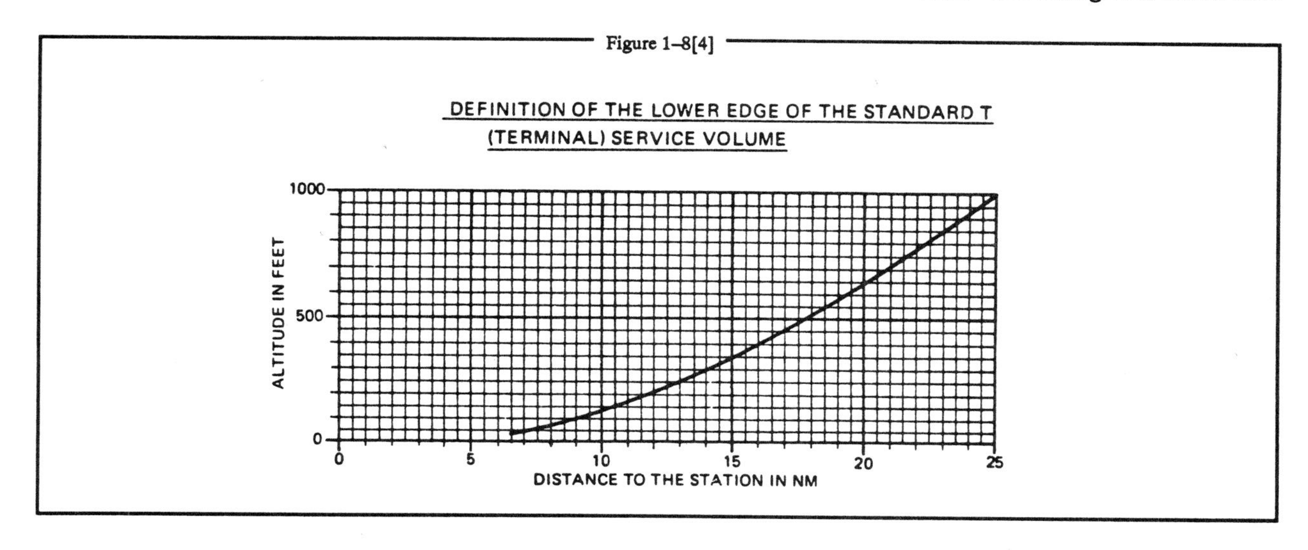

Figure 1–8[5]

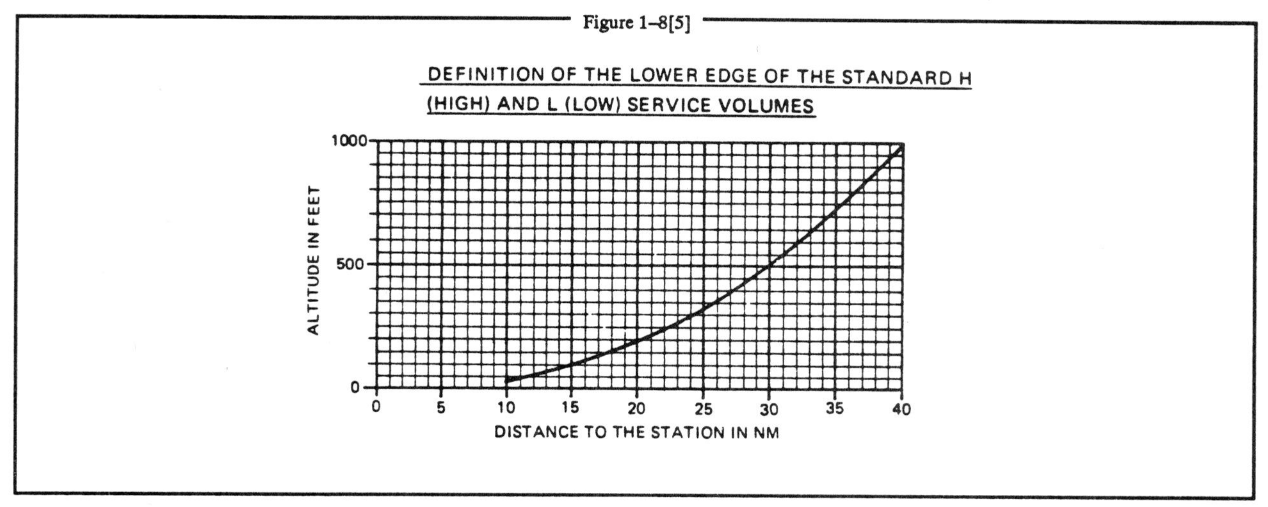

d. VOR/DME/TACAN STANDARD SERVICE VOLUMES (SSV)

Standard service volumes (SSVs) are graphically shown in Figure 1–8[1], Figure 1–8[2], Figure 1–8[3], Figure 1–8[4], and Figure 1–8[5]. The SSV of a station is indicated by using the class designator as a prefix to the station type designation.

EXAMPLE:

TVOR, LDME, and HVORTAC.

Within 25 NM, the bottom of the T service volume is defined by the curve in Figure 1–8[4]. Within 40 NM, the bottoms of the L and H service volumes are defined by the curve in Figure 1–8[5]. (See Table 1–8[1]).

e. NONDIRECTIONAL RADIO BEACON (NDB)

NDBs are classified according to their intended use. The ranges of NDB service volumes are shown in Table 1–8[2]. The distances (radius) are the same at all altitudes.

1–9. MARKER BEACON

a. Marker beacons serve to identify a particular location in space along an airway or on the approach to an instrument runway. This is done by means of a 75 MHz transmitter which transmits a directional signal to be received by aircraft flying overhead. These markers are generally used in conjunction with en route NAVAIDs and ILS as point designators.

b. There are three classes of en route marker beacons: Fan Marker (FM), Low Powered Fan Markers (LFM) and Z Markers. They transmit the letter "R" (dot dash dot) identification, or (if additional markers are in the same area) the letter "K," "P," "X," or "Z."

1. Class FMs are used to provide a positive identification of positions at definite points along

Table 1–8[1]

SSV CLASS DESIGNATOR	ALTITUDE AND RANGE BOUNDARIES
T (Terminal)	From 1000 feet above ground level (AGL) up to and including 12,000 feet AGL at radial distances out to 25 NM.
L (Low Altitude)	From 1000 feet AGL up to and including 18,000 feet AGL at radial distances out to 40 NM.
H (High Altitude)	From 1000 feet AGL up to and including 14,500 feet AGL at radial distances out to 40 NM. From 14,500 feet AGL up to and including 60,000 feet at radial distances out to 100 NM. From 18,000 feet AGL up to and including 45,000 feet AGL at radial distances out to 130 NM.

Table 1–8[2]

CLASS	DISTANCE (RADIUS)
Compass Locator	15 NM
MH	25 NM
H	50 NM*
HH	75 NM

* Service ranges of individual facilities may be less than 50 nautical miles (NM). Restrictions to service volumes are first published as a Notice to Airmen and then with the alphabetical listing of the NAVAID in the Airport/Facility Directory.

the airways. The transmitters have a power output of approximately 100 watts. Two types of antenna array are used with class FMs.

(a) The first type used, and generally referred to as the standard type, produces an elliptical shaped pattern, which, at an elevation of 1,000 feet above the station, is about 4 NM wide and 12 NM long. At 10,000 feet the pattern widens to about 12 NM wide and 35 NM long.

(b) The second array produces a dumbbell or boneshaped pattern, which, at the "handle", is about three miles wide at 1,000 feet. The boneshaped marker is preferred at approach control locations where "timed" approaches are used.

2. The class LFM or low powered FMs have a rated power output of 5 watts. The antenna array produces a circular pattern which appears elongated at right angles to the airway due to the directional characteristics of the aircraft receiving antenna.

3. The Station Location, or Z-Marker, was developed to meet the need for a positive position indicator for aircraft operating under instrument flight conditions to show the pilot when he was passing directly over a low frequency navigational aid. The marker consists of a 5 watt transmitter and a directional antenna array which is located on the range plot between the towers or the loop antennas.

1–9b3 NOTE.—ILS marker beacon information is included in 1–10 *INSTRUMENT LANDING SYSTEMS (ILS).*

1–10. INSTRUMENT LANDING SYSTEM (ILS)

a. GENERAL

1. The ILS is designed to provide an approach path for exact alignment and descent of an aircraft on final approach to a runway.

2. The ground equipment consists of two highly directional transmitting systems and, along the approach, three (or fewer) marker beacons. The directional transmitters are known as the localizer and glide slope transmitters.

3. The system may be divided functionally into three parts:

(a) Guidance information—localizer, glide slope

(b) Range information—marker beacon, DME

(c) Visual information—approach lights, touchdown and centerline lights, runway lights

4. Compass locators located at the Outer Marker (OM) or Middle Marker (MM) may be substituted for marker beacons. DME, when specified in the procedure, may be substituted for the OM.

5. Where a complete ILS system is installed on each end of a runway; (i.e. the approach end of Runway 4 and the approach end of Runway 22) the ILS systems are not in service simultaneously.

b. LOCALIZER

1. The localizer transmitter operates on one of 40 ILS channels within the frequency range of 108.10 to 111.95 MHz. Signals provide the pilot with course guidance to the runway centerline.

2. The approach course of the localizer is called the front course and is used with other functional parts, e.g., glide slope, marker beacons, etc. The localizer signal is transmitted at the far end of the runway. It is adjusted for a course width of (full scale fly-left to a full scale fly-right) of 700 feet at the runway threshold.

3. The course line along the extended centerline of a runway, in the opposite direction to the front course is called the back course.

CAUTION: Unless the aircraft's ILS equipment includes reverse sensing capability, when flying inbound on the back course it is necessary to steer the aircraft in the direction opposite the needle deflection when making corrections from off-course to on-course. This "flying away from the needle" is also required when flying outbound on the front course of the localizer. DO NOT USE BACK COURSE SIGNALS for approach unless a BACK COURSE APPROACH PROCEDURE is published for that particular runway and the approach is authorized by ATC.

4. Identification is in International Morse Code and consists of a three-letter identifier preceded by the letter I (..) transmitted on the localizer frequency.

EXAMPLE:
I-DIA

5. The localizer provides course guidance throughout the descent path to the runway threshold from a distance of 18 NM from the antenna between an altitude of 1,000 feet above the highest terrain along the course line and 4,500 feet above the elevation of the antenna site. Proper off-course indications are provided throughout the following angular areas of the operational service volume:

(a) To 10 degrees either side of the course along a radius of 18 NM from the antenna, and

Figure 1–10[1]

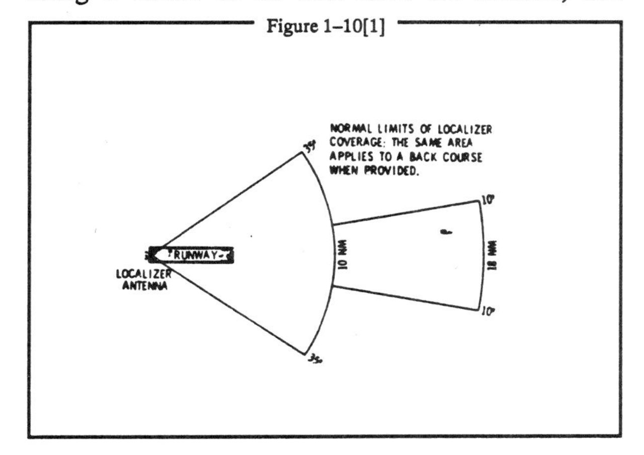

(b) From 10 to 35 degrees either side of the course along a radius of 10 NM. (See Figure 1–10[1])

6. Unreliable signals may be received outside these areas.

c. LOCALIZER-TYPE DIRECTIONAL AID

1. The Localizer-type Directional Aid (LDA) is of comparable use and accuracy to a localizer but is not part of a complete ILS. The LDA course usually provides a more precise approach course than the similar Simplified Directional Facility (SDF) installation, which may have a course width of 6 or 12 degrees. The LDA is not aligned with the runway. Straight-in minimums may be published where alignment does not exceed 30 degrees between the course and runway. Circling minimums only are published where this alignment exceeds 30 degrees.

d. GLIDE SLOPE/GLIDE PATH

1. The UHF glide slope transmitter, operating on one of the 40 ILS channels within the frequency range 329.15 MHz, to 335.00 MHz radiates its signals in the direction of the localizer front course. The term "glide path" means that portion of the glide slope that intersects the localizer.

CAUTION: False glide slope signals may exist in the area of the localizer back course approach which can cause the glide slope flag alarm to disappear and present unreliable glide slope information. Disregard all glide slope signal indications when making a localizer back course approach unless a glide slope is specified on the approach and landing chart.

2. The glide slope transmitter is located between 750 feet and 1,250 feet from the approach end of the runway (down the runway) and offset 250 to 650 feet from the runway centerline. It transmits a glide path beam 1.4 degrees wide. The signal provides descent information for navigation down to the lowest authorized decision height (DH) specified in the approved ILS approach procedure. The glidepath may not be suitable for navigation below the lowest authorized DH and any reference to glidepath indications below that height must be supplemented by visual reference to the runway environment. Glidepaths with no published DH are usable to runway threshold.

3. The glide path projection angle is normally adjusted to 3 degrees above horizontal so that it intersects the MM at about 200 feet and the OM at about 1,400 feet above the runway elevation. The glide slope is normally usable to the distance of 10 NM. However, at some locations, the glide slope has been certified for an extended service volume which exceeds 10 NM.

4. Pilots must be alert when approaching the glidepath interception. False courses and reverse sensing will occur at angles considerably greater than the published path.

5. Make every effort to remain on the indicated glide path (reference: FAR 91.129(d)(2)). Exercise caution: avoid flying below the glide path to assure obstacle/terrain clearance is maintained.

6. The published glide slope threshold crossing height (TCH) *DOES NOT* represent the height of the actual glide path on-course indication above the runway threshold. It is used as a reference for planning purposes which represents the height above the runway threshold that an aircraft's glide slope antenna should be, if that aircraft remains on a trajectory formed by the four-mile-to-middle marker glidepath segment.

7. Pilots must be aware of the vertical height between the aircraft's glide slope antenna and the main gear in the landing configuration and, at the DH, plan to adjust the descent angle accordingly if the published TCH indicates the wheel crossing height over the runway threshold may not be satisfactory. Tests indicate a comfortable wheel crossing height is approximately 20 to 30 feet, depending on the type of aircraft.

e. DISTANCE MEASURING EQUIPMENT (DME)

1. When installed with the ILS and specified in the approach procedure, DME may be used:

(a) In lieu of the OM.

(b) As a back course (BC) final approach fix (FAF).

(c) To establish other fixes on the localizer course.

2. In some cases, DME from a separate facility may be used within Terminal Instrument Procedures (TERPS) limitations:

(a) To provide ARC initial approach segments.

(b) As a FAF for BC approaches.

(c) As a substitute for the OM.

f. MARKER BEACON

1. ILS marker beacons have a rated power output of 3 watts or less and an antenna array designed to produce a elliptical pattern with dimensions, at 1,000 feet above the antenna, of approximately 2,400 feet in width and 4,200 feet in length. Airborne marker beacon receivers with a selective sensitivity feature should always be operated in the "low" sensitivity position for proper reception of ILS marker beacons.

2. Ordinarily, there are two marker beacons associated with an ILS, the OM and MM. Locations with a Category II and III ILS also have an Inner Marker (IM). When an aircraft passes over a marker, the pilot will receive the following indications: (See Table 1–10[1]).

Table 10[1]

MARKER	CODE	LIGHT
OM	– – –	BLUE
MM	. – . –	AMBER
IM	. . . .	WHITE
BC	. . . .	WHITE

(a) The OM normally indicates a position at which an aircraft at the appropriate altitude on the localizer course will intercept the ILS glide path.

(b) The MM indicates a position approximately 3,500 feet from the landing threshold. This is also the position where an aircraft on the glide path will be at an altitude of approximately 200 feet above the elevation of the touchdown zone.

(c) The inner marker (IM) will indicate a point at which an aircraft is at a designated decision height (DH) on the glide path between the MM and landing threshold.

3. A back course marker normally indicates the ILS back course final approach fix where approach descent is commenced.

g. COMPASS LOCATOR

1. Compass locator transmitters are often situated at the MM and OM sites. The transmitters have a power of less than 25 watts, a range of at least 15 miles and operate between 190 and 535 kHz. At some locations, higher powered radio beacons, up to 400 watts, are used as OM compass locators. These generally carry Transcribed Weather Broadcast (TWEB) information.

2. Compass locators transmit two letter identification groups. The outer locator transmits the first two letters of the localizer identification group, and the middle locator transmits the last two letters of the localizer identification group.

h. ILS FREQUENCY

1. See Table 1–10[2] for frequency pairs allocated for ILS.

i. ILS MINIMUMS

1. The lowest authorized ILS minimums, with all required ground and airborne systems components operative, are

(a) Category I—Decision Height (DH) 200 feet and Runway Visual Range (RVR) 2,400 feet (with touchdown zone and centerline lighting, RVR 1800 Category A, B, C; RVR 2000 Category D).

Table 1-10[2]

Localizer MHz	Glide Slope
108.10	334.70
108.15	334.55
108.3	334.10
108.35	333.95
108.5	329.90
108.55	329.75
108.7	330.50
108.75	330.35
108.9	329.30
108.95	329.15
109.1	331.40
109.15	331.25
109.3	332.00
109.35	331.85
109.50	332.60
109.55	332.45
109.70	333.20
109.75	333.05
109.90	333.80
109.95	333.65
110.1	334.40
110.15	334.25
110.3	335.00
110.35	334.85
110.5	329.60
110.55	329.45
110.70	330.20
110.75	330.05
110.90	330.80
110.95	330.65
111.10	331.70
111.15	331.55
111.30	332.30
111.35	332.15
111.50	332.9
111.55	332.75
111.70	333.5
111.75	333.35
111.90	331.1
111.95	330.95

(b) Category II—DH 100 feet and RVR 1,200 feet.

(c) Category IIIA—RVR 700 feet.

1-10i1c NOTE.—Special authorization and equipment are required for Category II and IIIA.

j. INOPERATIVE ILS COMPONENTS

1. Inoperative localizer: When the localizer fails, an ILS approach is not authorized.

2. Inoperative glide slope: When the glide slope fails, the ILS reverts to a nonprecision localizer approach.

1-10j2 NOTE.—Refer to the Inoperative Component Table in the U.S. Government Terminal Procedures Publication (TPP), for adjustments to minimums due to inoperative airborne or ground system equipment.

k. ILS COURSE DISTORTION

1. All pilots should be aware that disturbances to ILS localizer and glide slope courses may occur when surface vehicles or aircraft are operated near the localizer or glide slope antennas. Most ILS installations are subject to signal interference by either surface vehicles, aircraft or both. ILS CRITICAL AREAS are established near each localizer and glide slope antenna.

2. ATC issues control instructions to avoid interfering operations within ILS critical areas at controlled airports during the hours the Airport Traffic Control Tower (ATCT) is in operations as follows:

(a) Weather Conditions—Less than ceiling 800 feet and/or visibility 2 miles.

(1) LOCALIZER CRITICAL AREA—Except for aircraft that land, exit a runway, depart or miss approach, vehicles and aircraft are not authorized in or over the critical area when an arriving aircraft is between the ILS final approach fix and the airport. Additionally, when the ceiling is less than 200 feet and/or the visibility is RVR 2,000 or less, vehicle and aircraft operations in or over the area are not authorized when an arriving aircraft is inside the ILS MM.

(2) GLIDE SLOPE CRITICAL AREA—Vehicles and aircraft are not authorized in the area when an arriving aircraft is between the ILS final approach fix and the airport unless the aircraft has reported the airport in sight and is circling or side stepping to land on a runway other than the ILS runway.

(b) Weather Conditions—At or above ceiling 800 feet and/or visibility 2 miles.

(1) No critical area protective action is provided under these conditions

(2) If an aircraft advises the tower that an AUTOLAND or COUPLED approach will be conducted, an advisory will be promptly issued if a vehicle or aircraft will be in or over a critical area when the arriving aircraft is inside the ILS MM.

EXAMPLE:
GLIDE SLOPE SIGNAL NOT PROTECTED.

(3) Aircraft holding below 5000 feet between the outer marker and the airport may cause localizer signal variations for aircraft conducting the ILS Approach. Accordingly, such holding is not authorized when weather or visibility conditions are less than ceiling 800 feet and/or visibility 2 miles.

(4) Pilots are cautioned that vehicular traffic not subject to ATC may cause momentary deviation to ILS course or glide slope signals. Also, critical areas are not protected at uncontrolled airports or at airports with an operating control tower when weather or visibility conditions are above those requiring protective measures. Aircraft conducting coupled or autoland operations should be especially alert in monitoring automatic flight control systems. (See Figure 1-10[2].)

1-11. SIMPLIFIED DIRECTIONAL FACILITY (SDF)

a. The SDF provides a final approach course similar to that of the ILS localizer. It does not provide glide slope information. A clear understanding of the ILS localizer and the additional factors listed below completely describe the operational characteristics and use of the SDF.

b. The SDF transmits signals within the range of 108.10 to 111.95 MHz.

c. The approach techniques and procedures used in an SDF instrument approach are essentially the same as those employed in executing a standard localizer approach except the SDF course may not be aligned with the runway and the course may be wider, resulting in less precision.

d. Usable off—course indications are limited to 35 degrees either side of the course centerline. Instrument indications received beyond 35 degrees should be disregarded.

e. The SDF antenna may be offset from the runway centerline. Because of this, the angle of convergence between the final approach course and the runway bearing should be determined by reference to the instrument approach procedure chart. This angle is generally not more than 3 degrees. However, it should be noted that inasmuch as the approach course originates at the antenna site, an approach which is continued beyond the runway threshold will lead the aircraft to the SDF offset position rather than along the runway centerline.

f. The SDF signal is fixed at either 6 degrees or 12 degrees as necessary to provide maximum flyability and optimum course quality.

g. Identification consists of a three-letter identifier transmitted in Morse Code on the SDF frequency. The appropriate instrument approach chart will indicate the identifier used at a particular airport.

1-12. MICROWAVE LANDING SYSTEM (MLS)

a. GENERAL

1. The MLS provides precision navigation guidance for exact alignment and descent of aircraft on approach to a runway. It provides azimuth, elevation, and distance.

2. Both lateral and vertical guidance may be displayed on conventional course deviation indicators or incorporated into multipurpose cockpit displays. Range information can be displayed by conventional DME indicators and also incorporated into multipurpose displays.

3. The MLS initially supplements and will eventually replace ILS as the standard landing system in the United States for civil, military and international civil aviation. The transition plan assures duplicate ILS and MLS facilities where needed to protect current users of ILS. At international airports ILS service is protected to the year 1995.

4. The system may be divided into five functions:

(a) Approach azimuth.

(b) Back azimuth.

(c) Approach elevation.

(d) Range.

(e) Data communications.

5. The *standard configuration* of MLS ground equipment includes:

(a) An azimuth station to perform functions (a) and (e) above. In addition to providing azimuth navigation guidance, the station transmits basic data which consists of information associated directly with the operation of the landing system, as well as advisory data on the performance of the ground equipment.

(b) An elevation station to perform function (c).

(c) Precision Distance Measuring Equipment (DME/P) to perform function (d). The DME/P provides continuous range information that is compatible with standard navigation DME but has improved accuracy and additional channel capabilities.

6. MLS Expansion Capabilities—The standard configuration can be expanded by adding one or more of the following functions or characteristics.

(a) Back azimuth—Provides lateral guidance for missed approach and departure navigation.

(b) Auxiliary data transmissions—Provides *additional* data, including refined airborne positioning, meteorological information, runway status, and other supplementary information.

(c) Larger proportional guidance.

7. MLS identification is a four-letter designation starting with the letter M. It is transmitted in International Morse Code at least six times per minute by the approach azimuth (and back azimuth) ground equipment.

b. APPROACH AZIMUTH GUIDANCE

Figure 1–10[2]

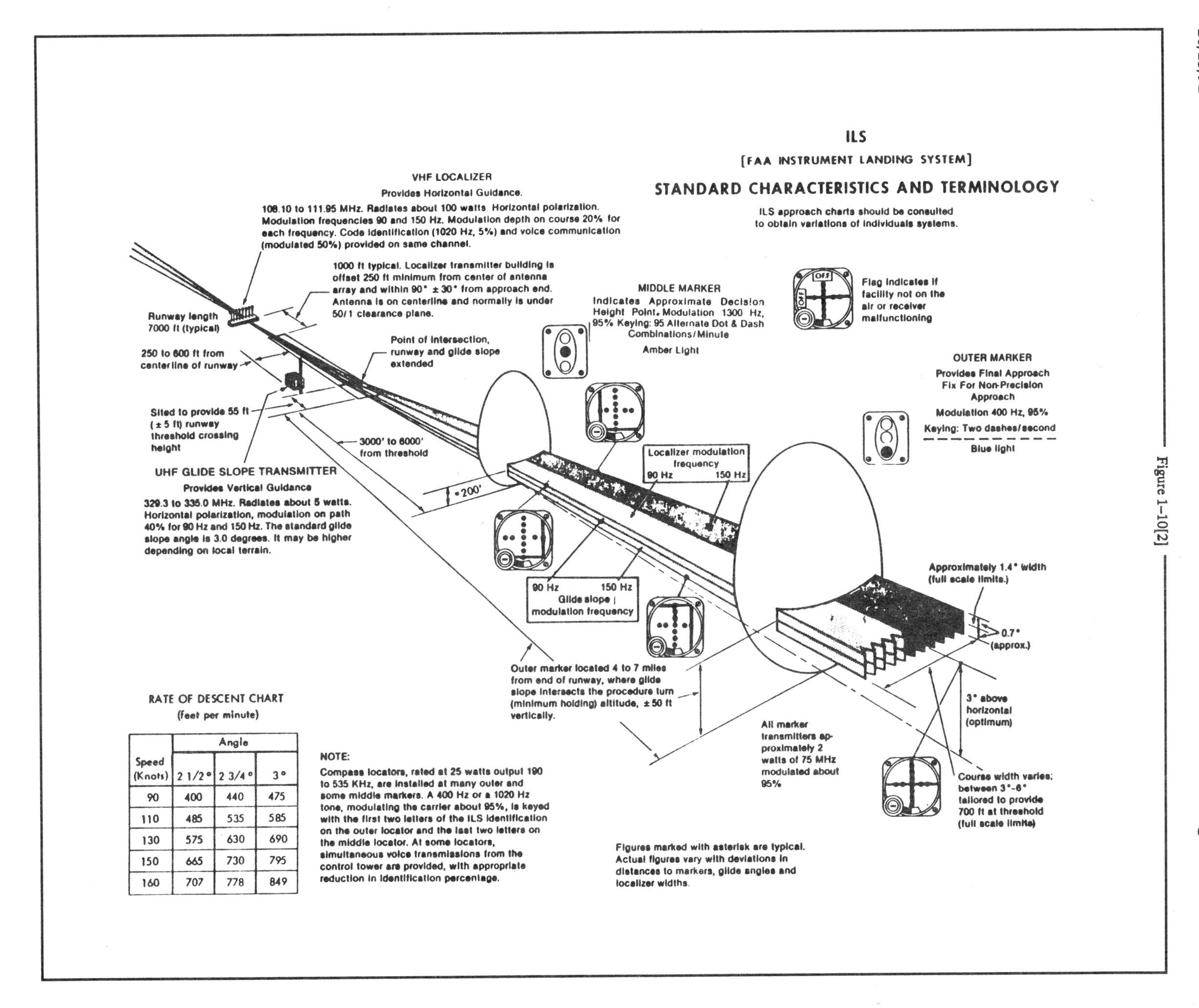

Speed (Knots)	Angle 2 1/2°	2 3/4°	3°
90	400	440	475
110	485	535	585
130	575	630	690
150	665	730	795
160	707	778	849

1. The azimuth station transmits MLS angle and data on one of 200 channels within the frequency range of 5031 to 5091 MHz. (See Table 1–12[1]) for MLS angle and data channeling, and (See Table 1–12[2]) for the DME.

Table 1–12[1]

CHANNEL NUMBER	FREQUENCY (MHz)	CHANNEL NUMBER	FREQUENCY (MHz)	CHANNEL NUMBER	FREQUENCY (MHz)	CHANNEL NUMBER	FREQUENCY (MHz)	CHANNEL NUMBER	FREQUENCY (MHz)
500	5031.0	540	5043.0	580	5055.0	620	5067.0	660	5079.0
501	5031.3	—	—	—	—	—	—	—	—
502	5031.6	—	—	—	—	—	—	—	—
503	5031.9	—	—	—	—	—	—	—	—
504	5032.2	—	—	—	—	—	—	—	—
505	5032.5	545	5044.5	585	5056.5	625	5068.5	665	5080.5
—	—	—	—	—	—	—	—	—	—
—	—	—	—	—	—	—	—	—	—
—	—	—	—	—	—	—	—	—	—
—	—	—	—	—	—	—	—	—	—
510	5034.0	550	5046.0	590	5058.0	630	5070.0	670	5982.0
—	—	—	—	—	—	—	—	—	—
—	—	—	—	—	—	—	—	—	—
—	—	—	—	—	—	—	—	—	—
—	—	—	—	—	—	—	—	—	—
515	5035.5	555	5047.5	595	5059.5	635	5071.5	675	5083.5
—	—	—	—	—	—	—	—	—	—
—	—	—	—	—	—	—	—	—	—
—	—	—	—	—	—	—	—	—	—
—	—	—	—	—	—	—	—	—	—
520	5037.0	560	5049.0	600	5061.0	640	5073.0	680	5085.0
—	—	—	—	—	—	—	—	—	—
—	—	—	—	—	—	—	—	—	—
—	—	—	—	—	—	—	—	—	—
—	—	—	—	—	—	—	—	—	—
525	5038.5	565	5050.5	605	5062.5	645	5074.5	685	5086.5
—	—	—	—	—	—	—	—	—	—
—	—	—	—	—	—	—	—	—	—
—	—	—	—	—	—	—	—	—	—
—	—	—	—	—	—	—	—	—	—
530	5040.0	570	5052.0	610	5064.0	650	5076.0	690	5088.0
—	—	—	—	—	—	—	—	—	—
—	—	—	—	—	—	—	—	—	—
—	—	—	—	—	—	—	—	—	—
—	—	—	—	—	—	—	—	—	—
535	5041.5	575	5053.5	615	5065.5	655	5077.5	695	5089.5
—	—	—	—	—	—	—	—	696	5089.8
—	—	—	—	—	—	—	—	697	5090.1
—	—	—	—	—	—	—	—	698	5090.4
—	—	—	—	—	—	—	—	699	5090.7

Table 1–12[2]

DME CHANNEL (NUMBER)	VHF CHANNEL (MHz)	C—BAND CHANNEL (MHz)	ANGLE CHANNEL (NUMBER)	INTERROGATOR FREQUENCY (MHz)	NON-PRECISION INTERROGATOR PULSE CODE (USEC)	PRECISION INTERROGATOR PULSE CODE (USEC)	TRANSPONDER FREQUENCY (MHz)	TRANSPONDER PULSE CODE (USEC)
1X				1025	12		962	12
1Y				1025	36		1088	30
2X				1026	12		963	12
2Y				1026	36		1089	30
3X				1027	12		964	12
3Y				1027	36		1090	30
4X				1028	12		965	12
4Y				1028	36		1091	30
5X				1029	12		966	12
5Y				1029	36		1092	30
6X				1030	12		967	12
6Y				1030	36		1093	30
7X				1031	12		968	12
7Y				1031	36		1094	30
8X				1032	12		969	12
8Y				1032	36		1095	30
9X				1033	12		970	12
9Y				1033	36		1096	30
10X				1034	12		971	12
10Y				1034	36		1097	30
11X				1035	12		972	12
11Y				1035	36		1098	30
12X				1036	12		973	12
12Y				1036	36		1099	30
13X				1037	12		974	12
13Y				1037	36		1100	30
14X				1038	12		975	12
14Y				1038	36		1101	30
15X				1039	12		976	12
15Y				1039	36		1102	30
16X				1040	12		977	12
16Y				1040	36		1103	30
17X	108.00	—	—	1041	12	—	978	12
17Y	108.05	5043.00	540	1041	36	42	1104	30
17Z	—	5043.30	541	1041	21	27	1104	15
18X	108.10	5031.00	500	1042	12	18	979	12
18W	—	5031.30	501	1042	24	33	979	24
18Y	108.15	5043.60	542	1042	36	42	1105	30
18Z	—	5043.90	543	1042	21	27	1105	15
19X	108.20	—	—	1043	12	—	980	12

2. The equipment is normally located about 1,000 feet beyond the stop end of the runway, but there is considerable flexibility in selecting sites. For example, for heliport operations the azimuth transmitter can be collocated with the elevation transmitter.

3. The azimuth coverage (See Figure 1–12[1].) extends:

(a) Laterally, at least 40 degrees on either side of the runway.

(b) In elevation, up to an angle of 15 degrees—and to at least 20,000 feet.

(c) In range, to at least 20 NM.

c. BACK AZIMUTH GUIDANCE.

1. The back azimuth transmitter is essentially the same as the approach azimuth transmitter. However, the equipment transmits at a somewhat lower data rate because the guidance accuracy requirements are not as stringent as for the landing approach. The equipment operates on the same frequency as the approach azimuth but at a different time in the transmission sequence.

2. The equipment is normally located about 1,000 feet in front of the approach end of the runway. On runways that have MLS service on both ends (e.g., Runway 9 and 27), the azimuth equipment can be switched in their operation from the approach azimuth to the back azimuth and vice versa, and thereby reduce the amount of equipment required.

3. The back azimuth provides coverage as follows:

(a) Laterally, at least 40 degrees on either side of the runway centerline.

(b) In elevation, up to an angle of 15 degrees.

(c) In range, to at least 7 NM from the runway stop end.(See Figure 1–12[1].)

1–12c3c NOTE.—The actual coverage is normally the same as for the approach azimuth.

d. ELEVATION GUIDANCE

1. The elevation station transmits signals on the same frequency as the azimuth station. A single frequency is time-shared between angle and data functions.

2. The elevation transmitter is normally located about 400 feet from the side of the runway between runway threshold and the touchdown zone.

Figure 1–12[1]

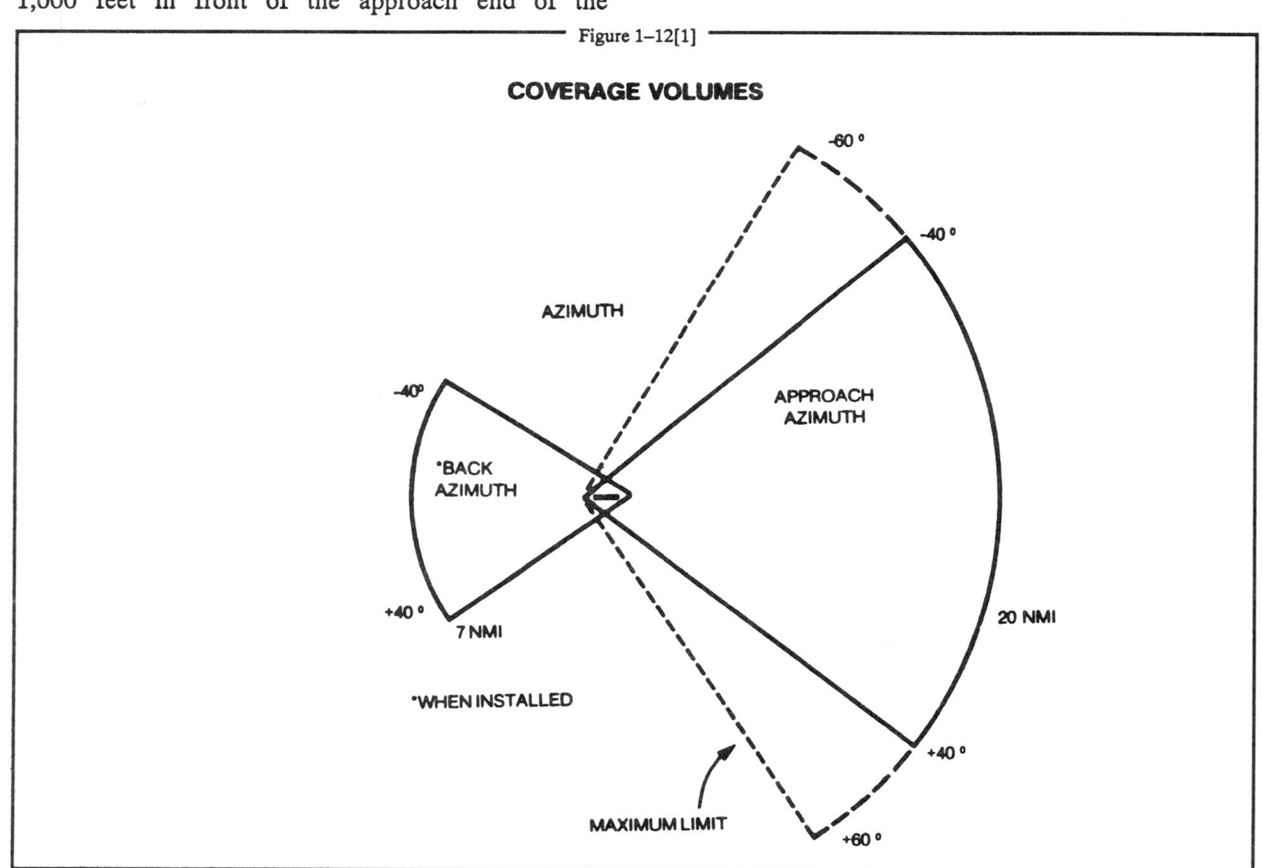

3. Elevation coverage is provided in the same airspace as the azimuth guidance signals:

(a) In elevation, to at least +15 degrees.

(b) Laterally, 40 degrees on either side of the runway centerline.

Figure 1–12[2]

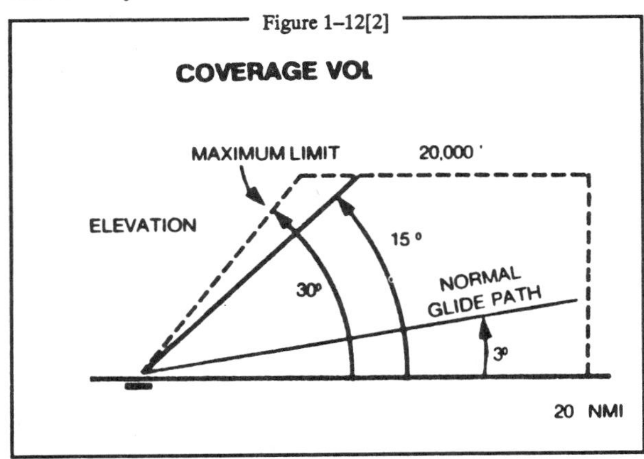

(c) In range, to at least 20 NM.(See Figure 1–12[2].)

e. RANGE GUIDANCE

1. The MLS Precision Distance Measuring Equipment (DME/P) functions the same as the navigation DME described in 1–7, but there are some technical differences. The beacon transponder operates in the frequency band 962 to 1105 MHz and responds to an aircraft interrogator. The MLS DME/P accuracy is improved to be consistent with the accuracy provided by the MLS azimuth and elevation stations.

2. A DME/P channel is paired with the azimuth and elevation channel. A complete listing of the 200 paired channels of the DME/P with the angle functions is contained in FAA Standard 022 (MLS Interoperability and Performance Requirements).

3. The DME/P is an integral part of the MLS and is installed at all MLS facilities unless a waiver is obtained. This occurs infrequently—and only at outlying, low density airports where marker beacons or compass locators are already in place.

f. DATA COMMUNICATIONS

1. The data transmission can include both basic and auxiliary data words. All MLS facilities transmit basic data. In the future facilities at some airports, including most high density airports, will also transmit auxilary data.

2. Coverage limits—MLS data are transmitted throughout the azimuth (and back azimuth when provided) coverage sectors.

3. Basic data content—Representative data include:

(a) Station identification.

(b) Exact locations of azimuth, elevation and DME/P stations (for MLS receiver processing functions).

(c) Ground equipment performance level.

(d) DME/P channel and status.

4. Auxiliary data content—Representative data include:

(a) 3–D locations of MLS equipment.

(b) Waypoint coordinates.

(c) Runway conditions.

(d) Weather (e.g., RVR, ceiling, altimeter setting, wind, wake vortex, wind shear).

g. OPERATIONAL FLEXIBILITY. The MLS has the capability to fulfill a variety of needs in the transition, approach, landing, missed approach and departure phases of flight. For example: Curved and segmented approaches; selectable glide path angles; accurate 3–D positioning of the aircraft in space; and the establishment of boundaries to ensure clearance from obstructions in the terminal area. While many of these capabilities are available to any MLS-equipped aircraft, the more sophisticated capabilities (such as curved and segmented approaches) are dependent upon the particular capabilities of the airborne equipment.

h. SUMMARY

1. Accuracy. The MLS provides precision three-dimensional navigation guidance accurate enough for all approach and landing maneuvers.

2. Coverage. Accuracy is consistent throughout the coverage volumes. (See Figure 1–12[3].)

Figure 1–12[3]

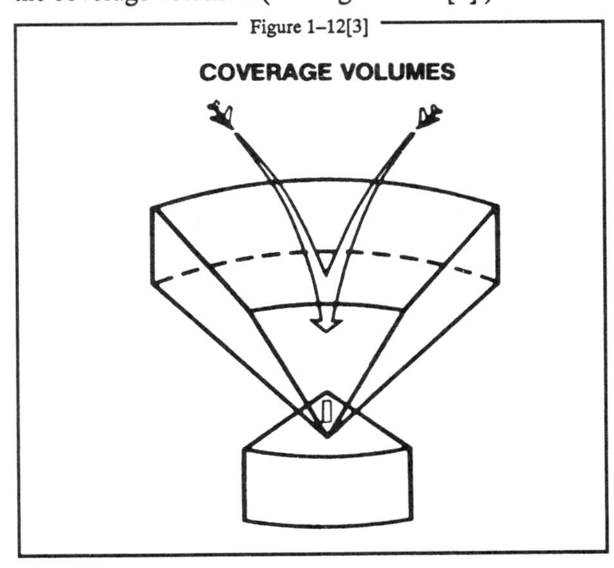

3. Environment. The system has low susceptibility to interference from weather conditions and airport ground traffic.

4. Channels. MLS has 200 channels—enough for any foreseeable need.

5. Data. The MLS transmits ground-air data messages associated with the systems operation.

6. Range information. Continuous range information is provided with an accuracy of about 100 feet.

1–13. INTERIM STANDARD MICROWAVE LANDING SYSTEM (ISMLS)

a. The ISMLS is designed to provide approach information similar to the ILS for an aircraft on final approach to a runway. The system provides both lateral and vertical guidance which is displayed on a conventional course deviation indicator or approach horizon. Operational performance and coverage areas are similar to the ILS system.

b. ISMLS operates in the C-band microwave frequency range (about 5000 MHz). ISMLS signals will not be received by unmodified VHF/UHF ILS receivers. Aircraft with ISMLS must be equipped with a C-band receiving antenna in addition to other special equipment mentioned below. The C-band antenna limits reception of the signal to an angle of about 50 degrees from the inbound course. An aircraft so equipped will not receive the ISMLS signal until flying a magnetic heading within 50 degrees either side of the inbound course. Because of this, ISMLS procedures are designed to restrict the use of the ISMLS signal until the aircraft is in position for the final approach. Transition to the ISMLS, holding and procedure turns at the ISMLS facility must be predicated on other navigation aids such as NDB, VOR, etc. Once established on the approach course inbound, the system can be flown the same as an ILS. No back course is provided.

c. The ISMLS consists of the following basic components:

1. C-Band (5000 MHz–5030 MHz) localizer.

2. C-Band (5220 MHz–5250 MHz) glide path.

3. VHF marker beacons (75 MHz).

4. A VHF/UHF ILS receiver modified to receive ISMLS signals.

5. C-Band antenna.

6. Converter unit.

7. A Microwave/ILS Mode Control.

d. Identification consists of a three letter Morse Code identifier preceded by the Morse Code for "M" (——) (e.g., M—STP). The "M" distinguishes this system from ILS which is preceded by the Morse Code for "I" (..) (e.g., I-STP).

e. Approaches published in conjunction with the ISMLS are identified as "MLS Rwy-(Interim)." The frequency displayed on the ISMLS approach chart is a VHF frequency. ISMLS frequencies are tuned by setting the receiver to the listed VHF frequencies. When the ISMLS mode is selected, receivers modified to accept ISMLS signals receive a paired C-band frequency that is processed by the receiver.

CAUTION: Pilots should not attempt to fly ISMLS procedures unless the aircraft is so equipped.

1–14. NAVAID IDENTIFIER REMOVAL DURING MAINTENANCE

During periods of routine or emergency maintenance, coded identification (or code and voice, where applicable) is removed from certain FAA NAVAIDs. Removal of identification serves as a warning to pilots that the facility is officially off the air for tune-up or repair and may be unreliable even though intermittent or constant signals are received.

1–14 NOTE.—During periods of maintenance VHF ranges may radiate a T—E—S—T code (– . ... –).

1–15. NAVAIDS WITH VOICE

a. Voice equipped en route radio navigational aids are under the operational control of either an FAA FSS or an approach control facility. Most NAVAIDs are remotely operated.

b. Unless otherwise noted on the chart, all radio navigation aids operate continuously except during interruptions for voice transmissions on the same frequencies where simultaneous transmission is not available, and during shutdowns for maintenance. Hours of operation of facilities not operating continuously are annotated on charts and in the Airport/Facility Directory.

1–16. USER REPORTS ON NAVAID PERFORMANCE

a. Users of the National Airspace System (NAS) can render valuable assistance in the early correction of NAVAID malfunctions by reporting their observations of undesirable NAVAID performance. Although NAVAID's are monitored by electronic detectors, adverse effects of electronic interference, new obstructions or changes in terrain near the NAVAID can exist without detection by the ground monitors. Some of the characteristics of malfunction or deteriorating performance which should be reported are: erratic course or bearing indications; intermittent, or full, flag alarm; garbled, missing or obviously improper coded identification; poor quality communications reception; or, in the case of frequency interference, an audible hum or tone accompanying radio communications or NAVAID identification.

b. Reporters should identify the NAVAID, location of the aircraft, time of the observation, type of aircraft and describe the condition observed; the type of receivers in use is also useful information. Reports can be made in any of the following ways:

1. Immediate report by direct radio communication to the controlling Air Route Traffic Control Center (ARTCC), Control Tower, or FSS. This method provides the quickest result.

2. By telephone to the nearest FAA facility.

3. By FAA Form 8000–7, Safety Improvement Report, a postage-paid card designed for this purpose. These cards may be obtained at FAA FSSs, General Aviation District Offices, Flight Standards District Offices, and General Aviation Fixed Base Operations.

c. In aircraft that have more than one receiver, there are many combinations of possible interference between units. This can cause either erroneous navigation indications or, complete or partial blanking out of the communications. Pilots should be familiar enough with the radio installation of the particular airplanes they fly to recognize this type of interference.

1–17. LORAN

a. LORAN—A service has been terminated in the U.S. coastal confluence region.

b. LORAN—C was developed to provide the Department of Defense with a radio navigation capability having longer range and much greater accuracy than its predecessor, LORAN—A. LORAN—C was subsequently selected by the U.S. Government as the radio navigation system for civil marine use in the U.S. coastal areas. Later, it was adopted as a supplemental means of navigation for civil aviation.

c. Operation

LORAN—C is a pulsed, hyperbolic system, operating in the 90–110 kHz frequency band. The system is based upon measurement of the difference in time of arrival of pulses of radio—frequency energy radiated by a group, or chain of transmitters which are separated by hundreds of miles. Within a chain, one station is designated as the Master station (M), and the other stations are designated as Secondary stations, Whiskey (W), X—ray (X), Yankee (Y), and Zulu (Z). Signals transmitted from the Secondary stations are synchronized with those from the Master station. The measurement of a time-difference (TD) is made by a receiver that achieves accuracy by comparing a zero crossing of a specified radio- frequency cycle within the pulses received from the Master and Secondary stations of a chain. Only groundwave signals are used for air navigation in the National Airspace System. Within the groundwave range, LORAN—C will provide the user, who uses an adequate receiver, with predictable accuracy of 0.25 nautical miles (2 drms) or better. All accuracy is dependent on the user's location within the signal coverage area of the chain of stations.

d. Signal coverage

The LORAN—C system provides signal coverage throughout the conterminous 48 states and much of Alaska. The so called "Mid-Continent Gap" was closed officially during May 1991.

e. Aviation uses

Note: The use of LORAN—C for navigation in the National Airspace System (NAS) has increased considerably in recent years. In 1990, LORAN—C installations in aircraft were estimated to be in excess of 100,000. Most of these installations are for VFR use only. Approximately 10 percent of the aircraft LORAN—C installations are approved for IFR use during en route and terminal operations. However, none of the installations are approved (as of August 1991) for use during nonprecision instrument approach operations. Several manufacturers of LORAN—C airborne equipment currently are seeking FAA certification of receiver models to be used for nonprecision instrument approach operations. Once approved, the approach model can also be used for en route and terminal use. Since LORAN—C receivers will be installed in aircraft with either of two types of IFR approval; (1) en route and terminal or (2) en route, terminal, and approach, or with a basic VFR capability only, pilots must be aware of the authorized operational approval level of the receiver installed in their aircraft. Approval information is contained in the Aircraft Flight Manual Supplement, on FAA Form 337 or in aircraft maintenance records, or possibly by a placard installed near or on the control panel. Pilots must familiarize themselves with the above referenced documents to verify the approval level of the LORAN—C receiver they are operating so the requirements of FAR 91.9 are met. All LORAN—C systems (both the receiver and its installation) installed in aircraft and used in the NAS must have FAA approval that is traceable through aircraft records.

f. Planned expansion

As part of the FAA's continued development of a national aviation LORAN—C system, 196 signal monitors were installed in VOR stations across the country to monitor the system. The monitors will provide data as the source of correction values published with approach charts to ensure the accuracy of LORAN—C during nonprecision approaches. The correction values will adjust for seasonal errors that degrade LORAN—C accuracy. By the end of 1992, the Automatic Blink System should be operational. Currently manually activated, signal Blink provides a warning to LORAN—C receivers that a LORAN—C signal is out of tolerance. On November 15, 1990, 10 public use LORAN—C approaches were approved for use. Additional approaches will be approved as soon as resources permit. The FAA has a list of 500 proposed LORAN—C approaches it is considering. The list was compiled when each state provided the FAA the names of the top 10 priority sites within each state.

g. Notices to Airmen (NOTAMs) are issued for LORAN—C chain or station outages. Domestic NOTAM (D)'s are issued under the identifier "LRN." International NOTAMs are issued under the KNMH series. Pilots may obtain these NOTAMs from FSS briefers upon request.

h. LORAN—C status information

Prerecorded telephone answering service messages pertaining to LORAN—C are available in Table 1–17[1] and Table 1–17[2].

Table 1–17[1]

RATE	CHAIN	TELEPHONE
5930	Canadian East Coast	709–454–3261*
7980	Southeast U.S.	904–569–5241
8970	Great Lakes	607–869–5395
9960	Northeast U.S.	607–869–5395

*St. Anthony, Newfoundland, Canada

Information can also be obtained directly from the office of the Coordinator of Chain Operations (COCO) for each chain. The following telephone numbers are for each COCO office:

Table 1–17[2]

RATE	CHAIN	TELEPHONE	LOCATION
4990	Central Pacific	808–247–5591	Kaneohe, HI
5930	Canadian East Coast	709–454–2392	St. Anthony, NF
5990	Canadian West Coast	604–666–0472	Vancouver, BC
7930	North Atlantic	011–44–1–409–4758	London, UK
7960	Gulf of Alaska	907–487–5583	Kodiak, AK
7970	Norwegian Sea	011–44–1–409–4758	London, UK
7980	Southeast U.S.	205–899–5225	Malone, FL
7990	Mediterranean Sea	011–44–1–409–4758	London, UK
8290	North Central U.S.	707–987–2911	Middletown, CA
8970	Great Lakes	607–869–5393	Seneca, NY
9610	South Central U.S.	205–899–5225	Malone, FL
9940	West Coast U.S.	707–987–2911	Middletown, CA
9960	Northeast U.S.	607–869–5393	Seneca, NY
9970	Northwest Pacific	415–437–3224	San Francisco, CA
9990	North Pacific	907–487–5583	Kodiak, AK

1–18. OMEGA AND OMEGA/VLF NAVIGATION SYSTEMS

a. OMEGA

1. OMEGA is a network of eight transmitting stations located throughout the world to provide worldwide signal coverage. These stations transmit in the Very Low Frequency (VLF) band. Because of the low frequency, the signals are receivable to ranges of thousands of miles. The stations are located in Norway, Liberia, Hawaii (USA), North Dakota (USA), La Reunion, Argentina, Australia, and Japan.

2. Presently each station transmits on four basic navigational frequencies: 10.2 kHz, 11.05 kHz, 11.3 kHz, and 13.6 kHz, in sequenced format. This time sequenced format prevents interstation signal interference. With eight stations and a silent .2–second interval between each transmission, the entire cycle repeats every 10 seconds.

3. In addition to the four basic navigational frequencies, each station transmits a unique navigation frequency. An OMEGA station is said to be operating in full format when the station transmits on the basic frequencies plus the unique frequency. Unique frequencies are presently assigned as follows: (See Table 1–18[1]).

Table 1–18[1]

STATION	LOCATION	FREQUENCY
Station A	Norway	12.1 kHz
Station B	Liberia	12.0 kHz
Station C	Hawaii	11.8 kHz
Station D	North Dakota	13.1 kHz
Station E	La Reunion	12.3 kHz
Station F	Argentina	12.9 kHz
Station G	Australia	13.0 kHz
Station H	Japan	12.8 kHz

b. VLF

1. The U.S. Navy operates a communications system in the VLF band. The stations are located worldwide and transmit at powers of 500–1000 kW. Some airborne OMEGA receivers have the capability to receive and process these VLF signals for navigation in addition to OMEGA signals. The VLF stations generally used for navigation are located in Australia, Japan, England, Hawaii and on the U.S. mainland in Maine, Washington state, and Maryland.

2. Although the Navy does not object to the use of VLF communications signals for navigation, the system is not dedicated to navigation. Signal format, transmission, and other parameters of the VLF system are subject to change at the Navy's discretion. The VLF communications stations are individually shut down for scheduled maintenance for a few hours each week. Regular NOTAM service regarding the VLF system or station status is not available. However, the Naval Observatory provides a taped message concerning phase differences, phase values, and shutdown information for both the VLF communications network and the OMEGA system (phone 202–653–1757).

c. Operational Use of OMEGA and OMEGA/VLF

1. The OMEGA navigation network is capable of providing consistent fixing information to an accuracy of plus or minus 2 NM depending upon the level of sophistication of the receiver/processing system. OMEGA signals are affected by propagation variables which may degrade fix accuracy. These variables include daily variation of phase velocity, polar cap absorption, and sudden solar activity. Daily compensation for variation within the receiver/processor, or occasional excessive solar activity and its effect on OMEGA, cannot be accurately forecast or anticipated. If an unusual amount of solar activity disturbs the OMEGA signal enlargement paths to any extent, the U.S. Coast Guard advises the FAA and an appropriate NOTAM is sent.

2. At 16 minutes past each hour, WWV (Fort Collins, Colorado) broadcasts a message concerning the status of each OMEGA station, signal irregularities, and other information concerning OMEGA. At 47 minutes past each hour, WWVH (Hawaii) broadcasts similar information. The U.S. Coast Guard provides a taped OMEGA status report (703–866–3801). NOTAMS concerning OMEGA are available through any FSS. OMEGA NOTAMs should be requested by OMEGA station name.

3. The FAA has recognized OMEGA and OMEGA/VLF systems as an additional means of en route IFR navigation in the conterminous United States and Alaska when approved in accordance with FAA guidance information. Use of OMEGA or OMEGA/VLF requires that all navigation equipment otherwise required by the Federal Aviation Regulations be installed and operating. When flying RNAV routes, VOR and DME equipment is required.

4. The FAA recognizes the use of the Naval VLF communications system as a supplement to OMEGA, but not the sole means of navigation.

1–19. VHF DIRECTION FINDER

a. The VHF Direction Finder (VHF/DF) is one of the common systems that helps pilots without their being aware of its operation. It is a ground based radio receiver used by the operator of the ground station. FAA facilities that provide VHF/DF service are identified in the Airport/Facility Directory.

b. The equipment consists of a directional antenna system and a VHF radio receiver.

c. The VHF/DF receiver display indicates the magnetic direction of the aircraft from the ground station each time the aircraft transmits.

d. DF equipment is of particular value in locating lost aircraft and in helping to identify aircraft on radar. (Reference.—Direction Finding Instrument Approach Procedure, paragraph 6–13).

1–20. INERTIAL NAVIGATION SYSTEM (INS)

A totally self-contained navigation system, comprised of gyros, accelerometers, and a navigation computer, which provides aircraft position and navigation information in response to signals resulting from inertial effects on system components, and does not require information from external references. INS is aligned with accurate position information prior to departure, and thereafter calculates its position as it progresses to the destination. By programming a series of waypoints, the system will navigate along a predetermined track. New waypoints can be inserted at any time if a revised routing is desired. INS accuracy is very high initially following alignment, and decays with time at the rate of about 1–2 nautical miles per hour. Position update alignment can be accomplished inflight using ground based references, and many INS systems now have sophisticated automatic update using dual DME and or VOR inputs. INS may be approved as the sole means of navigation or may be used in combination with other systems.

1–21. DOPPLER RADAR

A semiautomatic self-contained dead reckoning navigation system (radar sensor plus computer) which is not continuously dependent on information derived from ground based or external aids. The system employs radar signals to detect and measure ground speed and drift angle, using the aircraft compass system as its directional reference. Doppler is less accurate than INS or OMEGA however, and the use of an external reference is required for periodic updates if acceptable position accuracy is to be achieved on long range flights.

1–22. FLIGHT MANAGEMENT SYSTEM (FMS)

A computer system that uses a large data base to allow routes to be preprogrammed and fed into the system by means of a data loader. The system is constantly updated with respect to position accuracy by reference to conventional navigation aids. The sophisticated program and its associated data base insures that the most appropriate aids are automatically selected during the information update cycle.

1–23. GLOBAL POSITIONING SYSTEM (GPS)

A space-base radio positioning, navigation, and time-transfer system being developed by Department of Defense. When fully deployed, the system is intended to provide highly accurate position and velocity information, and precise time, on a continuous global basis, to an unlimited number of properly equipped users. The system will be unaffected by weather, and will provide a worldwide common grid reference system. The GPS concept is predicated upon accurate and continuous knowledge of the spatial position of each satellite in the system with respect to time and distance from a transmitting satellite to the user. The GPS receiver automatically selects appropriate signals from the satellites in view and translates these into a three-dimensional position, velocity, and time. Predictable system accuracy for civil users is projected to be 100 meters horizontally. Performance standards and certification criteria have not yet been established.

1–24 thru 1–30. RESERVED

Section 2. RADAR SERVICES AND PROCEDURES

1–31. RADAR

a. Capabilities

1. Radar is a method whereby radio waves are transmitted into the air and are then received when they have been reflected by an object in the path of the beam. *Range* is determined by measuring the time it takes (at the speed of light) for the radio wave to go out to the object and then return to the receiving antenna. The *direction* of a detected object from a radar site is determined by the position of the rotating antenna when the reflected portion of the radio wave is received.

2. More reliable maintenance and improved equipment have reduced radar system failures to a negligible factor. Most facilities actually have some components duplicated-one operating and another which immediately takes over when a malfunction occurs to the primary component.

b. Limitations

Figure 1–31[1]

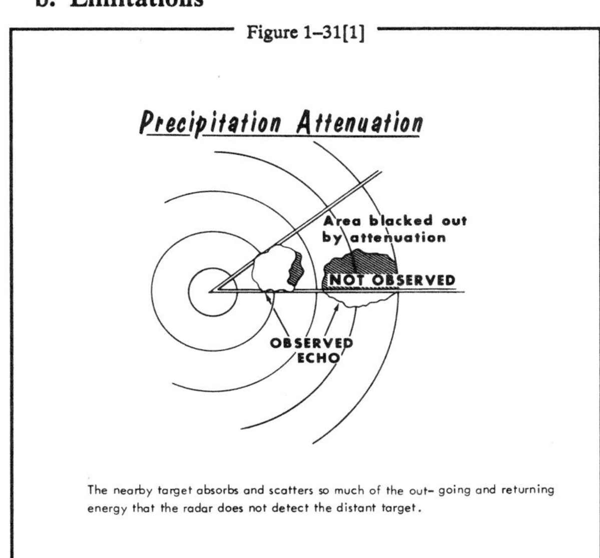

The nearby target absorbs and scatters so much of the out- going and returning energy that the radar does not detect the distant target.

1. It is very important for the aviation community to recognize the fact that there are limitations to radar service and that ATC controllers may not always be able to issue traffic advisories concerning aircraft which are not under ATC control and cannot be seen on radar. (See Figure 1–31[1].)

(a) The characteristics of radio waves are such that they normally travel in a continuous straight line unless they are:

(1) "Bent" by abnormal atmospheric phenomena such as temperature inversions;

(2) Reflected or attenuated by dense objects such as heavy clouds, precipitation, ground obstacles, mountains, etc.; or

(3) Screened by high terrain features.

(b) The bending of radar pulses, often called anomalous propagation or ducting, may cause many extraneous blips to appear on the radar operator's display if the beam has been bent toward the ground or may decrease the detection range if the wave is bent upward. It is difficult to solve the effects of anomalous propagation, but using beacon radar and electronically eliminating stationary and slow moving targets by a method called moving target indicator (MTI) usually negate the problem.

(c) Radar energy that strikes dense objects will be reflected and displayed on the operator's scope thereby blocking out aircraft at the same range and greatly weakening or completely eliminating the display of targets at a greater range. Again, radar beacon and MTI are very effectively used to combat ground clutter and weather phenomena, and a method of circularly polarizing the radar beam will eliminate some weather returns. A negative characteristic of MTI is that an aircraft flying a speed that coincides with the canceling signal of the MTI (tangential or "blind" speed) may not be displayed to the radar controller.

(d) Relatively low altitude aircraft will not be seen if they are screened by mountains or are below the radar beam due to earth curvature. The only solution to screening is the installation of strategically placed multiple radars which has been done in some areas.

(e) There are several other factors which affect radar control. The amount of reflective surface of an aircraft will determine the size of the radar return. Therefore, a small light airplane or a sleek jet fighter will be more difficult to see on radar than a large commercial jet or military bomber. Here again, the use of radar beacon is invaluable if the aircraft is equipped with an airborne transponder. All ARTCC radars in the conterminous U.S. and many airport surveillance radars have the capability to interrogate MODE C and display altitude information to the controller from appropriately equipped aircraft. However, there are a number of airport surveillance radars that are still two dimensional (range and azimuth) only and altitude information must be obtained from the pilot.

(f) At some locations within the ATC en route environment, secondary-radar-only (no primary radar) gap filler radar systems are used to give lower altitude radar coverage between two larger radar systems, each of which provides both primary and secondary radar coverage. In those geographical areas served by secondary-radar only, aircraft without transponders cannot be provided with radar service. Additionally, transponder equipped aircraft cannot be provided with radar advisories concerning primary targets and weather. (See Pilot/Controller Glossary, RADAR/Radio Detection and Ranging).

(g) The controllers' ability to advise a pilot flying on instruments or in visual conditions of his proximity to another aircraft will be limited if the unknown aircraft is not observed on radar, if no flight plan information is available, or if the volume of traffic and workload prevent his issuing traffic information. The controller's first priority is given to establishing vertical, lateral, or longitudinal separation between aircraft flying IFR under the control of ATC.

c. FAA radar units operate continuously at the locations shown in the Airport/Facility Directory, and their services are available to all pilots, both civil and military. Contact the associated FAA control tower or ARTCC on any frequency guarded for initial instructions, or in an emergency, any FAA facility for information on the nearest radar service.

1–32. AIR TRAFFIC CONTROL RADAR BEACON SYSTEM (ATCRBS)

a. The ATCRBS, sometimes referred to as secondary surveillance radar, consists of three main components:

1. Interrogator. Primary radar relies on a signal being transmitted from the radar antenna site and for this signal to be reflected or "bounced back" from an object (such as an aircraft). This reflected signal is then displayed as a "target" on the controller's radarscope. In the ATCRBS, the *Interrogator,* a ground based radar beacon transmitter-receiver, scans in synchronism with the primary radar and transmits discrete radio signals which repetitiously requests all transponders, on the mode being used, to reply. The replies received are then mixed with the primary returns and both are displayed on the same radarscope.

2. Transponder. This airborne radar beacon transmitter-receiver automatically receives the signals from the interrogator and selectively replies with a specific pulse group (code) only to those interrogations being received on the mode to which it is set. These replies are independent of, and much stronger than a primary radar return.

3. Radarscope. The radarscope used by the controller displays returns from both the primary radar system and the ATCRBS. These returns, called targets, are what the controller refers to in the control and separation of traffic.

b. The job of identifying and maintaining identification of primary radar targets is a long and tedious task for the controller. Some of the advantages of ATCRBS over primary radar are:

1. Reinforcement of radar targets.

2. Rapid target identification.

3. Unique display of selected codes.

c. A part of the ATCRBS ground equipment is the decoder. This equipment enables the controller to assign discrete transponder codes to each aircraft under his control. Normally only one code will be assigned for the entire flight. Assignments are made by the ARTCC computer on the basis of the National Beacon Code Allocation Plan. The equipment is also designed to receive MODE C altitude information from the aircraft.

1–32c NOTE—Refer to figures with explanatory legends for an illustration of the target symbology depicted on radar scopes in the NAS Stage A (en route), the ARTS III (terminal) systems, and other nonautomated (broadband) radar systems. (See Figure 1–32[1] and Figure 1–32[2].)

d. It should be emphasized that aircraft transponders greatly improve the effectiveness of radar systems. (Reference—Transponder Operation, paragraph 4–18).

1–33. SURVEILLANCE RADAR

a. Surveillance radars are divided into two general categories: Airport Surveillance Radar (ASR) and Air Route Surveillance Radar (ARSR).

1. ASR is designed to provide relatively short-range coverage in the general vicinity of an airport and to serve as an expeditious means of handling terminal area traffic through observation of precise aircraft locations on a radarscope. The ASR can also be used as an instrument approach aid.

2. ARSR is a long-range radar system designed primarily to provide a display of aircraft locations over large areas.

3. Center Radar Automated Radar Terminal Systems (ARTS) Processing (CENRAP) was developed to provide an alternative to a non-radar environment at terminal facilities should an Airport Surveillance Radar (ASR) fail or malfunction. CENRAP sends aircraft radar beacon target information to the ASR terminal facility equipped with ARTS. Procedures used for the separation of aircraft may increase under certain conditions when a facility is utilizing CENRAP because radar target information updates at a slower rate than the normal ASR radar. Radar services for VFR aircraft are also limited during CENRAP operations because of the

Figure 1–32[1]

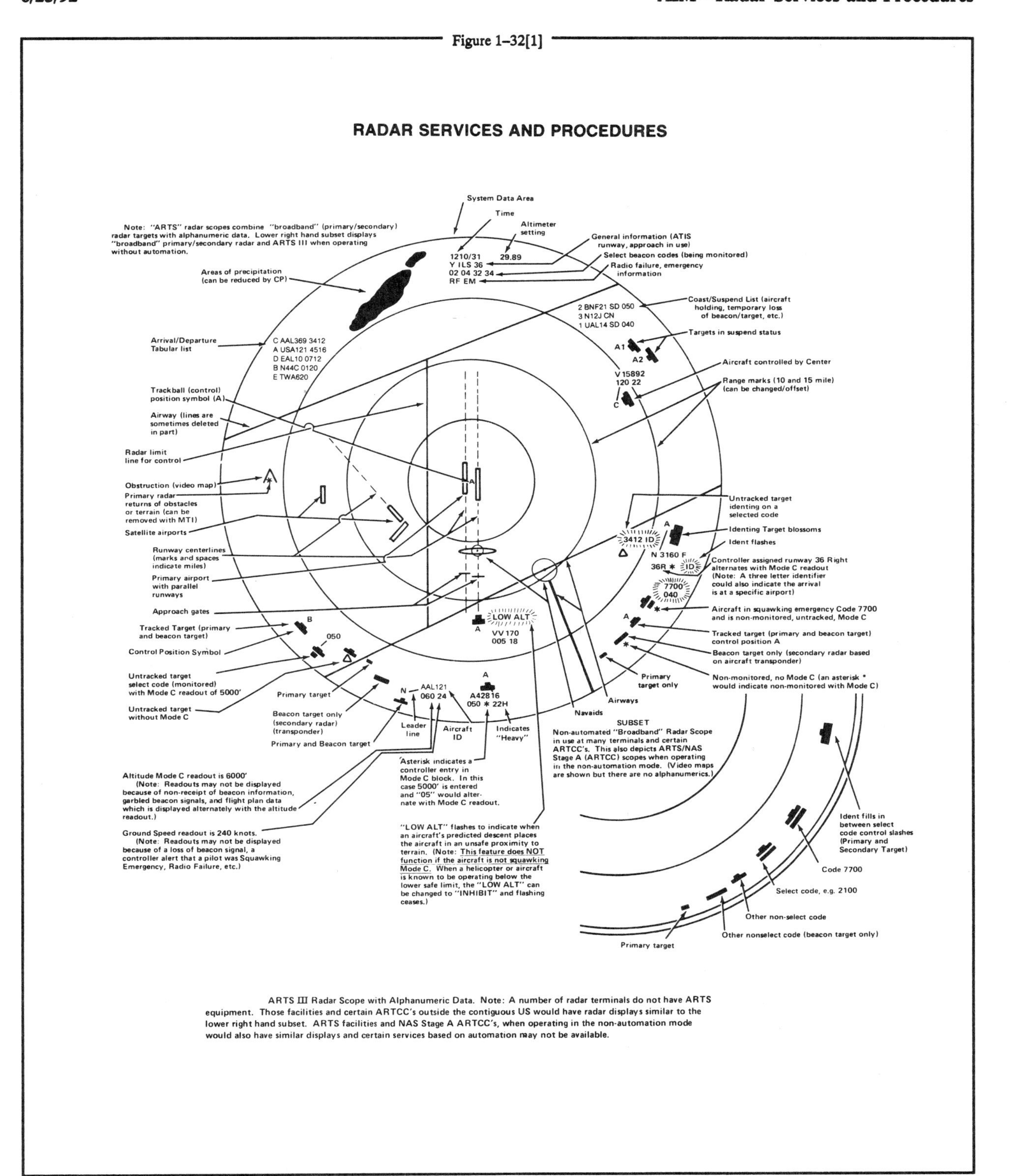

ARTS III Radar Scope with Alphanumeric Data. Note: A number of radar terminals do not have ARTS equipment. Those facilities and certain ARTCC's outside the contiguous US would have radar displays similar to the lower right hand subset. ARTS facilities and NAS Stage A ARTCC's, when operating in the non-automation mode would also have similar displays and certain services based on automation may not be available.

Figure 1–32[2]

RADAR SERVICES AND PROCEDURES

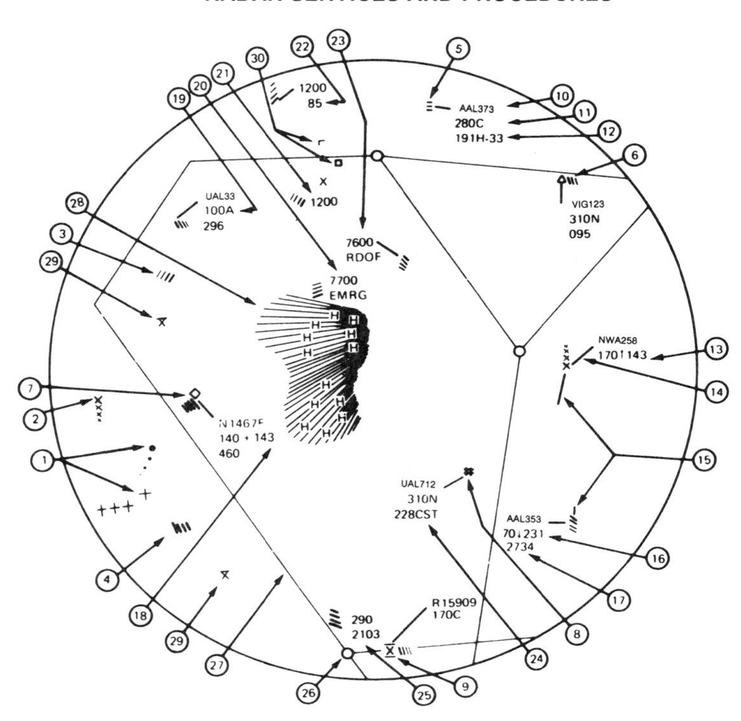

Target Symbols

1 Uncorrelated primary radar target + ●
2 *Correlated primary radar target X
3 Uncorrelated beacon target /
4 Correlated beacon target \
5 Identing beacon target ≡
(*Correlated means the association of radar data with the computer projected track of an identified aircraft)

Position Symbols

6 Free track (No flight plan tracking) △
7 Flat track (flight plan tracking) ◊
8 Coast (Beacon target lost) #
9 Present Position Hold ⧗

Data Block Information

10 *Aircraft Identification
11 *Assigned Altitude FL280, mode C altitude same or within ±200′ of asgnd altitude
12 *Computer ID #191, Handoff is to Sector 33 (0-33 would mean handoff accepted) (*Nr's 10, 11, 12 constitute a "full data block")
13 Assigned altitude 17,000′, aircraft is climbing, mode C readout was 14,300 when last beacon interrogation was received
14 Leader line connecting target symbol and data block
15 Track velocity and direction vector line (Projected ahead of target)
16 Assigned altitude 7000, aircraft is descending, last mode C readout (or last reported altitude was 100′ above FL230
17 Transponder code shows in full data block only when different than assigned code
18 Aircraft is 300′ above assigned altitude
19 Reported altitude (No mode C readout) same as assigned. An "N" would indicate no reported altitude)
20 Transponder set on emergency code 7700 (EMRG flashes to attract attention)
21 Transponder code 1200 (VFR) with no mode C
22 Code 1200 (VFR) with mode C and last altitude readout
23 Transponder set on Radio Failure code 7600, (RDOF flashes)
24 Computer ID #228, CST indicates target is in Coast status
25 Assigned altitude FL290, transponder code (These two items constitute a "limited data block")

Other symbols

26 Navigational Aid
27 Airway or jet route
28 Outline of weather returns based on primary radar (See Chapter 4, ARTCC Radar Weather Display. H's represent areas of high density precipitation which might be thunderstorms. Radial lines indicate lower density precipitation)
29 Obstruction
30 Airports Major: □ , Small: ┌

NAS Stage A Controllers View Plan Display. This figure illustrates the controller's radar scope (PVD) when operating in the full automation (RDP) mode, which is normally 20 hours per day. (Note: When not in automation mode, the display is similar to the broadband mode shown in the ARTS III Radar Scope figure. Certain ARTCC's outside the contiguous U.S. also operate in "broadband" mode.)

additional workload required to provide services to IFR aircraft.

b. Surveillance radars scan through 360 degrees of azimuth and present target information on a radar display located in a tower or center. This information is used independently or in conjunction with other navigational aids in the control of air traffic.

1–34. PRECISION APPROACH RADAR (PAR)

a. PAR is designed to be used as a *landing aid,* rather than an aid for sequencing and spacing aircraft. PAR equipment may be used as a primary landing aid, or it may be used to monitor other types of approaches. It is designed to display *range, azimuth* and *elevation* information.

b. Two antennas are used in the PAR array, one scanning a vertical plane, and the other scanning horizontally. Since the range is limited to 10 miles, azimuth to 20 degrees, and elevation to 7 degrees, only the final approach area is covered. Each scope is divided into two parts. The upper half presents altitude and distance information, and the lower half presents azimuth and distance.

1–35 thru 1–40. RESERVED

Chapter 2. AERONAUTICAL LIGHTING AND OTHER AIRPORT VISUAL AIDS

Section 1. AIRPORT LIGHTING AIDS

2–1. APPROACH LIGHT SYSTEMS (ALS)

a. Approach light systems provide the basic means to transition from instrument flight to visual flight for landing. Operational requirements dictate the sophistication and configuration of the approach light system for a particular runway.

b. Approach light systems are a configuration of signal lights starting at the landing threshold and extending into the approach area a distance of 2400–3000 feet for precision instrument runways and 1400–1500 feet for nonprecision instrument runways. Some systems include sequenced flashing lights which appear to the pilot as a ball of light traveling towards the runway at high speed (twice a second). (See Figure 2–1[1] and Figure 2–1[2].)

Figure 2–1[1]

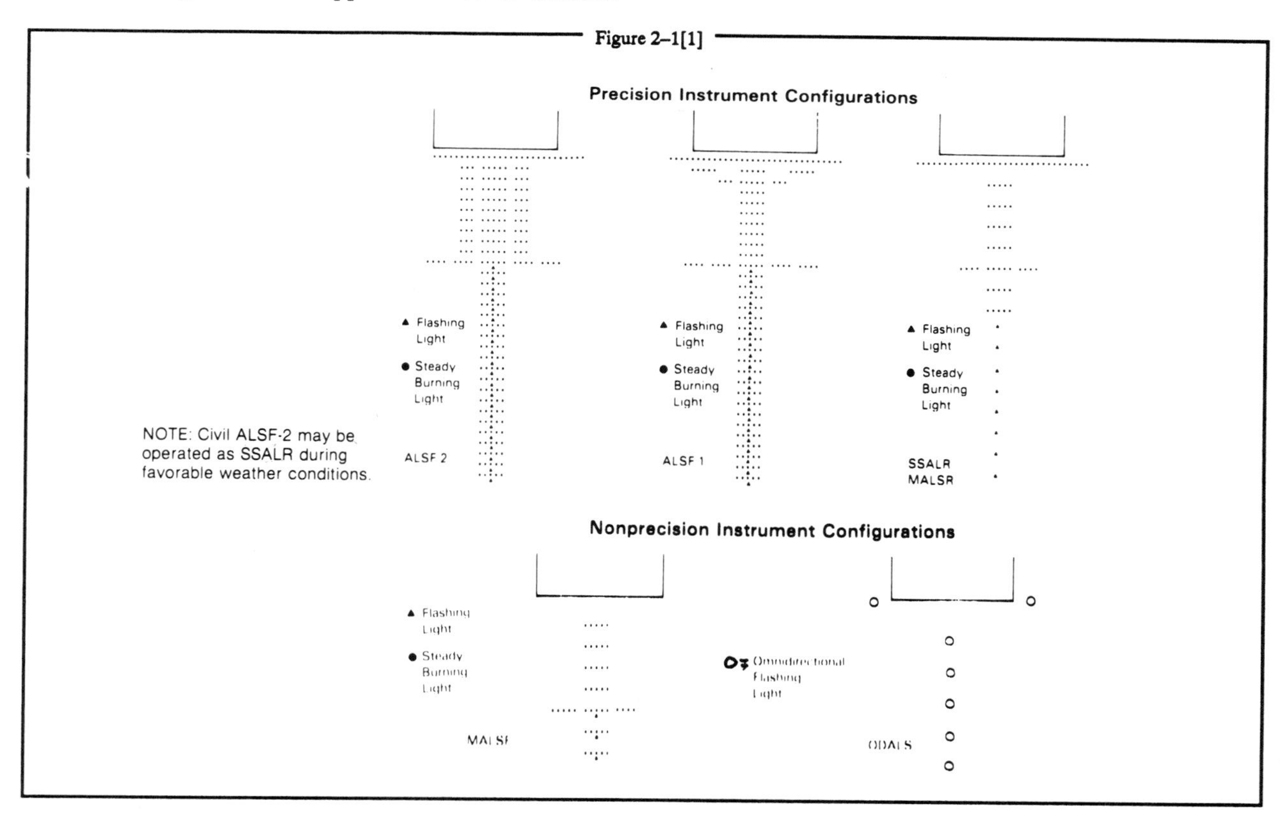

2–2. VISUAL GLIDESLOPE INDICATORS

a. Visual Approach Slope Indicator (VASI)

1. The VASI is a system of lights so arranged to provide visual descent guidance information during the approach to a runway. These lights are visible from 3–5 miles during the day and up to 20 miles or more at night. The visual glide path of the VASI provides safe obstruction clearance within plus or minus 10 degrees of the extended runway centerline and to 4 NM from the runway threshold. Descent, using the VASI, should not be initiated until the aircraft is visually aligned with the runway. Lateral course guidance is provided by the runway or runway lights.

2. VASI installations may consist of either 2, 4, 6, 12, or 16 lights units arranged in bars referred to as near, middle, and far bars. Most VASI installations consist of 2 bars, near and far, and may consist of 2, 4, or 12 light units. Some VASIs consist of three bars, near, middle, and far, which provide an additional visual glide path to accommodate high cockpit aircraft. This installation may consist of either 6 or 16 light units. VASI installations consisting of 2, 4, or 6 light units are located on one side of the runway, usually the left. Where the installation consists of 12 or 16 light units, the units are located on both sides of the runway.

3. Two-bar VASI installations provide one visual glide path which is normally set at 3 degrees. Three-bar VASI installations provide two visual glide paths. The lower glide path is provided by the near and middle bars and is normally set at 3 degrees while the upper glide path, provided by the middle and far bars, is normally 1/4 degree higher. This higher glide path is intended for use only by high cockpit aircraft to provide a sufficient threshold crossing height. Although normal glide path angles are three degrees, angles at some locations may be as high as 4.5 degrees to give proper obstacle clearance. Pilots of high performance aircraft are cautioned that use of VASI angles in excess of 3.5 degrees may cause an increase in runway length required for landing and rollout.

4. The basic principle of the VASI is that of color differentiation between red and white. Each light unit projects a beam of light having a white segment in the upper part of the beam and red segment in the lower part of the beam. The light units are arranged so that the pilot using the VASIs during an approach will see the combination of lights shown below.

5. For 2–bar VASI (4 light units) See Figure 2–2[1].

Figure 2–2[1]

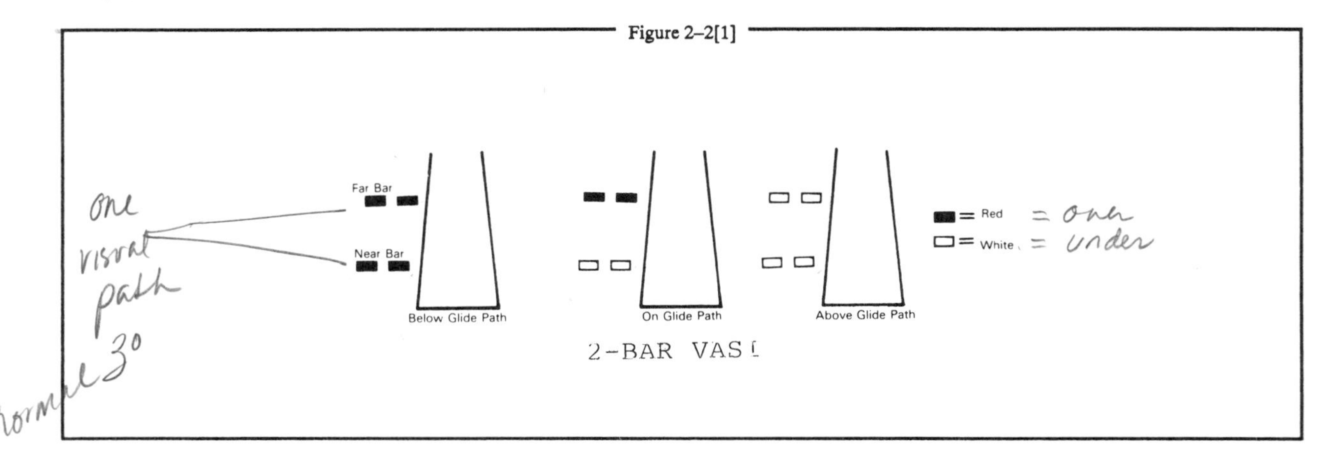

6. For 3–bar VASI (6 light units) See Figure 2–2[2].

Figure 2–2[2]

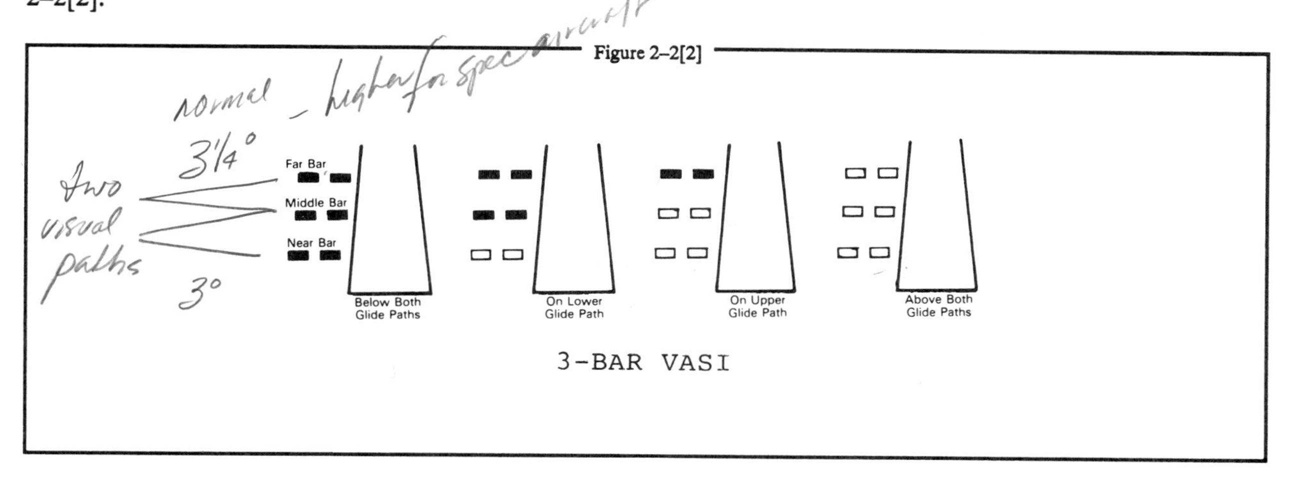

7. For other VASI configurations See Figure 2–2[3].

Figure 2–2[3]

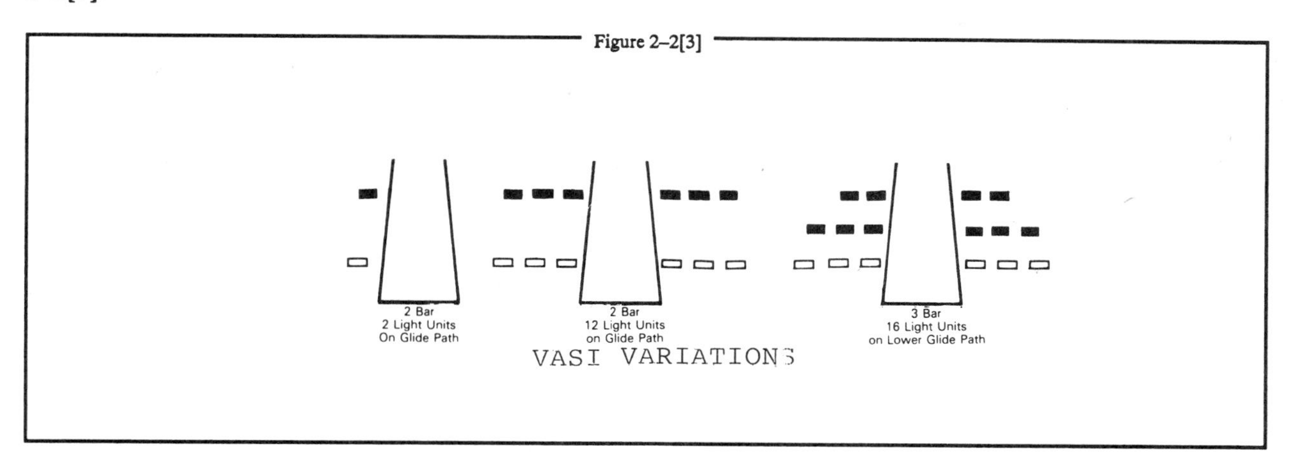

b. Precision Approach Path Indicator (PAPI)—The precision approach path indicator (PAPI) uses light units similar to the VASI but are installed in a single row of either two or four light units. These systems have an effective visual range of about 5 miles during the day and up to 20 miles at night. The row of light units is normally installed on the left side of the runway and the glide path indications are as depicted. (See Figure 2–2[4].)

Figure 2–2[4]

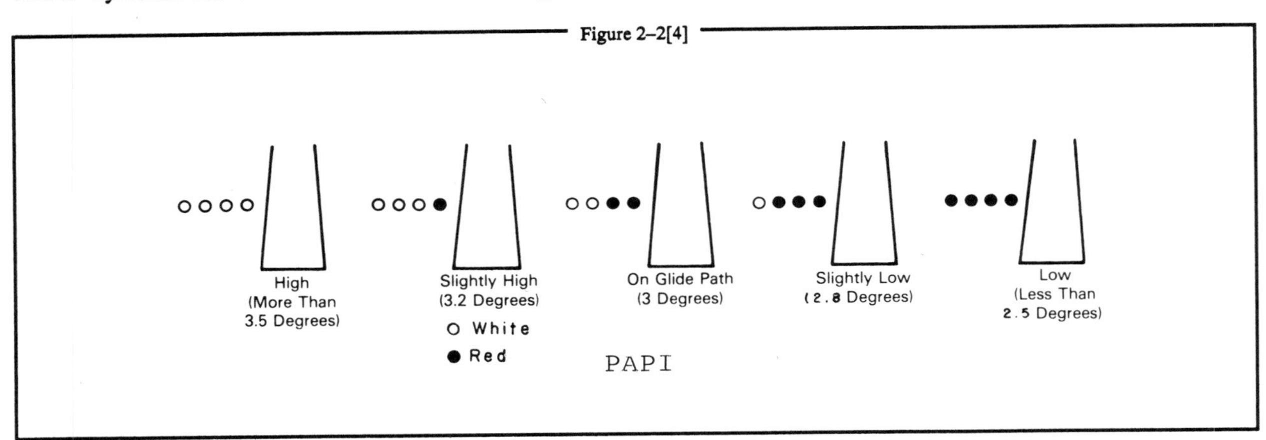

c. Tri-color Systems—Tri-color visual approach slope indicators normally consist of a single light unit projecting a three-color visual approach path into the final approach area of the runway upon which the indicator is installed. The below glide path indication is red, the above glide path indication is amber, and the on glide path indication is green. These types of indicators have a useful range of approximately one-half to one mile during the day and up to five miles at night depending upon the visibility conditions. (See Figure 2–2[5].)

Figure 2–2[5]

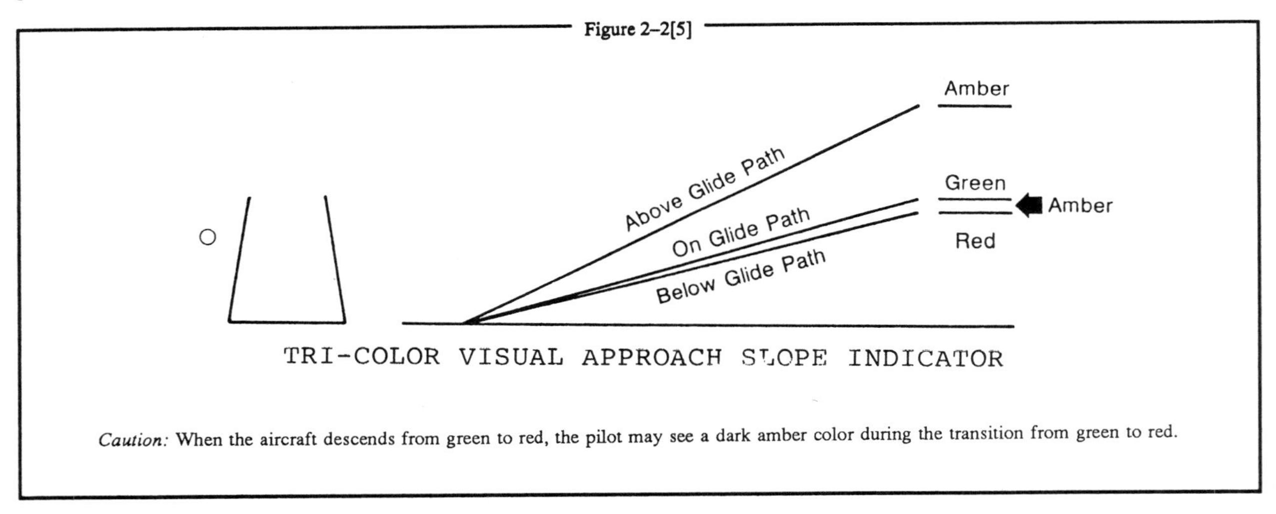

Caution: When the aircraft descends from green to red, the pilot may see a dark amber color during the transition from green to red.

d. Pulsating Systems—Pulsating visual approach slope indicators normally consist of a single light unit projecting a two-color visual approach path into the final approach area of the runway upon which the indicator is installed. The below glide path indication is normally pulsating red and the above glide path indication is normally pulsating white. The on glide path indication is a steady white light. The useful range of this system is about four miles during the day and up to ten miles at night. (See Figure 2–2[6].)

Figure 2–2[6]

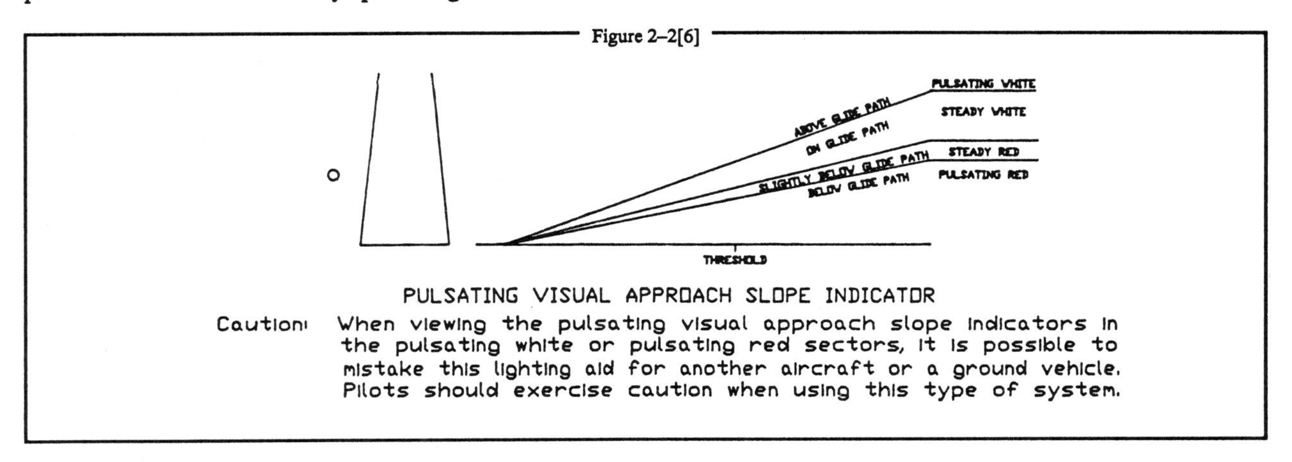

e. Alignment of Elements Systems—Alignment of elements systems are installed on some small general aviation airports and are a low-cost system consisting of painted plywood panels, normally black and white or fluorescent orange. Some of these systems are lighted for night use. The useful range of these systems is approximately three-quarter miles. To use the system the pilot positions his aircraft so the elements are in alignment. The glide path indications are shown in Figure 2–2[7].

Figure 2–2[7]

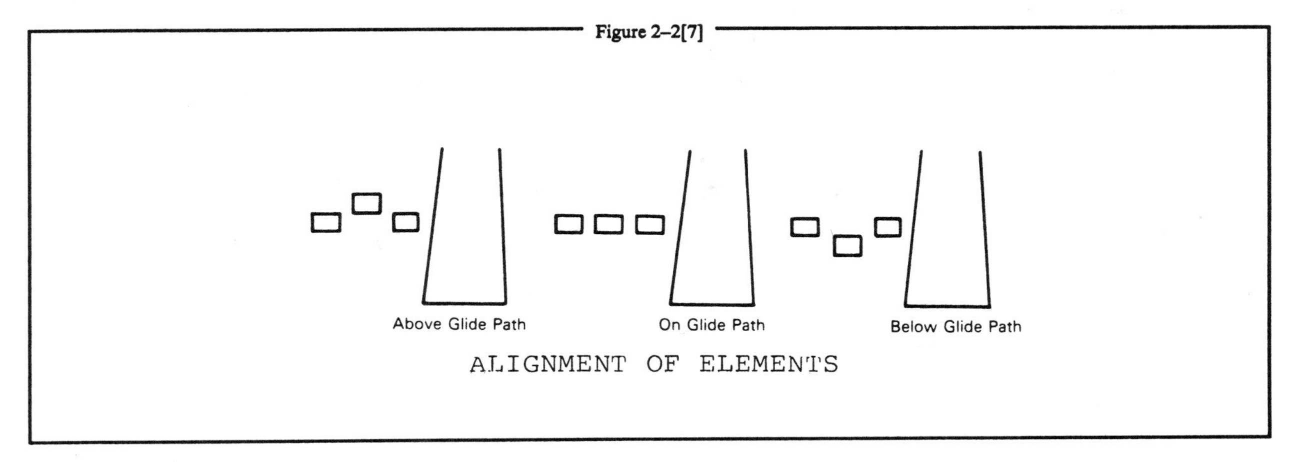

2–3. RUNWAY END IDENTIFIER LIGHTS (REIL)

REILs are installed at many airfields to provide rapid and positive identification of the approach end of a particular runway. The system consists of a pair of synchronized flashing lights located laterally on each side of the runway threshold. REILs may be either omnidirectional or unidirectional facing the approach area. They are effective for:

1. Identification of a runway surrounded by a preponderance of other lighting.

2. Identification of a runway which lacks contrast with surrounding terrain.

3. Identification of a runway during reduced visibility.

2–4. RUNWAY EDGE LIGHT SYSTEMS

a. Runway edge lights are used to outline the edges of runways during periods of darkness or restricted visibility conditions. These light systems are classified according to the intensity or brightness they are capable of producing: they are the High Intensity Runway Lights (HIRL), Medium Intensity Runway Lights (MIRL), and the Low Intensity Runway Lights (LIRL). The HIRL and MIRL systems have variable intensity controls, whereas the LIRLs normally have one intensity setting.

b. The runway edge lights are white, except on instrument runways amber replaces white on the last 2,000 feet or half the runway length, whichever is less, to form a caution zone for landings.

c. The lights marking the ends of the runway emit red light toward the runway to indicate the end of runway to a departing aircraft and emit green outward from the runway end to indicate the threshold to landing aircraft.

2–5. IN-RUNWAY LIGHTING

Touchdown zone lights and runway centerline lights are installed on some precision approach runways to facilitate landing under adverse visibility conditions. Taxiway turnoff lights may be added to expedite movement of aircraft from the runway.

a. Touchdown Zone Lighting (TDZL)—two rows of transverse light bars disposed symmetrically about the runway centerline in the runway touchdown zone. The system starts 100 feet from the landing threshold and extends to 3000 feet from the threshold or the midpoint of the runway, whichever is the lesser.

b. Runway Centerline Lighting (RCLS)—flush centerline lights spaced at 50–foot intervals beginning 75 feet from the landing threshold and extending to within 75 feet of opposite end of the runway.

c. Runway Remaining Lighting—centerline lighting systems in the final 3,000 feet as viewed from the takeoff or approach position. Alternate red and white lights are seen from the 3,000 foot points to the 1,000 foot points, and all red lights are seen for the last 1,000 feet of the runway. From the opposite direction, these lights are seen as white lights.

d. Taxiway turnoff lights—flush lights spaced at 50 foot intervals defining the curved path of aircraft travel from the runway centerline to a point on the taxiway. These lights are steady burning and emit green light.

2–6. CONTROL OF LIGHTING SYSTEMS

a. Operation of approach light systems and runway lighting is controlled by the control tower (ATCT). At some locations the FSS may control the lights where there is no control tower in operation.

b. Pilots may request that lights be turned on or off. Runway edge lights, in-pavement lights and approach lights also have intensity controls which may be varied to meet the pilots request. Sequenced flashing lights (SFL) may be turned on and off. Some sequenced flashing light systems also have intensity control.

2–7. PILOT CONTROL OF AIRPORT LIGHTING

a. Radio control of lighting is available at selected airports to provide airborne control of lights by keying the aircraft's microphone. Control of lighting systems is often available at locations without specified hours for lighting and where there is no control tower or FSS or when the tower or FSS is closed (locations with a part-time tower or FSS) or specified hours. All lighting systems which are radio controlled at an airport, whether on a single runway or multiple runways, operate on the same radio frequency. (See Table 2–7[1] and Table 2–7[2]).

Table 2–7[1]—RUNWAYS WITH APPROACH LIGHTS

Lighting System	No. of Int. Steps	Status During Nonuse Period	Intensity Step Selected Per No. of Mike Clicks		
			3 Clicks	5 Clicks	7 Clicks
Approach Lights (Med. Int.)	2	Off	Low	Low	High
Approach Lights (Med. Int.)	3	Off	Low	Med	High
MIRL	3	Off or Low	¤	¤	¤
HIRL	5	Off or Low	¤	¤	¤
VASI	2	Off	◊	◊	◊

NOTES:
¤ Predetermined intensity step.
◊ Low intensity for night use. High intensity for day use as determined by photocell control.

Table 2–7[2]—RUNWAYS WITHOUT APPROACH LIGHTS

Lighting System	No. of Int. Steps	Status During Nonuse Period	Intensity Step Selected Per No. of Mike Clicks		
			3 Clicks	5 Clicks	7 Clicks
MIRL	3	Off or Low	Low	Med.	High
HIRL	5	Off or Low	Step 1 or 2	Step 3	Step 5
LIRL	1	Off	On	On	On
VASI☆	2	Off	◊	◊	◊
REIL☆	1	Off	Off	On / Off	On
REIL☆	3	Off	Low	Med.	High

NOTES:
◊ Low intensity for night use. High intensity for day use as determined by photocell control.
☆ The control of VASI and/or REIL may be independent of other lighting systems.

b. With FAA approved systems, various combinations of medium intensity approach lights, runway lights, taxiway lights, VASI and/or REIL may be activated by radio control. On runways with both approach lighting and runway lighting (runway edge lights, taxiway lights, etc.) systems, the approach lighting system takes precedence for air-to-ground radio control over the runway lighting system which is set at a predetermined intensity step, based on expected visibility conditions. Runways without approach lighting may provide radio controlled intensity adjustments of runway edge lights. Other lighting systems, including VASI, REIL, and taxiway lights may be either controlled with the runway edge lights or controlled independently of the runway edge lights.

c. The control system consists of a 3–step control responsive to 7, 5, and/or 3 microphone clicks. This 3–step control will turn on lighting facilities capable of either 3–step, 2–step or 1–step operation. The 3–step and 2–step lighting facilities can be altered in intensity, while the 1–step cannot. All lighting is illuminated for a period of 15 minutes from the most recent time of activation and may not be extinguished prior to end of the 15 minute period (except for 1–step and 2–step REILs which may be turned off when desired by keying the mike 5 or 3 times respectively).

d. Suggested use is to always initially key the mike 7 times; this assures that all controlled lights are turned on to the maximum available intensity. If desired, adjustment can then be made, where the capability is provided, to a lower intensity (or the REIL turned off) by keying 5 and/or 3 times. Due to the close proximity of airports using the same frequency, radio controlled lighting receivers may be set at a low sensitivity requiring the aircraft to be relatively close to activate the system. Consequently, even when lights are on, always key mike as directed when overflying an airport of intended landing or just prior to entering the final segment of an approach. This will assure the aircraft is close enough to activate the system and a full 15 minutes lighting duration is available. Approved lighting systems may be activated by keying the mike (within 5 seconds) as indicated below: (See Table 2–7[3]).

Table 2–7[3]—RADIO CONTROL SYSTEM

Key Mike	*Function*
7 times within 5 seconds	Highest intensity available
5 times within 5 seconds	Medium or lower intensity (Lower REIL or REIL-off)
3 times within 5 seconds	Lowest intensity available (Lower REIL or REIL-off)

e. For all public use airports with FAA standard systems the Airport/Facility Directory contains the types of lighting, runway and the frequency that is used to activate the system. Airports with IAPs include data on the approach chart identifying the light system, the runway on which they are installed, and the frequency that is used to activate the system.

f. Where the airport is not served by an IAP, it may have either the standard FAA approved control system or an independent type system of different specification installed by the airport sponsor. The Airport/Facility Directory contains descriptions of pilot controlled lighting systems for each airport having other than FAA approved systems, and explains the type lights, method of control, and operating frequency in clear text.

2–8. AIRPORT (ROTATING) BEACONS

a. The airport beacon has a vertical light distribution to make it most effective from one to 10 degrees above the horizon; however, it can be seen well above and below this peak spread. The beacon may be an omnidirectional capacitor-discharge device, or it may rotate at a constant speed which produces the visual effect of flashes at regular intervals. Flashes may be one or two colors alternately. The total number of flashes are:

1. 12 to 30 per minute for beacons marking airports, landmarks, and points on Federal airways.

2. 30 to 60 per minute for beacons marking heliports.

b. The colors and color combinations of beacons are:

1. White and Green—Lighted land airport

2. *Green alone—Lighted land airport

3. White and Yellow—Lighted water airport

4. *Yellow alone—Lighted water airport

5. Green, Yellow, and White—Lighted heliport

2–8b5 NOTE—*Green alone or yellow alone is used only in connection with a white-and-green or white-and-yellow beacon display, respectively.

c. Military airport beacons flash alternately white and green, but are differentiated from civil beacons by dualpeaked (two quick) white flashes between the green flashes.

d. In control zones, operation of the airport beacon during the hours of daylight often indicates that the ground visibility is less than 3 miles and/or the ceiling is less than 1,000 feet. ATC clearance in accordance with FAR 91 is required for landing, takeoff and flight in the traffic pattern. Pilots should not rely solely on the operation of the airport beacon to indicate if weather conditions are IFR or VFR. At some locations with operating control towers, ATC personnel turn the beacon on or off when controls are in the tower. At many airports the airport beacon is turned on by a photoelectric cell or time clocks and ATC personnel can not control them. There is no regulatory requirement for daylight operation and it is the pilot's responsibility to comply with proper preflight planning as required by FAR 91.103.

2–9. TAXIWAY LIGHTS

a. Taxiway Edge Lights.—Taxiway edge lights are used to outline the edges of taxiways during periods of darkness or restricted visibility conditions. These fixtures emit blue light.

2–9a NOTE—At most major airports these lights have variable intensity settings and may be adjusted at pilot request or when deemed necessary by the controller.

b. Taxiway Centerline Lights.—Taxiway centerline lights are used to facilitate ground traffic under low visibility conditions. They are located along the taxiway centerline in a straight line on straight portions, on the centerline of curved portions, and along designated taxiing paths in portions of runways, ramp, and apron areas. Taxiway centerline lights are steady burning and emit green light.

2–10. RESERVED

Section 2. AIR NAVIGATION AND OBSTRUCTION LIGHTING

2–11. AERONAUTICAL LIGHT BEACONS

a. An aeronautical light beacon is a visual NAVAID displaying flashes of white and/or colored light to indicate the location of an airport, a heliport, a landmark, a certain point of a Federal airway in mountainous terrain, or an obstruction. The light used may be a rotating beacon or one or more flashing lights. The flashing lights may be supplemented by steady burning lights of lesser intensity.

b. The color or color combination displayed by a particular beacon and/or its auxiliary lights tell whether the beacon is indicating a landing place, landmark, point of the Federal airways, or an obstruction. Coded flashes of the auxiliary lights, if employed, further identify the beacon site.

2–12. CODE BEACONS AND COURSE LIGHTS

a. CODE BEACONS

1. The code beacon, which can be seen from all directions, is used to identify airports and landmarks and to mark obstructions. The number of code beacon flashes are:

(a) **Green coded flashes** not exceeding 40 flashes or character elements per minute, or constant flashes 12 to 15 per minute, for identifying land airports.

(b) **Yellow coded flashes** not exceeding 40 flashes or character elements per minute, or constant flashes 12 to 15 per minute, for identifying water airports.

(c) **Red flashes,** constant rate, 12 to 40 flashes per minute for marking hazards.

b. COURSE LIGHTS

1. The course light, which can be seen clearly from only one direction, is used only with rotating beacons of the Federal Airway System: two course lights, back to back, direct coded flashing beams of light in either direction along the course of airway.

2–12b1 NOTE—Airway beacons are remnants of the "lighted" airways which antedated the present electronically equipped Federal Airways System. Only a few of these beacons exist today to mark airway segments in remote mountain areas. Flashes in Morse Code identify the beacon site.

2–13. OBSTRUCTION LIGHTS

a. Obstructions are marked/lighted to warn airmen of their presence during daytime and nighttime conditions. They may be marked/lighted in any of the following combinations:

1. Aviation Red Obstruction Lights. Flashing aviation red beacons and steady burning aviation red lights during nighttime operation. Aviation orange and white paint is used for daytime marking.

2. High Intensity White Obstruction Lights. Flashing high intensity white lights during daytime with reduced intensity for twilight and nighttime operation. When this type system is used, the marking of structures with red obstruction lights and aviation orange and white paint may be omitted.

3. Dual Lighting. A combination of flashing aviation red beacons and steady burning aviation red lights for nighttime operation and flashing high intensity white lights for daytime operation. Aviation orange and white paint may be omitted.

b. High intensity flashing white lights are being used to identify some supporting structures of overhead transmission lines located across rivers, chasms, gorges, etc. These lights flash in a middle, top, lower light sequence at approximately 60 flashes per minute. The top light is normally installed near the top of the supporting structure, while the lower light indicates the approximate lower portion of the wire span. The lights are beamed towards the companion structure and identify the area of the wire span.

c. High intensity, flashing white lights are also employed to identify tall structures, such as chimneys and towers, as obstructions to air navigation. The lights provide a 360 degree coverage about the structure at 40 flashes per minute and consist of from one to seven levels of lights depending upon the height of the structure. Where more than one level is used the vertical banks flash simultaneously.

2–14 thru 2–20. RESERVED

Section 3. AIRPORT MARKING AIDS AND SIGNS

2–21. GENERAL

a. Airport marking aids and signs provide information that is useful to a pilot during landing, takeoff, and taxiing. The markings and signs described in this section reflect the FAA recommended standards.

b. Uniformity in airport marking and signs from one airport to another enhances safety and improves efficiency. Pilots are encouraged to work with the operators of the airports they use to achieve the marking and sign standards described in this section.

c. Pilots who encounter ineffective, incorrect, or confusing marking or signs on an airport should make the operator of the airport aware of the problem. These situations may also be reported under the Aviation Safety Reporting Program as described in paragraph 7–81). Pilots may also report these situations to the FAA regional airports division.

d. The FAA has issued revised standards on airport signs. The information contained in this section of the AIM is based upon these revised sign standards. Airports certificated under Federal Aviation Regulation Part FAR 139 are expected to be in compliance with the revised standards by January 1, 1994. Until that date, signs on these certificated airports may not be in accordance with the information provided herein. Airports not subject to certification under Part 139 may not be in compliance with these standards.

2–22. AIRPORT MARKING AIDS

a. Markings for runways on airports and STOLports are white. Markings defining the landing area on a heliport are also white except for hospital heliports which use a red ''H'' on a white cross. Markings for taxiways, closed areas, hazardous areas, and holding positions (even if they are on a runway) are yellow.

2–22 NOTE.—Refer to ADVISORY CIRCULAR–150/5340–1 Marking of Paved Areas on Airports for detailed airport marking information.

b. Runway designators.—Runway numbers and letters are determined from the approach direction. The runway number is the whole number nearest one-tenth the magnetic azimuth of the centerline of the runway, measured clockwise from the magnetic north. The letter, or letters, differentiate between left (L), right (R), or center (C), parallel runways, as applicable:

1. For two parallel runways ''L'' ''R''

2. For three parallel runways ''L'' ''C'' ''R''

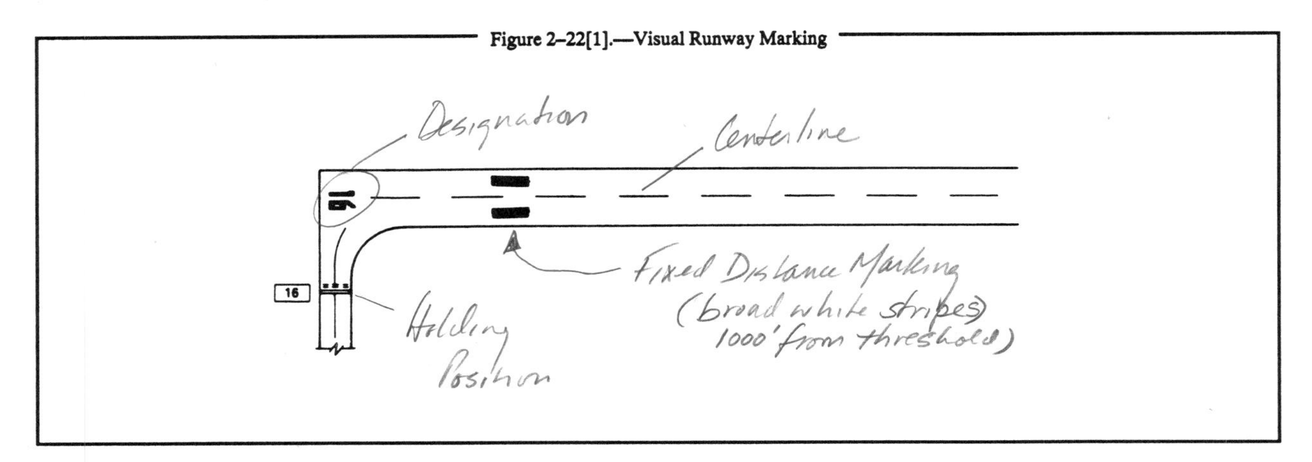

Figure 2–22[1].—Visual Runway Marking

c. Visual Runway Marking.—Used for operations under Visual Flight Rules. (See Figure 2–22[1]).

1. Centerline marking.

2. Designation marking.

3. Threshold marking on runways used or intended to be used by international commercial air transport.

4. Fixed distance marking (on runways 4,000 feet (1200 m) or longer used by jet aircraft).

5. Holding position markings for taxiway/runway intersections.

6. Holding position markings at runway/runway intersections when runways are normally used for "land, hold short operations" or taxiing.

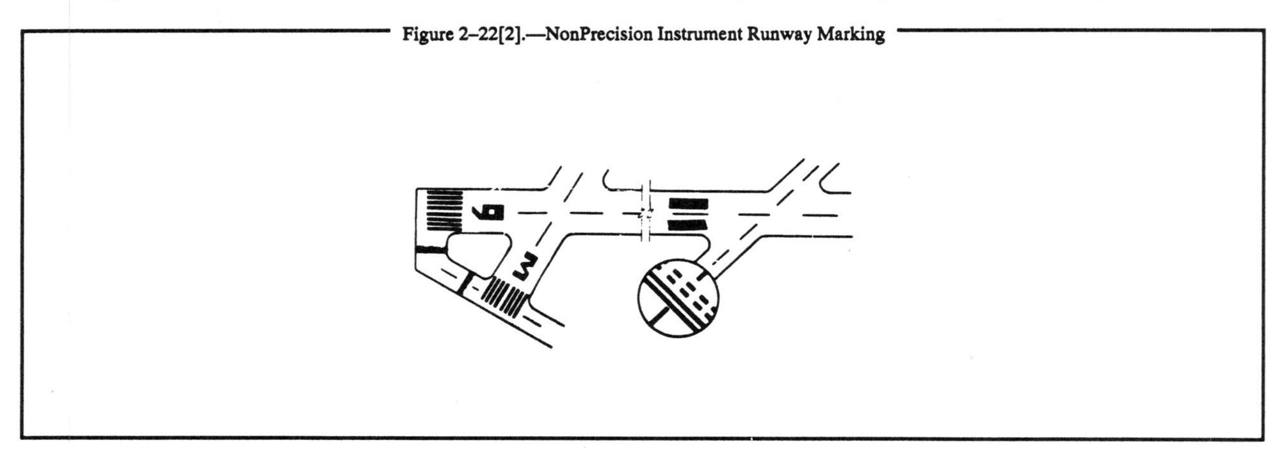

Figure 2–22[2].—NonPrecision Instrument Runway Marking

d. NonPrecision Instrument Runway Marking.—Used on runways served by a nonvisual navigation aid and intended for landings under instrument weather conditions. (See Figure 2–22[2]).

1. Centerline marking.

2. Designation marking.

3. Threshold marking.

4. Fixed distance marking (on runways 4,000 feet (1200 m) or longer used by jet aircraft).

5. Holding position markings for taxiway/runway intersections and instrument landing system (ILS) critical areas.

6. Holding position markings at runway/runway intersections when runways are normally used for "land, hold short operations" or taxiing.

Figure 2–22[3].—Precision Instrument Runway Marking

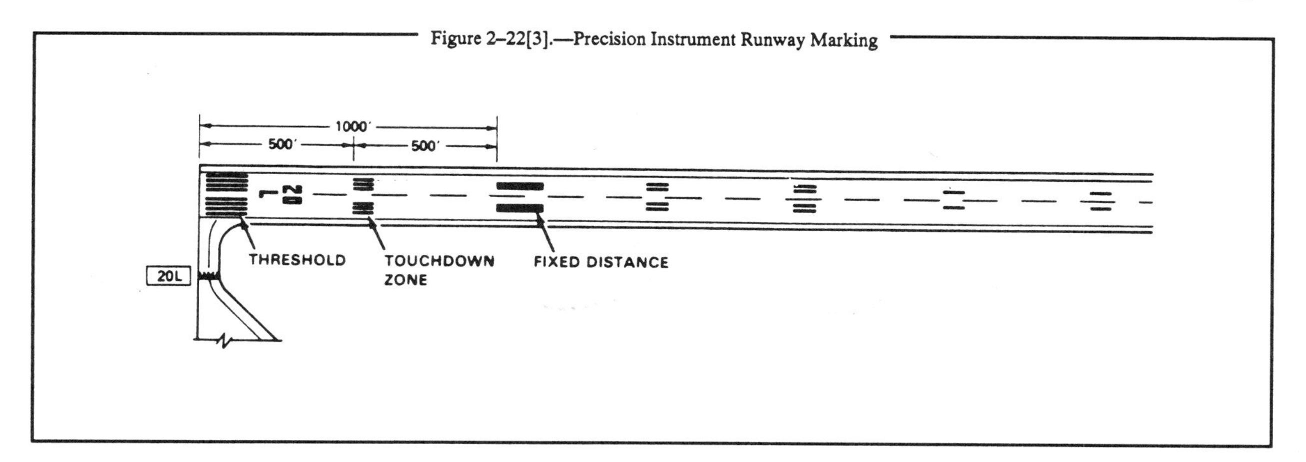

e. Precision Instrument Runway Marking.— Used on runways served by nonvisual precision approach aids and on runways having special operational requirements. (See Figure 2–22[3]).

1. Centerline marking.
2. Designation marking.
3. Threshold marking.
4. Fixed distance marking.
5. Touchdown zone marking.
6. Side stripes.
7. Holding position markings for taxiway/runway intersections and instrument landing system (ILS) critical areas.
8. Holding position markings at runway/runway intersections when runways are normally used for ''land, hold short operations'' or taxiing.

f. Threshold.—The designated beginning of the runway that is available and suitable for the landing of aircraft.

2–22f NOTE.—Sometimes constructionst or maintenance activities require the threshold to be relocated towards the departure end of the runway. In these cases, a NOTAM should be issued by the airport operator identifying the portion of the runway that is closed, e.g., First 2,000 feet of Runway 24 closed. Because the duration of the relocation can vary from a few hours to several months, methods for identifying the relocated threshold vary. One common practice is to use a ten feet wide white threshold bar across the width of the runway. Although the runway lights in the area between the old threshold and relocated threshold will not be illuminated, the runway markings in this area may or may not be obliterated, removed, or covered.

Figure 2–22[4].—Displaced Threshold

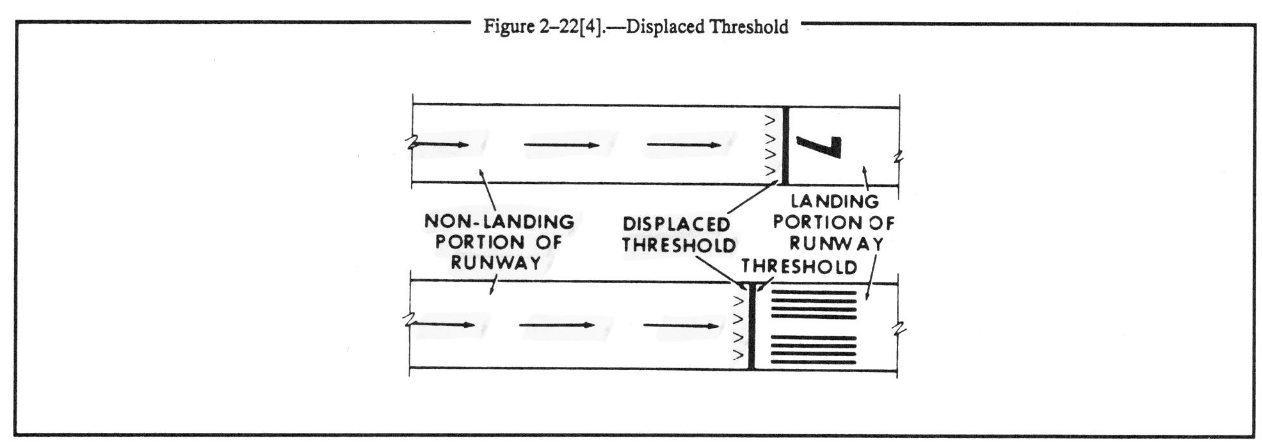

g. Displaced Threshold.—A threshold located at a point on the runway other than the designated beginning of the runway. A ten feet white threshold bar is located across the width of the runway at the displaced threshold. White arrows are located along the centerline in the area between the beginning of the runway and displaced threshold. White arrow heads are located across the width of the runway just prior to the threshold bar.(See Figure 2–22[4]).

Figure 2–22[5].—Paved Areas Beyond the Runway End

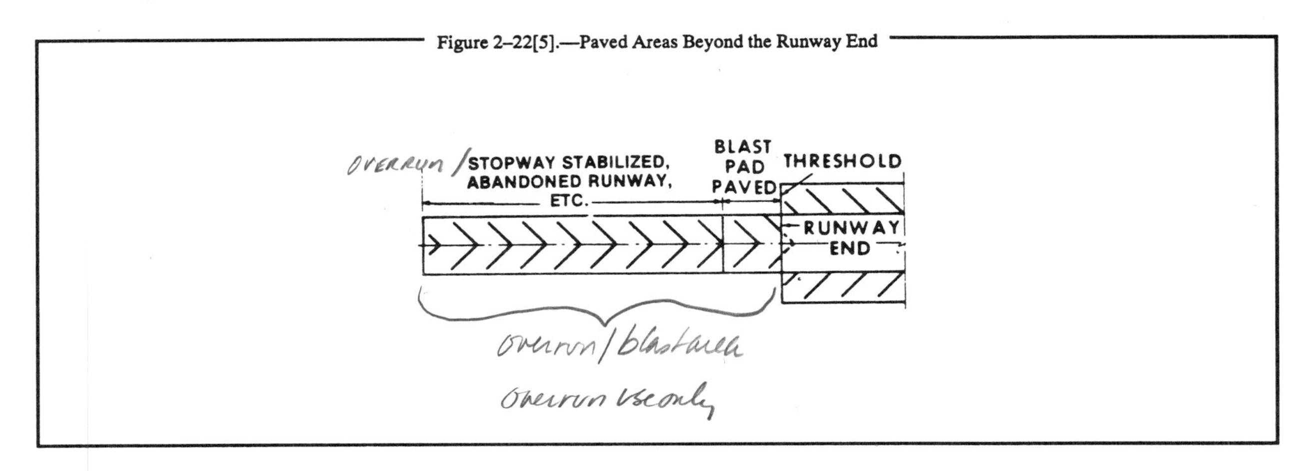

h. Paved Areas Beyond the Runway End.—Any paved area beyond the runway end that is not intended to be use as a runway in the opposite direction or as a taxiway is marked with yellow chevrons across the width of the pavement. Blast pads, stopways, and abandoned sections of runway are marked in this manner. (See Figure 2–22[5]).

i. Closed Runway.—A runway which is unusable and may be hazardous even though it may appear usable. (See Figure 2–22[6]).

Figure 2–22[6].—Closed Runway

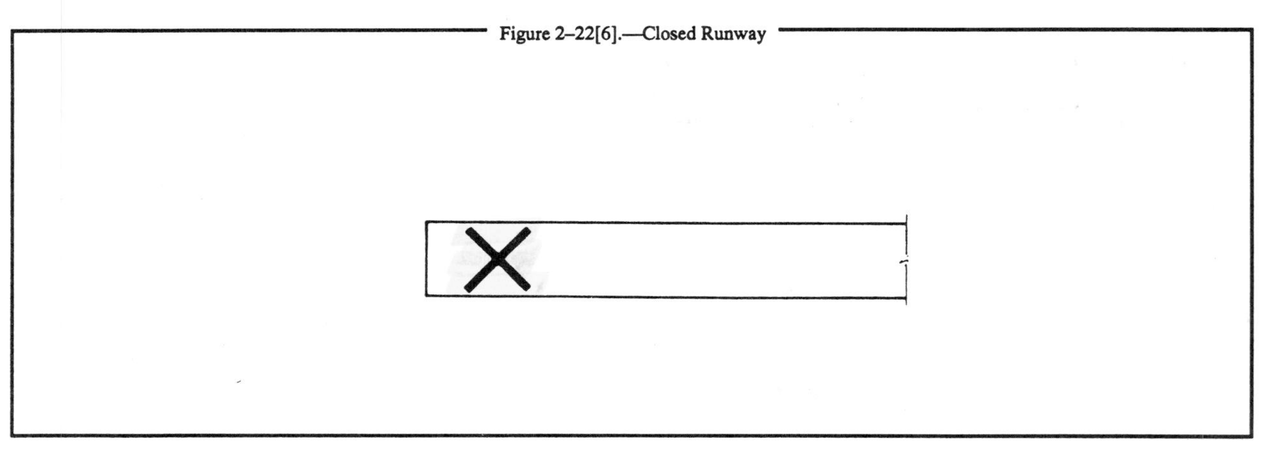

j. Fixed Distance Marking.—The fixed distance marking identifies the aiming point for the pilot of a landing aircraft. The marking consists of a broad white stripe located on each side of the runway centerline approximately 1,000 feet from the landing threshold. (See Figure 2–22[1], Figure 2–22[2], and Figure 2–22[3].)

Figure 2–22[7].—Short Take Off and Landing

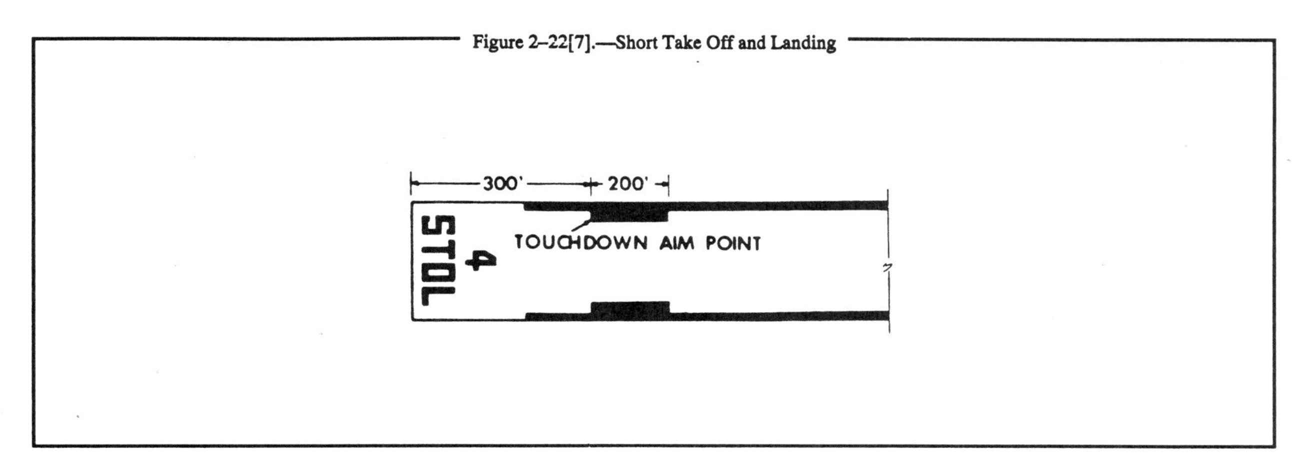

k. STOL (Short Take Off and Landing) Runway.—In addition to the normal runway number marking, the letters STOL are painted on the approach end of the runway and a touchdown aim point is shown. (See Figure 2–22[7]).

l. Taxiway Marking.—The taxiway centerline is marked with a continuous yellow line. When the taxiway edge is marked, two continuous yellow lines spaced six inches apart are used.

m. Holding Position Markings.—There are three types of holding position markings that may be encountered on an airport. The following is a description of each of these markings and their applications:

1. Holding Position Markings for Taxiway/Runway Intersections, Taxiways Located in Runway Approach Areas, and Runway/Runway Intersections consist of four yellow lines—two solid and two dashed, spaced six inches apart and extending across the width of the taxiway or runway. An example of a taxiway/runway intersection is shown. (See Figure 2–22[8]). The solid lines are always on the side where the aircraft is to hold. These markings are installed on runways only if the runway is normally used by air traffic control for "land, hold short" operations or taxiing operations and have operational significance only for those two types of operations. A sign with a white inscription on a red background is installed adjacent to these holding position markings.

2–22m1 NOTE 1.—Yellow hold position markings are being placed on runways prior to the intersection with another runway, when the runway is normally used by air traffic control for taxiing aircraft or for "Land, Hold Short" operations. Pilots receiving instructions "Clear to Land, Runway XX" from Air Traffic Control are authorized to use the entire landing length of the runway and should disregard any holding position markings located on the runway. Pilots receiving and accepting instructions "Clear to Land Runway XX, Hold Short of Runway YY" from Air Traffic Control must either exit Runway XX prior to Runway YY or stop prior to Runway YY.

2–22m1 NOTE 2.—When instructed by ATC "HOLD SHORT OF (runway, or runway approach)" the pilot should stop so no part of his aircraft extends beyond the holding position marking. When approaching the holding position marking, a pilot should not cross the marking without ATC clearance at a controlled airport or without making sure of adequate separation from other aircraft at uncontrolled airports. An aircraft exiting a runway is not clear until all parts of the aircraft have crossed the applicable holding position marking.

Figure 2–22[8].—Holding Position Markings: ILS Critical Areas

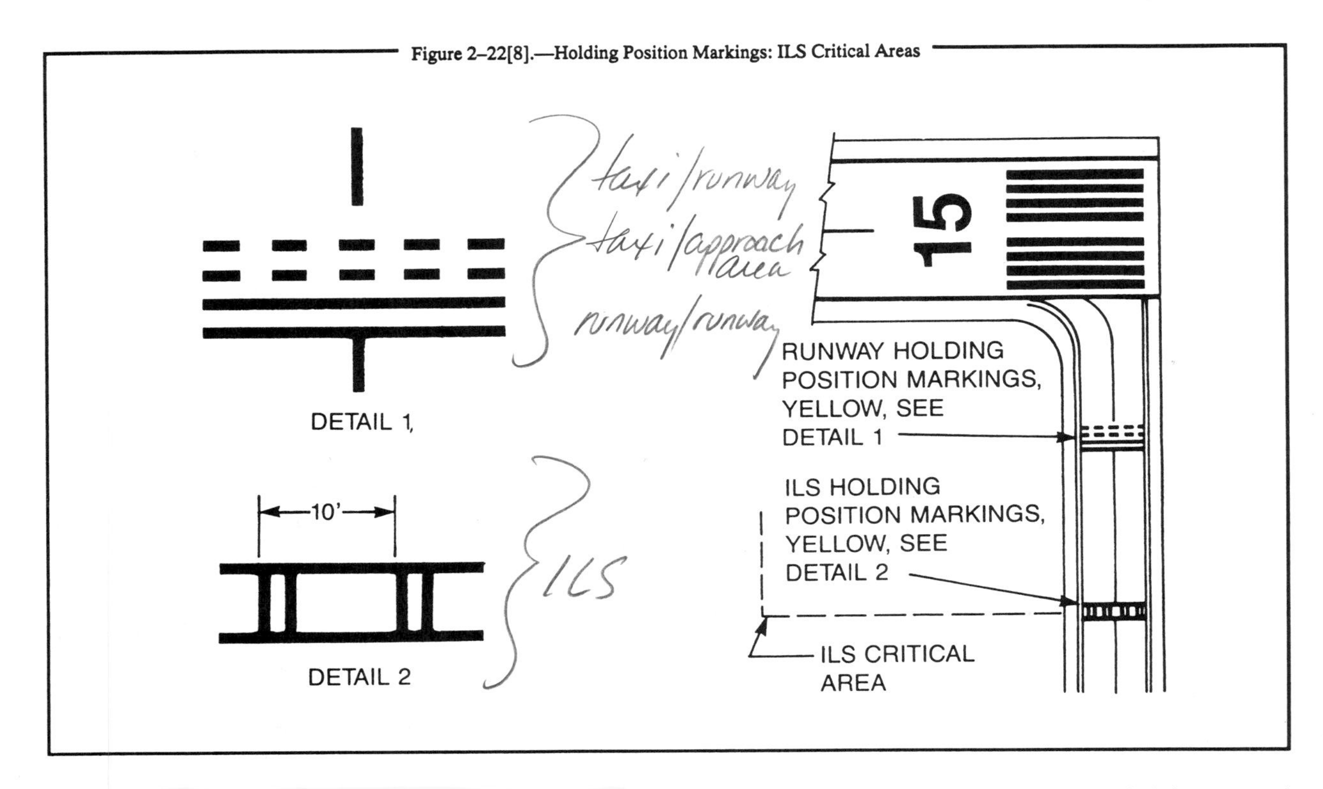

2. Holding Position Markings for ILS Critical Areas consist of two yellow solid lines spaced two feet apart connected by pairs of solid lines spaced ten feet apart extending across the width of the taxiway as shown. (See Figure 2–22[8]). A sign with an inscription "ILS" in white on a red background is installed adjacent to these hold position markings.

2–22m2 NOTE.—When the ILS critical area is being protected (See 1–10k, ILS COURSE DISTORTION) the pilot should stop so no part of his aircraft extends beyond the holding position marking. When approaching the holding position marking, a pilot should not cross the marking without ATC clearance. ILS critical area is not clear until all parts of the aircraft have crossed the applicable holding position marking.

Figure 2–22[9].—Holding Position Markings: Taxiway/Taxiway Intersections

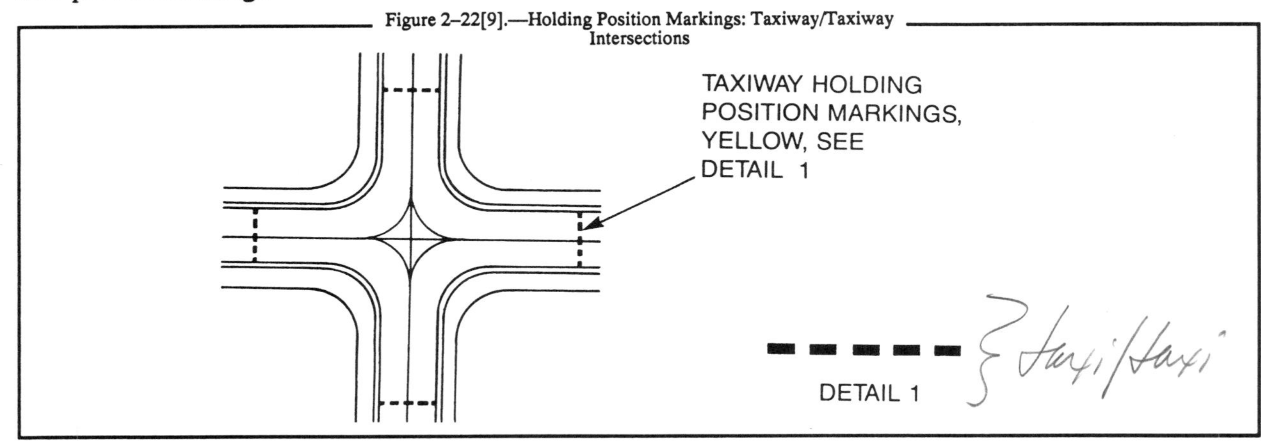

3. Holding Position Markings for Taxiway/ Taxiway Intersections consist of a single dashed line extending across the width of the taxiway as shown. (See Figure 2–22[9]). They are installed on taxiways where air traffic control normally holds aircraft short of a taxiway intersection.

2–22m3 NOTE.—When instructed by ATC "HOLD SHORT OF (taxiway)" the pilot should stop so no part of his aircraft extends beyond the holding position marking. When approaching the holding position marking, a pilot should not cross the marking without ATC clearance at a controlled airport or without making sure of adequate separation from other aircraft at uncontrolled airports.

Figure 2–22[10].—Helicopter Landing Areas

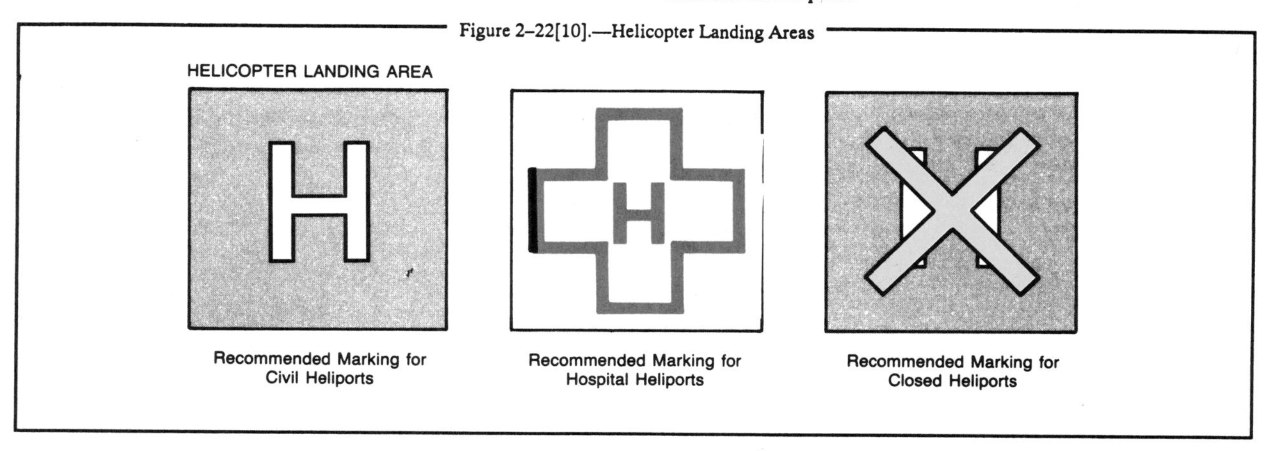

n. Helicopter Landing Areas.—The markings illustrated in Figure 2–22[10] are used to identify the landing and takeoff area at a public use heliport and hospital heliport. The letter "H" in the markings is oriented to align with the intended direction of approach. Figure 2–22[10] also depicts the markings for a closed heliport.

2–23. AIRPORT SIGNS

There are six types of signs installed on airfields: mandatory instruction signs, location signs, direction signs, destination signs, information signs, and runway distance remaining signs. The characteristics and use of these signs are discussed in paragraphs 2–24 through 2–29.

2–23 NOTE.—Refer to Advisory Circular 150/5340–18C, Standards for Airport Sign Systems for detailed information on airport signs.

2–24. MANDATORY INSTRUCTION SIGNS

a. These signs have a red background with a white inscription and are used to denote:

1. An entrance to a runway or critical area and;

2. Areas where an aircraft is prohibited from entering.

Typical mandatory signs and applications are:

b. Runway Holding Position Sign. This sign is located at the holding position on taxiways that intersect a runway or on runways that intersect other runways. The inscription on the sign contains the designation of the intersecting runway as shown in Figure 2–24[1]. The runway numbers on the sign are arranged to correspond to the respective runway threshold. For example, "15–33" indicates that the threshold for Runway 15 is to the left and the threshold for Runway 33 is to the right.

Figure 2–24[1].—Runway Holding Position Sign

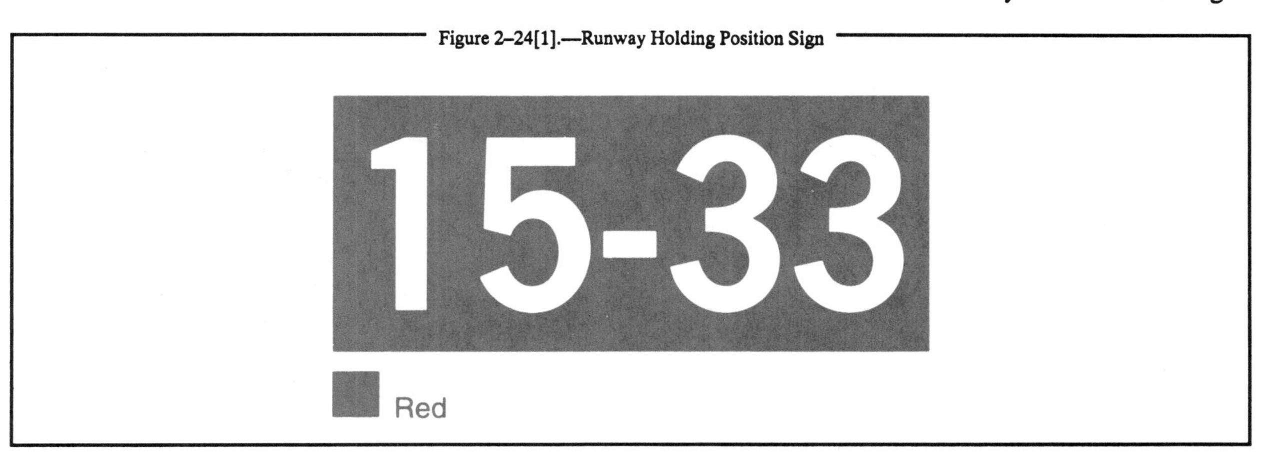

1. On taxiways that intersect the beginning of the takeoff runway, only the designation of the takeoff runway may appear on the sign as shown in Figure 2–24[2], while all other signs will have the designation of both runway directions.

Figure 2–24[2].—Holding Position Sign at Beginning of Takeoff Runway

2. If the sign is located on a taxiway that intersects the intersection of two runways, the designations for both runways will be shown on the sign along with arrows showing the approximate alignment of each runway as shown in Figure 2–24[3]. In addition to showing the approximate runway alignment, the arrow indicates the direction **to the threshold of the runway** whose designation is immediately next to the arrow.

Figure 2–24[3].—Holding Position Sign for a Taxiway that Intersects the Intersection of Two Runways

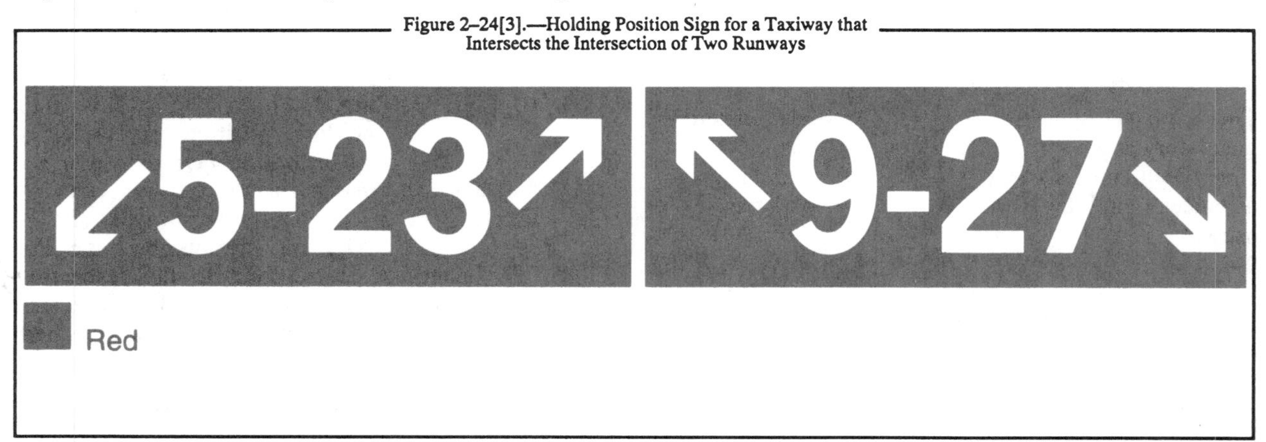

3. A runway holding position sign on a taxiway will be installed adjacent to holding position markings on the taxiway pavement. On runways, holding position markings will be located only on the runway pavement adjacent to the sign, if the runway is normally used by air traffic control for "Land, Hold Short" operations or as a taxiway. The holding position markings are described in Paragraph 2–22m(1).

c. Runway Approach Area Holding Position Sign. At some airports, it is necessary to hold an aircraft on a taxiway located in the approach or departure area for a runway so that the aircraft does not interfere with operations on that runway. In these situations a sign with the designation of the approach end of the runway followed by a "dash" (—) and letters "APCH" will be located at the holding position on the taxiway. Holding position markings in accordance with Paragraph 2–22m(1) will be located on the taxiway pavement. An example of this sign is shown in Figure 2–24[4]. In this example, the sign may protect the approach to Runway 15 and/or the departure for Runway 33.

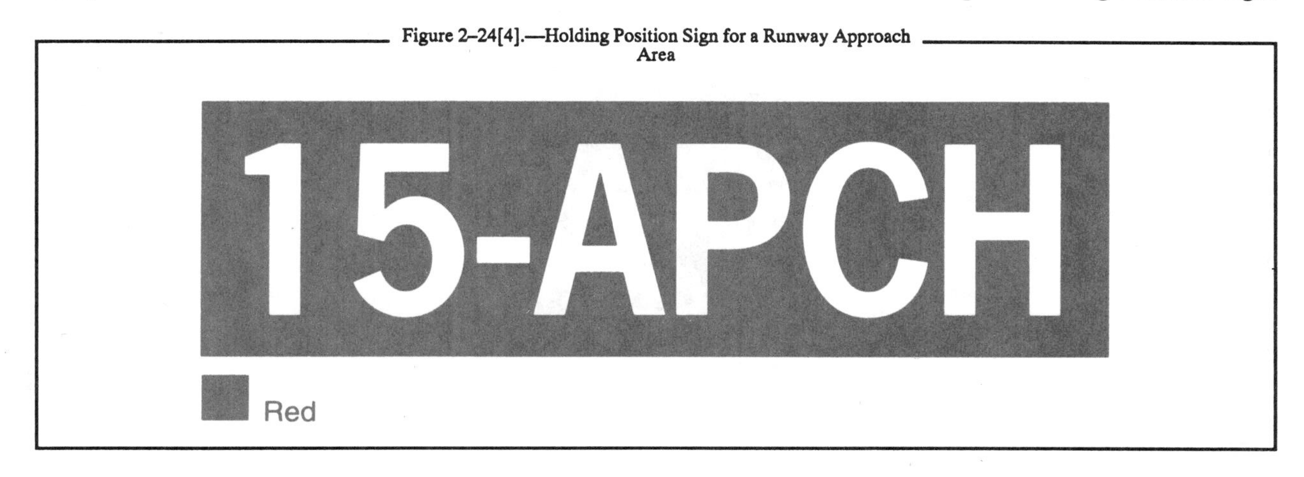

Figure 2–24[4].—Holding Position Sign for a Runway Approach Area

d. ILS Critical Area Holding Position Sign. At some airports, when the instrument landing system is being used, it is necessary to hold an aircraft on a taxiway at a location other than the holding position described in Paragraph 2–22m(1). In these situations the holding position sign for these operations will have the inscription "ILS" and be located adjacent to the holding position marking on the taxiway described in Paragraph 2–22m(2). An example of this sign is shown in Figure 2–24[5].

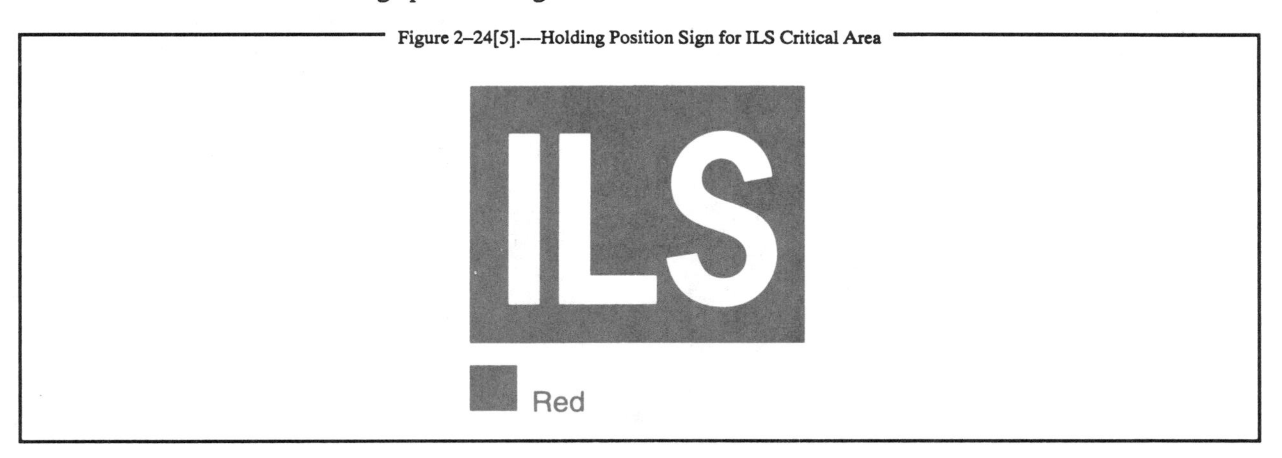

Figure 2–24[5].—Holding Position Sign for ILS Critical Area

e. No Entry Sign. This sign, shown in Figure 2–24[6], prohibits an aircraft from entering an area. Typically, this sign would be located on a taxiway intended to be used in only one direction or at the intersection of vehicle roadways with runways, taxiways or aprons where the roadway may be mistaken as a taxiway or other aircraft movement surface.

Figure 2–24[6].—Sign Prohibiting Aircraft Entry into an Area

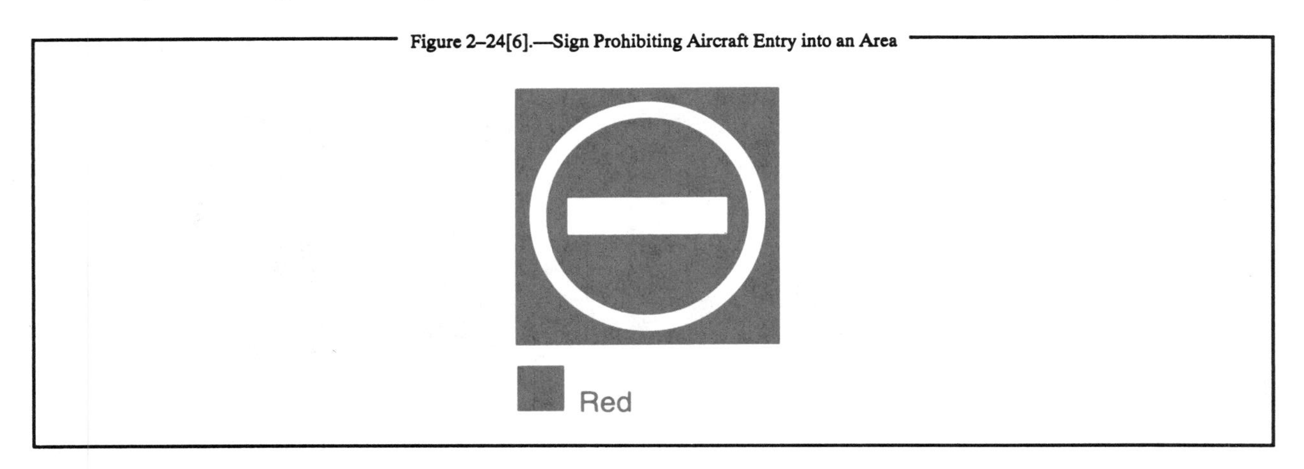

2–24 NOTE.—The holding position sign provides the pilot with a visual cue as to the location of the holding position marking. The operational significance of holding position markings are described in the notes for Paragraph 2–22m.

2–25. LOCATION SIGNS

a. Location signs are used to identify either a taxiway or runway on which the aircraft is located. Other location signs provide a visual cue to pilots to assist them in determining when they have exited an area. The various location signs are described below.

b. Taxiway Location Sign. This sign has a black background with a yellow inscription and yellow border as shown in Figure 2–25[1]. The inscription is the designation of the taxiway on which the aircraft is located. These signs are installed along taxiways either by themselves or in conjunction with direction signs (See Figure 2–26[1]) or runway holding position signs (See Figure 2–25[2]).

Figure 2–25[1].—Taxiway Location Sign

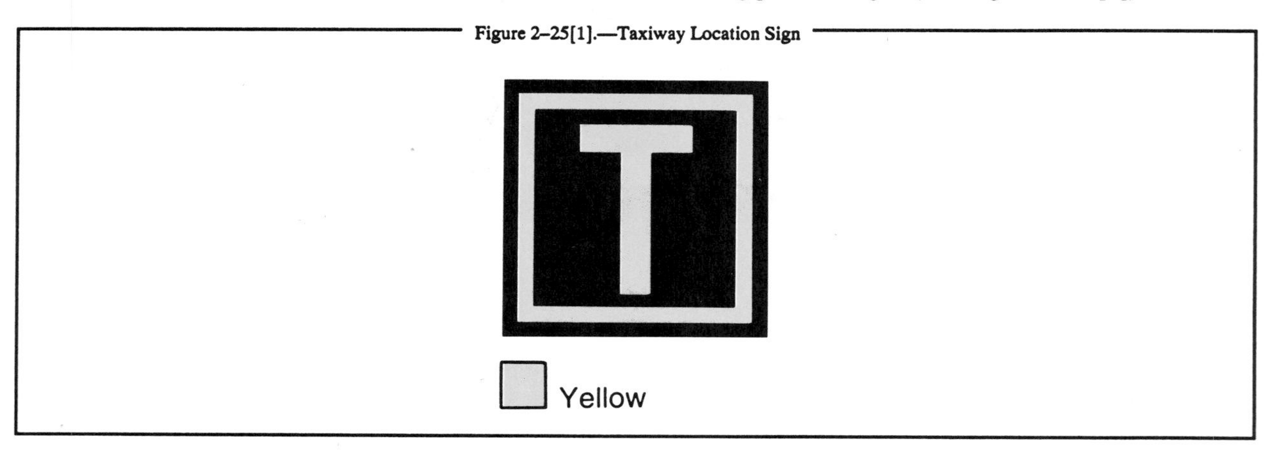

Figure 2–25[2].—Taxiway Location Sign Collocated with Runway Holding Position Sign

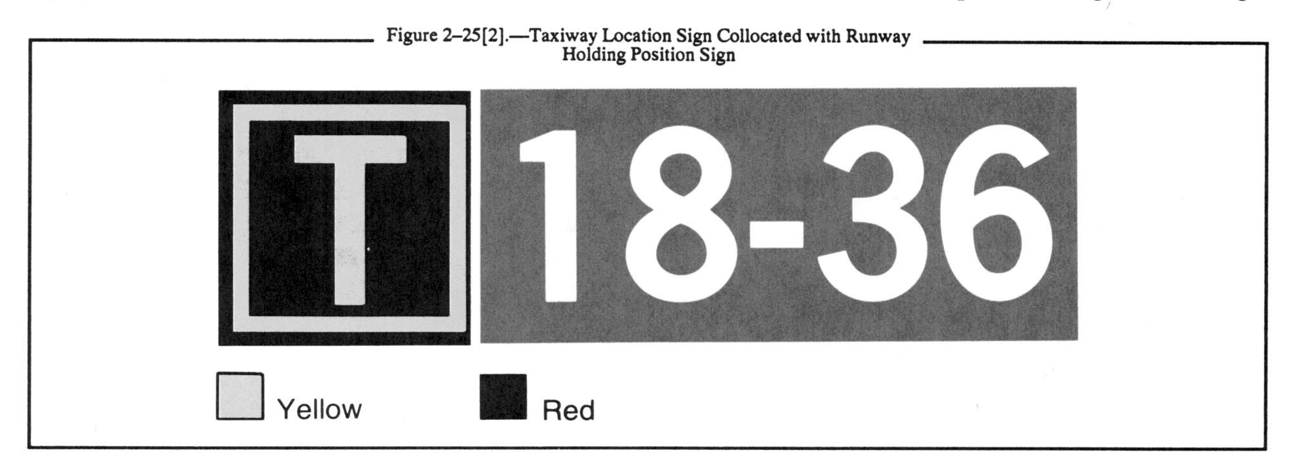

c. Runway Location Sign. This sign has a black background with a yellow inscription and yellow border as shown in Figure 2–25[3]. The inscription is the designation of the runway on which the aircraft is located. These signs are intended to complement the information available to pilots through their magnetic compass and typically are installed where the proximity of two or more runways to one another could cause pilots to be confused as to which runway they are on.

Figure 2–25[3].—Runway Location Sign

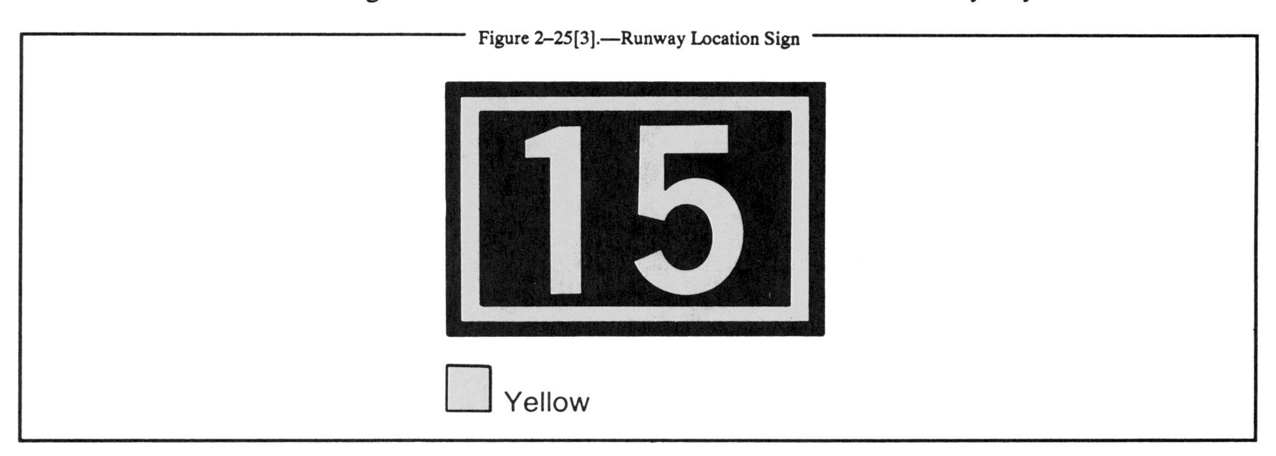

d. Runway Boundary Sign. This sign has a yellow background with a black inscription with a graphic depicting the pavement holding position marking as shown in Figure 2–25[4]. This sign, which faces the runway and is visible to the pilot exiting the runway, is located adjacent to the holding position marking on the pavement. The sign is intended to provide pilots with another visual cue which they can use as a guide in deciding when they are "clear of the runway."

Figure 2–25[4].—Runway Boundary Sign

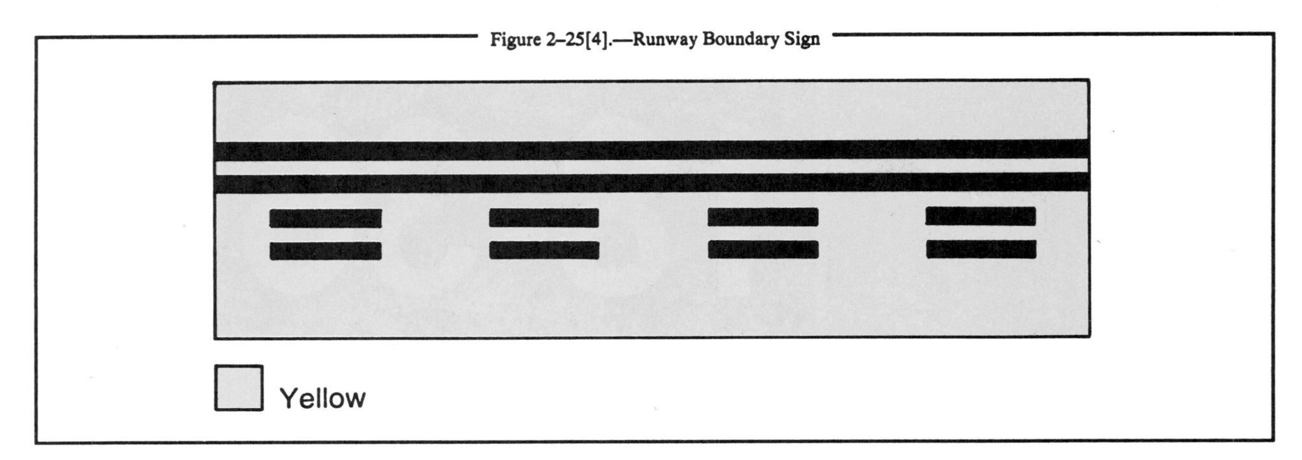

e. ILS Critical Area Boundary Sign. This sign has a yellow background with a black inscription with a graphic depicting the ILS pavement holding position marking as shown in Figure 2–25[5]. This sign is located adjacent to the ILS holding position marking on the pavement and can seen by pilots leaving the critical area. The sign is intended to provide pilots with another visual cue which they can use as a guide in deciding when they are "clear of the ILS critical area."

Figure 2–25[5].—ILS Critical Area Boundary Sign

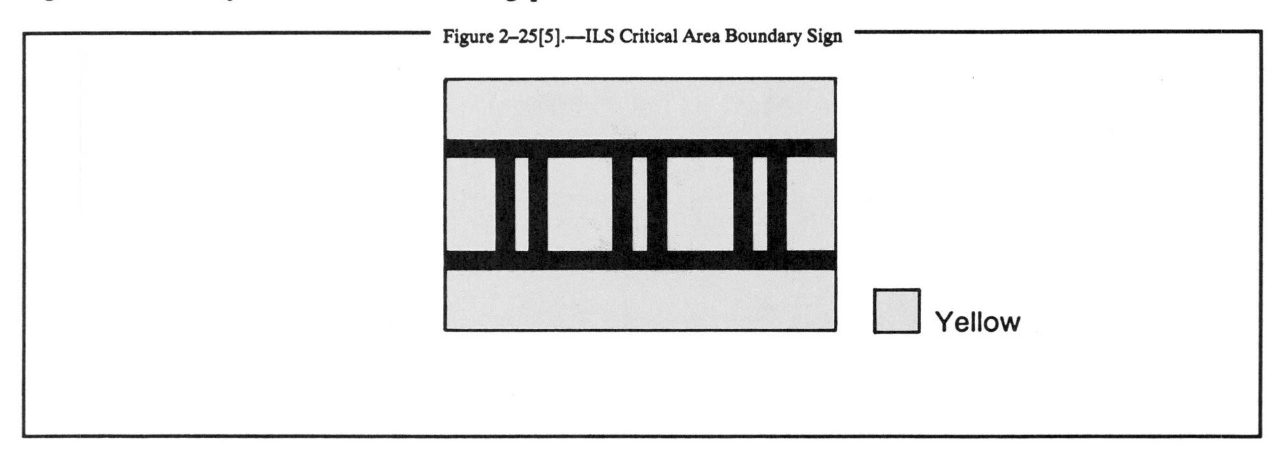

2–26. DIRECTION SIGNS

a. Direction signs have a yellow background with a black inscription. The inscription identifies the designation(s) of the intersecting taxiway(s) leading out of intersection that a pilot would normally be expected to turn onto or hold short of. Each designation is accompanied by an arrow indicating the direction of the turn.

b. Except as noted in subparagraph e, each taxiway designation shown on the sign is accompanied by only one arrow. When more than one taxiway designation is shown on the sign each designation and its associated arrow is separated from the other taxiway designations by either a vertical message divider or a taxiway location sign as shown in Figure 2–26[1].

Figure 2–26[1].—Direction Sign Array with Location Sign on Far Side of Intersection

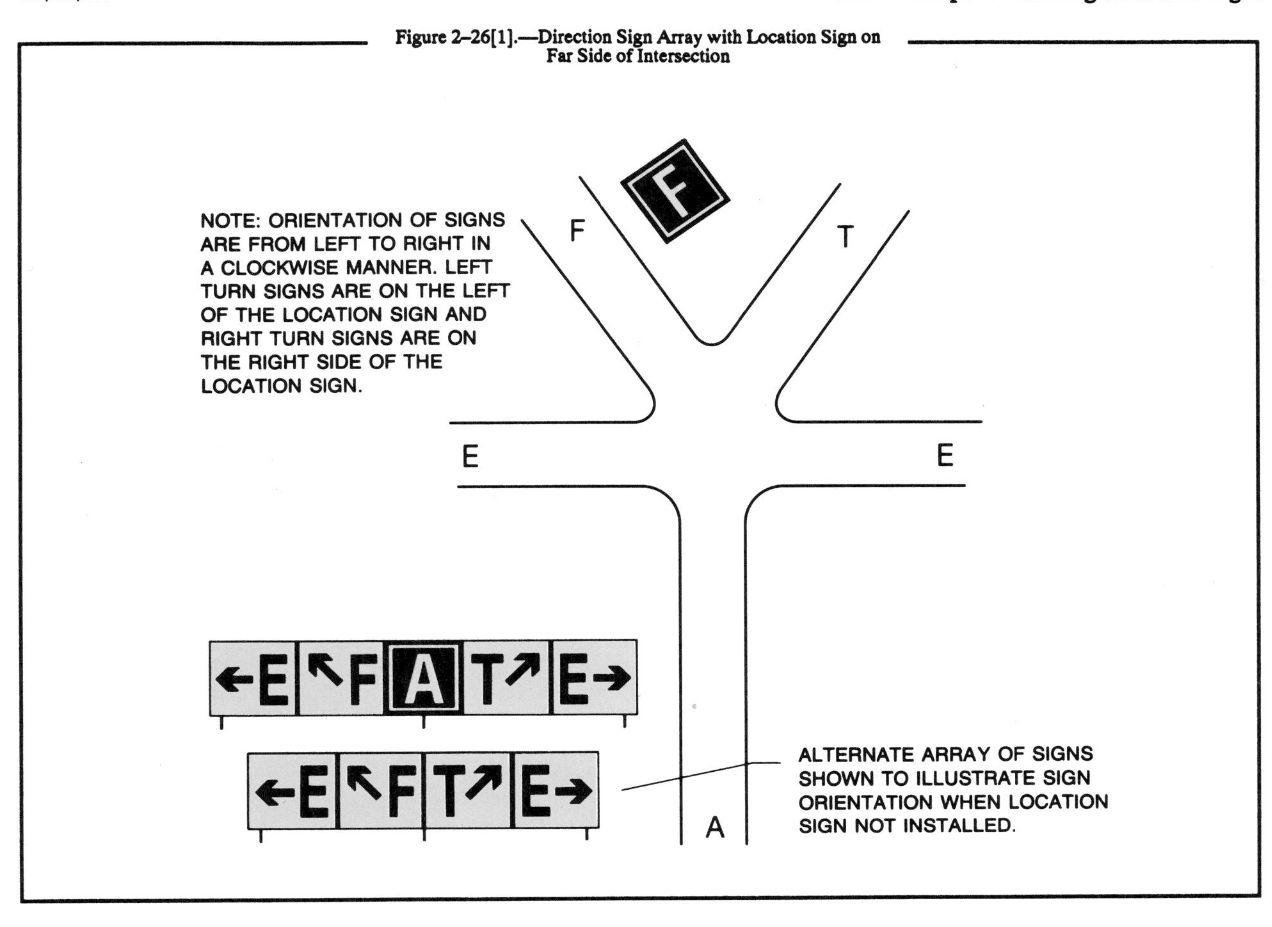

c. Direction signs are normally located on the left prior to the intersection. When used on a runway to indicate an exit, the sign is located on the same side of the runway as the exit. Figure 2–26[2] shows a direction sign used to indicate a runway exit.

Figure 2–26[2].—Direction Sign for Runway Exit

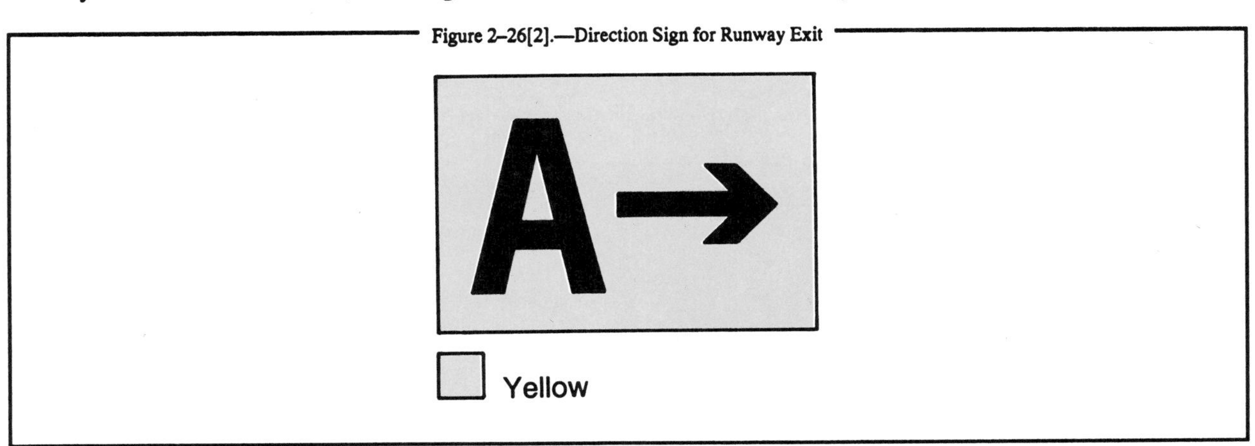

d. The taxiway designations and their associated arrows on the sign are arranged clockwise starting from the first taxiway on the pilot's left, see Figure 2–26[1].

e. If a location sign is located with the direction signs, it is placed so that the designations for all turns to the left will be to the left of the location sign; the designations for continuing straight ahead or for all turns to the right would be located to the right of the location sign. See Figure 2–26[1].

f. When the intersection is comprised of only one crossing taxiway, it is permissible to have two arrows associated with the crossing taxiway as shown in Figure 2–26[3]. In this case, the location sign is located to the left of the direction sign.

Figure 2–26[3].—Direction Sign Array for Simple Intersection

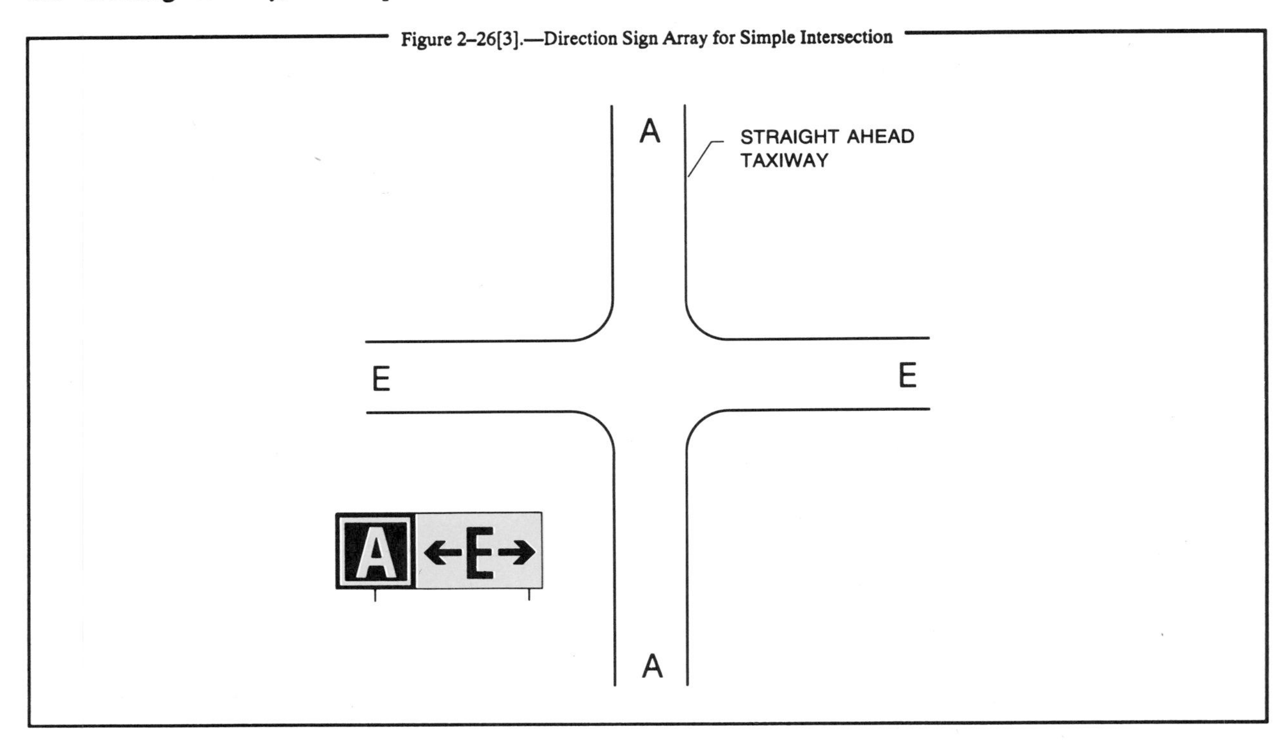

2–27. DESTINATION SIGNS

a. Destination signs also have a yellow background with a black inscription indicating a destination on the airport. These signs always have an arrow showing the direction of the taxiing route to that destination. Figure 2–27[1] is an example of a typical destination sign. When the arrow on the destination sign indicates a turn, the sign is located prior to the intersection.

Figure 2–27[1].—Destination Sign for Military Area

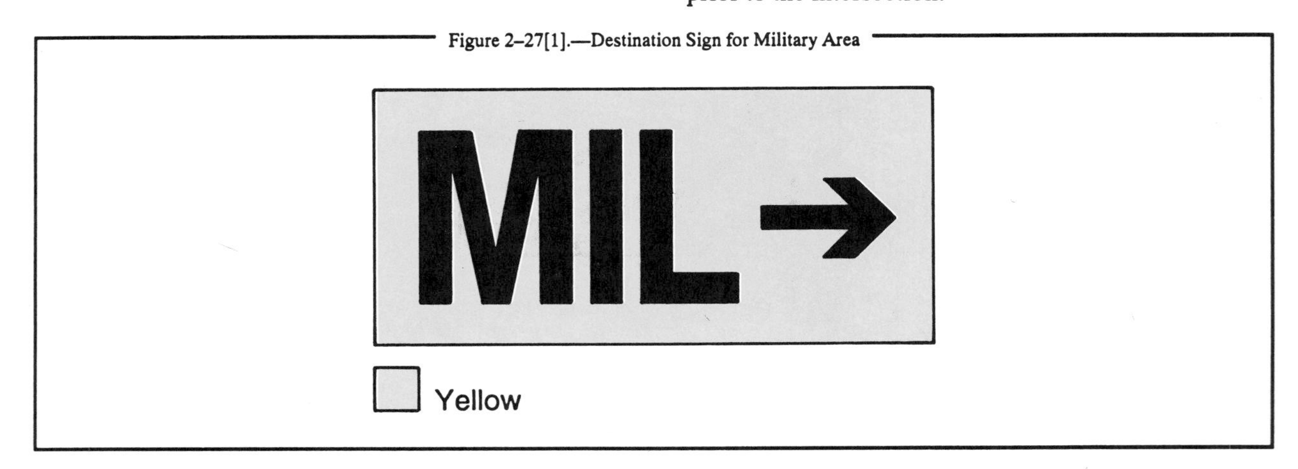

b. Destinations commonly shown on these types of signs include runways, aprons, terminals, military areas, civil aviation areas, cargo areas, international areas, and fixed base operators. An abbreviation may be used as the inscription on the sign for some of these destinations.

c. When the inscription for two or more destinations having a common taxiing route are placed on a sign, the destinations are separated by a "dot" (•) and one arrow would be used as shown in Figure 2–27[2]. When the inscription on a sign contains two or more destinations having different taxiing routes, each destination will be accompanied by an arrow and will be separated from the other destinations on the sign with a vertical black message divider as shown in Figure 2–27[3].

Figure 2–27[2].—Destination Sign for Common Taxiing Route to Two Runways

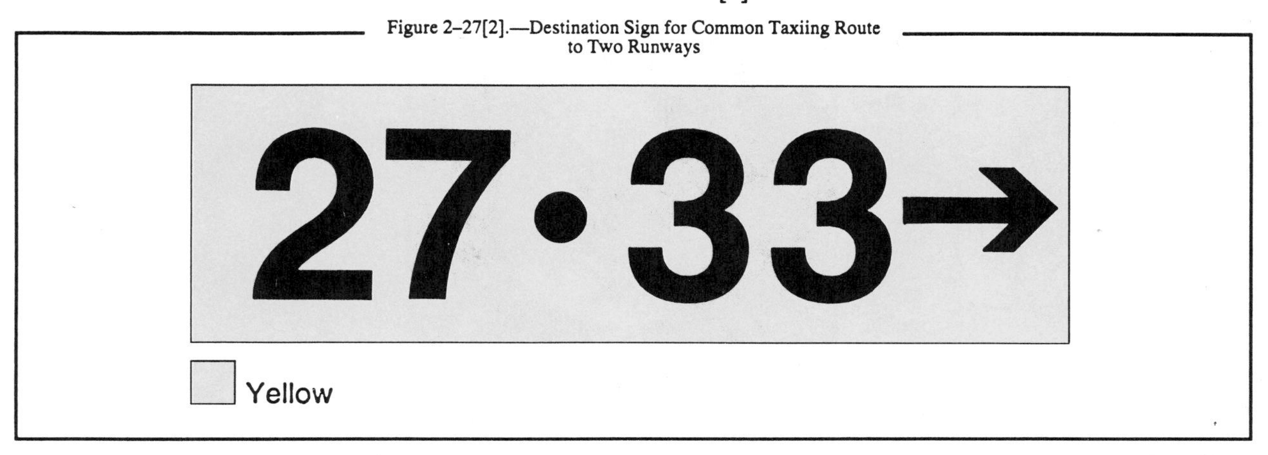

Figure 2–27[3].—Destination Sign for Different Taxiing Routes to Two Runways

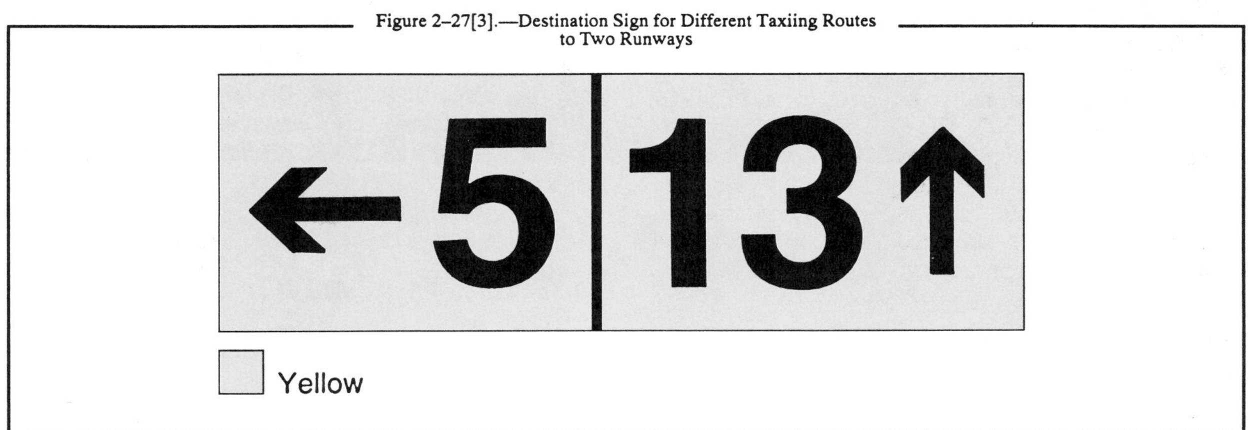

2–28. INFORMATION SIGNS

Information signs have a yellow background with a black inscription. They are used to provide the pilot with information on such things as areas that cannot be seen from the control tower, applicable radio frequencies, and noise abatement procedures. The airport operator determines the need, size, and location for these signs.

2–29. RUNWAY DISTANCE REMAINING SIGNS

Runway distance remaining signs have a black background with a white numeral inscription and may be installed along one or both side(s) of the runway. The number on the signs indicates the distance (in thousands of feet) of landing runway remaining. The last sign, i.e., the sign with the numeral "1," will be located at least 950 feet from the runway end. Figure 2–29[1] shows an example of a runway distance remaining sign.

Figure 2–29[1].—Runway Distance Remaining Sign Indicating 3000 feet of Runway Remaining

2–30. AIRCRAFT ARRESTING DEVICES

a. Certain airports are equipped with a means of rapidly stopping military aircraft on a runway. This equipment, normally referred to as EMERGENCY ARRESTING GEAR, generally consists of pendant cables supported over the runway surface by rubber "donuts". Although most devices are located in the overrun areas, a few of these arresting systems have cables stretched over the operational areas near the ends of a runway.

b. Arresting cables which cross over a runway require special markings on the runway to identify the cable location. These markings consist of 10 feet diameter solid circles painted "identification yellow", 30 feet on center, perpendicular to the runway centerline across the entire runway width. Additional details are contained in ADVISORY CIRCULAR–150/5220–9, Aircraft Arresting Systems for Joint/Civil Military Airports.

2–30b NOTE.—Aircraft operations on the runway are NOT restricted by the installation of aircraft arresting devices.

2–31 thru 2–40. RESERVED

SPECIAL NOTICE

AIRSPACE RECLASSIFICATION AND OTHER CHARTING CHANGES

Beginning with this edition, and continuing for at least two additional editions, this insert will be printed in the AIM. As indicated by the title, ''AIRSPACE RECLASSIFICATION AND CHARTING CHANGES FOR VFR PRODUCTS,'' the figures depict the charting changes associated with airspace reclassification and other independent changes. The charting insert has been prepared by the National Ocean Service (NOS) in cooperation with the Department of Defense (DOD) and the Federal Aviation Administration (FAA).

The actual charting changes will begin to appear on October 15, 1992 and be ccompleted on or before the effective date of September 16, 1993. Also beginning on October 15, 1992, a charting insert will accompany each visual (Sectional, Terminal, and World) chart. Of particualar interest to the reader is the scheduled publication dates for Sectional Aeronautical Charts. Every pilot is encourage to check the publication date for his or her area; For example, a chart user in the Kansas City area would not get a chart with all the changes until December 10, 1992.

New FAA Airspace Classifications

Current Airspace Classification

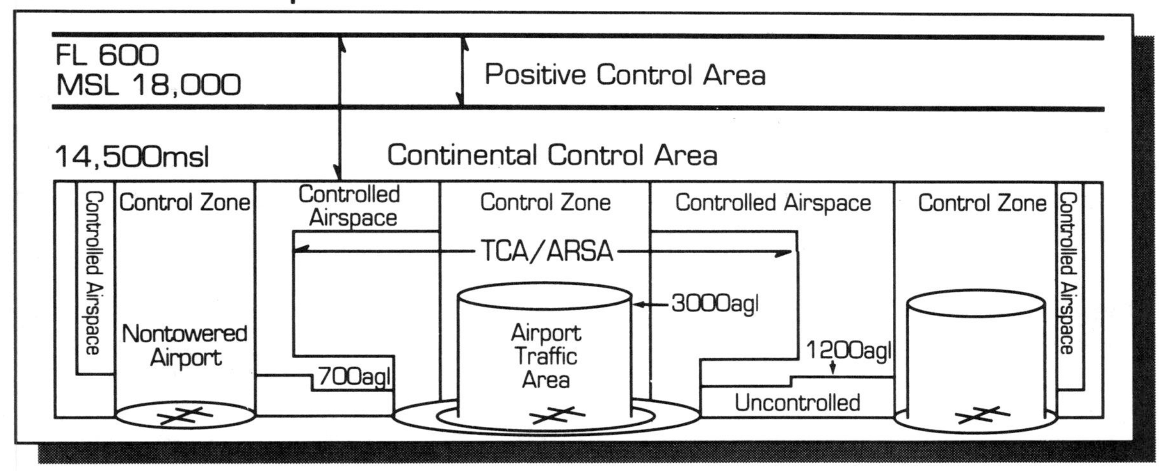

New Airspace Classification

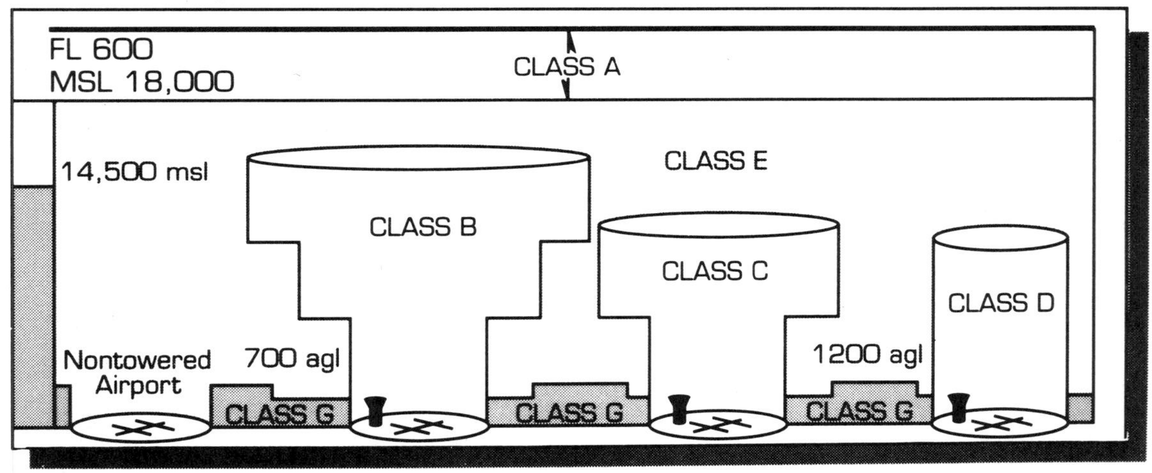

msl - mean sea level
agl - above ground level
FL - flight level

AIRSPACE RECLASSIFICATION
AND CHARTING CHANGES FOR VFR PRODUCTS

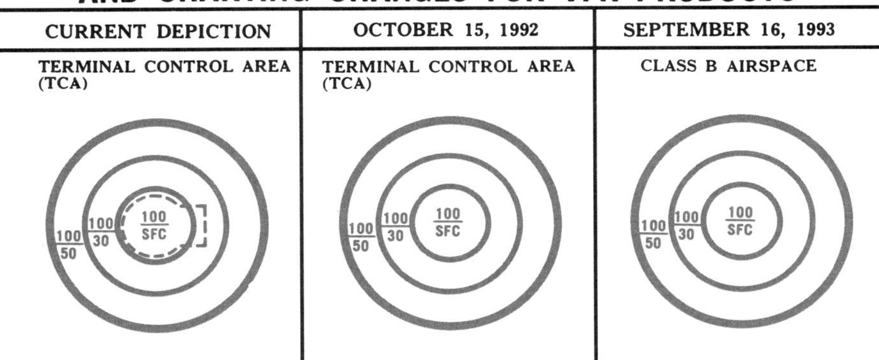

There are no charting symbology changes. However, beginning on October 15, 1992, the control zones associated with any airport in the Terminal Control Area (TCA) surface area will cease to be charted. The control zones will continue to legally exist until the TCAs become Class B airspace on September 16, 1993. One rule change, effective with Class B airspace, is that the cloud clearance criteria for VFR flight operations is changed to "clear of clouds."

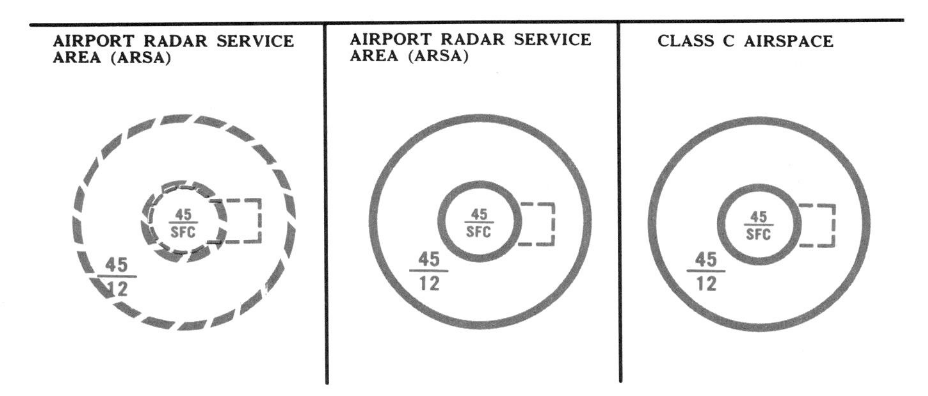

Beginning on October 15, 1992, the solid magenta line formerly used for Terminal Radar Service Areas (TRSAs) is used for Airport Radar Service Areas (ARSAs). This change will be completed on sectional and terminal area charts by April 1, 1993. Also beginning on October 15, 1992, the control zones associated with any airport in the ARSA surface area will coincide with the ARSA surface area and cease to be charted. The ARSAs will become Class C airspace on September 16, 1993. Any extension of a control zone that exceeds an ARSA surface area will be depicted with a magenta segmented line. The magenta segmented line denotes controlled airspace extending upward from the surface to the overlying or adjacent controlled airspace. Such extensions will become Class E airspace on September 16, 1993. There are no operating rule changes, pilots may continue to operate VFR underneath the ARSA/Class C shelf without contacting air traffic control. On April 2, 1992, TRSAs began to be depicted with a solid black line. The interim conversion will be complete October 15, 1992. TRSAs, as entities, will not become an airspace class on September 16, 1993.

Published at Washington, D.C.
U.S. Department of Commerce
National Oceanic and Atmospheric Administration
National Ocean Service

Page 1

CURRENT DEPICTION	OCTOBER 15, 1992	SEPTEMBER 16, 1993
CONTROL ZONE	CONTROL ZONE WITH TOWER	CLASS D AIRSPACE
	26	26
	CONTROL ZONE WITH TOWER AND EXTENSION WITHOUT COMMUNICATIONS REQUIREMENT	CLASS D AIRSPACE WITH ASSOCIATED CLASS E AIRSPACE
CONTROL ZONES ARE SHOWN ON ALL VFR CHARTS	26	26

Those control zones with an operating control tower will continue to be depicted with a blue segmented line. The vertical limit, in AMSL, is now charted in hundreds of feet. The airport traffic area (ATA) and its communications requirement with air traffic control remains until September 16, 1993 at which time, a like communications requirement within the Class D airspace is established.

Arrival extensions will either be charted as part of the basic surface area with the blue segmented symbology or as a separate surface area indicated by the magenta segmented line. Communications with air traffic control are not required within the airspace encompassed by the magenta lines, which will be Class E airspace.

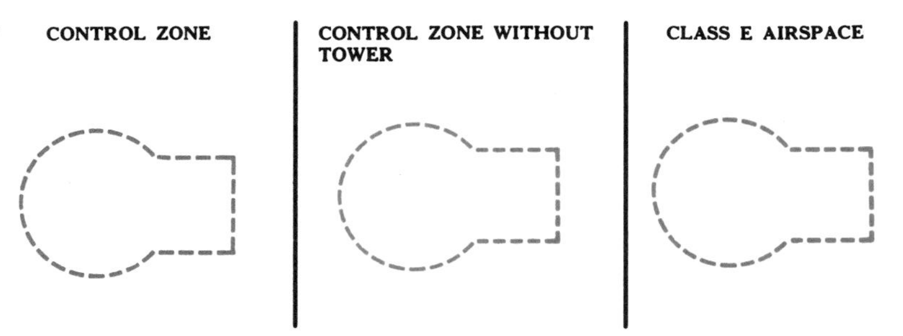

Control zones without an operating control tower will now be depicted with a magenta segmented line which denotes controlled airspace extending upward from the surface to the overlying or adjacent controlled airspace; therefore, the vertical limit is not depicted. There are no operating rule changes and such control zones become Class E airspace on September 16, 1993.

CONTROL ZONES WILL NOT BE CHARTED ON WORLD AERONAUTICAL CHARTS (WACs) PRINTED OCTOBER 15, 1992 AND AFTER.

CURRENT DEPICTION	OCTOBER 15, 1992	SEPTEMBER 16, 1993

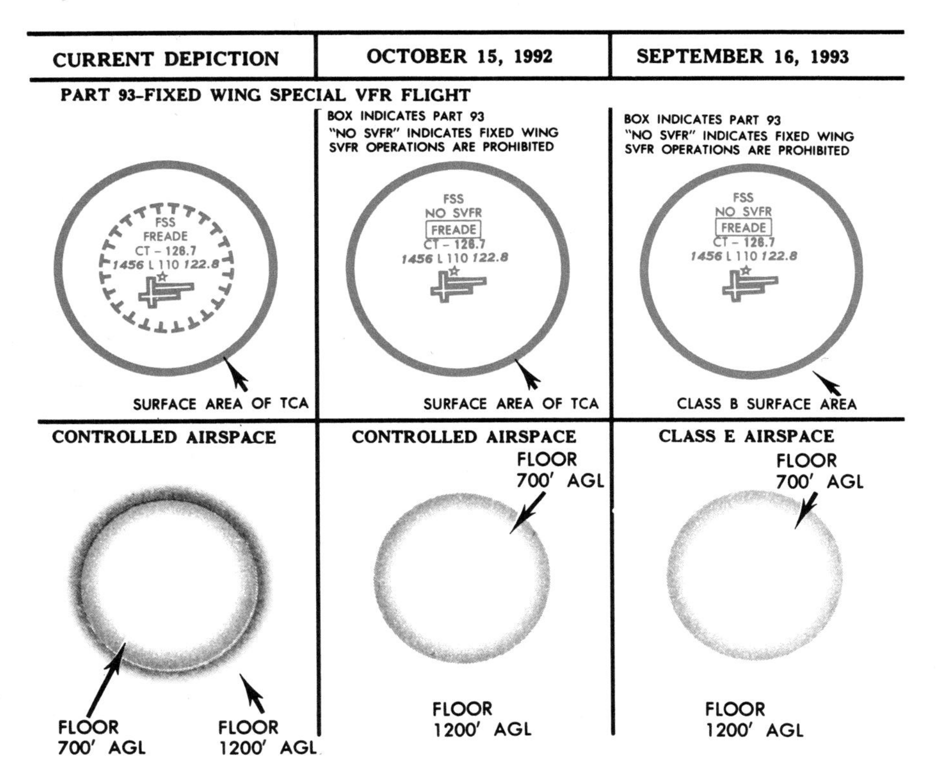

Effective October 15, 1992, the blue vignette (light-blue shaded) line will not be used to depict the 1200 foot or above airspace, unless it abuts uncontrolled airspace. Where the outer edge of the 700 foot transition area (magenta vignette) ends, the 1200 foot or greater area, automatically begins. Effective September 16, 1993, these areas become Class E airspace extending upward from other than the surface. There are no operating rule changes.

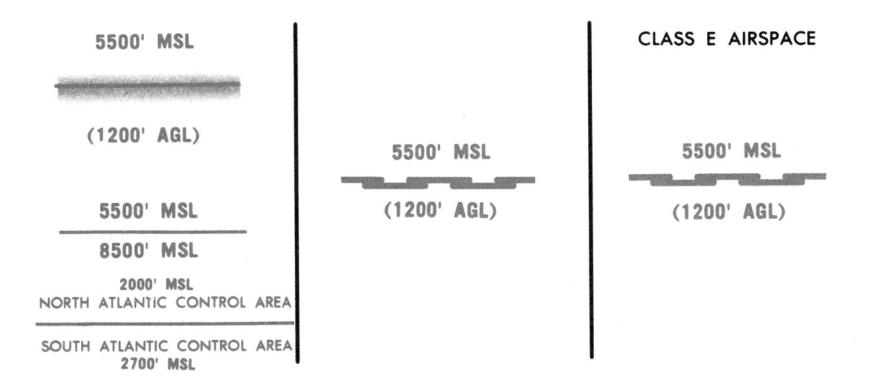

This is a new symbology which will be used to depict the boundary of controlled airspace (Class E) with floors other than 700 feet or 1200 feet. This symbology will also be used to distinguish the floors of the domestic offshore areas and the offshore control areas beyond 12 NM of the U.S. coast.

WACs DIFFERENTIATE BETWEEN HORIZONTAL LIMITS OF AIRSPACE, NOT VERTICAL LIMITS. WACs WILL CONTINUE TO USE THE SOLID BLUE LINE FOR DIFFERENTIATING CONTROL AREAS.

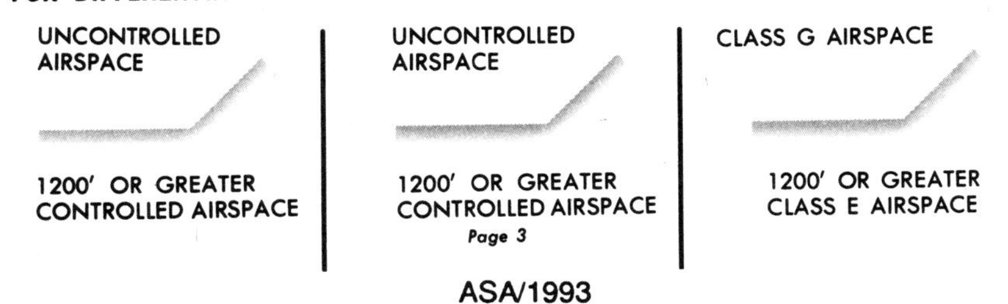

Page 3

CURRENT DEPICTION	OCTOBER 15, 1992	SEPTEMBER 16, 1993
AIR DEFENSE IDENTIFICATION ZONE	AIR DEFENSE IDENTIFICATION ZONE	AIR DEFENSE IDENTIFICATION ZONE

The Air Defense Identification Zone (ADIZ) is changing symbology size and color. The ADIZ is not a type of airspace nor will it become a class of airspace.

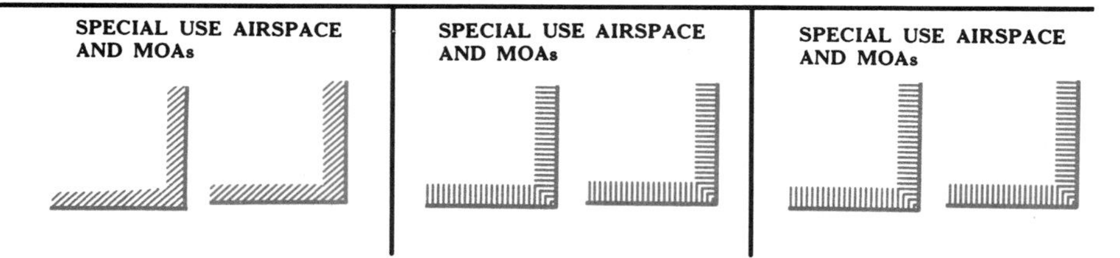

AN IFR INSERT WILL BE AVAILABLE PRIOR TO THE SEPTEMBER 16, 1993 EFFECTIVE DATE OF THE AIRSPACE RECLASSIFICATION.

USE OF OBSOLETE CHARTS FOR NAVIGATION IS DANGEROUS. Aeronautical information changes rapidly, and it is vitally important that pilots check the effective dates on each aeronautical chart to be used. Obsolete charts should be discarded and replaced by current editions. To make certain a chart is current, refer to the next scheduled edition date.

Major changes will occur to the VFR aeronautical charts begining with the October 15, 1992 editions. Listed below are the first scheduled dates for sectional charts depicting these changes. These dates are subject to change.

SECTIONAL AERONAUTICAL CHARTS

Albuquerque ..... Nov 12, 1992
Anchorage .... Dec 10, 1992
Atlanta .... Apr 1, 1993
Bethel .... Aug 19, 1993
Billings .... Apr 1, 1993
Brownsville .... Dec 10, 1992
Cape Lisburne .... Sep 16, 1993
Charlotte .... Feb 4, 1993
Cheyenne .... Mar 4, 1993
Chicago .... Oct 15, 1992
Cincinnati .... Jan 7, 1993
Cold Bay .... Mar 4, 1993
Dallas-Ft Worth .... Jan 7, 1993
Dawson .... Nov 12, 1992
Denver .... Feb 4, 1993
Detroit .... Oct 15, 1992
Dutch Harbor ...... Apr 1, 1993
El Paso .... Mar 4, 1993
Fairbanks .... Jan 7, 1993
Great Falls .... Feb 4, 1993
Green Bay .... Jan 7, 1993
Halifax .... Apr 1, 1993
Hawaiian Is ..... Nov 12, 1992
Houston .... Oct 15, 1992
Jacksonville .... Mar 4, 1993
Juneau .... Apr 29, 1993
Kansas City .... Dec 10, 1992
Ketchikan .... Apr 29, 1993
Klamath Falls ..... Oct 15, 1992
Kodiak .... Mar 4, 1993
Lake Huron .... Oct 15, 1992
Las Vegas .... Apr 1, 1993
Los Angeles .... Jun 7, 1993
McGrath .... Aug 19, 1993
Memphis .... Oct 15, 1992
Miami .... Mar 4, 1993
Montreal .... Apr 1, 1993
New Orleans ..... Dec 10, 1992
New York .... Dec 10, 1992
Nome .... Jul 22, 1993
Omaha .... Mar 4, 1993
Phoenix .... Nov 12, 1992
Point Barrow ..... Nov 12, 1992
Saint Louis .... Nov 12, 1992
Salt Lake City ..... Nov 12, 1992
San Antonio ...... Dec 10, 1992
San Francisco ..... Oct 15, 1992
Seattle .... Jan 7, 1993
Seward .... Dec 10, 1992
Twin Cities .... Feb 4, 1993
Washington .... Mar 4, 1993
West Aleutian ...... Apr 1, 1993
Whitehorse .... May 27, 1993
Wichita .... Nov 12, 1992

Your local agent may place a special order for you, or you may order directly from NOAA at the following address:

NOAA Distribution Branch, N/CG33
National Ocean Service
Riverdale, Maryland 20737-1199

Telephone orders are also accepted by the NOAA Distribution Branch.

Individual Orders 301-436-6990
Subscriptions 301-436-6993

Chapter 3. AIRSPACE
Section 1. GENERAL

3–1. GENERAL

Because of the nature of some operations, restrictions are required for safety reasons. The complexity or density of aircraft movements in other airspace areas may result in additional aircraft and pilot requirements for operation within such airspace. It is important that pilots be familiar with the operational requirements for the various airspace segments.

3–2. GENERAL DIMENSIONS OF AIRSPACE SEGMENTS

Refer to Federal Aviation Regulations (FARs) for specific dimensions, exceptions, geographical areas covered, exclusions, specific transponder or equipment requirements, and flight operations. In the following illustration, arrows ending near but not touching reference lines mean up to or down to, but not including, the referenced altitude. (See Figure 3–2[1].)

Figure 3–2[1]

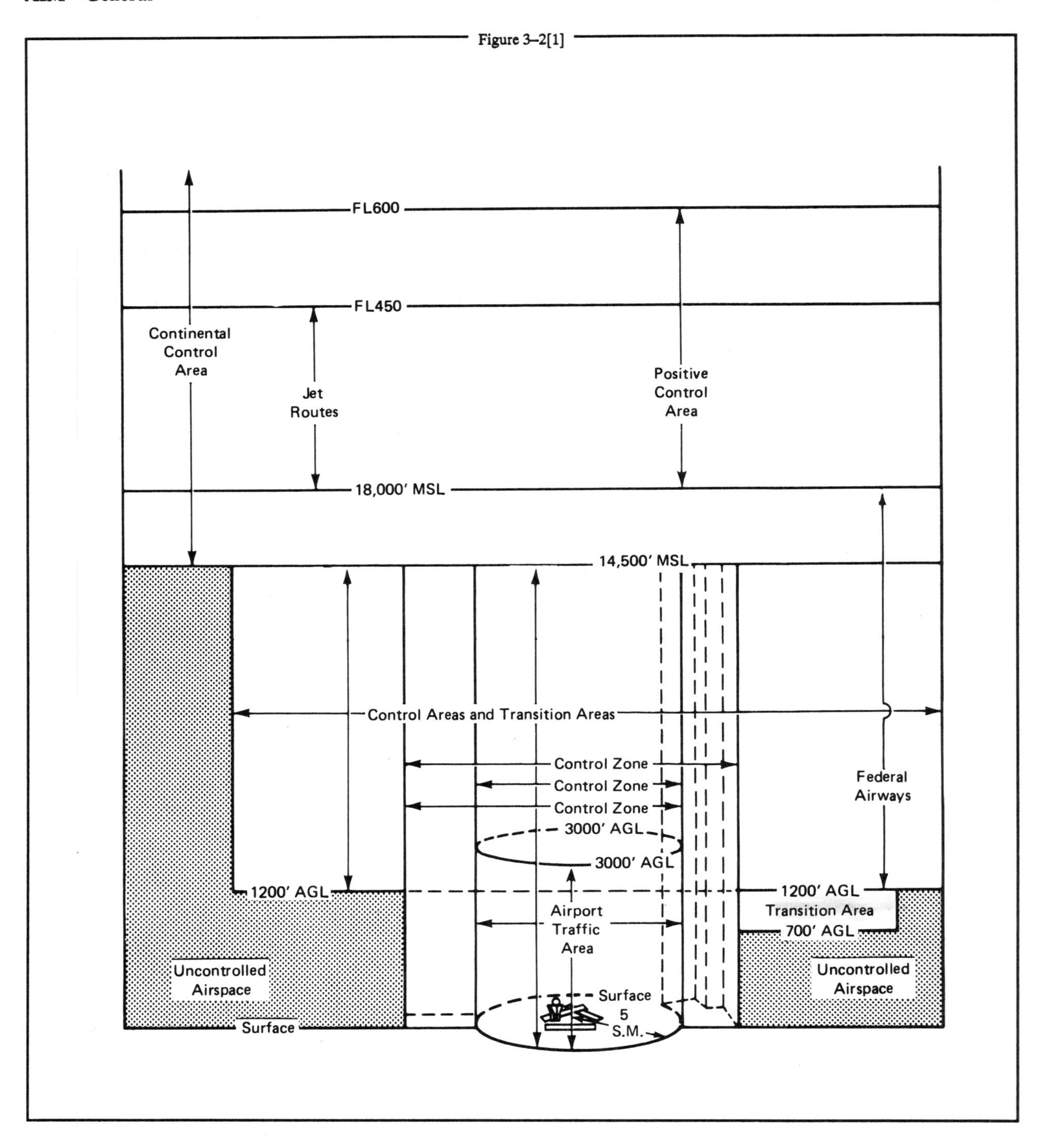

3–3 thru 3–10. RESERVED

Section 2. UNCONTROLLED AIRSPACE

3–11. GENERAL

Uncontrolled airspace is that portion of the airspace that has not been designated as a Control Zone, Airport Radar Service Area, Terminal Control Area, Transition Area, Control Area, Continental Control Area, or Positive Control Area.

3–12. VFR REQUIREMENTS

Rules governing VFR flight have been adopted to assist the pilot in meeting his responsibility to see and avoid other aircraft. Minimum weather conditions and distance from clouds required for VFR flight are contained in these rules. (FAR 91.155.)

3–13. IFR REQUIREMENTS

FARs specify the pilot and aircraft equipment requirements for IFR flight. Pilots are reminded that in addition to altitude or flight level requirements, FAR 91.177 includes a requirement to remain at least 1,000 feet (2,000 feet in designated mountainous terrain) above the highest obstacle within a horizontal distance of 4 nautical miles from the course to be flown.

3–14. MINIMUM VFR VISIBILITY AND DISTANCE FROM CLOUDS

(See Table 3–14[1]).

Table 3–14[1]

BASIC VFR WEATHER MINIMUMS

Altitude	*Flight Visibility*	*Distance From Clouds*
1,200 feet or less above the surface—		
Within controlled airspace:	3 statute miles	500 feet below. 1,000 feet above. 2,000 feet horizontal.
Outside controlled airspace:		
Day: (except as provided in section 91.155(b))	1 statute mile	Clear of clouds.
Night: (except as provided in section 91.155(b))	3 statute miles	500 feet below. 1,000 feet above. 2,000 feet horizontal.
More than 1,200 feet above the surface but less than 10,000 feet MSL—		
Within controlled airspace:	3 statute miles	500 feet below. 1,000 feet above. 2,000 feet horizontal.
Outside controlled airspace:		
Day:	1 statute mile	500 feet below. 1,000 feet above. 2,000 feet horizontal.
Night:	3 statute miles	500 feet below. 1,000 feet above. 2,000 feet horizontal.
More than 1,200 feet above the surface and at or above 10,000 feet MSL—.	5 statute miles	1,000 feet below. 1,000 feet above. 1 mile horizontal.

The following operations may be conducted outside of controlled airspace below 1,200 feet above the surface:

1. *Helicopter.* When the visibility is less than 1 mile during day hours or less than 3 miles during night hours, a helicopter may be operated clear of clouds if operated at a speed that allows the pilot adequate opportunity to see any air traffic or obstruction in time to avoid a collision.

2. *Airplane.* When the visibility is less than 3 miles but not less than 1 mile during night hours, an airplane may be operated clear of clouds if operated in an airport traffic pattern within one-half mile of the runway.

3–15. ALTITUDES AND FLIGHT LEVELS

(See Table 3–15[1] and Table 3–15[2]).

Table 3–15[1]

VFR CRUISING ALTITUDES AND FLIGHT LEVELS

If your magnetic course (ground track) is:	*And you are more than 3,000 feet above the surface but below 18,000 feet MSL, fly:*	*And you are above 18,000 feet MSL to FL 290 (except within Positive Control Area, FAR 71.193), fly:*	*And you are above FL 290 (except within Positive Control Area, FAR 71.193), fly 4,000 foot intervals:*
0° to 179°	Odd thousands MSL, plus 500 feet (3,500, 5,500, 7,500, etc.).	Odd Flight Levels plus 500 feet (FL 195, FL 215, FL 235, etc.).	Beginning at FL 300 (FL 300, 340, 380, etc.).
180° to 359°	Even thousands MSL, plus 500 feet (4,500, 6,500, 8,500, etc.).	Even Flight Levels plus 500 feet (FL 185, FL 205, FL 225, etc.).	Beginning at FL 320 (FL 320, 360, 400, etc.).

Table 3–15[2]

IFR ALTITUDES AND FLIGHT LEVELS—UNCONTROLLED AIRSPACE

If your magnetic course (ground track) is:	*And you are below 18,000 feet MSL, fly:*	*And you are at or above 18,000 feet MSL but below FL 290, fly:*	*And you are at or above FL 290, fly 4,000 foot intervals:*
0° to 179°	Odd thousands MSL, (3,000, 5,000, 7,000, etc.).	Odd Flight Levels, FL 190, 210, 230, etc.).	Beginning at FL 290, (FL 290, 330, 370, etc.).
180° to 359°	Even thousands MSL, (2,000, 4,000, 6,000, etc.).	Even Flight Levels (FL 180, 200, 220, etc.).	Beginning at FL 310, (FL 310, 350, 390, etc.).

3–16 thru 3–20. RESERVED

Section 3. CONTROLLED AIRSPACE

3–21. GENERAL

Controlled airspace consists of those areas designated as Continental Control Area, Control Area, Positive Control Area, Control Zones, Terminal Control Areas, Airport Radar Service Area and Transition Areas, within which some or all aircraft may be subject to ATC. Safety, users' needs, and volume of flight operations are some of the factors considered in the designation of controlled airspace. When so designated, the airspace is supported by ground to air communications, navigation aids, and air traffic services.

3–22. CONTINENTAL CONTROL AREA

a. The Continental Control Area consists of the airspace of the 48 contiguous States, the District of Columbia and Alaska, excluding the Alaska peninsula west of longitude 160 degrees 00 minutes 00 seconds W, at and above 14,500 feet MSL, but does not include:

1. airspace less than 1,500 feet above the surface of the earth; or

2. Prohibited and Restricted areas, other than the Restricted areas listed in FAR 71 Subpart D.

3–23. CONTROL AREAS

Control Areas consist of the airspace designated as Colored Federal airways, VOR Federal airways, Additional Control Areas, and Control Area Extensions, but do not include the Continental Control Area. Unless otherwise designated, Control Areas also include the airspace between a segment of a main VOR airway and its associated alternate segments. The vertical extent of the various categories of airspace contained in Control Areas is defined in FAR 71.

3–24. POSITIVE CONTROL AREA

Positive Control Area is airspace so designated as positive control area in FAR 71.193. This area includes specified airspace within the conterminous U.S. from 18,000 feet to and including FL 600, excluding Santa Barbara Island, Farallon Island, and that portion south of latitude 25 degrees 04 minutes north. In Alaska, it includes the airspace over the State of Alaska from 18,000 feet to and including FL 600, but not including the airspace less than 1,500 feet above the surface of the earth and the Alaskan Peninsula west of longitude 160 degrees 00 minutes west. Rules for operating in Positive Control Area are found in FAR 91.135 and FAR 91.215.

3–25. TRANSITION AREAS

a. Transition Areas are designated to contain IFR operations in controlled airspace during portions of the terminal operation and while transitioning between the terminal and en route environments.

b. Transition Areas are controlled airspace extending upward from 700 feet or more above the surface when designated in conjunction with an airport for which an instrument approach procedure has been prescribed; or from 1,200 feet or more above the surface when designated in conjunction with airway route structures or segments. Unless specified otherwise, Transition Areas terminate at the base of overlying controlled airspace.

3–26. CONTROL ZONES

a. Control Zones are regulatory in nature and established as controlled airspace. They extend upward from the surface and terminate at the base of the Continental Control Area. Control Zones that do not underlie the Continental Control Area have no upper limit. A Control Zone is based on a primary airport but may include one or more airports and is normally a circular area within a radius of 5 statute miles around the primary airport, except that it may include extensions necessary to include instrument departure and arrival paths.

b. Some basic requirements for designating a control zone are communications and weather observation reporting:

1. Communications capability with aircraft which normally operate within the control zone must exist down to the runway surface of the primary airport. Communications may be either direct from the ATC facility having jurisdiction over the control zone or by rapid relay through other communications facilities which are acceptable to that ATC facility.

2. Federally certificated weather observers take hourly and special weather observations at the primary airport in the control zone during the times and dates a control zone is designated. The required weather observations must be forwarded expeditiously to the ATC facility having jurisdiction over the control zones.

c. Control Zones are depicted on charts (for example, on the Sectional Charts the Control Zone is outlined by a broken blue line). If a Control Zone is effective only during certain hours of the day (a part-time Control Zone as prescribed in the regulation) it will be reflected on the charts. A typical Control Zone is depicted in General

Dimensions of Airspace Segments, paragraph 3–2. (Reference—Special VFR Clearances, paragraph 4–85).

3–27. TERMINAL CONTROL AREA (TCA)

a. Terminal Control Area (TCA) consist of controlled airspace extending upward from the surface or higher to specified altitudes, within which **all aircraft** are subject to the operating rules and pilot/equipment requirements specified in FAR 91. Each TCA location includes at least one primary airport around which the TCA is located. Descriptions of TCAs can be found in FAR 71.

Table 3–27[1]

These areas are currently designated as TCA's and are depicted on VFR Terminal Area Charts.

Atlanta	Minneapolis
Boston	New Orleans
Charlotte	New York
Chicago	Philadelphia
Cleveland	Phoenix
Dallas	Pittsburgh
Denver	Orlando
Detroit	St. Louis
Honolulu	Salt Lake City
Houston	San Diego
Kansas City	San Francisco
Las Vegas	Seattle
Los Angeles	Tampa
Memphis	Washington, DC
Miami	

b. TCAs are charted on Sectional, World Aeronautical, En route Low Altitude, DOD FLIP and Terminal Area Charts. (See Table 3–27[1].)

3–28. TERMINAL CONTROL AREA (TCA) OPERATION

a. Operating Rules and Pilot/Equipment Requirements. REGARDLESS OF WEATHER CONDITIONS, AN ATC AUTHORIZATION IS REQUIRED PRIOR TO OPERATING WITHIN A TCA. Pilots should not request an authorization to operate within a TCA unless the requirements of FAR 91.215 and FAR 91.131 are met. Included among these requirements are:

1. Unless otherwise authorized by ATC, aircraft must be equipped with an operable two-way radio capable of communicating with ATC on appropriate frequencies for that terminal control area.

2. person may takeoff or land a civil aircraft at an airport within a TCA or operate a civil aircraft within a TCA unless:

(a) pilot-in-command holds at least a private pilot certificate; or,

(b) aircraft is operated by a student pilot who has met the requirements of FAR 61.95; however,

(c) the following TCA primary airports, no person may takeoff or land a civil aircraft unless the pilot-in-command holds **at least** a private pilot certificate:

Atlanta Hartsfield Airport, GA
Boston Logan Airport, MA
Chicago O'Hare Intl Airport, IL
Dallas/Fort Worth Intl Airport, TX
Los Angeles Intl Airport, CA
Miami Intl Airport, FL
Newark Intl Airport, NJ
New York Kennedy Airport, NY
New York La Guardia Airport, NY
San Francisco Intl Airport, CA
Washington National Airport, DC
Andrews Air Force Base, MD

3. Unless otherwise authorized by ATC, each person operating a large turbine engine-powered airplane to or from a primary airport shall operate at or above the designated floors while within the lateral limits of the TCA.

4. Unless otherwise authorized by ATC, each aircraft must be equipped as follows:

(a) Two-way radio capable of communications with ATC on appropriate frequencies for that area.

(b) IFR operations, an operable VOR or TACAN receiver.

(c) Unless otherwise authorized by ATC, an operable radar beacon transponder with automatic altitude reporting equipment.

3–28a4c NOTE—ATC may, upon notification, immediately authorize a deviation from the altitude reporting equipment requirement; however, a request for a deviation from the 4096 transponder equipment requirement must be submitted to the controlling ATC facility at least one hour before the proposed operation. (Reference—Transponder Operation, paragraph 4–18).

b. Flight Procedures.

1. *IFR Flights.* Aircraft within the TCA are required to operate in accordance with current IFR procedures. A clearance for a visual approach to a primary airport is not authorization for turbine powered airplanes to operate below the designated floors of the TCA.

2. *VFR Flights.*

(a) ARRIVING AIRCRAFT MUST OBTAIN AUTHORIZATION PRIOR TO ENTERING A TCA AND MUST CONTACT ATC ON THE APPROPRIATE FREQUENCY, and in relation to geographical fixes shown on local charts. Although a pilot may be operating beneath the floor of the TCA on initial contact, communications with ATC should be established in relation to the points indicated for spacing and sequencing purposes.

(b) Departing aircraft require a clearance to depart the TCA and should advise the clearance delivery position of their intended altitude and route of flight. ATC will normally advise VFR aircraft when leaving the geographical limits of the TCA. Radar service is not automatically terminated with this advisory unless specifically stated by the controller.

(c) Aircraft not landing or departing the primary airport may obtain ATC clearance to transit the TCA when traffic conditions permit and provided the requirements of FAR 91.131 are met. Such VFR aircraft are encouraged, to the extent possible, to operate at altitudes above or below the TCA or transit through established VFR corridors. Pilots operating in VFR corridors are urged to use frequency 122.750 MHZ for the exchange of aircraft position information.

(d) VFR non-TCA aircraft are cautioned against operating too closely to TCA boundaries, especially where the floor of the TCA is 3,000 feet or less or where VFR cruise altitudes are at or near the floor of higher levels. Observance of this precaution will reduce the potential for encountering a TCA aircraft operating at TCA floor altitudes. Additionally, VFR non-TCA aircraft are encouraged to utilize the VFR Planning Chart as a tool for planning flight in proximity to a TCA. Charted VFR Flyway Planning charts are published on the back of the existing VFR Terminal Area Charts.

c. ATC Clearances and Separation. AN ATC AUTHORIZATION IS REQUIRED TO ENTER AND OPERATE WITHIN A TCA. VFR pilots are provided sequencing and separation from other aircraft while operating within a TCA. Reference—Terminal Radar Programs for VFR Aircraft, paragraph 4–16.

3–28c1 NOTE— Separation and sequencing of VFR aircraft will be suspended in the event of a radar outage as this service is dependent on radar. The pilot will be advised that the service is not available and issued wind, runway information and the time or place to contact the tower.

3–28c2 NOTE— Separation of VFR aircraft will be suspended during CENRAP operations. Traffic advisories and sequencing to the primary airport will be provided on a workload permitting basis. The pilot will be advised when CENRAP is in use.

1. VFR aircraft are separated from all VFR/IFR aircraft which weigh 19,000 pounds or less by a minimum of:

(a) Target resolution, or

(b) 500 feet vertical separation, or

(c) Visual separation.

2. VFR aircraft are separated from all VFR/IFR aircraft which weight more than 19,000 and turbojets by no less than:

(a) 1 ½ miles lateral separation, or

(b) 500 feet vertical separation, or

(c) Visual separation.

3. This program is not to be interpreted as relieving pilots of their responsibilities to see and avoid other traffic operating in basic VFR weather conditions, to adjust their operations and flight path as necessary to preclude serious wake encounters, to maintain appropriate terrain and obstruction clearance, or to remain in weather conditions equal to or better than the minimums required by FAR 91.155. Approach control should be advised and a revised clearance or instruction obtained when compliance with an assigned route, heading and/or altitude is likely to compromise pilot responsibility with respect to terrain and obstruction clearance, vortex exposure, and weather minimums.

4. ATC may assign altitudes to VFR aircraft that do not conform to FAR 91.159. "Resume Appropriate VFR Altitudes" will be broadcast when the altitude assignment is no longer needed for separation or when leaving the TCA. Pilots must return to an altitude that conforms to FAR 91.159 as soon as practicable.

3–29. IFR ALTITUDES AND FLIGHT LEVELS

Pilots operating IFR within controlled airspace will fly at an altitude/flight level assigned by ATC.

When operating IFR within controlled airspace with an altitude assignment of **VFR ON TOP,** flight is to be conducted at an appropriate VFR altitude which is not below the minimum IFR altitude for the route. (Reference—Altitudes and Flight Levels, paragraph 3–15). **VFR ON TOP** is not permitted in certain airspace areas, such as positive control airspace, certain Restricted Areas, etc. Consequently, IFR flights operating **VFR ON TOP** will avoid such airspace.

3–30. SUSPENSION OF THE TRANSPONDER AND AUTOMATIC ALTITUDE REPORTING EQUIPMENT REQUIREMENTS FOR OPERATIONS IN THE VICINITY OF DESIGNATED MODE C VEIL AIRPORTS

a. Pursuant to Special Federal Aviation Regulation (SFAR) No. 62, effective until December 30, 1993, aircraft operating in the vicinity of the airports listed below are excluded from the Mode C transponder equipment requirements of Federal Aviation Regulation FAR 91.215(b)(2). The exclusion from the Mode C transponder equipment requirement only applies to those operations at or below the altitude specified for each airport that are: (1) within a 2–nautical mile radius of a listed airport: and (2) along a direct route between that airport and the outer boundary of the Mode C veil. The routing must be consistent with established traffic patterns, noise abatement procedures, and safety. The designation of altitudes for each airport is not intended to supersede the provisions of FAR 91.119, Minimum Safe Altitudes. Routings to and from each airport are intentionally unspecified to permit the pilot, complying with FAR 91.119, to avoid operating over obstructions, noise-sensitive areas, etc. Further, should the pilot of an aircraft intending to operate into or out of an airport listed in the SFAR determine that the operation at or below the specified altitude is unsafe due to meteorological conditions, aircraft operating characteristics, or other factors, the pilot should seek relief from the Mode C transponder requirement via the ATC authorization process.

b. Effective until December 30, 1993. Airports at which the Mode C transponder equipment requirements of FAR 91.215(b)(2) do not apply.

1. Atlanta TCA Mode C veil. (See Table 3–30[1].)

Table 3–30[1]

Airport Name	Airport ID	Altitude (AGL)
Air Acres Airport, Woodstock, GA	5GA4	1,500
B & L Strip Airport, Hollonville, GA	GA29	1,500
Camfield Airport, McDonough, GA	GA36	1,500
Cobb County-McCollum Field Airport, Marietta, GA	RYY	1,500
Covington Municipal Airport, Covington, GA	9A1	1,500
Diamond R Ranch Airport, Villa Rica, GA	3GA5	1,500
Dresden Airport, Newnan, GA	GA79	1,500
Eagles Landing Airport, Williamson, GA	5GA3	1,500
Fagundes Field Airport, Haralson, GA	6GA1	1,500
Gable Branch Airport, Haralson, GA	5GA0	1,500
Georgia Lite Flite Ultralight Airport, Acworth, GA	31GA	1,500
Griffin-Spalding County Airport, Griffin, GA	6A2	1,500
Howard Private Airport, Jackson, GA	GA02	1,500
Newnan Coweta County Airport, Newnan, GA	CCO	1,500
Peach State Airport, Williamson, GA	3GA7	1,500
Poole Farm Airport, Oxford, GA	2GA1	1,500
Powers Airport, Hollonville, GA	GA31	1,500
S & S Landing Strip Airport, Griffin, GA	8GA6	1,500
Shade Tree Airport, Hollonville, GA	GA73	1,500

2. Boston TCA Mode C veil. (See Table 3–30[2].)

Table 3–30[2]

Airport Name	Airport ID	Altitude (AGL)
Berlin Landing Area Airport, Berlin, MA	MA19	2,500
Hopedale Industrial Park Airport, Hopedale, MA	1B6	2,500
Larson's SPB, Tyngsboro, MA	MA74	2,500
Moore AAF, Ayer/Fort Devens, MA	AYE	2,500
New England Gliderport, Salem, NH	NH29	2,500
Plum Island Airport, Newburyport, MA	2B2	2,500
Plymouth Municipal Airport, Plymouth, MA	PYM	2,500
Taunton Municipal Airport, Taunton, MA	TAN	2,500
Unknown Field Airport, Southborough, MA	1MA5	2,500

3. Charlotte TCA Mode C veil. (See Table 3–30[3].)

Table 3–30[3]

Airport Name	Airport ID	Altitude (AGL)
Arant Airport, Wingate, NC	1NC6	2,500
Bradley Outernational Airport, China Grove, NC	NC29	2,500
Chester Municipal Airport, Chester, SC	9A6	2,500
China Grove Airport, China Grove, NC	76A	2,500
Goodnight's Airport, Kannapolis, NC	2NC8	2,500
Knapp Airport, Marshville, NC	3NC4	2,500
Lake Norman Airport, Mooresville, NC	14A	2,500
Lancaster County Airport, Lancaster, SC	LKR	2,500
Little Mountain Airport, Denver, NC	66A	2,500
Long Island Airport, Long Island, NC	NC26	2,500
Miller Airport, Mooresville, NC	8A2	2,500
U S Heliport, Wingate, NC	NC56	2,500
Unity Aerodrome Airport, Lancaster, SC	SC76	2,500
Wilhelm Airport, Kannapolis, NC	6NC2	2,500

4. Chicago TCA Mode C veil. (See Table 3–30[4].)

Table 3–30[4]

Airport Name	Airport ID	Altitude (AGL)
Aurora Municipal Airport, Chicago/Aurora, IL	ARR	1,200
Donald Alfred Gade Airport, Antioch, IL	IL11	1,200
Dr. Joseph W. Esser Airport, Hampshire, IL	7IL6	1,200
Flying M. Farm Airport, Aurora, IL	IL20	1,200
Fox Lake SPB, Fox Lake, IL	IS03	1,200
Graham SPB, Crystal Lake, IL	IS79	1,200
Herbert C. Mass Airport, Zion, IL	IL02	1,200
Landings Condominium Airport, Romeoville, IL	C49	1,200
Lewis University Airport, Romeoville, IL	LOT	1,200
Mc Henry Farms Airport, Mc Henry, IL	44IL	1,200
Olson Airport, Plato Center, IL	LL53	1,200
Redeker Airport, Milford, IL	IL85	1,200
Reid RLA Airport, Gilberts, IL	6IL6	1,200
Shamrock Beef Cattle Farm Airport, Mc Henry, IL	49LL	1,200
Sky Soaring Airport, Union, IL	55LL	1,200
Waukegan Regional Airport, Waukegan, IL	UGN	1,200
Wormley Airport, Oswego, IL	85LL	1,200

5. Cleveland TCA Mode C veil. (See Table 3–30[5].)

Table 3–30[5]

Airport Name	Airport ID	Altitude (AGL)
Akron Fulton International Airport, Akron, OH	AKR	1,300
Bucks Airport, Newbury, OH	40OH	1,300
Derecsky Airport, Auburn Center, OH	6OI0	1,300
Hannum Airport, Streetsboro, OH	69OH	1,300
Kent State University Airport, Kent, OH	1G3	1,300
Lost Nation Airport, Willoughby, OH	LNN	1,300
Mills Airport, Mantua, OH	OH06	1,300
Portage County Airport, Ravenna, OH	29G	1,300
Stoney's Airport, Ravenna, OH	OI32	1,300
Wadsworth Municipal Airport, Wadsworth, OH	3G3	1,300

6. Dallas/Fort Worth TCA Mode C veil. (See Table 3–30[6].)

Table 3–30[6]

Airport Name	Airport ID	Altitude (AGL)
Beggs Ranch/Aledo Airport, Aledo, TX	TX15	1,800
Belcher Airport, Sanger, TX	TA25	1,800
Bird Dog Field Airport, Krum, TX	TA48	1,800
Boe-Wrinkle Airport, Azle, TX	28TS	1,800
Flying V Airport, Sanger, TX	71XS	1,800
Graham Ranch Airport, Celina, TX	TX44	1,800
Haire Airport, Bolivar, TX	TX33	1,800
Hartlee Field Airport, Denton, TX	1F3	1,800
Hawkin's Ranch Strip Airport, Rhome, TX	TA02	1,800
Horseshoe Lake Airport, Sanger, TX	TE24	1,800
Ironhead Airport, Sanger, TX	T58	1,800
Kezer Air Ranch Airport, Springtown, TX	61F	1,800
Lane Field Airport, Sanger, TX	58F	1,800
Log Cabin Airport, Aledo, TX	TX16	1,800
Lone Star Airpark Airport, Denton, TX	T32	1,800
Rhome Meadows Airport, Rhome, TX	TS72	1,800
Richards Airport, Krum, TX	TA47	1,800
Tallows Field Airport, Celina, TX	79TS	1,800
Triple S Airport, Aledo, TX	42XS	1,800
Warshun Ranch Airport, Denton, TX	4TA1	1,800
Windy Hill Airport, Denton, TX	46XS	1,800
Aero Country Airport, McKinney, TX	TX05	1,400
Bailey Airport, Midlothian, TX	7TX8	1,400
Bransom Farm Airport, Burleson, TX	TX42	1,400
Carroll Air Park Airport, De Soto, TX	F66	1,400
Carroll Lake-View Airport, Venus, TX	70TS	1,400
Eagle's Nest Estates Airport, Ovilla, TX	2T36	1,400
Flying B Ranch Airport, Ovilla, TX	TS71	1,400
Lancaster Airport, Lancaster, TX	LNC	1,400
Lewis Farm Airport, Lucas, TX	6TX1	1,400
Markum Ranch Airport, Fort Worth, TX	TX79	1,400
McKinney Municipal Airport, McKinney, TX	TKI	1,400
O'Brien Airpark Airport, Waxahachie, TX	F25	1,400
Phil L. Hudson Municipal Airport, Mesquite, TX	HQZ	1,400
Plover Heliport, Crowley, TX	82Q	1,400
Venus Airport, Venus, TX	75TS	1,400

7. Denver TCA Mode C veil. (See Table 3–30[7].)

Table 3–30[7]

Airport Name	Airport ID	Altitude (AGL)
Athanasiou Valley Airport, Blackhawk, CO	CO07	1,200
Boulder Municipal Airport, Boulder, CO	1V5	1,200
Bowen Farms No. 2 Airport, Strasburg, CO	3CO5	1,200
Carrera Airpark Airport, Mead, CO	93CO	1,200
Cartwheel Airport, Mead, CO	0CO8	1,200
Colorado Antique Field Airport, Niwot, CO	8CO7	1,200
Comanche Airfield Airport, Strasburg, CO	3CO6	1,200
Comanche Livestock Airport, Strasburg, CO	59CO	1,200
Flying J Ranch Airport, Evergreen, CO	27CO	1,200
Frederick-Firestone Air Strip Airport, Frederick, CO	CO58	1,200
Frontier Airstrip Airport, Mead, CO	84CO	1,200
Hoy Airstrip Airport, Bennett, CO	76CO	1,200
J & S Airport, Bennett, CO	CD14	1,200
Kugel-Strong Airport, Platteville, CO	27V	1,200
Land Airport, Keenesburg, CO	CO82	1,200
Lindys Airpark Airport, Hudson, CO	7CO3	1,200
Marshdale STOL, Evergreen, CO	CO52	1,200
Meyer Ranch Airport, Conifer, CO	5CO6	1,200
Parkland Airport, Erie, CO	7CO0	1,200
Pine View Airport, Elizabeth, CO	02V	1,200
Platte Valley Airport, Hudson, CO	18V	1,200
Rancho De Aereo Airport, Mead, CO	05CO	1,200
Spickard Farm Airport, Byers, CO	5CO4	1,200
Vance Brand Airport, Longmont, CO	2V2	1,200
Yoder Airstrip Airport, Bennett, CO	CD09	1,200

8. Detroit TCA Mode C veil. (See Table 3–30[8].)

Table 3–30[8]

Airport Name	Airport ID	Altitude (AGL)
Al Meyers Airport, Tecumseh, MI	3TE	1,400
Brighton Airport, Brighton, MI	45G	1,400
Cackleberry Airport, Dexter, MI	2MI9	1,400
Erie Aerodome Airport, Erie, MI	05MI	1,400
Ham-A-Lot Field Airport, Petersburg, MI	MI48	1,400
Merillat Airport, Tecumseh, MI	34G	1,400
Rossettie Airport, Manchester, MI	75G	1,400
Tecumseh Products Airport, Tecumseh, MI	0D2	1,400

9. Honolulu TCA Mode C veil. (See Table 3–30[9].)

Table 3–30[9]

Airport Name	Airport ID	Altitude (AGL)
Dillingham Airfield Airport, Mokuleia, HI	HDH	2,500

10. Houston TCA Mode C veil. (See Table 3–30[10].)

Table 3–30[10]

Airport Name	Airport ID	Altitude (AGL)
Ainsworth Airport, Cleveland, TX	0T6	1,200
Biggin Hill Airport, Hockley, TX	0TA3	1,200
Cleveland Municipal Airport, Cleveland, TX	6R3	1,200
Fay Ranch Airport, Cedar Lane, TX	0T2	1,200
Freeman Property Airport, Katy, TX	61T	1,200
Gum Island Airport, Dayton, TX	3T6	1,200
Harbican Airpark Airport, Katy, TX	9XS9	1,200
Harold Freeman Farm Airport, Katy, TX	8XS1	1,200
Hoffpauir Airport, Katy, TX	59T	1,200
Horn-Katy Hawk International Airport, Katy, TX	57T	1,200
Houston-Hull Airport, Houston, TX	SGR	1,200
Houston-Southwest Airport, Houston, TX	AXH	1,200
King Air Airport, Katy, TX	55T	1,200
Lake Bay Gall Airport, Cleveland, TX	0T5	1,200
Lake Bonanza Airport, Montgomery, TX	33TA	1,200
R W J Airpark Airport, Baytown, TX	54TX	1,200
Westheimer Air Park Airport, Houston, TX	5TA4	1,200

11. Kansas City TCA Mode C veil. (See Table 3–30[11].)

Table 3–30[11]

Airport Name	Airport ID	Altitude (AGL)
Amelia Earhart Airport, Atchison, KS	K59	1,000
Booze Island Airport, St. Joseph,MO	64MO	1,000
Cedar Air Park Airport, Olathe, KS	51K	1,000
D'Field Airport, McLouth, KS	KS90	1,000
Dorei Airport, McLouth, KS	K69	1,000
East Kansas City Airport, Grain Valley, MO	3GV	1,000
Excelsior Springs Memorial Airport, Excelsior Springs, MO	3EX	1,000
Flying T Airport, Oskaloosa, KS	7KS0	1,000
Hermon Farm Airport, Gardner, KS	KS59	1,000
Hillside Airport, Stilwell, KS	63K	1,000
Independence Memorial Airport, Independence, MO	3IP	1,000
Johnson County Executive Airport, Olathe, KS	OJC	1,000
Johnson County Industrial Airport, Olathe, KS	IXD	1,000
Kimray Airport, Plattsburg, MO	7MO7	1,000
Lawrence Municipal Airport, Lawrence, KS	LWC	1,000
Martins Airport, Lawson, MO	21MO	1,000
Mayes Homestead Airport, Polo, MO	37MO	1,000
McComas-Lee's Summit Municipal Airport, Lee's Summit, MO	K84	1,000
Mission Road Airport, Stilwell, KS	64K	1,000
Northwood Airport, Holt, MO	2MO2	1,000
Plattsburg Airpark Airport, Plattsburg, MO	MO28	1,000
Richards-Gebaur Airport, Kansas City, MO	GVW	1,000
Rosecrans Memorial Airport, St. Joseph, MO	STJ	1,000
Runway Ranch Airport, Kansas City, MO	2MO9	1,000
Sheller's Airport, Tonganoxie, KS	11KS	1,000
Shomin Airport, Oskaloosa, KS	0KS1	1,000
Stonehenge Airport, Williamstown, KS	71KS	1,000
Threshing Bee Airport, McLouth, KS	41K	1,000

12. Las Vegas TCA Mode C veil. (See Table 3–30[12].)

Table 3–30[12]

Airport Name	Airport ID	Altitude (AGL)
Sky Ranch Estates Airport, Sandy Valley, NV	3L2	2,500

13. Memphis TCA Mode C veil. (See Table 3–30[13].)

Table 3–30[13]

Airport Name	Airport ID	Altitude (AGL)
Bernard Manor Airport, Earle, AR	65M	2,500
Holly Springs-Marshall County Airport, Holly Springs, MS	M41	2,500
Mc Neely Airport, Earle, AR	M63	2,500
Price Field Airport, Joiner, AR	80M	2,500
Tucker Field Airport, Hughes, AR	78M	2,500
Tunica Airport, Tunica, MS	30M	2,500
Tunica Municipal Airport, Tunica, MS	M97	2,500

14. Minneapolis TCA Mode C veil. (See Table 3–30[14].)

Table 3–30[14]

Airport Name	Airport ID	Altitude (AGL)
Belle Plaine Airport, Belle Plaine, MN	7Y7	1,200
Carleton Airport, Stanton, MN	SYN	1,200
Empire Farm Strip Airport, Bongards, MN	MN15	1,200
Flying M Ranch Airport, Roberts, WI	78WI	1,200
Johnson Airport, Rockford, MN	MY86	1,200
River Falls Airport, River Falls, WI	Y53	1,200
Rusmar Farms Airport, Roberts, WI	WS41	1,200
Waldref SPB, Forest Lake, MN	9Y6	1,200
Ziermann Airport, Mayer, MN	MN71	1,200

15. New Orleans TCA Mode C veil. (See Table 3–30[15].)

Table 3–30[15]

Airport Name	Airport ID	Altitude (AGL)
Bollinger SPB, Larose, LA	L38	1,500
Clovelly Airport, Cut Off, LA	LA09	1,500

16. New York TCA Mode C veil. (See Table 3–30[16].)

Table 3–30[16]

Airport Name	Airport ID	Altitude (AGL)
Allaire Airport, Belmar/Farmingdale, NJ	BLM	2,000
Cuddihy Landing Strip Airport, Freehold, NJ	NJ60	2,000
Ekdahl Airport, Freehold, NJ	NJ59	2,000
Fla-Net Airport, Netcong, NJ	0NJ5	2,000
Forrestal Airport, Princeton, NJ	N21	2,000
Greenwood Lake Airport, West Milford, NJ	4N1	2,000
Greenwood Lake SPB, West Milford, NJ	6NJ7	2,000
Lance Airport, Whitehouse Station, NJ	6NJ8	2,000
Mar Bar L Farms, Englishtown, NJ	NJ46	2,000
Peekskill SPB, Peekskill, NY	7N2	2,000
Peters Airport, Somerville, NJ	4NJ8	2,000
Princeton Airport, Princeton/Rocky Hill, NJ	39N	2,000
Solberg-Hunterdon Airport, Readington, NJ	N51	2,000

17. Orlando TCA Mode C veil. (See Table 3–30[17].)

Table 3–30[17]

Airport Name	Airport ID	Altitude (AGL)
Arthur Dunn Air Park Airport, Titusville, FL	X21	1,400
Space Center Executive Airport, Titusville, FL	TIX	1,400

18. Philadelphia TCA Mode C veil. (See Table 3–30[18].)

Table 3–30[18]

Airport Name	Airport ID	Altitude (AGL)
Ginns Airport, West Grove, PA	78N	1,000
Hammonton Municipal Airport, Hammonton, NJ	N81	1,000
Li Calzi Airport, Bridgeton, NJ	N50	1,000
New London Airport, New London, PA	N01	1,000
Wide Sky Airpark Airport, Bridgeton, NJ	N39	1,000

19. Phoenix TCA Mode C veil. (See Table 3–30[19].)

Table 3–30[19]

Airport Name	Airport ID	Altitude (AGL)
Ak Chin Community Airfield Airport, Maricopa, AZ	E31	2,500
Boulais Ranch Airport, Maricopa, AZ	9E7	2,500
Estrella Sailport, Maricopa, AZ	E68	2,500
Hidden Valley Ranch Airport, Maricopa, AZ	AZ17	2,500
Millar Airport, Maricopa, AZ	2AZ4	2,500
Pleasant Valley Airport, New River, AZ	AZ05	2,500
Serene Field Airport, Maricopa, AZ	AZ31	2,500
Sky Ranch Carefree Airport, Carefree, AZ	E18	2,500
Sycamore Creek Airport, Fountain Hills, AZ	0AS0	2,500
University of Arizona, Maricopa Agricultural Center Airport, Maricopa, AZ	3AZ2	2,500

20. St. Louis TCA Mode C veil. (See Table 3–30[20].)

Table 3–30[20]

Airport Name	Airport ID	Altitude (AGL)
Blackhawk Airport, Old Monroe, MO	6MO0	1,000
Lebert Flying L Airport, Lebanon, IL	3H5	1,000
Shafer Metro East Airport, St. Jacob, IL	3K6	1,000
Sloan's Airport, Elsberry, MO	0MO8	1,000
Wentzville Airport, Wentzville, MO	MO50	1,000
Woodliff Airpark Airport, Foristell, MO	98MO	1,000

21. Salt Lake City TCA Mode C veil. (See Table 3–30[21].)

Table 3–30[21]

Airport Name	Airport ID	Altitude (AGL)
Bolinder Field-Tooele Valley Airport, Tooele, UT	TVY	2,500
Cedar Valley Airport, Cedar Fort, UT	UT10	2,500
Morgan County Airport, Morgan, UT	42U	2,500
Tooele Municipal Airport, Tooele, UT	U26	2,500

22. Seattle TCA Mode C veil. (See Table 3–30[22].)

Table 3–30[22]

Airport Name	Airport ID	Altitude (AGL)
Firstair Field Airport, Monroe, WA	WA38	1,500
Gower Field Airport, Olympia, WA	6WA2	1,500
Harvey Field Airport, Snohomish, WA	S43	1,500

23. Tampa TCA Mode C veil. (See Table 3–30[23].)

Table 3–30[23]

Airport Name	Airport ID	Altitude (AGL)
Hernando County Airport, Brooksville, FL	BKV	1,500
Lakeland Municipal Airport, Lakeland, FL	LAL	1,500
Zephyrhills Municipal Airport, Zephyrhills, FL	ZPH	1,500

Effective until the establishment of the Washington Tri-Area TCA or December 30, 1993, whichever occurs first:

24. Washington TCA Mode C veil. (See Table 3–30[24].)

Table 3–30[24]

Airport Name	Airport ID	Altitude (AGL)
Barnes Airport, Lisbon, MD	MD47	2,000
Bay Bridge Airport, Stevensville, MD	W29	2,000
Castle Marina Airport, Chester, MD	0W6	2,000
Davis Airport, Laytonsville, MD	W50	2,000
Fremont Airport, Kemptown,MD	MD41	2,000
Kentmorr Airpark Airport, Stevensville, MD	3W3	2,000
Montgomery County Airpark Airport, Gaithersburg, MD	GAI	2,000
Waredaca Farm Airport, Brookeville, MD	MD16	2,000
Aqua-Land/Cliffton Skypark Airport, Newburg, MD	2W8	1,000
Buds Ferry Airport, Indian Head, MD	MD39	1,000
Burgess Field Airport, Riverside, MD	3W1	1,000
Chimney View Airport, Fredericksburg, VA	5VA5	1,000
Holly Springs Farm Airport, Nanjemoy, MD	MD55	1,000
Lanseair Farms Airport, La Plata, MD	MD97	1,000
Nyce Airport, Mount Victoria, MD	MD84	1,000
Parks Airpark Airport, Nanjemoy, MD	MD54	1,000
Pilots Cove Airport, Tompkinsville, MD	MD06	1,000
Quantico MCAF, Quantico, VA	NYG	1,000
Stewart Airport, St. Michaels, MD	MD64	1,000
U S Naval Weapons Center, Dahlgren Lab Airport, Dahlgren, VA	NDY	1,000

Effective upon the establishment of the Washington Tri-Area TCA;

25. Washington Tri-Area Mode C veil. (See Table 3–30[25].)

Table 3–30[25]

Airport Name	Airport ID	Altitude (AGL)
Albrecht Airstrip Airport, Long Green, MD	MD48	2,000
Armacost Farms Airport, Hampstead, MD	MD38	2,000
Barnes Airport, Lisbon, MD	MD47	2,000
Bay Bridge Airport, Stevensville, MD	W29	2,000
Carroll County Airport, Westminster, MD	W54	2,000
Castle Marina Airport, Chester, MD	0W6	2,000
Clearview Airpark Airport, Westminster, MD	2W2	2,000
Davis Airport, Laytonsville, MD	W50	2,000
Fallston Airport, Fallston, MD	W42	2,000
Faux-Burhans Airport, Frederick, MD	3MD0	2,000
Forest Hill Airport, Forest Hill, MD	MD31	2,000
Fort Detrick Helipad Heliport, Fort Detrick (Frederick), MD	MD32	2,000
Frederick Municipal Airport, Frederick, MD	FDK	2,000
Fremont Airport, Kemptown, MD	MD41	2,000
Good Neighbor Farm Airport, Unionville, MD	MD74	2,000
Happy Landings Farm Airport, Unionville, MD	MD73	2,000
Harris Airport, Still Pond, MD	MD69	2,000
Hybarc Farm Airport, Chestertown, MD	MD19	2,000
Kennersley Airport, Church Hill, MD	MD23	2,000
Kentmorr Airpark Airport, Stevensville, MD	3W3	2,000
Montgomery County Airpark Airport, Gaithersburg, MD	GAI	2,000
Phillips AAF, Aberdeen, MD	APG	2,000
Pond View Private Airport, Chestertown, MD	0MD4	2,000
Reservoir Airport, Finksburg, MD	1W8	2,000
Scheeler Field Airport, Chestertown, MD	0W7	2,000
Stolcrest STOL, Urbana, MD	MD75	2,000
Tinsley Airstrip Airport, Butler, MD	MD17	2,000
Walters Airport, Mount Airy, MD	0MD6	2,000
Waredaca Farm Airport, Brookeville, MD	MD16	2,000
Weide AAF, Edgewood Arsenal, MD	EDG	2,000
Woodbine Gliderport, Woodbine, MD	MD78	2,000
Wright Field Airport, Chestertown, MD	MD11	2,000
Aviacres Airport, Warrenton, VA	3VA2	1,500
Birch Hollow Airport, Hillsboro, VA	W60	1,500
Flying Circus Aerodrome Airport, Warrenton, VA	3VA3	1,500
Fox Acres Airport, Warrenton, VA	15VA	1,500
Hartwood Airport, Somerville, VA	8W8	1,500
Horse Feathers Airport, Midland, VA	53VA	1,500
Krens Farm Airport, Hillsboro, VA	14VA	1,500
Scott Airpark Airport, Lovettsville, VA	VA61	1,500
The Grass Patch Airport, Lovettsville, VA	VA62	1,500
Walnut Hill Airport, Calverton, VA	58VA	1,500
Warrenton Air Park Airport, Warrenton, VA	9W0	1,500
Warrenton-Fauquier Airport, Warrenton, VA	W66	1,500
Whitman Strip Airport, Manassas, VA	0V5	1,500
Aqua-Land/Cliffton Skypark Airport, Newburg, MD	2W8	1,000
Buds Ferry Airport, Indian Head, MD	MD39	1,000
Burgess Field Airport, Riverside, MD	3W1	1,000
Chimney View Airport, Fredericksburg, VA	5VA5	1,000
Holly Springs Farm Airport, Nanjemoy, MD	MD55	1,000
Lanseair Farms Airport, La Plata, MD	MD97	1,000
Nyce Airport, Mount Victoria, MD	MD84	1,000
Parks Airpark Airport, Nanjemoy, MD	MD54	1,000
Pilots Cove Airport, Tompkinsville, MD	MD06	1,000
Quantico MCAF, Quantico, VA	NYG	1,000
Stewart Airport, St. Michaels, MD	MD64	1,000
U S Naval Weapons Center, Dahlgren Lab Airport, Dahlgren, VA	NDY	1,000

3–31. AIRPORT RADAR SERVICE AREA (ARSA)

a. An Airport Radar Service Area (ARSA) consists of controlled airspace extending upward from the surface or higher to specified altitudes, within which all aircraft are subject to the operating rules and pilot and equipment requirements specified in FAR 91. ARSAs are described in FAR 71. (See FAR 71.14.)

b. Dimensions

1. *ARSA* (A basic standard design with minor site specific variations.) The ARSA airspace consists of two circles, both centered on the primary/ARSA airport. The inner circle has a radius of 5NM. The outer circle has a radius of 10NM. The airspace of the inner circle extends from the surface of the ARSA up to 4,000 feet above that airport. The airspace area between the 5 and 10NM rings begins at a height 1,200 feet AGL and extends to the same altitude cap as the inner circle.

2. *OUTER AREA* The normal radius will be 20NM, with some variations based on site specific requirements. The outer area extends outward from the primary/ARSA airport and extends from the lower limits of radar/radio coverage up to the ceiling of the approach control's delegated airspace, excluding the ARSA and other airspace as appropriate.

c. ARSAs are charted on Sectional Charts, and some Terminal Control Area Charts.

1. Airport Radar Service Areas by State. (See Table 3–31[1].)

These areas are currently designated as ARSA's and are depicted on Sectional Charts. The table will be updated as additional ARSA's are implemented. Pilots should consult current sectional charts and NOTAM's for the latest information on services available. Pilots should be aware that some ARSA's underlie or are adjacent to TCA's.

Table 3–31[1]

STATE/CITY	AIRPORT
ALABAMA.	
Birmingham	Municipal
Huntsville	Madison Co-Carl T Jones Fld
Mobile	Bates Field
ALASKA.	
Anchorage	International
ARIZONA.	
Davis-Monthan	AFB
Tucson	International
ARKANSAS.	
Little Rock	Adams Field
CALIFORNIA.	
Beale	AFB
Burbank	Burbank - Glendale - Pasadena
Castle	AFB
Fresno	Air Terminal
McClellan	AFB
March	AFB
Mather	AFB
Monterey	Peninsula
Norton	AFB
Oakland	Metropolitan Int
Ontario	International
Sacramento	Metropolitan
San Jose	International
Santa Ana	El Toro MCAS
Santa Ana	John Wayne/Orange County
Santa Barbara	Municipal
COLORADO.	
Colorado Springs	Municipal Airport
CONNECTICUT.	
Windsor Locks	Bradley Int
DELAWARE.	
DISTRICT OF COLUMBIA.	
FLORIDA.	
Daytona Beach	Regional
Fort Lauderdale	Hollywood Int
Fort Myers	SW Florida Regional
Jacksonville	International
Palm Beach	International
Pensacola	NAS
Pensacola	Regional
Sarasota	Bradenton
Tallahassee	Municipal
Whiting	NAS
GEORGIA.	
Columbus	Metropolitan
Savannah	International
HAWAII.	
Kahului	International
IDAHO.	
Boise	Air Terminal
ILLINOIS.	

Table 3–31[1]—CONTINUED

STATE/CITY	AIRPORT
Champaign	U of Illinois Willard
Chicago	Midway
Moline	Quad City
Peoria	Greater Peoria
Springfield	Capital
INDIANA.	
Evansville	Evansville Dress Regional
Fort Wayne	Municipal
Indianapolis	International
South Bend	Michiana Regional
IOWA.	
Cedar Rapids	Municipal
Des Moines	International
KANSAS.	
Wichita	Mid-Continent
KENTUCKY.	
Covington	Greater Cincinnati International
Lexington	Blue Grass Field
Louisville	Standiford Field
LOUISIANA.	
Barksdale	Air Force Base
Baton Rouge	BTR Metro, Ryan Field
Lafayette	Regional
Shreveport	Regional
MAINE.	
Portland	International Jetport
MARYLAND.	
MASSACHUSETTS.	
MICHIGAN.	
Flint	Bishop
Grand Rapids	Kent County
Lansing	Capital City
MINNESOTA.	
MISSISSIPPI.	
Columbus	Air Force Base
Jackson	Allen C Thompson Field
MISSSOURI.	
MONTANA.	
NEBRASKA.	
Lincoln	Municipal
Omaha	Eppley Airfield
Offutt	Air Force Base
NEVADA.	
Reno	Cannon Int
NEW HAMPSHIRE.	
Manchester	Manchester
NEW JERSEY.	
Atlantic City	City
NEW MEXICO.	
Albuquerque	International
NEW YORK.	

Table 3–31[1]—CONTINUED

STATE/CITY	AIRPORT
Albany	County
Buffalo	Greater Buff. Int
Islip	Long Island MacArthur
Rochester	Rochester-Monroe Co
Syracuse	Hancock Int
NORTH CAROLINA.	
Fayetteville	Municipal/Grannis Field
Greensboro	Highpoint-Winston-Salem Regional
Pope	Air Force Base
Raleigh	Raleigh-Durham
NORTH DAKOTA.	
OHIO.	
Akron	Akron-Canton Reg
Columbus	Port Columbus Int
Dayton	James M. Cox Int
Toledo	Express
OKLAHOMA.	
Oklahoma City	Will Rogers World
Tinker	Air Force Base
Tulsa	International
OREGON.	
Portland	International
PENNSYLVANIA.	
Allentown	Bethlehem-Easton
PUERTO RICO.	
San Juan	International
RHODE ISLAND.	
Providence	Theodore Francis Green St.
SOUTH CAROLINA.	
Charleston	Air Force Base / International
Columbia	Metropolitan
Greer	Greenville-Spartanburg
Shaw	Air Force Base
SOUTH DAKOTA.	
TENNESSEE.	
Chattanooga	Lovell Field
Knoxville	McGhee Tyson
Nashville	Metropolitan
TEXAS.	
Abilene	Municipal
Amarillo	International
Austin	Robert Mueller Mun
Corpus Christi	International
Dyess	Air Force Base
El Paso	International
Harlingen	Rio Grande Valley
Houston	William P. Hobby
Laughlin	Air Force Base
Lubbock	International
Midland	International
San Antonio	International
UTAH.	
VERMONT.	
Burlington	International
VIRGINIA.	
Richmond	Richard Evelyn Byrd Inter.
Norfolk	International
Roanoke	Regional / Woodrum Field
WASHINGTON.	
Fairchild	Air Force Base
Spokane	International

Table 3–31[1]—CONTINUED

STATE/CITY	AIRPORT
Whidbey Island	Naval Air Station
WEST VIRGINIA.	
Charleston	Yeager
WISCONSIN.	
Green Bay	Austin Straubel Field
Madison	Dane County Reg
Milwaukee	General Mitchell Int
WYOMING.	

3–32. AIRPORT RADAR SERVICE AREA (ARSA) OPERATION

a. Operating Rules and Pilot/Equipment Requirements

1. **Pilot Certification:** No specific certification required.

2. **Equipment:**

(a) Two-Way Radio, and

(b) For additional information: Reference—Transponder Operation, paragraph 4–18(f).

3. **Arrivals and Transitions:** Two-way radio communication must be established with the ATC facility having jurisdiction over the ARSA prior to entry and thereafter as instructed by ATC.

4. **Departures:**

(a) Primary or Satellite Airport with an Operating Control Tower: Two-way radio communication must be established and maintained with the control tower in accordance with Title 14 Part 91.87 [91.129] and thereafter as instructed by ATC.

(b) Satellite Airports without an Operating Control Tower: Two-way radio communication must be established as soon as practicable after departing with the ATC facility having jurisdiction over the ARSA and thereafter, as instructed by ATC.

5. **Traffic Patterns:** Pilots must comply with FAA arrival or departure traffic patterns.

6. **Ultralight Vehicles:** Ultralight vehicle operations are not permitted in an ARSA unless otherwise authorized by the ATC Facility having jurisdiction over the ARSA. (FAR 103)

7. **Parachute Jumps:** Parachute jumps are not permitted in an ARSA except under the terms of an ATC authorization issued by the ATC facility having jurisdiction over the ARSA. (FAR 105)

b. ATC Services (See Figure 3–32[1].)

1. **Within The ARSA:**

(a) Sequencing of all arriving aircraft to the primary/ARSA airport.

(b) Standard IFR separation between IFR aircraft.

Figure 3–32[1]

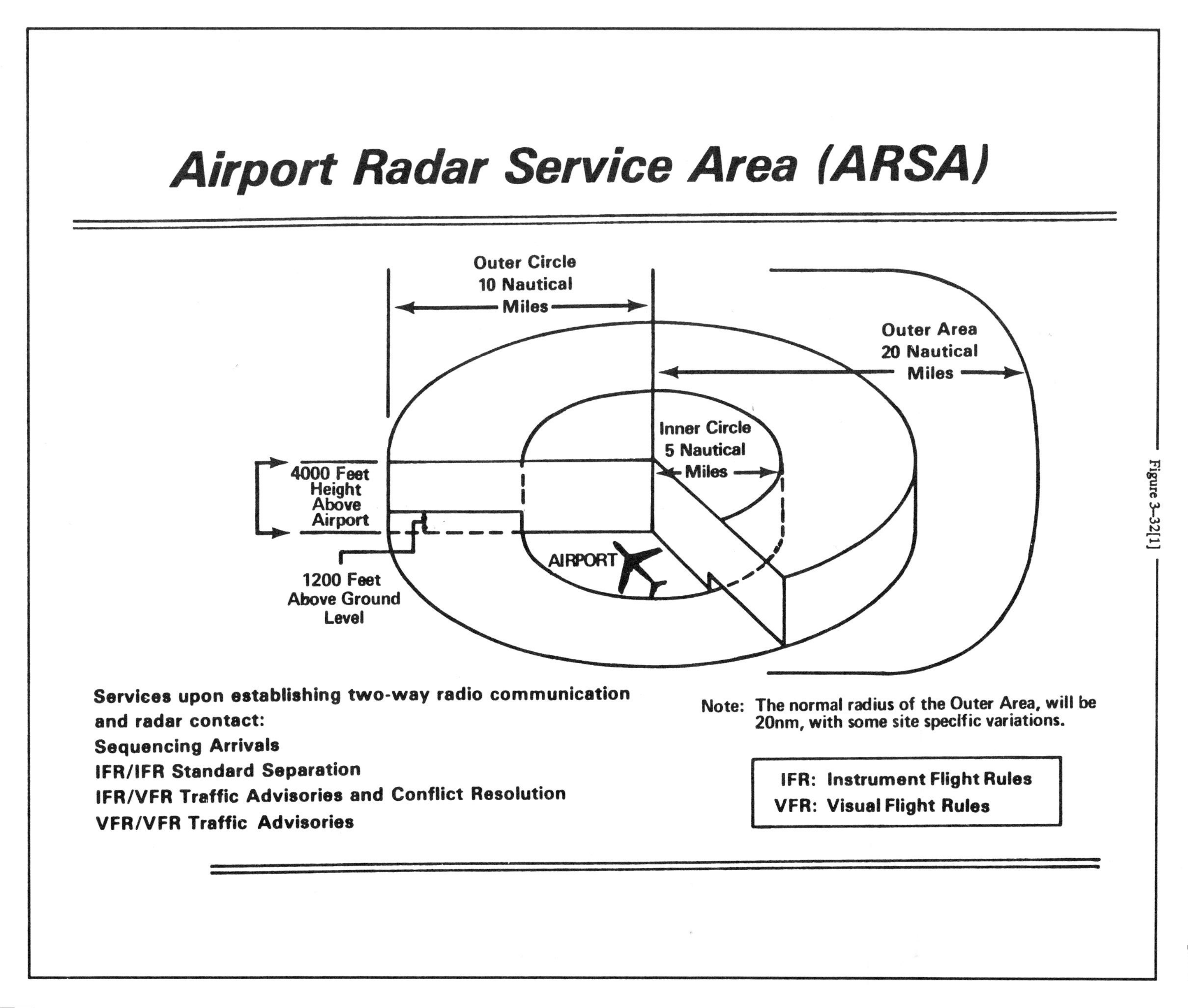
Airport Radar Service Area (ARSA)
Outer Circle
10 Nautical
Miles
Outer Area
20 Nautical
Miles
Inner Circle
5 Nautical
Miles
4000 Feet
Height
Above
Airport
AIRPORT
1200 Feet
Above Ground
Level
Services upon establishing two-way radio communication and radar contact:
Sequencing Arrivals
IFR/IFR Standard Separation
IFR/VFR Traffic Advisories and Conflict Resolution
VFR/VFR Traffic Advisories
Note: The normal radius of the Outer Area, will be 20nm, with some site specific variations.
IFR: Instrument Flight Rules
VFR: Visual Flight Rules

(c) Between IFR and VFR aircraft — traffic advisories and conflict resolution so that radar targets do not touch, or 500 feet vertical separation.

(d) Between VFR aircraft — traffic advisories and as appropriate, safety alerts.

2. Within The Outer Area:

(a) The same services are provided for aircraft operating within the outer area, as within the ARSA, *when two-way communication and radar contact is established.*

(b) While pilot participation in this area is strongly encouraged, it is not a VFR requirement.

3. Beyond The Outer Area:

(a) Standard IFR separation.

(b) Basic Radar Service.

(c) Stage II/Stage III Service, where appropriate.

(d) Safety Alert, as appropriate.

c. Air traffic control radar is required to provide ARSA services. ARSA services may not be available or may be limited during radar outages.

1. Separation and sequencing of VFR aircraft will be suspended during a radar outage. The pilot will be advised that ARSA service is not available and issued wind, runway information and the time or location to contact the tower.

2. Separation of VFR aircraft will be suspended during CENRAP operations. Traffic advisories and sequencing to the primary airport will be provided on a workload permitting basis. The pilot will be advised when CENRAP is in use.

d. While pilot participation is required within the ARSA, it is voluntary within the outer area and can be discontinued, within the outer area, at the pilots request.

e. ARSA services will be provided in the outer area unless the pilot requests termination of the service.

f. Service provided beyond the outer area will be on a workload permitting basis and can be terminated by the controller if workload dictates.

g. In some locations an ARSA may overlap the airport traffic area of a secondary airport. In order to allow that control tower to provide service to aircraft, portions of the overlapping ARSA may be procedurally excluded when the secondary airport tower is in operation. Aircraft operating in these procedurally excluded areas will only be provided airport traffic control services when in communication with the secondary airport tower. ARSA service to aircraft inbound to these airports will be discontinued when the aircraft is instructed to contact the tower.

h. Aircraft departing secondary controlled airports will not receive ARSA service until they have been radar identified and two-way communication has been established with the ARSA facility.

i. ARSA service to aircraft proceeding to a satellite airport will be terminated at a sufficient distance to allow time to change to the appropriate tower or advisory frequency.

j. Some ARSA facilities shut down for portions of the night. When this occurs, the effective hours of the ARSA will be the same as the operating hours of the serving facility.

k. This program is not to be interpreted as relieving pilots of their responsibilities to see and avoid other traffic operating in basic VFR weather conditions, to adjust their operations and flight path as necessary to preclude serious wake encounters, to maintain appropriate terrain and obstruction clearance, or to remain in weather conditions equal to or better than the minimums required by FAR 91.155. Whenever compliance with an assigned route, heading and/or altitude is likely to compromise pilot responsibility respecting terrain and obstruction clearance, vortex exposure, and weather minimums, approach control should be so advised and a revised clearance or instruction obtained.

l. Pilots of arriving aircraft should contact the ARSA facility on the publicized frequency and give their position, altitude, radar beacon code (if transponder equipped), destination, and request ARSA service. Radio contact should be initiated far enough from the ARSA boundary to preclude entering the ARSA before radio communication is established.

m. If the controller responds to a radio call with, ''(aircraft callsign) standby,'' radio communications have been established and the pilot can enter the ARSA. If workload or traffic conditions prevent immediate provision of ARSA services, the controller will inform the pilot to remain outside the ARSA until conditions permit the services to be provided.

EXAMPLE:

''(AIRCRAFT CALLSIGN) REMAIN OUTSIDE THE ARSA AND STANDBY.''

3–32m NOTE— It is important to understand that if the controller responds to the initial radio call without using the aircraft callsign, radio communications have **not** been established and the pilot may not enter the ARSA.

EXAMPLE:

''AIRCRAFT CALLING DULLES APPROACH CONTROL, STANDBY.''

3–33 thru 3–40. RESERVED

Section 4. SPECIAL USE AIRSPACE

3–41. GENERAL

Special use airspace consists of that airspace wherein activities must be confined because of their nature, or wherein limitations are imposed upon aircraft operations that are not a part of those activities, or both. Except for Controlled Firing Areas, special use airspace areas are depicted on aeronautical charts.

3–42. PROHIBITED AREA

Prohibited Areas contain airspace of defined dimensions identified by an area on the surface of the earth within which the flight of aircraft is prohibited. Such areas are established for security or other reasons associated with the national welfare. These areas are published in the Federal Register and are depicted on aeronautical charts.

3–43. RESTRICTED AREA

a. Restricted Areas contain airspace identified by an area on the surface of the earth within which the flight of aircraft, while not wholly prohibited, is subject to restrictions. Activities within these areas must be confined because of their nature or limitations imposed upon aircraft operations that are not a part of those activities or both. Restricted areas denote the existence of unusual, often invisible, hazards to aircraft such as artillery firing, aerial gunnery, or guided missiles. Penetration of Restricted Areas without authorization from the using or controlling agency may be extremely hazardous to the aircraft and its occupants. Restricted Areas are published in the Federal Register and constitute FAR 73.

b. ATC facilities apply the following procedures when aircraft are operating on an IFR clearance (including those cleared by ATC to maintain VFR-ON-TOP) via a route which lies within joint-use restricted airspace.

1. If the restricted area is not active and has been released to the controlling agency (FAA), the ATC facility will allow the aircraft to operate in the restricted airspace without issuing specific clearance for it to do so.

2. If the restricted area is active and has not been released to the controlling agency (FAA), the ATC facility will issue a clearance which will ensure the aircraft avoids the restricted airspace unless it is on an approved altitude reservation mission or has obtained its own permission to operate in the airspace and so informs the controlling facility.

3–43b2 NOTE—The above apply only to joint-use restricted airspace and not to prohibited and nonjoint-use airspace. For the latter categories, the ATC facility will issue a clearance so the aircraft will avoid the restricted airspace unless it is on an approved altitude reservation mission or has obtained its own permission to operate in the airspace and so informs the controlling facility.

c. Restricted airspace is depicted on the En Route Chart appropriate for use at the altitude or flight level being flown. For joint-use restricted areas, the name of the controlling agency is shown on these charts. For all prohibited areas and nonjoint-use restricted areas, unless otherwise requested by the using agency, the phrase "NO A/G" is shown.

3–44. WARNING AREA

Warning Areas are airspace which may contain hazards to nonparticipating aircraft in international airspace. Warning Areas are established beyond the 3 mile limit. Though the activities conducted within Warning Areas may be as hazardous as those in Restricted Areas, Warning Areas cannot be legally designated as Restricted Areas because they are over international waters. Penetration of Warning Areas during periods of activity may be hazardous to the aircraft and its occupants. Official descriptions of Warning Areas may be obtained on request to the FAA, Washington, D.C.

3–45. MILITARY OPERATIONS AREAS (MOA)

a. MOAs consist of airspace of defined vertical and lateral limits established for the purpose of separating certain military training activities from IFR traffic. Whenever a MOA is being used, nonparticipating IFR traffic may be cleared through a MOA if IFR separation can be provided by ATC. Otherwise, ATC will reroute or restrict nonparticipating IFR traffic.

b. Most training activities necessitate acrobatic or abrupt flight maneuvers. Military pilots conducting flight in Department of Defense aircraft within a designated and active military operations area (MOA) are exempted from the provisions of FAR 91.303(c) and (d) which prohibit acrobatic flight within Federal airways and control zones.

c. Pilots operating under VFR should exercise extreme caution while flying within a MOA when military activity is being conducted. The activity status (active/inactive) of MOA's may change frequently. Therefore, pilots should contact any FSS within 100 miles of the area to obtain accurate real-time information concerning the MOA hours of operation. Prior to entering an active MOA, pilots should contact the controlling agency for traffic advisories.

d. MOA's are depicted on Sectional, VFR Terminal, Area and Low Altitude En Route Charts.

3–46. ALERT AREA

Alert Areas are depicted on aeronautical charts to inform nonparticipating pilots of areas that may contain a high volume of pilot training or an unusual type of aerial activity. Pilots should be particularly alert when flying in these areas. All activity within an Alert Area shall be conducted in accordance with FARs, without waiver, and pilots of participating aircraft as well as pilots transiting the area shall be equally responsible for collision avoidance. Information concerning these areas may be obtained upon request to the FAA, Washington, D.C.

3–47. CONTROLLED FIRING AREAS

Controlled Firing Areas contain activities which, if not conducted in a controlled environment, could be hazardous to nonparticipating aircraft. The distinguishing feature of the Controlled Firing Area, as compared to other special use airspace, is that its activities are suspended immediately when spotter aircraft, radar, or ground lookout positions indicate an aircraft might be approaching the area. There is no need to chart Controlled Firing Areas since they do not cause a nonparticipating aircraft to change its flight path.

3–48 thru 3–60. RESERVED

Section 5. OTHER AIRSPACE AREAS

3–61. AIRPORT TRAFFIC AREAS

a. Unless otherwise specifically designated (FAR 93), Airport traffic areas consist of the airspace within a horizontal radius of 5 statute miles from the geographical center of any airport at which a control tower is operating, extending from the surface up to, but not including, an altitude of 3,000 feet above the elevation of the airport.

b. FAR 91.127 requires that unless a pilot is landing at or taking off from an airport in the airport traffic area or authorized otherwise by ATC, *the pilot must avoid the area.* Generally, it is the pilot's responsibility to obtain any necessary authorization. Pilots operating under IFR or receiving radar services from an ATC facility are not expected to obtain their own authorizations through each area. Rather, the ATC facility providing the service will coordinate with the appropriate control towers for the approval to transit each area.

c. FAR 91.129 requires that unless otherwise authorized by ATC, a pilot operating to or from an airport served by an operating control tower must maintain two-way radio communications with the control tower while in the airport traffic area which includes the movement areas of that airport. FAR 91.117 sets the maximum indicated airspeed for operations in an airport traffic area for at 200 knots. airport traffic areas are indicated on aeronautical charts by the blue airport symbol, but the actual boundary is not depicted. (Reference—General Dimensions of Airspace Segments, paragraph 3–2).

3–62. AIRPORT ADVISORY AREA

a. The airport advisory area is the area within 10 statute miles of an airport where a control tower is not operating but where a FSS is located. At such locations, the FSS provides advisory service to arriving and departing aircraft. (Reference—Traffic Advisory Practices at Airports Where a Tower is Not in Operation, paragraph 4–8).

b. It is not mandatory that pilots participate in the Airport Advisory Service program, but it is strongly recommended that they do.

3–63. MILITARY TRAINING ROUTES (MTR)

a. National security depends largely on the deterrent effect of our airborne military forces. To be proficient, the military services must train in a wide range of airborne tactics. One phase of this training involves "low level" combat tactics. The required maneuvers and high speeds are such that they may occasionally make the see-and-avoid aspect of VFR flight more difficult without increased vigilance in areas containing such operations. In an effort to ensure the greatest practical level of safety for all flight operations, the MTR program was conceived.

b. The MTRs program is a joint venture by the FAA and the Department of Defense (DOD). MTR routes are mutually developed for use by the military for the purpose of conducting low-altitude, high-speed training. The routes above 1,500 feet above ground level (AGL) are developed to be flown, to the maximum extent possible, under IFR. The routes at 1,500 feet AGL and below are generally developed to be flown under Visual Flight Rules (VFR).

c. Generally, MTRs are established below 10,000 feet MSL for operations at speeds in excess of 250 knots. However, route segments may be defined at higher altitudes for purposes of route continuity. For example, route segments may be defined for descent, climbout, and mountainous terrain. There are IFR and VFR routes as follows:

1. IFR Military Training Routes—IR: Operations on these routes are conducted in accordance with IFRs regardless of weather conditions.

2. VFR Military Training Routes—VR: Operations on these routes are conducted in accordance with VFRs except, flight visibility shall be 5 miles or more; and flights shall not be conducted below a ceiling of less than 3,000 feet AGL.

d. Military training routes will be identified and charted as follows:

1. Route identification.

(a) MTRs with no segment above 1,500 feet AGL shall be identified by four number characters; e.g., IR1206, VR1207.

(b) MTRs that include one or more segments above 1,500 feet AGL shall be identified by three number characters; e.g., IR206, VR207.

(c) Alternate IR/VR routes or route segments are identified by using the basic/principal route designation followed by a letter suffix, e.g., IR008A, VR1007B, etc.

2. Route charting.

(a) IFR Low Altitude En Route Chart—This chart will depict all IR routes and all VR routes that accommodate operations above 1,500 feet AGL.

(b) VFR Planning Chart—This chart will depict routes (military training activities such as IR and VR regardless of altitude), MOAs, and restricted, warning and alert areas.

(c) Area Planning (AP/1B) Chart (DOD Flight Information Publication—FLIP). This chart is published by the DOD primarily for military users and contains detailed information on both IR and VR routes. (Reference—Auxiliary Charts, paragraph 9–6).

e. The FLIP contains charts and narrative descriptions of these routes. This publication is available to the general public by single copy or annual subscription from the DIRECTOR, DMACSC, Attention: DOCP, Washington, DC 20315–0020. This DOD FLIP is available for pilot briefings at FSS and many airports.

f. Nonparticipating aircraft are not prohibited from flying within an MTR; however, extreme vigilance should be exercised when conducting flight through or near these routes. Pilots should contact FSSs within 100 NM of a particular MTR to obtain current information or route usage in their vicinity. Information available includes times of scheduled activity, altitudes in use on each route segment, and actual route width. Route width varies for each MTR and can extend several miles on either side of the charted MTR centerline. Route width information for IR and VR MTR's is also available in the FLIP AP/1B along with additional MTR (SR/AR) information. When requesting MTR information, pilots should give the FSS their position, route of flight, and destination in order to reduce frequency congestion and permit the FSS specialist to identify the MTR routes which could be a factor.

3–64. TEMPORARY FLIGHT RESTRICTIONS

a. General—This paragraph describes the types of conditions under which the FAA may impose temporary flight restrictions. It also explains which FAA elements have been delegated authority to issue a temporary flight restrictions NOTAM and lists the types of responsible agencies/offices from which the FAA will accept requests to establish temporary flight restrictions. The FAR is explicit as to what operations are prohibited, restricted, or allowed in a temporary flight restrictions area. Pilots are responsible to comply with FAR 91.137 when conducting flight in an area where a temporary flight restrictions area is in effect, and should check appropriate NOTAMs during flight planning.

b. The purpose for establishing a temporary flight restrictions area is to:

(1) Protect persons and property in the air or on the surface from an existing or imminent hazard associated with an incident on the surface when the presence of low flying aircraft would magnify, alter, spread, or compound that hazard (FAR 91.137(a)(1));

(2) Provide a safe environment for the operation of disaster relief aircraft (FAR 91.137(a)(2));

(3) Or prevent an unsafe congestion of sightseeing aircraft above an incident or event which may generate a high degree of public interest (FAR 91.137(a)(3)).

c. Except for hijacking situations, when the provisions of FAR 91.137(a)(1) or (a)(2) are necessary, a temporary flight restrictions area will only be established by or through the area manager at the Air Route Traffic Control Center (ARTCC) having jurisdiction over the area concerned. A temporary flight restrictions NOTAM involving the conditions of FAR 91.137(a)(3) will be issued at the direction of the regional air traffic division manager having oversight of the airspace concerned. When hijacking situations are involved, a temporary flight restrictions area will be implemented through the FAA Washington Headquarters Office of Civil Aviation Security. The appropriate FAA air traffic element, upon receipt of such a request, will establish a temporary flight restrictions area under FAR 91.137(a)(1).

d. The FAA accepts recommendations for the establishment of a temporary flight restrictions area under FAR 91.137(a)(1) from military major command headquarters, regional directors of the Office of Emergency Planning, Civil Defense State Directors, State Governors, or other similar authority. For the situations involving FAR 91.137(a)(2), the FAA accepts recommendations from military commanders serving as regional, subregional, or Search and Rescue (SAR) coordinators; by military commanders directing or coordinating air operations associated with disaster relief; or by civil authorities directing or coordinating organized relief air operations (includes representatives of the Office of Emergency Planning, U.S. Forest Service, and State aeronautical agencies). Appropriate authorities for a temporary flight restrictions establishment under FAR 91.137(a)(3) are any of those listed above or by State, county, or city government entities.

e. The type of restrictions issued will be kept to a minimum by the FAA consistent with achievement of the necessary objective. Situations which warrant the extreme restrictions of FAR 91.137(a)(1) include, but are not limited to: toxic gas leaks or spills, flammable agents, or fumes which if fanned by rotor or propeller wash could endanger persons or property on the surface, or if entered by an aircraft could endanger persons or property in the air; imminent volcano eruptions which could endanger airborne aircraft and occupants; nuclear accident or incident; and hijackings. Situations which warrant the restrictions associated with FAR

91.137(a)(2) include: forest fires which are being fought by releasing fire retardants from aircraft; and aircraft relief activities following a disaster (earthquake, tidal wave, flood, etc.). FAR 91.137(a)(3) restrictions are established for events and incidents that would attract an unsafe congestion of sightseeing aircraft.

f. The amount of airspace needed to protect persons and property or provide a safe environment for rescue/relief aircraft operations is normally limited to within 2,000 feet above the surface and within a two nautical mile radius. Incidents occurring within an airport traffic area or terminal control area (TCA) will normally be handled through existing procedures and should not require the issuance of temporary flight restrictions NOTAM.

g. The FSS nearest the incident site is normally the "coordination facility." When FAA communications assistance is required, the designated FSS will function as the primary communications facility for coordination between emergency control authorities and affected aircraft. The ARTCC may act as liaison for the emergency control authorities if adequate communications cannot be established between the designated FSS and the relief organization. For example, the coordination facility may relay authorizations from the on-scene emergency response official in cases where news media aircraft operations are approved at the altitudes used by relief aircraft.

h. ATC may authorize operations in a temporary flight restrictions area under its own authority only when flight restrictions are established under FAR 91.137(a)(2) and (a)(3) and only when such operations are conducted under instrument flight rules (IFR). The appropriate ARTCC/air traffic control tower manager will, however, ensure that such authorized flights do not hamper activities or interfere with the event for which restrictions were implemented. However, ATC will not authorize local IFR flights into the temporary flight restrictions area.

i. To preclude misunderstanding, the implementing NOTAM will contain specific and formatted information. The facility establishing a temporary flight restrictions area will format a NOTAM beginning with the phrase "FLIGHT RESTRICTIONS" followed by: the location of the temporary flight restrictions area; the effective period; the area defined in statute miles; the altitudes affected; the FAA coordination facility and commercial telephone number; the reason for the temporary flight restrictions; the agency directing any relief activities and its commercial telephone number; and other information considered appropriate by the issuing authority.

EXAMPLE FAR 91.137(a)(1):

The following NOTAM prohibits all aircraft operations except those specified in the NOTAM.

FLIGHT RESTRICTIONS MATTHEWS, VIRGINIA, EFFECTIVE IMMEDIATELY UNTIL 1200 GMT JANUARY 20, 1987. PURSUANT TO FAR 91.137(a)(1) TEMPORARY FLIGHT RESTRICTIONS ARE IN EFFECT. RESCUE OPERATIONS IN PROGRESS. ONLY RELIEF AIRCRAFT OPERATIONS UNDER THE DIRECTION OF THE DEPARTMENT OF DEFENSE ARE AUTHORIZED IN THE AIRSPACE AT AND BELOW 5,000 FEET MSL WITHIN A TWO MILE RADIUS OF LASER AFB, MATTHEWS, VIRGINIA. COMMANDER, LASER AFB, IN CHARGE (897) 946–5543. STEENSON FSS IS THE FAA COORDINATION FACILITY (792) 555–6141.

EXAMPLE FAR 91.137(a)(2):

The following NOTAM permits the on-site emergency response official to authorize media aircraft operations below the altitudes used by the relief aircraft.

FLIGHT RESTRICTIONS 25 MILES EAST OF BRANSOME, IDAHO, EFFECTIVE IMMEDIATELY UNTIL 2359 JANUARY 20, 1987. PURSUANT TO FAR 91.137(a)(2) TEMPORARY FLIGHT RESTRICTIONS ARE IN EFFECT WITHIN A FOUR MILE RADIUS OF THE INTERSECTION OF COUNTY ROADS 564 and 315 AT AND BELOW 3,500 FEET MSL TO PROVIDE A SAFE ENVIRONMENT FOR FIRE FIGHTING AIRCRAFT OPERATIONS. DAVIS COUNTY SHERIFF'S DEPARTMENT (792) 555–8122 IS IN CHARGE OF ON-SCENE EMERGENCY RESPONSE ACTIVITIES. GLIVINGS FSS (792) 555–1618 IS THE FAA COORDINATION FACILITY.

EXAMPLE FAR 91.137(a)(3):

The following NOTAM prohibits sightseeing aircraft operations.

FLIGHT RESTRICTIONS BROWN, TENNESSEE, DUE TO OLYMPIC ACTIVITY. EFFECTIVE 1100 GMT JUNE 18, 1987, UNTIL 0200 GMT JULY 19, 1987. PURSUANT TO FAR 91.137(a)(3) TEMPORARY FLIGHT RESTRICTIONS ARE IN EFFECT WITHIN A THREE MILE RADIUS OF THE BASSETT SPORTS

COMPLEX AT AND BELOW 2,500 FEET MSL. NORTON FSS (423) 555–6742 IS THE FAA COORDINATION FACILITY.

3–65. FLIGHT LIMITATIONS/PROHIBITIONS

a. Flight Limitations in the proximity of Space Flight Operations are designated in a Notice to Airman (NOTAM). FAR 91.143 provides protection from potentially hazardous situations for pilots and space flight crews and costly delays of shuttle operations.

b. Flight Restrictions in the proximity of Presidential and Other Parties are put into effect because numerous aircraft and large assemblies of persons may be attracted to areas to be visited or traveled by the President or Vice President, heads of foreign states, and other public figures. Such conditions may create a hazard to aircraft engaged in air commerce and to persons and property on the ground. In addition, responsible agencies of the United States Government may determine that certain regulatory actions should be taken in the interest of providing protection for these public figures. FAR 91.141 provides for the issuance of a regulatory NOTAM to establish flight restrictions where required in such cases.

3–66. PARACHUTE JUMP AIRCRAFT OPERATIONS

a. Procedures relating to parachute jump areas are contained in FAR 105. Tabulations of parachute jump areas in the U.S. are contained in the Airport/Facility Directory.

b. Pilots of aircraft engaged in parachute jump operations are reminded that all reported altitudes must be with reference to mean sea level, or flight level, as appropriate, to enable ATC to provide meaningful traffic information.

c. Parachute operations in the vicinity of an airport without an operating control tower—There is no substitute for alertness while in the vicinity of an airport. It is essential that pilots conducting parachute operations be alert, look for other traffic, and exchange traffic information as recommended in 4–8 *TRAFFIC ADVISORY PRACTICES AT AIRPORTS WITHOUT OPERATING CONTROL TOWERS*. In addition, pilots should avoid releasing parachutes while in an airport traffic pattern when there are other aircraft in that pattern. Pilots should make appropriate broadcasts on the designated Common Traffic Advisory Frequency (CTAF), and monitor that CTAF until all parachute activity has terminated or the aircraft has left the area. Prior to commencing a jump operation, the pilot should broadcast the aircraft's altitude and position in relation to the airport, the approximate relative time when the jump will commence and terminate, and listen to the position reports of other aircraft in the area.

3–67 thru 3–80. RESERVED

Chapter 4. AIR TRAFFIC CONTROL

Section 1. SERVICES AVAILABLE TO PILOTS

4–1. AIR ROUTE TRAFFIC CONTROL CENTERS

Centers are established primarily to provide Air Traffic Service to aircraft operating on IFR flight plans within controlled airspace, and principally during the en route phase of flight.

4–2. CONTROL TOWERS

Towers have been established to provide for a safe, orderly and expeditious flow of traffic on and in the vicinity of an airport. When the responsibility has been so delegated, towers also provide for the separation of IFR aircraft in the terminal areas (Reference—Approach Control, paragraph 5–43).

4–3. FLIGHT SERVICE STATIONS

FSSs are air traffic facilities which provide pilot briefings, en route communications and VFR search and rescue services, assist lost aircraft and aircraft in emergency situations, relay ATC clearances, originate Notices to Airmen, broadcast aviation weather and NAS information, receive and process IFR flight plans, and monitor NAVAIDS. In addition, at selected locations FSSs provide En Route Flight Advisory Service (Flight Watch), take weather observations, issue airport advisories, and advise Customs and Immigration of transborder flights.

4–4. RECORDING AND MONITORING

a. Calls to air traffic control (ATC) facilities (ARTCCs, Towers, FSSs, Central Flow, and Operations Centers) over radio and ATC operational telephone lines (lines used for operational purposes such as controller instructions, briefings, opening and closing flight plans, issuance of IFR clearances and amendments, counter hijacking activities, etc.) may be monitored and recorded for operational uses such as accident investigations, accident prevention, search and rescue purposes, specialist training and evaluation, and technical evaluation and repair of control and communications systems.

b. Where the public access telephone is recorded, a beeper tone is not required. In place of the "beep" tone the FCC has substituted a mandatory requirement that persons to be recorded be given notice they are to be recorded and give consent. Notice is given by this entry, consent to record is assumed by the individual placing a call to the operational facility.

4–5. COMMUNICATIONS RELEASE OF IFR AIRCRAFT LANDING AT AN AIRPORT NOT BEING SERVED BY AN OPERATING TOWER

Aircraft operating on an IFR flight plan, landing at an airport not being served by a tower will be advised to change to the airport advisory frequency when direct communications with ATC is no longer required. Towers and centers do not have nontower airport traffic and runway in use information. The instrument approach may not be aligned with the runway in use; therefore, if the information has not already been obtained, pilots should make an expeditious change to the airport advisory frequency when authorized. (Reference—Advance Information on Instrument Approach, paragraph 5–44).

4–6. PILOT VISITS TO AIR TRAFFIC FACILITIES

Pilots are encouraged to visit air traffic facilities (Towers, Centers and FSSs) and participate in "Operation Raincheck". Operation Raincheck is a program designed to familiarize pilots with the ATC system, its functions, responsibilities and benefits. On rare occasions, facilities may not be able to approve a visit because of ATC workload or other reasons. It is therefore requested that pilots contact the facility prior to the visit and advise of the number of persons in the group, the time and date of the proposed visit and the primary interest of the group. With this information available, the facility can prepare an itinerary and have someone available to guide the group through the facility.

4–7. APPROACH CONTROL SERVICE FOR VFR ARRIVING AIRCRAFT

a. Numerous approach control facilities have established programs for arriving VFR aircraft to contact approach control for landing information. This information includes: wind, runway, and altimeter setting at the airport of intended landing. This information may be omitted if contained in the ATIS broadcast and the pilot states the appropriate ATIS code.

4–7a NOTE—Pilot use of "Have Numbers" does not indicate receipt of the ATIS broadcast. In addition, the controller will provide traffic advisories on a workload permitting basis.

b. Such information will be furnished upon initial contact with concerned approach control facility. The pilot will be requested to change to the *tower* frequency at a predetermined time or point, to receive further landing information.

c. Where available, use of this procedure will not hinder the operation of VFR flights by requiring excessive spacing between aircraft or devious routing.

d. Compliance with this procedure is not mandatory but pilot participation is encouraged. (Reference—Terminal Radar Programs for VFR Aircraft, paragraph 4–16).

4–7d NOTE—Approach control services for VFR aircraft are normally dependent on air traffic control radar. These services are not available during periods of a radar outage. Approach control services for VFR aircraft are limited when CENRAP is in use.

4–8. TRAFFIC ADVISORY PRACTICES AT AIRPORTS WITHOUT OPERATING CONTROL TOWERS

(See Table 4–8[1].)

Table 4–8[1]

SUMMARY OF RECOMMENDED COMMUNICATION PROCEDURES

	FACILITY AT AIRPORT	FREQUENCY USE	COMMUNICATION/BROADCAST PROCEDURES		
			OUTBOUND	INBOUND	PRACTICE INSTRUMENT APPROACH
1.	UNICOM (No Tower or FSS)	Communicate with UNICOM station on published CTAF frequency (122.7, 122.8, 122.725, 122.975, or 123.0). If unable to contact UNICOM station, use self-announce procedures on CTAF.	Before taxiing and before taxiing on the runway for departure.	10 miles out. Entering downwind, base, and final. Leaving the runway.	
2.	No Tower, FSS, or UNICOM	Self-announce on MULTICOM frequency 122.9.	Before taxiing and before taxiing on the runway for departure.	10 miles out. Entering downwind, base, and final. Leaving the runway.	Departing final approach fix (name) or on final approach segment inbound.
3.	No Tower in operation, FSS open	Communicate with FSS on CTAF frequency.	Before taxiing and before taxiing on the runway for departure.	10 miles out. Entering downwind, base, and final. Leaving the runway.	Approach completed/ terminated.
4.	FSS closed (No Tower)	Self-announce on CTAF.	Before taxiing and before taxiing on the runway for departure.	10 miles out. Entering downwind, base, and final. Leaving the runway.	
5.	Tower or FSS not in operation	Self-announce on CTAF.	Before taxiing and before taxiing on the runway for departure.	10 miles out. Entering downwind, base, and final. Leaving the runway.	

a. Airport Operations Without Operating Control Tower

1. There is no substitute for alertness while in the vicinity of an airport. It is essential that pilots be alert and look for other traffic and exchange traffic information when approaching or departing an airport without an operating control tower. This is of particular importance since other aircraft may not have communication capability or, in some cases, pilots may not communicate their presence or intentions when operating into or out of such airports. To achieve the greatest degree of safety, it is essential that all radio-equipped aircraft transmit/receive on a common frequency identified for the purpose of airport advisories.

2. An airport may have a full or part-time tower or Flight Service Station (FSS) located on the airport, a full or part-time UNICOM station or no aeronautical station at all. There are three ways for pilots to communicate their intention and obtain airport/traffic information when operating at an airport that does not have an operating tower: by communicating with an FSS, a UNICOM operator, or by making a self-announce broadcast.

b. Communicating on a Common Frequency.

1. The key to communicating at an airport without an operating control tower is selection of the correct common frequency. The acronym *CTAF* which stands for Common Traffic Advisory Frequency, is synonymous with this program. A CTAF is a frequency designated for the purpose of carrying out airport advisory practices while operating to or from an airport without an operating control tower. The CTAF may be a UNICOM, MULTICOM, FSS, or tower frequency and is identified in appropriate aeronautical publications.

2. The CTAF frequency for a particular airport is contained in the Airport/Facility Directory (A/FD), Alaska Supplement, Alaska Terminal Publication, Instrument Approach Procedure Charts, and Standard Instrument Departure (SID) charts. Also, the CTAF frequency can be obtained by contacting any FSS. Use of the appropriate CTAF, combined with a visual alertness and application of the following recommended good operating practices, will enhance safety of flight into and out of all uncontrolled airports.

c. Recommended Traffic Advisory Practices.

1. Pilots of inbound traffic should monitor and communicate as appropriate on the designated CTAF from 10 miles to landing. Pilots of departing aircraft should monitor/communicate on the appropriate frequency from start-up, during taxi, and until 10 miles from the airport unless the FARs or local procedures require otherwise.

2.Pilots of aircraft conducting other than arriving or departing operations at altitudes normally used by arriving and departing aircraft should monitor/communicate on the appropriate frequency while within 10 miles of the airport unless required to do otherwise by the FAR's or local procedures. Such operations include parachute jumping/dropping, (Reference—Parachute Jump Aircraft Operations, paragraph 3–66, enroute, practicing maneuvers, etc.

d. Airport Advisory Service Provided by an FSS.

1. Airport Advisory Service (AAS) is a service provided by an FSS physically located on an airport which does not have a control tower or where the tower is operated on a part-time basis. The CTAF for FSSs which provide this service will be disseminated in appropriate aeronautical publications.

2. In communicating with a CTAF FSS, establish two-way communications before transmitting outbound/inbound intentions or information. An inbound aircraft should report approximately 10 miles from the airport, reporting altitude and aircraft type, location relative to the airport, state whether landing or overflight, and request airport advisory. Departing aircraft should state the aircraft type, full identification number, type of flight planned, i.e., VFR or IFR and the planned destination or direction of flight. Report before taxiing and before taxiing on the runway for departure. If communications with a UNICOM are necessary after initial report to FSS, return to FSS frequency for traffic update.

(a) Inbound

EXAMPLE:
VERO BEACH RADIO, CENTURION SIX NINER DELTA DELTA IS TEN MILES SOUTH, TWO THOUSAND, LANDING VERO BEACH. REQUEST AIRPORT ADVISORY.

(b) Outbound

EXAMPLE:
VERO BEACH RADIO, CENTURION SIX NINER DELTA DELTA, READY TO TAXI, VFR, DEPARTING TO THE SOUTHWEST. REQUEST AIRPORT ADVISORY.

3. A CTAF FSS provides wind direction and velocity, favored or designated runway, altimeter setting, known traffic, notices to airmen, airport taxi routes, airport traffic pattern information, and instrument approach procedures. These elements are varied so as to best serve the current traffic situation. Some airport managers have specified that under certain wind or other conditions designated runways be used. Pilots should advise the FSS of the runway they intend to use.

CAUTION: All aircraft in the vicinity of an airport may not be in communication with the FSS.

e. Information Provided by Aeronautical Advisory Stations (UNICOM).

1. UNICOM is a nongovernment air/ground radio communication station which may provide airport information at public use airports where there is no tower or FSS.

2. On pilot request, UNICOM stations may provide pilots with weather information, wind direction, the recommended runway, or other necessary information. If the UNICOM frequency is designated as the CTAF, it will be identified in appropriate aeronautical publications.

3. Should AAS by an FSS or, Aeronautical Advisory Station (UNICOM) be unavailable, wind and weather information may be obtainable from nearby controlled airports via Automatic Terminal Information Service (ATIS) or Automated Weather Observing System (AWOS) frequency.

f. Self-Announce Position and/or Intentions.

1. General. *Self-announce* is a procedure whereby pilots broadcast their position or intended flight activity or ground operation on the designated CTAF. This procedure is used primarily at airports which do not have an FSS on the airport. The self-announce procedure should also be used if a pilot is unable to communicate with the FSS on the designated CTAF.

2. If an airport has a tower and it is temporarily closed, or operated on a part-time basis and there is no FSS on the airport or the FSS is closed, use the CTAF to self-announce your position or intentions.

3. Where there is no tower, FSS, or UNICOM station on the airport, use MULTICOM frequency 122.9 for self-announce procedures. Such airports will be identified in appropriate aeronautical information publications.

4. Practice Approaches. Pilots conducting practice instrument approaches should be particularly alert for other aircraft that may be departing in the opposite direction. When conducting any practice approach, regardless of its direction relative to other airport operations, pilots should make announcements on the CTAF as follows:

(a) departing the final approach fix, inbound (non precision approach) or departing the Outer Marker or fix used in lieu of the outer marker, inbound (precision approach);

(b) established on the final approach segment or immediately upon being released by ATC;

(c) upon completion or termination of the approach; and

(d) upon executing the missed approach procedure.

5. Departing aircraft should always be alert for arrival aircraft coming from the opposite direction.

6. Recommended Self-Announce Phraseologies: It should be noted that aircraft operating to or from another nearby airport may be making self-announce broadcasts on the same UNICOM or MULTICOM frequency. To help identify one airport from another, the airport name should be spoken at the beginning and end of each self-announce transmission.

(a) Inbound

EXAMPLE:

STRAWN TRAFFIC, APACHE TWO TWO FIVE ZULU, (POSITION), (ALTITUDE), (DESCENDING) OR ENTERING DOWNWIND/BASE/FINAL (AS APPROPRIATE) RUNWAY ONE SEVEN FULL STOP, TOUCH-AND-GO, STRAWN.

STRAWN TRAFFIC APACHE TWO TWO FIVE ZULU CLEAR OF RUNWAY ONE SEVEN STRAWN.

(b) Outbound

EXAMPLE:

STRAWN TRAFFIC, QUEEN AIR SEVEN ONE FIVE FIVE BRAVO (LOCATION ON AIRPORT) TAXIING TO RUNWAY TWO SIX STRAWN.

STRAWN TRAFFIC, QUEEN AIR SEVEN ONE FIVE FIVE BRAVO DEPARTING RUNWAY TWO SIX. "DEPARTING THE PATTERN TO THE (DIRECTION), CLIMBING TO (ALTITUDE) STRAWN."

(c) Practice Instrument Approach

EXAMPLE:

STRAWN TRAFFIC, CESSNA TWO ONE FOUR THREE QUEBEC (NAME—FINAL APPROACH FIX) INBOUND DESCENDING THROUGH (ALTITUDE) PRACTICE (TYPE) APPROACH RUNWAY THREE FIVE STRAWN.

STRAWN TRAFFIC, CESSNA TWO ONE FOUR THREE QUEBEC PRACTICE (TYPE) APPROACH COMPLETED OR TERMINATED RUNWAY THREE FIVE STRAWN.

g. UNICOM Communications Procedures.

1. In communicating with a UNICOM station, the following practices will help reduce frequency congestion, facilitate a better understanding of pilot intentions, help identify the location of aircraft in the traffic pattern, and enhance safety of flight:

(a) Select the correct UNICOM frequency.

(b) State the identification of the UNICOM station you are calling in each transmission.

(c) Speak slowly and distinctly.

(d) Report approximately 10 miles from the airport, reporting altitude, and state your aircraft type, aircraft identification, location relative to the airport, state whether landing or overflight, and request wind information and runway in use.

(e) Report on downwind, base, and final approach.

(f) Report leaving the runway.

2. Recommended UNICOM Phraseologies:

(a) Inbound

EXAMPLE:

FREDERICK UNICOM CESSNA EIGHT ZERO ONE TANGO FOXTROT 10 MILES SOUTHEAST DESCENDING THROUGH (ALTITUDE) LANDING FREDERICK, REQUEST WIND AND RUNWAY INFORMATION FREDERICK.

FREDERICK TRAFFIC CESSNA EIGHT ZERO ONE TANGO FOXTROT ENTERING DOWNWIND/BASE/FINAL (AS APPROPRIATE) FOR RUNWAY ONE NINER (FULL STOP/ TOUCH-AND-GO) FREDERICK.

FREDERICK TRAFFIC CESSNA EIGHT ZERO ONE TANGO FOXTROT CLEAR OF RUNWAY ONE NINER FREDERICK.

(b) Outbound

EXAMPLE:

FREDERICK UNICOM CESSNA EIGHT ZERO ONE TANGO FOXTROT (LOCATION ON AIRPORT) TAXIING

TO RUNWAY ONE NINER, REQUEST WIND AND TRAFFIC INFORMATION FREDERICK.

FREDERICK TRAFFIC CESSNA EIGHT ZERO ONE TANGO FOXTROT DEPARTING RUNWAY ONE NINER. "REMAINING IN THE PATTERN" OR "DEPARTING THE PATTERN TO THE (DIRECTION) (AS APPROPRIATE)" FREDERICK.

4–9. IFR APPROACHES/GROUND VEHICLE OPERATIONS

a. IFR Approaches.

When operating in accordance with an IFR clearance and ATC approves a change to the advisory frequency, make an expeditious change to the CTAF and employ the recommended traffic advisory procedures.

b. Ground Vehicle Operation.

Airport ground vehicles equipped with radios should monitor the CTAF frequency when operating on the airport movement area and remain clear of runways/taxiways being used by aircraft. Radio transmissions from ground vehicles should be confined to safety-related matters.

c. Radio Control of Airport Lighting Systems.

Whenever possible, the CTAF will be used to control airport lighting systems at airports without operating control towers. This eliminates the need for pilots to change frequencies to turn the lights on and allows a continuous listening watch on a single frequency. The CTAF is published on the instrument approach chart and in other appropriate aeronautical information publications. For further details concerning radio controlled lights, see AC 150/5340–27.

4–10. DESIGNATED UNICOM/MULTICOM FREQUENCIES

a. Communications Between Aircraft.

CAUTION—The Federal Communications Commission (FCC) requires an aircraft station license to operate on UNICOM/MULTICOM frequencies and usage must be in accordance with Part 87 of the FCC Rules (See Section 87.29 regarding license applications). Misuse of these frequencies may result in either the imposition of fines and/or revocation/suspension of FCC aircraft station license.

b. Frequency Use.

1. The following listing depicts UNICOM and MULTICOM frequency uses as designated by the Federal Communications Commission (FCC). (See Table 4–10[1].)

Table 4–10[1]

USE	FREQUENCY
Airports without an operating control tower.	122.700, 122.725, 122.800, 122.975, 123.000, 123.050, 123.075
(MULTICOM FREQUENCY) Activities of a temporary, seasonal, emergency nature or search and rescue, as well as, airports with no tower, FSS, or UNICOM..	122.900
(MULTICOM FREQUENCY) Forestry management and fire suppression, fish and game management and protection, and environmental monitoring and protection..	122.925
Airports with a control tower or FSS on airport.	122.950

4–10b1 NOTE—In some areas of the country, frequency interference may be encountered from nearby airports using the same UNICOM frequency. Where there is a problem, UNICOM operators are encouraged to develop a "least interference" frequency assignment plan for airports concerned using the frequencies designated for airports without operating control towers. UNICOM licensees are encouraged to apply for UNICOM 25 kHz spaced channel frequencies. Due to the extremely limited number of frequencies with 50 kHz channel spacing, 25 kHz channel spacing should be implemented. UNICOM licensees may then request FCC to assign frequencies in accordance with the plan, which FCC will reeview and consider for approval.

4–10b1 NOTE—Wind direction and runway information may not be available on UNICOM frequency 122.950.

2. The following listing depicts other frequency uses as designated by the Federal Communications Commission (FCC). (See Table 4–10[2]).

Table 4–10[2]

USE	FREQUENCY
Air-to-air communications & private airports (not open to the public).	122.750 122.850
Air-to-air communications (general aviation helicopters).	123.025
Aviation instruction, Glider, Hot Air Balloon (**not to be used for advisory service**).	123.300, 123.500

4–11. USE OF UNICOM FOR ATC PURPOSES

UNICOM service may be used for air traffic control purposes, only under the following circumstances:

1. Revision to proposed departure time.

2. Takeoff, arrival, or flight plan cancellation time.

3. ATC clearance, provided arrangements are made between the ATC facility and the UNICOM licensee to handle such messages.

4–12. AUTOMATIC TERMINAL INFORMATION SERVICE (ATIS)

a. ATIS is the continuous broadcast of recorded noncontrol information in selected high activity terminal areas. Its purpose is to improve controller effectiveness and to relieve frequency congestion by automating the repetitive transmission of essential but routine information. Pilots are urged to cooperate in the ATIS program as it relieves frequency congestion on approach control, ground control, and local control frequencies. The Airport Facility/Directory indicates airports for which ATIS is provided.

b. ATIS information includes the time of the latest weather sequence, ceiling, visibility (if the weather is above a ceiling/sky condition of 5,000 feet and the visibility is 5 miles or more, inclusion of the ceiling/sky condition, visibility, and obstructions to vision in the ATIS message is optional), obstructions to visibility, temperature, dew point (if available), wind direction (magnetic) and velocity, altimeter, other pertinent remarks, instrument approach and runways in use is continuously broadcast on the voice feature of a TVOR/VOR/VORTAC located on or near the airport, or in a discrete VHF/UHF frequency. The departure runway will only be given if different from the landing runway except at locations having a separate ATIS for departure. Where VFR arrival aircraft are expected to make initial contact with approach control, this fact and the appropriate frequencies may be broadcast on ATIS. Pilots of aircraft arriving or departing the terminal area can receive the continuous ATIS broadcasts at times when cockpit duties are least pressing and listen to as many repeats as desired. ATIS broadcasts shall be updated upon the receipt of any official weather, regardless of content change and reported values. A new recording will also be made when there is a change in other pertinent data such as runway change, instrument approach in use, etc.

EXAMPLE:

DULLES INTERNATIONAL INFORMATION SIERRA. 1300 ZULU WEATHER. MEASURED CEILING THREE THOUSAND OVERCAST. VISIBILITY THREE, SMOKE. TEMPERATURE SIX EIGHT. WIND THREE FIVE ZERO AT EIGHT. ALTIMETER TWO NINER NINER TWO. ILS RUNWAY ONE RIGHT APPROACH IN USE. LANDING RUNWAY ONE RIGHT AND LEFT. DEPARTURE RUNWAY THREE ZERO. ARMEL VORTAC OUT OF SERVICE. ADVISE YOU HAVE SIERRA.

c. Pilots should listen to ATIS broadcasts whenever ATIS is in operation.

d. Pilots should notify controllers on initial contact that they have received the ATIS broadcast by repeating the alphabetical code word appended to the broadcast.

EXAMPLE:

"INFORMATION SIERRA RECEIVED."

e. When the pilot acknowledges that he has received the ATIS broadcast, controllers may omit those items contained in the broadcast if they are current. Rapidly changing conditions will be issued by ATC and the ATIS will contain words as follows:

EXAMPLE:

"LATEST CEILING/VISIBILITY/ALTIMETER/WIND/(other conditions) WILL BE ISSUED BY APPROACH CONTROL/TOWER."

4–12e NOTE—The absence of a sky condition or ceiling and/or visibility on ATIS indicates a sky condition or ceiling of 5,000 feet or above and visibility of 5 miles or more. A remark may be made on the broadcast, "The weather is better than 5000 and 5," or the existing weather may be broadcast.

f. Controllers will issue pertinent information to pilots who do not acknowledge receipt of a broadcast or who acknowledge receipt of a broadcast which is not current.

g. To serve frequency limited aircraft, FSSs are equipped to transmit on the omnirange frequency at most en route VORs used as ATIS voice outlets. Such communication interrupts the ATIS broadcast. Pilots of aircraft equipped to receive on other FSS frequencies are encouraged to do so in order that these override transmissions may be kept to an absolute minimum.

h. While it is a good operating practice for pilots to make use of the ATIS broadcast where it is available, some pilots use the phrase "Have Numbers" in communications with the control tower. Use of this phrase means that the pilot has received wind, runway, and altimeter information ONLY and the tower does not have to repeat this information. It does not indicate receipt of the ATIS broadcast and should never be used for this purpose.

4–13. RADAR TRAFFIC INFORMATION SERVICE

This is a service provided by radar ATC facilities. Pilots receiving this service are advised of any radar target observed on the radar display which may be in such proximity to the position of their aircraft or its intended route of flight that it warrants their attention. This service is not intended to relieve the pilot of his responsibility for continual vigilance to see and avoid other aircraft.

a. Purpose of the Service:

1. The issuance of traffic information as observed on a radar display is based on the principle of assisting and advising a pilot that a particular radar target's position and track indicates it may intersect or pass in such proximity to his intended flight path that it warrants his attention. This is to alert the pilot to the traffic so that he can be on the lookout for it and thereby be in a better position to take appropriate action should the need arise.

2. Pilots are reminded that the surveillance radar used by ATC does not provide altitude information unless the aircraft is equipped with MODE C and the Radar Facility is capable of displaying altitude information.

b. Provisions of the Service:

1. Many factors, such as limitations of the radar, volume of traffic, controller workload and communications frequency congestion, could prevent the controller from providing this service. The controller possesses complete discretion for determining whether he is able to provide or continue to provide this service in a specific case. His reason against providing or continuing to provide the service in a particular case is not subject to question nor need it be communicated to the pilot. In other words, the provision of this service is entirely dependent upon whether the controller believes he is in a position to provide it. Traffic information is routinely provided to all aircraft operating on IFR Flight Plans except when the pilot advises he does not desire the service, or the pilot is operating within Positive Controlled Airspace. Traffic information may be provided to flights not operating on IFR Flight Plans when requested by pilots of such flights.

4–13b1 NOTE—Radar ATC facilities normally display and monitor both primary and secondary radar when it is available, except that secondary radar may be used as the sole display source in Positive Control Airspace (PCA), and under some circumstances outside of PCA (beyond primary coverage and in en route areas where only secondary is available). Secondary radar may also be used outside PCA as the sole display source when the primary radar is temporarily unusable or out of service. Pilots in contact with the affected ATC facility are normally advised when a temporary outage occurs; i.e., "primary radar out of service; traffic advisories available on transponder aircraft only." This means simply that only the aircraft which have transponders installed and in use will be depicted on ATC radar indicators when the primary radar is temporarily out of service.

2. When receiving VFR radar advisory service, pilots should monitor the assigned frequency at all times. This is to preclude controllers' concern for radio failure or emergency assistance to aircraft under his jurisdiction. VFR radar advisory service does not include vectors away from conflicting traffic unless requested by the pilot. When advisory service is no longer desired, advise the controller before changing frequencies and then change your transponder code to 1200, if applicable. Pilots should also inform the controller when changing VFR cruising altitude. Except in programs where radar service is automatically terminated, the controller will advise the aircraft when radar is terminated.

4–13b2 NOTE—Participation by VFR pilots in formal programs implemented at certain terminal locations constitutes pilot request. This also applies to participating pilots at those locations where arriving VFR flights are encouraged to make their first contact with the tower on the approach control frequency.

c. Issuance of Traffic Information—Traffic information will include the following concerning a target which may constitute traffic for an aircraft that is:

1. Radar identified:

(a) Azimuth from the aircraft in terms of the 12 hour clock, or

(b) When rapidly maneuvering civil test or military aircraft prevent accurate issuance of traffic as in (a) above, specify the direction from an aircraft's position in terms of the eight cardinal compass points (N, NE, E, SE, S, SW, W, NW). This method shall be terminated at the pilot's request.

(c) Distance from the aircraft in nautical miles;

(d) Direction in which the target is proceeding; and

(e) Type of aircraft and altitude if known.

EXAMPLE:

Traffic 10 o'clock, 3 miles, west-bound (type aircraft and altitude, if known, of the observed traffic). The altitude may be known, by means of MODE C, but not verified with the pilot for accuracy. (To be valid for separation purposes by ATC, the accuracy of MODE C readouts must be verified. This is usually accomplished upon initial entry into the radar system by a comparison of the readout to pilot stated altitude, or the field elevation in the case of continuous readout being received from an aircraft on the airport.) When necessary to issue traffic advisories containing unverified altitude information, the controller will issue the advisory in the same manner as if it were verified due to the accuracy of these readouts. The pilot may upon receipt of traffic information, request a vector (heading) to avoid such traffic. The vector will be provided to the extent possible as determined by the controller provided the aircraft to be vectored is within the airspace under the jurisdiction of the controller.

2. Not radar identified:

(a) Distance and direction with respect to a fix;

(b) Direction in which the target is proceeding; and

(c) Type of aircraft and altitude if known.

EXAMPLE:

Traffic 8 miles south of the airport northeastbound, (type aircraft and altitude if known).

d. The examples depicted in the following figures point out the possible error in the position of this traffic when it is necessary for a pilot to apply drift correction to maintain this track. This error could also occur in the event a change in course is made at the time radar traffic information is issued.

Figure 4–13[1]

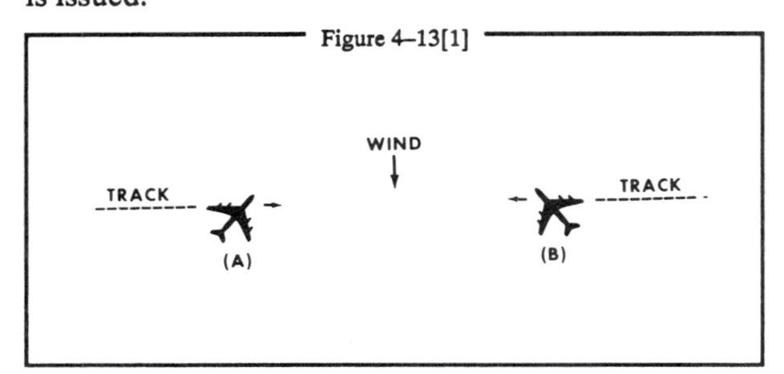

1. In Figure 4–13[1] traffic information would be issued to the pilot of aircraft "A" as 12 o'clock. The actual position of the traffic as seen by the pilot of aircraft "A" would be 2 o'clock. Traffic information issued to aircraft "B" would also be given as 12 o'clock, but in this case, the pilot of "B" would see his traffic at 10 o'clock.

Figure 4–13[2]

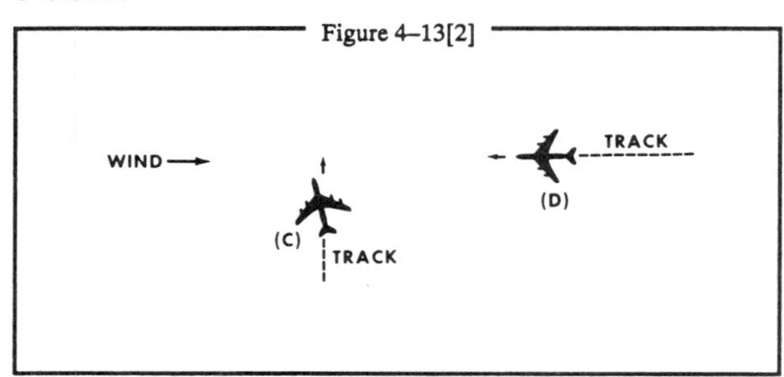

2. In Figure 4–13[2] traffic information would be issued to the pilot of aircraft "C" as 2 o'clock. The actual position of the traffic as seen by the pilot of aircraft "C" would be 3 o'clock. Traffic information issued to aircraft "D" would be at an 11 o'clock position. Since it is not necessary for the pilot of aircraft "D" to apply wind correction (crab) to make good his track, the actual position of the traffic issued would be correct. Since the radar controller can only observe aircraft track (course) on his radar display, he must issue traffic advisories accordingly, and pilots should give due consideration to this fact when looking for reported traffic.

4–14. SAFETY ALERT

A safety alert will be issued to pilots of aircraft being controlled by ATC if the controller is aware the aircraft is at an altitude which, in the controller's judgment, places the aircraft in unsafe proximity to terrain, obstructions or other aircraft. The provision of this service is contingent upon the capability of the controller to have an awareness of a situation involving unsafe proximity to terrain, obstructions and uncontrolled aircraft. The issuance of a safety alert cannot be mandated, but it can be expected on a reasonable, though intermittent basis. Once the alert is issued, it is solely the pilot's prerogative to determine what course of action, if any, he will take. This procedure is intended for use in time critical situations where aircraft safety is in question. Noncritical situations should be handled via the normal traffic alert procedures.

a. Terrain or Obstruction Alert:

1. The controller will immediately issue an alert to the pilot of an aircraft under his control when he recognizes that the aircraft is at an altitude which, in his judgment, may be in unsafe proximity to terrain/obstructions. The primary method of detecting unsafe proximity is through MODE C automatic altitude reports.

EXAMPLE:

LOW ALTITUDE ALERT, CHECK YOUR ALTITUDE IMMEDIATELY. THE, as appropriate, MEA/MVA/MOCA IN YOUR AREA IS (altitude) or, if past the final approach fix (non precision approach) or the Outer Marker or fix used in lieu of the outer marker (precision approach), THE, as appropriate, MDA/DH (if known) is (altitude).

2. Terminal ARTS IIA, III, AND IIIA facilities have an automated function which, if operating, alerts the controller when a tracked MODE C equipped aircraft under his control is below or is predicted to be below a predetermined Minimum Safe Altitude. This function, called Minimum Safe Altitude Warning (MSAW), is designed solely as a controller aid in detecting potentially unsafe aircraft proximity to terrain/ obstructions. The ARTS IIA, III, and IIIA facility will, when MSAW is operating, provide MSAW monitoring for all aircraft with an operating MODE C altitude encoding transponder that are tracked by the system and are:

(a) Operating on an IFR flight plan, or

(b) Operating VFR and have requested MSAW monitoring.

(3) Terminal AN/TPX–42A (number beacon decoder system) facilities have an automated function called Low Altitude Alert System (LAAS). Although not as sophisticated as MSAW, LAAS alerts the controller when a MODE C transponder equipped aircraft operating on an IFR flight plan is below a predetermined Minimum Safe Altitude.

4–14a3 NOTE—Pilots operating VFR may request MSAW or LAAS monitoring if their aircraft are equipped with MODE C transponders.

EXAMPLE:
APACHE THREE THREE PAPA REQUEST MSAW/LAAS.

b. Aircraft Conflict Alert:

1. The controller will immediately issue an alert to the pilot of an aircraft under his control if he is aware of another aircraft which is not under his control is at an altitude which, in the controller's judgment, places both aircraft in unsafe proximity to each other. With the alert, when feasible, the controller will offer the pilot the position of the traffic if time permits and an alternate course(s) of action. Any alternate course(s) of action the controller may recommend to the pilot will be predicated only on other traffic under his control.

EXAMPLE:
AMERICAN THREE, TRAFFIC ALERT, (position of traffic, if time permits), ADVISE YOU TURN RIGHT/LEFT HEADING (degrees) AND/OR CLIMB/DESCEND TO (altitude) IMMEDIATELY.

4-15. RADAR ASSISTANCE TO VFR AIRCRAFT

a. Radar equipped FAA ATC facilities provide radar assistance and navigation service (vectors) to VFR aircraft provided the aircraft can communicate with the facility, are within radar coverage, and can be radar identified.

b. Pilots should clearly understand that authorization to proceed in accordance with such radar navigational assistance does not constitute authorization for the pilot to violate FARs. In effect, assistance provided is on the basis that navigational guidance information issued is advisory in nature and the job of flying the aircraft safely, remains with the pilot.

c. In many cases, the controller will be unable to determine if flight into instrument conditions will result from his instructions. To avoid possible hazards resulting from being vectored into IFR conditions, pilots should keep the controller advised of the weather conditions in which he is operating and along the course ahead.

d. Radar navigation assistance (vectors) may be initiated by the controller when one of the following conditions exist:

1. The controller suggests the vector and the pilot concurs.

2. A special program has been established and vectoring service has been advertised.

3. In the controller's judgment the vector is necessary for air safety.

e. Radar navigation assistance (vectors) and other radar derived information may be provided in response to pilot requests. Many factors, such as limitations of radar, volume of traffic, communications frequency, congestion, and controller workload could prevent the controller from providing it. The controller has complete discretion for determining if he is able to provide the service in a particular case. His decision not to provide the service in a particular case is not subject to question.

4-16. TERMINAL RADAR PROGRAMS FOR VFR AIRCRAFT

a. Basic Radar Service

1. In addition to the use of radar for the control of IFR aircraft, all commissioned radar facilities provide traffic advisories and limited vectoring (on a workload permitting basis) to VFR aircraft.

4-16a1 NOTE—When the STAGE services were developed, two basic radar services (traffic advisories and limited vectoring) were identified as STAGE I. This definition became, over the years, unnecessary. Therefore, the term STAGE I has been eliminated from use in the field and the handbooks. These basic services will still be provided by all commissioned terminal radar facilities whether they are STAGE II, III, or neither.

2. Vectoring service may be provided when requested by the pilot or with pilot concurrence when suggested by ATC.

3. Pilots of arriving aircraft should contact approach control on the publicized frequency and give their position, altitude, radar beacon code (if transponder equipped), destination, and request traffic information.

4. Approach control will issue wind and runway, except when the pilot states "Have Numbers" or this information is contained in the ATIS broadcast and the pilot indicates he has received the ATIS information. Traffic information is provided on a workload permitting basis. Approach control will specify the time or place at which the pilot is to contact the tower on local control frequency for further landing information. Upon being told to contact the tower, radar service is automatically terminated.

b. Stage II Service (Radar Advisory and Sequencing for VFR Aircraft)

1. This service has been implemented at certain terminal locations (See locations listed in the Airport/Facility Directory). The purpose of the service is to adjust the flow of arriving VFR and IFR aircraft into the traffic pattern in a safe and orderly manner and to provide radar traffic information to departing VFR aircraft. Pilot participation is urged but it is not mandatory.

2. Pilots of arriving VFR aircraft should initiate radio contact on the publicized frequency with approach control when approximately 25 miles from the airport at which Stage II services are being provided. On initial contact by VFR aircraft, approach control will assume that Stage II service is requested.

Approach control will provide the pilot with wind and runway (except when the pilot states "Have Numbers" or that he has received the ATIS information), routings, etc., as necessary for proper sequencing with other participating VFR and IFR traffic en route to the airport. Traffic information will be provided on a workload permitting basis. If an arriving aircraft does not want the service, the pilot should state "NEGATIVE STAGE II" or make a similar comment, on initial contact with approach control.

3. After radar contact is established, the pilot may use pilot navigation to enter the traffic pattern or, depending on traffic conditions, may be instructed by ATC to fly specific headings to position the flight behind a preceding aircraft in the approach sequence. When a flight is positioned behind the preceding aircraft and the pilot reports having that aircraft in sight, the pilot will be instructed to follow the preceding aircraft. THE ATC INSTRUCTION TO FOLLOW THE PRECEDING AIRCRAFT DOES NOT AUTHORIZE THE PILOT TO COMPLY WITH ANY ATC CLEARANCE OR INSTRUCTION ISSUED TO THE PRECEDING AIRCRAFT. If other "nonparticipating" or "local" aircraft are in the traffic pattern, the tower will issue a landing sequence. Radar service will be continued to the runway.

4. Standard radar separation will be provided between IFR aircraft until such time as the aircraft is sequenced and the pilot sees the traffic he is to follow. Standard radar separation between VFR or between VFR and IFR aircraft will not be provided.

5. Pilots of departing VFR aircraft are encouraged to request radar traffic information by notifying ground control on initial contact with their request and proposed direction of flight.

EXAMPLE:

XRAY GROUND CONTROL, NOVEMBER ONE EIGHT SIX, READY TO TAXI, VFR SOUTHBOUND, HAVE INFORMATION BRAVO AND REQUEST RADAR TRAFFIC INFORMATION.

4–16b5 NOTE—Following takeoff, the tower will advise when to contact departure control.

6. Pilots of aircraft transiting the area and in radar contact/communication with approach control will receive traffic information on a controller workload permitting basis. Pilots of such aircraft should give their position, altitude, radar beacon code (if transponder equipped), destination, and/or route of flight.

c. Stage III Service (Radar Sequencing and Separation Service for VFR Aircraft).

1. This service has been implemented at certain terminal locations. The service is advertised in the Airport/Facility Directory. The purpose of this service is to provide separation between all participating VFR aircraft and all IFR aircraft operating within the airspace defined as the Terminal Radar Service Area (TRSA). Pilot participation is urged but it is not mandatory.

2. If any aircraft does not want the service, the pilot should state "NEGATIVE STAGE III" or make a similar comment, on initial contact with approach control or ground control, as appropriate.

3. TRSAs are depicted on sectional aeronautical charts and listed in the Airport/Facility Directory.

4. While operating within a TRSA, pilots are provided Stage III service and separation as prescribed in this paragraph. In the event of a radar outage, separation and sequencing of VFR aircraft will be suspended as this service is dependent on radar. The pilot will be advised that the service is not available and issued wind, runway information, and the time or place to contact the tower. Traffic information will be provided on a workload permitting basis.

5. Visual separation is used when prevailing conditions permit and it will be applied as follows:

(a) When a VFR flight is positioned behind a preceding aircraft and the pilot reports having that aircraft in sight, the pilot will be instructed by ATC to follow the preceding aircraft. THE ATC INSTRUCTION TO FOLLOW THE PRECEDING AIRCRAFT DOES NOT AUTHORIZE THE PILOT TO COMPLY WITH ANY ATC CLEARANCE OR INSTRUCTION ISSUED TO THE PRECEDING AIRCRAFT. Radar service will be continued to the runway.

(b) When an IFR aircraft is being sequenced with other traffic and the pilot reports seeing the aircraft that the pilot is to follow, the pilot may be instructed by ATC to follow it and will be cleared for a "visual approach." THE ATC INSTRUCTION TO FOLLOW THE PRECEDING AIRCRAFT DOES NOT AUTHORIZE THE PILOT TO COMPLY WITH ANY ATC CLEARANCE OR INSTRUCTION ISSUED TO THE PRECEDING AIRCRAFT.

(c) If other "nonparticipating" or "local" aircraft are in the traffic pattern, the tower will issue a landing sequence.

(d) Departing VFR aircraft may be asked if they can visually follow a preceding departure out of the TRSA. If the pilot concurs, he will be directed to follow it until leaving the TRSA.

6. Until visual separation is obtained, standard vertical or radar separation will be provided.

(a) 1000 feet vertical separation may be used between IFR aircraft.

(b) 500 feet vertical separation may be used between VFR aircraft, or between a VFR and an IFR aircraft.

(c) Radar separation varies depending on size of aircraft and aircraft distance from the radar antenna. The minimum separation used will be 1½ miles for most VFR aircraft under 12,500 pounds GWT. If being separated from larger aircraft, the minimum is increased appropriately.

7. Pilots operating VFR under Stage III in a TRSA—

(a) Must maintain an altitude when assigned by ATC unless the altitude assignment is to maintain at or below a specified altitude. ATC may assign altitudes for separation that do not conform to FAR 91.159. When the altitude assignment is no longer needed for separation or when leaving the TRSA, the instruction will be broadcast, "RESUME APPROPRIATE VFR ALTITUDES." Pilots must then return to an altitude that conforms to FAR 91.159 as soon as practicable.

(b) When not assigned an altitude, the pilot should coordinate with ATC prior to any altitude change.

8. Within the TRSA, traffic information on observed but unidentified targets will, to the extent possible, be provided all IFR and participating VFR aircraft. At the request of the pilot, he will be vectored to avoid the observed traffic, insofar as possible, provided the aircraft to be vectored is within the airspace under the jurisdiction of the controller.

9. Departing aircraft should inform ATC of their intended destination and/or route of flight and proposed cruising altitude.

10. ATC will normally advise participating VFR aircraft when leaving the geographical limits of the TRSA. Radar service is not automatically terminated with this advisory unless specifically stated by the controller.

d. PILOT RESPONSIBILITY: THESE PROGRAMS ARE NOT TO BE INTERPRETED AS RELIEVING PILOTS OF THEIR RESPONSIBILITIES TO SEE AND AVOID OTHER TRAFFIC OPERATING IN BASIC VFR WEATHER CONDITIONS, TO ADJUST THEIR OPERATIONS AND FLIGHT PATH AS NECESSARY TO PRECLUDE SERIOUS WAKE ENCOUNTERS, TO MAINTAIN APPROPRIATE TERRAIN AND OBSTRUCTION CLEARANCE, OR TO REMAIN IN WEATHER CONDITIONS EQUAL TO OR BETTER THAN THE MINIMUMS REQUIRED BY FAR 91.155. WHENEVER COMPLIANCE WITH AN ASSIGNED ROUTE, HEADING AND/OR ALTITUDE IS LIKELY TO COMPROMISE PILOT RESPONSIBILITY RESPECTING TERRAIN AND OBSTRUCTION CLEARANCE, VORTEX EXPOSURE, AND WEATHER MINIMUMS, APPROACH CONTROL SHOULD BE SO ADVISED AND A REVISED CLEARANCE OR INSTRUCTION OBTAINED.

e. ATC services for VFR aircraft participating in terminal radar programs are dependent on air traffic control radar. Services for VFR aircraft are not available during periods of a radar outage and are limited during CENRAP operations. The pilot will be advised when VFR services are limited or not available.

4–16e NOTE—The Terminal Control Areas (TCA) and Airport Radar Service Area (ARSA) are areas of regulated airspace. The absence of ATC radar does not negate the requirement of an ATC clearance to enter a TCA or two-way radio contact with ATC to enter an ARSA.

4–17. TOWER EN ROUTE CONTROL (TEC)

a. TEC is an ATC program to provide a service to aircraft proceeding to and from metropolitan areas. It links designated approach control areas by a network of identified routes made up of the existing airway structure of the National Airspace System. The FAA initiated an expanded TEC program to include as many facilities as possible. The program's intent is to provide an overflow resource in the low altitude system which would enhance ATC services. A few facilities have historically allowed turbojets to proceed between certain city pairs, such as Milwaukee and Chicago, via tower en route and these locations may continue this service. However, the expanded TEC program will be applied, generally, for nonturbojet aircraft operating at and below 10,000 feet. The program is entirely within the approach control airspace of multiple terminal facilities. Essentially, it is for relatively short flights. Participating pilots are encouraged to use TEC for flights of two hours duration or less. If longer flights are planned, extensive coordination may be required within the multiple complex which could result in unanticipated delays.

b. Pilots requesting TEC are subject to the same delay factor at the destination airport as other aircraft in the ATC system. In addition, departure and en route delays may occur depending upon individual facility workload. When a major metropolitan airport is incurring significant delays, pilots in the TEC program may want to consider an alternative airport experiencing no delay.

c. There are no unique requirements upon pilots to use the TEC program. Normal flight plan filing procedures will ensure proper flight plan processing. Pilots should include the acronym "TEC" in the remarks section of the flight plan when requesting tower en route.

d. All approach controls in the system may not operate up to the maximum TEC altitude of 10,000 feet. IFR flight may be planned to any satellite airport in proximity to the major primary airport via the same routing.

4–18. TRANSPONDER OPERATION

a. GENERAL

1. Pilots should be aware that proper application of transponder operating procedures will provide both VFR and IFR aircraft with a higher degree of safety in the environment where high-speed closure rates are possible. Transponders substantially increase the capability of radar to see an aircraft and the MODE C feature enables the controller to quickly determine where potential traffic conflicts may exist. Even VFR pilots who are not in contact with ATC will be afforded greater protection from IFR aircraft and VFR aircraft which are receiving traffic advisories. Nevertheless, pilots should never relax their visual scanning vigilance for other aircraft.

2. Air Traffic Control Radar Beacon System (ATCRBS) is similar to and compatible with military coded radar beacon equipment. Civil MODE A is identical to military MODE 3.

3. Civil and military transponders should be adjusted to the "on" or normal operating position as late as practicable prior to takeoff and to "off" or "standby" as soon as practicable after completing landing roll, unless the change to "standby" has been accomplished previously at the request of ATC. IN ALL CASES, WHILE IN CONTROLLED AIRSPACE EACH PILOT OPERATING AN AIRCRAFT EQUIPPED WITH AN OPERABLE ATC TRANSPONDER MAINTAINED IN ACCORDANCE WITH FAR 91.413 **SHALL** OPERATE THE TRANSPONDER, INCLUDING MODE C IF INSTALLED, ON THE APPROPRIATE CODE OR AS ASSIGNED BY ATC. IN UNCONTROLLED AIRSPACE, THE TRANSPONDER SHOULD BE OPERATING WHILE AIRBORNE UNLESS OTHERWISE REQUESTED BY ATC.

4. If a pilot on an IFR flight cancels his IFR flight plan prior to reaching his destination, he should adjust his transponder according to VFR operations.

5. If entering a U.S. domestic control area from outside the U.S., the pilot should advise on first radio contact with a U.S. radar ATC facility that such equipment is available by adding "transponder" to the aircraft identification.

6. It should be noted by all users of the ATC Transponders that the coverage they can expect is limited to "line of sight." Low altitude or aircraft antenna shielding by the aircraft itself may result in reduced range. Range can be improved by climbing to a higher altitude. It may be possible to minimize antenna shielding by locating the antenna where dead spots are only noticed during abnormal flight attitudes.

b. TRANSPONDER CODE DESIGNATION

1. For ATC to utilize one or a combination of the 4096 discrete codes FOUR DIGIT CODE DESIGNATION will be used, e.g., code 2100 will be expressed as TWO ONE ZERO ZERO. Due to the operational characteristics of the rapidly expanding automated air traffic control system, THE LAST TWO DIGITS OF THE SELECTED TRANSPONDER CODE SHOULD ALWAYS READ "00" UNLESS SPECIFICALLY REQUESTED BY ATC TO BE OTHERWISE.

c. AUTOMATIC ALTITUDE REPORTING (MODE C)

1. Some transponders are equipped with a MODE C automatic altitude reporting capability. This system converts aircraft altitude in 100 foot increments to coded digital information which is transmitted together with MODE C framing pulses to the interrogating radar facility. The manner in which transponder panels are designed differs, therefore, a pilot should be thoroughly familiar with the operation of his transponder so that ATC may realize its full capabilities.

2. Adjust transponder to reply on the MODE A/3 code specified by ATC and, if equipped, to reply on MODE C with altitude reporting *capability activated* unless deactivation is directed by ATC or unless the installed aircraft equipment has not been tested and calibrated as required by FAR 91.217. If deactivation is required by ATC, turn off the altitude reporting feature of your transponder. An instruction by ATC to "STOP ALTITUDE SQUAWK, ALTITUDE DIFFERS (number of feet) FEET," may be an indication that your transponder is transmitting incorrect altitude information or that you have an incorrect altimeter setting. While an incorrect altimeter setting has no effect on the MODE C altitude information transmitted by your transponder (transponders are preset at 29.92), it would cause you to fly at an actual altitude different from your assigned altitude. When a controller indicates that an altitude readout is invalid, the pilot should initiate a check to verify that the aircraft altimeter is set correctly.

3. Pilots of aircraft with operating MODE C altitude reporting transponders should report exact altitude or Flight Level to the nearest hundred foot increment when establishing initial contact with an ATC facility. Exact altitude or flight level reports on initial contact provide ATC with information that is required prior to using MODE C altitude information for separation purposes. This will significantly reduce altitude verification requests.

d. TRANSPONDER IDENT FEATURE

1. The transponder shall be operated only as specified by ATC. Activate the "IDENT" feature only upon request of the ATC controller.

e. CODE CHANGES

1. When making routine code changes, pilots should avoid inadvertent selection of codes 7500, 7600 or 7700 thereby causing momentary false alarms at automated ground facilities. For example when switching from code 2700 to code 7200, switch first to 2200 then to 7200, NOT to 7700 and then 7200. This procedure applies to nondiscrete code 7500 and all discrete codes in the 7600 and 7700 series (i.e. 7600–7677, 7700–7777) which will trigger special indicators in automated facilities. Only nondiscrete code 7500 will be decoded as the hijack code.

2. Under no circumstances should a pilot of a civil aircraft operate the transponder on Code 7777. This code is reserved for military interceptor operations.

3. Military pilots operating VFR or IFR within restricted/warning areas should adjust their transponders to code 4000 unless another code has been assigned by ATC.

f. MODE C TRANSPONDER REQUIREMENTS

1. Specific details concerning requirements to carry and operate Mode C transponders, as well as exceptions and ATC authorized deviations from the requirements are found in FAR 91.215 and FAR 99.12.

. In general, the FAR requires aircraft to be equipped with Mode C transponders when operating:

(a) at or above 10,000 feet MSL over the 48 contiguous states or the District of Columbia, excluding that airspace below 2,500 feet AGL;

(b) within 30 miles of a TCA primary airport, below 10,000 feet MSL. Balloons, gliders, and aircraft not equipped with an engine driven electrical system are excepted from the above requirements when operating below the floor of Positive Control Area and/or; outside of a TCA and below the ceiling of the TCA (or 10,000 feet MSL, whichever is lower);

(c) within and above all ARSA's, up to 10,000 feet MSL;

(d) within 10 miles of certain designated airports, excluding that airspace which is both outside the Airport Traffic Area and below 1,200 feet AGL. Balloons, gliders and aircraft not equipped with an engine driven electrical system are excepted from this requirement.

3. FAR 99.12 requires all aircraft flying into, within, or across the contiguous U.S. ADIZ be equipped with a Mode C or Mode S transponder. Balloons, gilders and aircraft not equipped with an engine driven electrical system are excepted from this requirement.

4. Pilots shall ensure that their aircraft transponder is operating on an appropriate ATC assigned VFR/IFR code and MODE C when operating in such airspace. If in doubt about the operational status of either feature of your transponder while airborne, contact the nearest ATC facility or FSS and they will advise you what facility you should contact for determining the status of your equipment.

5. In-flight requests for "immediate" deviation from the transponder requirement may be approved by controllers only when the flight will continue IFR or when weather conditions prevent VFR descent and continued VFR flight in airspace not affected by the FAR. All other requests for deviation should be made by contacting the nearest Flight Service or Air Traffic facility in person or by telephone. The nearest ARTCC will normally be the controlling agency and is responsible for coordinating requests involving deviations in other ARTCC areas.

g. TRANSPONDER OPERATION UNDER VISUAL FLIGHT RULES (VFR)

1. Unless otherwise instructed by an Air Traffic Control Facility, adjust Transponder to reply on MODE 3/A code 1200 regardless of altitude.

2. Adjust transponder to reply on MODE C, with altitude reporting *capability activated* if the aircraft is so equipped, unless deactivation is directed by ATC or unless the installed equipment has not been tested and calibrated as required by FAR 91.217. If deactivation is required and your transponder is so designed, turn off the altitude reporting switch and continue to transmit MODE C framing pulses. If this capability does not exist, turn off MODE C.

h. RADAR BEACON PHRASEOLOGY

Air traffic controllers, both civil and military, will use the following phraseology when referring to operation of the Air Traffic Control Radar Beacon System (ATCRBS). Instructions by ATC refer only to MODE A/3 or MODE C operation and do not affect the operation of the transponder on other MODEs.

1. SQUAWK (number)—Operate radar beacon transponder on designated code in MODE A/3.

2. IDENT—Engage the "IDENT" feature (military I/P) of the transponder.

3. SQUAWK (number) and IDENT—Operate transponder on specified code in MODE A/3 and engage the "IDENT" (military I/P) feature.

4. SQUAWK STANDBY—Switch transponder to standby position.

5. SQUAWK LOW/NORMAL—Operate transponder on low or normal sensitivity as specified. Transponder is operated in "NORMAL" position unless ATC specifies "LOW", ("ON" is used instead of "NORMAL" as a master control label on some types of transponders.)

6. SQUAWK ALTITUDE—Activate MODE C with automatic altitude reporting.

7. STOP ALTITUDE SQUAWK—Turn off altitude reporting switch and continue transmitting MODE C framing pulses. If your equipment does not have this capability, turn off MODE C.

8. STOP SQUAWK (mode in use)—Switch off specified mode. (Used for military aircraft when the controller is unaware of military service requirements for the aircraft to continue operation on another MODE.)

9. STOP SQUAWK—Switch off transponder.

10. SQUAWK MAYDAY—Operate transponder in the emergency position (MODE A Code 7700 for civil transponder. MODE 3 Code 7700 and emergency feature for military transponder.)

11. SQUAWK VFR—Operate radar beacon transponder on code 1200 in the MODE A/3, or other appropriate VFR code.

4–19. HAZARDOUS AREA REPORTING SERVICE

a. Selected FSSs provide flight monitoring where regularly traveled VFR routes cross large bodies of water, swamps, and mountains. This service is provided for the purpose of expeditiously alerting Search and Rescue facilities when required. (See Figure 4–19[1].)

1. When requesting the service either in person, by telephone or by radio, pilots should be prepared to give the following information—type of aircraft, altitude, indicated airspeed, present position, route of flight, heading.

2. Radio contacts are desired at least every 10 minutes. If contact is lost for more than 15 minutes, Search and Rescue will be alerted. Pilots are responsible for canceling their request for service when they are outside the service area boundary. Pilots experiencing two-way radio failure are expected to land as soon as practicable and cancel their request for the service. The accompanying illustration titled "LAKE, ISLAND, MOUNTAIN AND SWAMP REPORTING SERVICE" depicts the areas and the FSS facilities involved in this program.

b. LONG ISLAND SOUND REPORTING SERVICE

The New York and Bridgeport AFSSs provide Long Island Sound Reporting service on request for aircraft traversing Long Island Sound.

1. When requesting the service pilots should ask for SOUND REPORTING SERVICE and should be prepared to provide the following appropriate information:

(a) Type and color of aircraft,

(b) The specific route and altitude across the sound including the shore crossing point,

(c) The overwater crossing time,

(d) Number of persons on board,

(e) True air speed.

2. Radio contacts are desired at least every 10 minutes, however, for flights of shorter duration a midsound report is requested. If contact is lost for more than 15 minutes Search and Rescue will be alerted. Pilots are responsible for canceling their request for the Long Island Sound Reporting Service when outside the service area boundary. Aircraft experiencing radio failure will be expected to land as soon as practicable and cancel their request for the service.

3. COMMUNICATIONS: Primary communications—pilots are to transmit on 122.1 MHz and listen on one of the following VOR frequencies:

(a) NEW YORK AFSS CONTROLS:

(1) Hampton RCO (FSS transmits and receives on 122.6 MHz).

(2) Calverton VORTAC (FSS transmits on 117.2 and receives on standard FSS frequencies).

(3) Kennedy VORTAC (FSS transmits on 115.9 and receives on 122.1 MHz).

(b) BRIDGEPORT AFSS CONTROLS:

(1) Madison VORTAC (FSS transmits on 110.4 and receives on 122.1 MHz).

(2) Groton VOR (FSS transmits on 111.8 and receives on 122.1 MHz).

(3) Bridgeport VOR (FSS transmits on 108.8 and receives on 122.1 MHz).

c. BLOCK ISLAND REPORTING SERVICE

Within the Long Island Reporting Service, the New York FSS also provides an additional service for aircraft operating between Montauk Point and Block Island. When requesting this service, pilots should ask for BLOCK ISLAND REPORTING SERVICE and should be prepared to provide the

Figure 4–19[1]

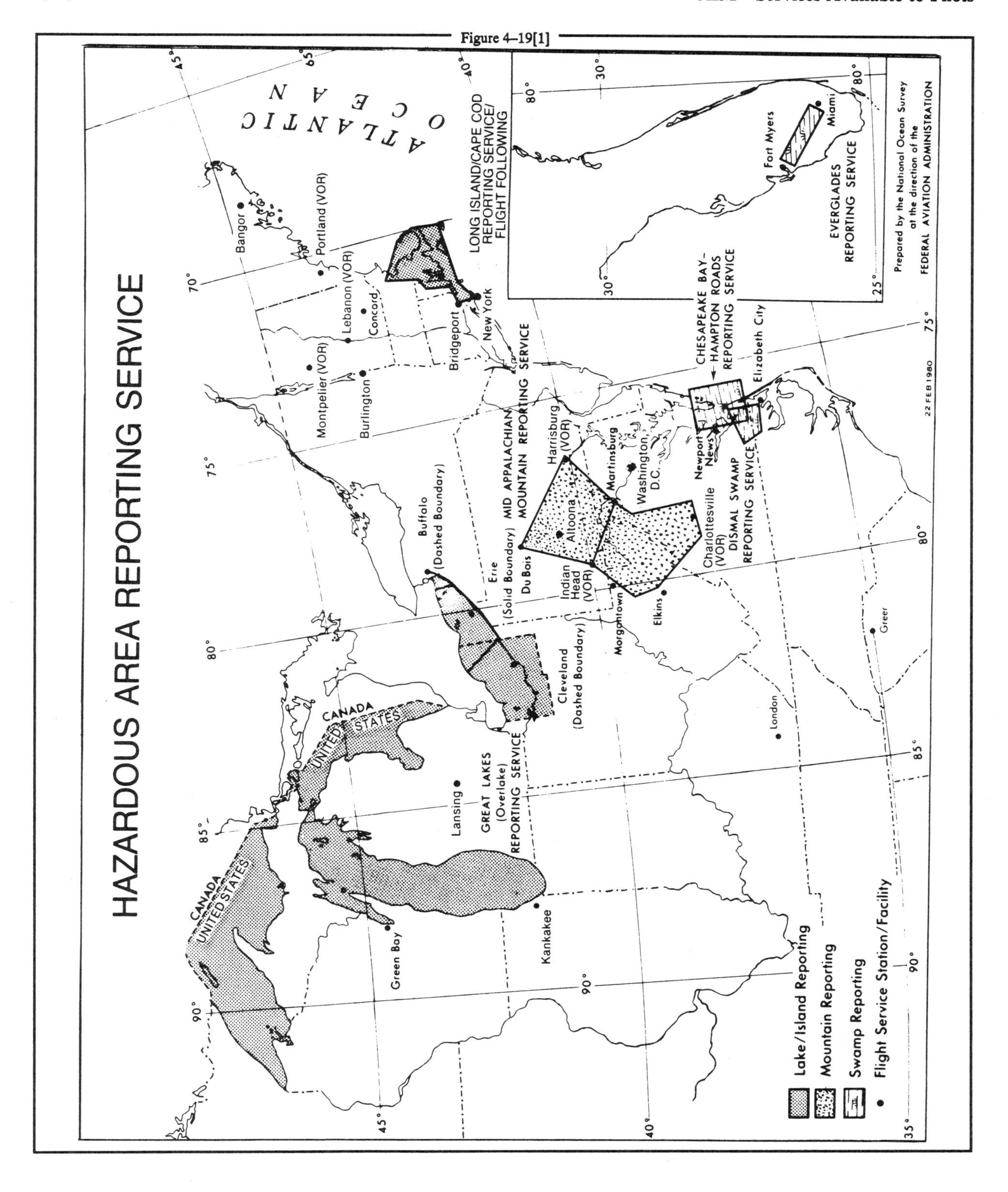
HAZARDOUS AREA REPORTING SERVICE
ATLANTIC OCEAN
LONG ISLAND/CAPE COD REPORTING SERVICE/ FLIGHT FOLLOWING
EVERGLADES REPORTING SERVICE
Fort Myers
Miami
Prepared by the National Ocean Survey at the direction of the FEDERAL AVIATION ADMINISTRATION
CHESAPEAKE BAY– HAMPTON ROADS REPORTING SERVICE
Elizabeth City
Newport News
DISMAL SWAMP REPORTING SERVICE
MID APPALACHIAN MOUNTAIN REPORTING SERVICE
Bangor
Portland (VOR)
Lebanon (VOR)
Concord
Montpelier (VOR)
Burlington
Bridgeport
New York
Harrisburg (VOR)
Martinsburg
Washington, D.C.
Altoona
Du Bois
Indian Head (VOR)
Charlottesville (VOR)
Morgantown
Elkins
Buffalo (Dashed Boundary)
Erie (Solid Boundary)
Cleveland (Dashed Boundary)
GREAT LAKES (Overlake) REPORTING SERVICE
Lansing
Kankakee
Green Bay
London
Greer
CANADA
UNITED STATES
22 FEB 1980
Lake/Island Reporting
Mountain Reporting
Swamp Reporting
Flight Service Station/Facility

same flight information as required for the Long Island Sound Reporting Service.

1. A minimum of three position reports are mandatory for this service; these are:

(a) Reporting leaving either Montauk Point or Block Island.

(b) Midway report.

(c) Report when over either Montauk Point or Block Island. At this time, the overwater service is canceled.

2. COMMUNICATIONS: Pilots are to transmit and receive on 122.6 MHz.

4–19c2 NOTE—Pilots are advised that 122.6 MHz is a remote receiver located at the Hampton VORTAC site and designed to provide radio coverage between Hampton and Block Island. Flights proceeding beyond Block Island may contact the Bridgeport AFSS by transmitting on 122.1 MHz and listening on Groton VOR (TMU) frequency 111.8 MHz.

d. CAPE COD AND ISLANDS RADAR OVERWATER FLIGHT FOLLOWING

In addition to normal VFR radar advisory services, traffic permitting, Cape Approach Control provides a radar overwater flight following service for aircraft traversing the Cape Cod and adjacent Island area. Pilots desiring this service may contact Cape RAPCON on 118.2 MHz.

1. Pilots requesting this service should be prepared to give the following information:

(a) type and color of aircraft,

(b) altitude,

(c) position and heading,

(d) route of flight, and

(e) true airspeed.

2. For best radar coverage, pilots are encouraged to fly at 1,500 feet MSL or above.

3. Pilots are responsible for canceling their request for overwater flight following when they are over the mainland and/or outside the service area boundary.

4–20. SPECIAL TRAFFIC MANAGEMENT PROGRAMS

a. Special procedures may be established when a location requires special traffic handling to accommodate above normal traffic demand (e.g., the Indianapolis 500, the Super Bowl) or reduced airport capacity (e.g., airport runway/taxiway closures for airport construction). The special procedures may remain in effect until the problem has been resolved or until local traffic management procedures can handle the situation and a need for special handling no longer exists.

b. The Airport Reservations Office (ARO) has been established to monitor the operation of the high density rule required by FAR 93, Subpart K. The ARO receives and processes all IFR requests for operations at designated high density traffic airports and allots reservations on a "first come, first serve" basis determined at the time the request is received at the office. Standby lists are not maintained. The toll free number to obtain a slot is 1–800–322–1212. For telephones without Touch-Tone service, the number is (202) 267–5312.

c. The high density airports are: John F. Kennedy International Airport, La Guardia Airport, Chicago O'Hare International Airport, and Washington National Airport.

1. Reservations for John F. Kennedy International Airport are required between 3 p.m. and 7:59 p.m. local time.

2. Reservations at Chicago O'Hare International Airport are required between 6:45 a.m. and 9:15 p.m. local time.

3. Reservations for La Guardia Airport and Washington National Airport are required between 6 a.m. and 11:59 p.m. local time.

d. Requests for IFR reservations will be accepted 48 hours prior to the proposed time of operation at the affected airport. An exception to the 48–hour limitation is made for holidays.

4–21 thru 4–30. RESERVED

Section 2. RADIO COMMUNICATIONS PHRASEOLOGY AND TECHNIQUES

4–31. GENERAL

a. Radio communications are a critical link in the ATC system. The link can be a strong bond between pilot and controller or it can be broken with surprising speed and disastrous results. Discussion herein provides basic procedures for new pilots and also highlights safe operating concepts for all pilots.

b. The single, most important thought in pilot-controller communications is understanding. It is essential, therefore, that pilots acknowledge each radio communication with ATC by using the appropriate aircraft call sign. Brevity is important, and contacts should be kept as brief as possible, but the controller must know what you want to do before he can properly carry out his control duties. And you, the pilot, must know exactly what he wants you to do. Since concise phraseology may not always be adequate, use whatever words are necessary to get your message across. Pilots are to maintain vigilance in monitoring air traffic control radio communications frequencies for potential traffic conflicts with their aircraft especially when operating on an active runway and/or when conducting a final approach to landing.

c. All pilots will find the Pilot/Controller Glossary very helpful in learning what certain words or phrases mean. Good phraseology enhances safety and is the mark of a professional pilot. Jargon, chatter, and "CB" slang have no place in ATC communications. The Pilot/Controller Glossary is the same glossary used in the ATC controller's handbook. We recommend that it be studied and reviewed from time to time to sharpen your communication skills.

4–32. RADIO TECHNIQUE

a. *Listen* before you transmit. Many times you can get the information you want through ATIS or by monitoring the frequency. Except for a few situations where some frequency overlap occurs, if you hear someone else talking, the keying of your transmitter will be futile and you will probably jam their receivers causing them to repeat their call. If you have just changed frequencies, pause, listen, and make sure the frequency is clear.

b. *Think* before keying your transmitter. Know what you want to say and if it is lengthy; e.g., a flight plan or IFR position report, jot it down.

c. The microphone should be very close to your lips and after pressing the mike button, a slight pause may be necessary to be sure the first word is transmitted. Speak in a normal, conversational tone.

d. When you release the button, wait a few seconds before calling again. The controller or FSS specialist may be jotting down your number, looking for your flight plan, transmitting on a different frequency, or selecting his transmitter to your frequency.

e. Be alert to the sounds *or the lack of sounds* in your receiver. Check your volume, recheck your frequency, and *make sure that your microphone is not stuck* in the transmit position. Frequency blockage can, and has, occurred for extended periods of time due to unintentional transmitter operation. This type of interference is commonly referred to as a "stuck mike," and controllers may refer to it in this manner when attempting to assign an alternate frequency. If the assigned frequency is completely blocked by this type of interference, use the procedures described for en route IFR radio frequency outage to establish or reestablish communications with ATC.

f. Be sure that you are within the performance range of your radio equipment and the ground station equipment. Remote radio sites do not always transmit and receive on all of a facility's available frequencies, particularly with regard to VOR sites where you can hear but not reach a ground station's receiver. Remember that higher altitudes increases the range of VHF "line of sight" communications.

4–33. CONTACT PROCEDURES

a. Initial Contact—

1. The terms *initial contact* or *initial callup* means the first radio call you make to a given facility or the first call to a different controller or FSS specialist within a facility. Use the following format:

(a) Name of the facility being called;

(b) Your *full* aircraft identification as filed in the flight plan or as discussed under Aircraft Call Signs below;

(c) The type of message to follow or your request if it is short, and

(d) The word "Over" if required.

EXAMPLE:
"NEW YORK RADIO, MOONEY THREE ONE ONE ECHO."

EXAMPLE:
"COLUMBIA GROUND, CESSNA THREE ONE SIX ZERO FOXTROT, I-F-R MEMPHIS."

EXAMPLE:

"MIAMI CENTER, BARON FIVE SIX THREE HOTEL, REQUEST V-F-R TRAFFIC ADVISORIES."

2. Many FSS's are equipped with RCO's and can transmit on the same frequency at more than one location. The frequencies available at specific locations are indicated on charts above FSS communications boxes. To enable the specialist to utilize the correct transmitter, advise the location and the frequency on which you expect a reply.

EXAMPLE:

St. Louis FSS can transmit on frequency 122.3 at either Farmington, MO, or Decatur, IL. If you are in the vicinity of Decatur, your callup should be "SAINT LOUIS RADIO, PIPER SIX NINER SIX YANKEE, RECEIVING DECATUR ONE TWO TWO POINT THREE."

3. If radio reception is reasonably assured, inclusion of your request, your position or altitude, and the phrase "Have Numbers" or "Information Charlie received" (for ATIS) in the initial contact helps decrease radio frequency congestion. Use discretion, and do not overload the controller with information he does not need. If you do not get a response from the ground station, recheck your radios or use another transmitter, but keep the next contact short.

EXAMPLE:

"ATLANTA CENTER, DUKE FOUR ONE ROMEO, REQUEST V-F-R TRAFFIC ADVISORIES, TWENTY NORTHWEST ROME, SEVEN THOUSAND FIVE HUNDRED, OVER."

b. Initial Contact When your Transmitting and Receiving Frequencies are Different—

1. If you are attempting to establish contact with a ground station and you are receiving on a different frequency than that transmitted, indicate the VOR name or the frequency on which you expect a reply. Most FSS's and control facilities can transmit on several VOR stations in the area. Use the appropriate FSS call-sign as indicated on charts.

EXAMPLE:

New York FSS transmits on the Kennedy, the Hampton, and the Calverton VORTAC's. If you are in the Calverton area, your callup should be "NEW YORK RADIO, CESSNA THREE ONE SIX ZERO FOXTROT, RECEIVING CALVERTON V-O-R, OVER."

2. If the chart indicates FSS frequencies above the VORTAC or in the FSS communications boxes, transmit or receive on those frequencies nearest your location.

3. When unable to establish contact and you wish to call *any* ground station, use the phrase "ANY RADIO (tower) (station), GIVE CESSNA THREE ONE SIX ZERO FOXTROT A CALL ON (frequency) OR (V-O-R)." If an emergency exists or you need assistance, so state.

c. Subsequent Contacts and Responses to Callup from a Ground Facility—

Use the same format as used for the initial contact except you should state your message or request with the callup in one transmission. The ground station name and the word "Over" may be omitted if the message requires an obvious reply and there is no possibility for misunderstandings. *You should acknowledge all callups or clearances* unless the controller or FSS specialist advises otherwise. There are some occasions when the controller must issue time-critical instructions to other aircraft, and he may be in a position to observe your response, either visually or on radar. If the situation demands your response, take appropriate action or immediately advise the facility of any problem. Acknowledge with your aircraft identification and one of the words "Wilco," "Roger," " Affirmative," " Negative," or other appropriate remarks; e.g., "PIPER TWO ONE FOUR LIMA, ROGER." If you have been receiving services; e.g., VFR traffic advisories and you are leaving the area or changing frequencies, advise the ATC facility and terminate contact.

d. Acknowledgement of Frequency Changes—

1. When advised by ATC to change frequencies, acknowledge the instruction. If you select the new frequency without an acknowledgement, the controller's workload is increased because he has no way of knowing whether you received the instruction or have had radio communications failure.

2. At times, a controller/specialist may be working a sector with multiple frequency assignments. In order to eliminate unnecessary verbiage and to free the controller/specialist for higher priority transmissions, the controller/specialist may request the pilot "(Identification), change to my frequency 123.4." This phrase should alert the pilot that he is only changing frequencies, not controller/specialist, and that initial callup phraseology may be abbreviated.

EXAMPLE:

"UNITED TWO TWENTY-TWO ON ONE TWO THREE POINT FOUR."

e. Compliance with Frequency Changes—

When instructed by ATC to change frequencies, select the new frequency as soon as possible unless instructed to make the change at a specific time, fix, or altitude. A delay in making the change could result in an untimely receipt of important information. If you are instructed to make the frequency change at a specific time, fix, or altitude, monitor the frequency you are on until reaching the specified time, fix, or altitudes unless instructed otherwise by ATC. (Reference—ARTCC Communications, paragraph 5–31).

4–34. AIRCRAFT CALL SIGNS

a. Precautions in the Use of Call Signs—

1. Improper use of call signs can result in pilots executing a clearance intended for another aircraft. Call signs should *never be abbreviated on an initial contact or at any time when other aircraft call signs have similar numbers/sounds or identical letters/number;* e.g., Cessna 6132F, Cessna 1622F, Baron 123F, Cherokee 7732F, etc.

EXAMPLE:
Assume that a controller issues an approach clearance to an aircraft at the bottom of a holding stack and an aircraft with a similar call sign (at the top of the stack) acknowledges the clearance with the last two or three numbers of his call sign. If the aircraft at the bottom of the stack did not hear the clearance and intervene, flight safety would be affected, and there would be no reason for either the controller or pilot to suspect that anything is wrong. This kind of "human factors" error can strike swiftly and is extremely difficult to rectify.

2. Pilots, therefore, must be certain that aircraft identification is complete and clearly identified before taking action on an ATC clearance. ATC specialists will not abbreviate call signs of air carrier or other civil aircraft having authorized call signs. ATC specialists may initiate abbreviated call signs of other aircraft by using the *prefix and the last three digits/letters* of the aircraft identification after communications are established. The pilot may use the abbreviated call sign in subsequent contacts with the ATC specialist. When aware of similar/identical call signs, ATC specialists will take action to minimize errors by emphasizing certain numbers/letters, by repeating the entire call sign, by repeating the prefix, or by asking pilots to use a different call sign temporarily. Pilots should use the phrase "VERIFY CLEARANCE FOR (your complete call sign)" if doubt exists concerning proper identity.

3. Civil aircraft pilots should state the aircraft type, model or manufacturer's name, followed by the digits/letters of the registration number. When the aircraft manufacturer's name or model is stated, the prefix "N" is dropped; e.g., Aztec Two Four Six Four Alpha.

EXAMPLE:
BONANZA SIX FIVE FIVE GOLF.

EXAMPLE:
BREEZY SIX ONE THREE ROMEO EXPERIMENTAL (omit "Experimental" after initial contact).

4. Air Taxi or other commercial operators *not* having FAA authorized call signs should prefix their normal identification with the phonetic word "Tango."

EXAMPLE:
TANGO AZTEC TWO FOUR SIX FOUR ALPHA.

5. Air carriers and commuter air carriers having FAA authorized call signs should identify themselves by stating the complete call sign (using group form for the numbers) and the word "heavy" if appropriate.

EXAMPLE:
UNITED TWENTY-FIVE HEAVY.

EXAMPLE:
MIDWEST COMMUTER SEVEN ELEVEN.

6. Military aircraft use a variety of systems including serial numbers, word call signs, and combinations of letters/numbers. Examples include Army Copter 48931, Air Force 61782, REACH 31792, Pat 157, Air Evac 17652, Navy Golf Alfa Kilo 21, Marine 4 Charlie 36, etc.

b. Air Ambulance Flights—

Because of the priority afforded air ambulance flights in the ATC system, extreme discretion is necessary when using the term "LIFEGUARD." It is only intended for those missions of an urgent medical nature and to be utilized only for that portion of the flight requiring expeditious handling. When requested by the pilot, necessary notification to expedite ground handling of patients, etc., is provided by ATC; however, when possible, this information should be passed in advance through nonATC communications systems.

1. Civilian air ambulance flights responding to medical emergencies (first call to an accident scene, carrying patients, organ donors, organs, or other urgently needed lifesaving medical material) will be expedited by ATC when necessary. When expeditious handling is necessary, add the word "LIFEGUARD" in the remarks section of the flight plan. In radio communications, use the call sign "LIFEGUARD" followed by the aircraft registration letters/numbers.

2. Similar provisions have been made for the use of "AIR EVAC" and "MED EVAC" by military air ambulance flights, except that these military flights will receive priority handling only when specifically requested.

EXAMPLE:
LIFEGUARD TWO SIX FOUR SIX.

3. Air carrier and Air Taxi flights responding to medical emergencies will also be expedited by ATC when necessary. The nature of these medical emergency flights usually concerns the transportation of urgently needed lifesaving medical materials or vital organs. IT IS IMPERATIVE THAT THE COMPANY/PILOT DETERMINE, BY THE NATURE/URGENCY OF THE SPECIFIC MEDICAL CARGO, IF PRIORITY ATC ASSISTANCE IS REQUIRED. Pilots shall ensure that the word "LIFEGUARD" is included in the remarks section of the flight plan and use the call sign "LIFEGUARD" followed by the company name and flight number for all transmissions when expedi-

tious handling is required. It is important for ATC to be aware of "LIFEGUARD" status, and it is the pilot's responsibility to ensure that this information is provided to ATC.

EXAMPLE:

LIFEGUARD DELTA THIRTY-SEVEN.

c. Student Pilots Radio Identification—

1. The FAA desires to help the student pilot in acquiring sufficient practical experience in the environment in which he will be required to operate. To receive additional assistance while operating in areas of concentrated air traffic, a student pilot need only identify himself as a student pilot during his initial call to an FAA radio facility.

EXAMPLE:

DAYTON TOWER, THIS IS FLEETWING ONE TWO THREE FOUR, STUDENT PILOT.

2. This special identification will alert FAA ATC personnel and enable them to provide the student pilot with such extra assistance and consideration as he may need. This procedure is not mandatory.

4–35. DESCRIPTION OF INTERCHANGE OR LEASED AIRCRAFT

a. Controllers issue traffic information based on familiarity with airline equipment and color/markings. When an air carrier dispatches a flight using another company's equipment and the pilot does not advise the terminal ATC facility, the possible confusion in aircraft identification can compromise safety.

b. Pilots flying an "interchange" or "leased" aircraft not bearing the colors/markings of the company operating the aircraft should inform the terminal ATC facility on first contact the name of the operating company and trip number, followed by the company name as displayed on the aircraft, and aircraft type.

EXAMPLE:

AIR CAL THREE ELEVEN, UNITED (INTERCHANGE/LEASE), BOEING SEVEN TWO SEVEN.

4–36. GROUND STATION CALL SIGNS

Pilots, when calling a ground station, should begin with the name of the facility being called followed by the type of the facility being called as indicated in the Table 4–36[1].

Table 4–36[1]

Facility	*Call Sign*
Airport UNICOM	"Shannon UNICOM"
FAA Flight Service Station	"Chicago Radio"
FAA Flight Service Station (En Route Flight Advisory Service (Weather))	"Seattle Flight Watch"
Airport Traffic Control Tower	"Augusta Tower"
Clearance Delivery Position (IFR)	"Dallas Clearance Delivery"
Ground Control Position in Tower	"Miami Ground"
Radar or Nonradar Approach Control Position	"Oklahoma City Approach"
Radar Departure Control Position	"St. Louis Departure"
FAA Air Route Traffic Control Center	"Washington Center"

4–37. PHONETIC ALPHABET

The International Civil Aviation Organization (ICAO) phonetic alphabet is used by FAA personnel when communications conditions are such that the information cannot be readily received without their use. ATC facilities may also request pilots to use phonetic letter equivalents when aircraft with similar sounding identifications are receiving communications on the same frequency. Pilots should use the phonetic alphabet when identifying their aircraft during initial contact with air traffic control facilities. Additionally, use the phonetic equivalents for single letters and to spell out groups of letters or difficult words during adverse communications conditions. (See Table 4–37[1].)

Table 4–37[1]

CHARACTER	MORSE CODE	TELEPHONY	PHONIC (PRONUNCIATION)
A	● —	Alfa	(AL-FAH)
B	— ● ● ●	Bravo	(BRAH-VOH)
C	— ● — ●	Charlie	(CHAR-LEE) or (SHAR-LEE)
D	— ● ●	Delta	(DELL-TAH)
E	●	Echo	(ECK-OH)
F	● ● — ●	Foxtrot	(FOKS-TROT)
G	— — ●	Golf	(GOLF)
H	● ● ● ●	Hotel	(HOH-TEL)
I	● ●	India	IN-DEE-AH)
J	● — — —	Juliett	(JEW-LEE-ETT)
K	— ● —	Kilo	(KEY-LOH)
L	● — ● ●	Lima	(LEE-MAH)
M	— —	Mike	(MIKE)
N	— ●	November	(NO-VEM-BER)
O	— — —	Oscar	(OSS-CAH)
P	● — — ●	Papa	(PAH-PAH)
Q	— — ● —	Quebec	(KEH-BECK)
R	● — ●	Romeo	(ROW-ME-OH)
S	● ● ●	Sierra	(SEE-AIR-RAH)
T	—	Tango	(TANG-GO)
U	● ● —	Uniform	(YOU-NEE-FORM) or (OO-NEE-FORM)
V	● ● ● —	Victor	(VIK-TAH)
W	● — —	Whiskey	(WISS-KEY)
X	— ● ● —	Xray	(ECKS-RAY)
Y	— ● — —	Yankee	(YANG-KEY)
Z	— — ● ●	Zulu	(ZOO-LOO)
1	● — — — —	One	(WUN)
2	● ● — — —	Two	(TOO)
3	● ● ● — —	Three	(TREE)
4	● ● ● ● —	Four	(FOW-ER)
5	● ● ● ● ●	Five	(FIFE)
6	— ● ● ● ●	Six	(SIX)
7	— — ● ● ●	Seven	(SEV-EN)
8	— — — ● ●	Eight	(AIT)
9	— — — — ●	Nine	(NIN-ER)
0	— — — — —	Zero	(ZEE-RO)

4–38. FIGURES

a. Figures indicating hundreds and thousands in round number, as for ceiling heights, and upper wind levels up to 9,900 shall be spoken in accordance with the following:

EXAMPLE:

Table 4–38[1]

500	FIVE HUNDRED

EXAMPLE:

Table 4–38[2]

4,500	FOUR THOUSAND FIVE HUNDRED

b. Numbers above 9,900 shall be spoken by separating the digits preceding the word "thousand."

EXAMPLE:

Table 4–38[3]

10,000	ONE ZERO THOUSAND

EXAMPLE:

Table 4–38[4]

13,500	ONE THREE THOUSAND FIVE HUNDRED

c. Transmit airway or jet route numbers as follows:

EXAMPLE:

Table 4–38[5]

V12	VICTOR TWELVE

EXAMPLE:

Table 4–38[6]

J533	J FIVE THIRTY-THREE

d. All other numbers shall be transmitted by pronouncing each digit.

EXAMPLE:

Table 4–38[7]

10	ONE ZERO

e. When a radio frequency contains a decimal point, the decimal point is spoken as "POINT."

EXAMPLE:

Table 4–38[8]

122.1	ONE TWO TWO POINT ONE

4–38e NOTE—ICAO Procedures require the decimal point be spoken as "DECIMAL," and FAA will honor such usage by military aircraft and all other aircraft required to use ICAO Procedures.

4–39. ALTITUDES AND FLIGHT LEVELS

a. Up to but not including 18,000 feet MSL, state the separate digits of the thousands plus the hundreds if appropriate.

EXAMPLE:

Table 4–39[1]

12,000	ONE TWO THOUSAND

EXAMPLE:

Table 4–39[2]

12,500	ONE TWO THOUSAND FIVE HUNDRED

b. At and above 18,000 feet MSL (FL 180), state the words "flight level" followed by the separate digits of the flight level.

EXAMPLE:

Table 4–39[3]

190	FLIGHT LEVEL ONE NINER ZERO

4–40. DIRECTIONS

The three digits of bearing, course, heading, or wind direction should always be magnetic. The word "true" must be added when it applies.

EXAMPLE:

Table 4–40[1]

(Magnetic course) 005	ZERO ZERO FIVE

EXAMPLE:

Table 4–40[2]

(True course) 050	ZERO FIVE ZERO TRUE

EXAMPLE:

Table 4–40[3]

(Magnetic bearing) 360	THREE SIX ZERO

EXAMPLE:

Table 4–40[4]

(Magnetic heading) 100	ONE ZERO ZERO

EXAMPLE:

Table 4–40[5]

(Wind direction) 220	TWO TWO ZERO

4–41. SPEEDS

The separate digits of the speed followed by the word "KNOTS." Except, controllers may omit the word "KNOTS" when using speed adjustment procedures; e.g., "REDUCE/INCREASE SPEED TO TWO FIVE ZERO."

EXAMPLES:

Table 4–41[1]

(Speed) 250	TWO FIVE ZERO KNOTS
(Speed) 190	ONE NINER ZERO KNOTS

The separate digits of the MACH Number preceded by "MACH."

EXAMPLES:

Table 4–41[2]

(Mach number) 1.5	MACH ONE POINT FIVE
(Mach number) 0.64	MACH POINT SIX FOUR
(Mach number) 0.7	MACH POINT SEVEN

4–42. TIME

a. FAA uses Coordinated Universal Time (UTC) for all operations. The term "Zulu" is used when ATC procedures require a reference to UTC.

EXAMPLE:

Table 4–42[1]

0920	ZERO NINER TWO ZERO

b. To Convert from Standard Time to Coordinated Universal Time:

Table 4–42[2]

Eastern Standard Time	Add 5 hours
Central Standard Time	Add 6 hours
Mountain Standard Time	Add 7 hours
Pacific Standard Time	Add 8 hours
Alaska Standard Time	Add 9 hours
Hawaii Standard Time	Add 10 hours

4–42b NOTE—For Daylight Time, subtract 1 hour.

c. The 24–hour clock system is used in radiotelephone transmissions. The hour is indicated by the first two figures and the minutes by the last two figures.

EXAMPLE:

Table 4–42[3]

0000	ZERO ZERO ZERO ZERO

EXAMPLE:

Table 4–42[4]

0920	ZERO NINER TWO ZERO

d. Time may be stated in minutes only (two figures) in radio telephone communications when no misunderstanding is likely to occur.

e. Current time in use at a station is stated in the nearest quarter minute in order that pilots may use this information for time checks. Fractions of a quarter minute less than 8 seconds are stated as the preceding quarter minute; fractions of a quarter minute of 8 seconds or more are stated as the succeeding quarter minute.

EXAMPLE:

Table 4–42[5]

0929:05	TIME, ZERO NINER TWO NINER

EXAMPLE:

Table 4–42[6]

0929:10	TIME, ZERO NINER TWO NINER AND ONE-QUARTER

4–43. COMMUNICATIONS WITH TOWER WHEN AIRCRAFT TRANSMITTER OR RECEIVER OR BOTH ARE INOPERATIVE

a. Arriving Aircraft—

1. Receiver inoperative-If you have reason to believe your receiver is inoperative, remain outside or above the airport traffic area until the direction and flow of traffic has been determined; then, advise the tower of your type aircraft, position, altitude, intention to land, and request that you be controlled with light signals. (Reference—Traffic Control Light Signals, paragraph 4–63). When you are approximately 3 to 5 miles from the airport, advise the tower of your position and join the airport traffic pattern. From this point on, watch the tower for light signals. Thereafter, if a complete pattern is made, transmit your position downwind and/or turning base leg.

2. Transmitter inoperative—Remain outside or above the airport traffic area until the direction and flow of traffic has been determined; then, join the airport traffic pattern. Monitor the primary local control frequency as depicted on Sectional Charts for landing or traffic information, and look for a light signal which may be addressed to your aircraft. During hours of daylight, acknowledge tower transmissions or light signals by rocking your wings. At night, acknowledge by blinking the landing or navigation lights. To acknowledge tower transmissions during daylight hours, hovering helicopters will turn in the direction of the controlling facility and flash the landing light. While in flight, helicopters should show their acknowledgement of receiving a transmission by making shallow banks in opposite directions. At night, helicopters will acknowledge receipt of transmissions by flashing either the landing or the search light.

3. Transmitter and receiver inoperative—Remain outside or above the airport traffic area until the direction and flow of traffic has been determined; then, join the airport traffic pattern and maintain visual contact with the tower to receive light signals. Acknowledge light signals as noted above.

b. Departing Aircraft—If you experience radio failure prior to leaving the parking area, make every effort to have the equipment repaired. If you are unable to have the malfunction repaired, call the tower by telephone and request authorization to depart without two-way radio communications. If tower authorization is granted, you will be given departure information and requested to monitor the tower frequency or watch for light signals as appropriate. During daylight hours, acknowledge tower transmissions or light signals by moving the ailerons or rudder. At night, acknowledge by blinking the landing or navigation lights. If radio malfunction occurs after departing the parking area, watch the tower for light signals or monitor tower frequency.

4–43b NOTE—Refer to FAR 91.129 and FAR 91.125.

4–44. COMMUNICATIONS FOR VFR FLIGHTS

a. FSS's are allocated frequencies for different functions; for example, 122.0 MHz is assigned as the En Route Flight Advisory Service frequency at selected FSS's. In addition, certain FSS's provide Airport Advisory Service on 123.6 MHz. Other FSS frequencies are listed in the Airport/Facility Directory. If you are in doubt as to what frequency to use, 122.2 MHz is assigned to the majority of FSS's as a common en route simplex frequency.

4–44a NOTE—1In order to expedite communications, state the frequency being used and the aircraft location during initial callup.

EXAMPLE:

DAYTON RADIO, THIS IS NOVEMBER ONE TWO THREE FOUR FIVE ON ONE TWO TWO POINT TWO, OVER SPRINGFIELD V-O-R, OVER.

b. Certain VOR voice channels are being utilized for recorded broadcasts; i.e., ATIS, HIWAS, etc. These services and appropriate frequencies are listed in the Airport/Facility Directory. On VFR flights, pilots are urged to monitor these frequencies. When in contact with a control facility, notify the controller if you plan to leave the frequency to monitor these broadcasts.

4–45 thru 4–50. RESERVED

Section 3. AIRPORT OPERATIONS

4–51. GENERAL

Increased traffic congestion, aircraft in climb and descent attitudes, and pilots preoccupation with cockpit duties are some factors that increase the hazardous accident potential near the airport. The situation is further compounded when the weather is marginal-that is, just meeting VFR requirements. Pilots must be particularly alert when operating in the vicinity of an airport. This section defines some rules, practices, and procedures that pilots should be familiar with and adhere to for safe airport operations.

4–52. TOWER CONTROLLED AIRPORTS

a. When operating at an airport where traffic control is being exercised by a control tower, pilots are required to maintain two-way radio contact with the tower while operating within the Airport Traffic Area unless the tower authorizes otherwise. Initial callup should be made about 15 miles from the airport. Unless there is a good reason to leave the tower frequency before exiting the airport traffic area, it is a good operating practice to remain on the tower frequency for the purpose of receiving traffic information. In the interest of reducing tower frequency congestion, pilots are reminded that it is not necessary to request permission to leave the tower frequency once outside of the airport traffic area.

b. When necessary, the tower controller will issue clearances or other information for aircraft to generally follow the desired flight path (traffic patterns) when flying in the airport traffic area/Control Zone and the proper taxi routes when operating on the ground. If not otherwise authorized or directed by the tower, pilots of fixed-wing aircraft approaching to land must circle the airport to the left. Pilots approaching to land in a helicopter must avoid the flow of fixed-wing traffic. However, in all instances, an appropriate clearance must be received from the tower before landing.

Figure 4–52[1]

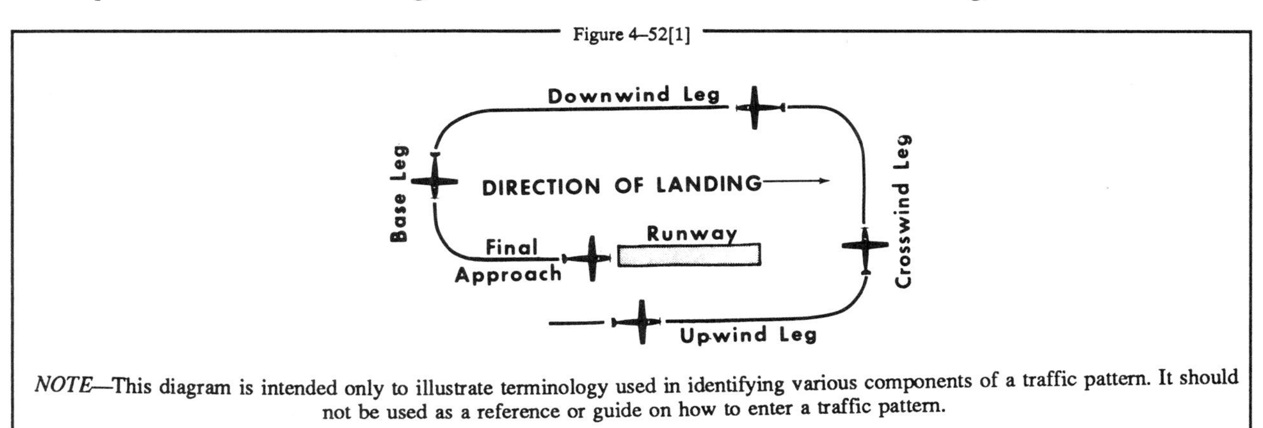

NOTE—This diagram is intended only to illustrate terminology used in identifying various components of a traffic pattern. It should not be used as a reference or guide on how to enter a traffic pattern.

NOTE—This diagram is intended only to illustrate terminology used in identifying various components of a traffic pattern.

It should not be used as a reference or guide on how to enter a traffic pattern.

c. The following terminology for the various components of a traffic pattern has been adopted as standard for use by control towers and pilots (See Figure 4–52[1]):

1. Upwind leg—A flight path parallel to the landing runway in the direction of landing.

2. Crosswind leg—A flight path at right angles to the landing runway off its takeoff end.

3. Downwind leg—A flight path parallel to the landing runway in the opposite direction of landing.

4. Base leg—A flight path at right angles to the landing runway off its approach end and extending from the downwind leg to the intersection of the extended runway centerline.

5. Final approach—A flight path in the direction of landing along the extended runway centerline from the base leg to the runway.

d. Many towers are equipped with a tower radar display. The radar uses are intended to enhance the effectiveness and efficiency of the local control, or tower, position. They are not intended to provide radar services or benefits to pilots except as they may accrue through a more efficient tower operation. The four basic uses are:

1. To determine an aircraft's exact location—This is accomplished by radar identifying the VFR aircraft through any of the techniques available to a radar position, such as having the aircraft *squawk ident.* Once identified, the aircraft's position and spatial relationship to other aircraft can be quickly determined, and standard instructions regarding VFR operation in the Airport Traffic Area will be issued. Once initial radar identification of a VFR aircraft has been established and the appropriate instructions have been issued, radar monitoring may be discontinued; the reason being that the local controller's primary means of surveillance in VFR conditions is visually scanning the airport and local area.

2. To provide radar traffic advisories—Radar traffic advisories may be provided to the extent that the local controller is able to monitor the radar display. Local control has primary control responsibilities to the aircraft operating on the runways, which will normally supersede radar monitoring duties.

3. To provide a direction or suggested heading—The local controller may provide pilots flying VFR with generalized instructions which will facilitate operations; e.g.,"PROCEED SOUTHWEST BOUND, ENTER A RIGHT DOWNWIND RUNWAY THREE ZERO," or provide a suggested heading to establish radar identification or as an advisory aid to navigation; e.g., "SUGGESTED HEADING TWO TWO ZERO, FOR RADAR IDENTIFICATION." In both cases, the instructions are advisory aids to the pilot flying VFR and are not radar vectors. PILOTS HAVE COMPLETE DISCRETION REGARDING ACCEPTANCE OF THE SUGGESTED HEADINGS OR DIRECTIONS AND HAVE SOLE RESPONSIBILITY FOR SEEING AND AVOIDING OTHER AIRCRAFT.

4. To provide information and instructions to aircraft operating within the Airport Traffic Area—In an example of this situation, the local controller would use the radar to advise a pilot on an extended downwind when to turn base leg.

4–52d4 NOTE—The above tower radar applications are intended to augment the standard functions of the local control position. There is no controller requirement to maintain constant radar identification. In fact, such a requirement could compromise the local controller's ability to visually scan the airport and local area to meet FAA responsibilities to the aircraft operating on the runways and within the airport traffic area. Normally, pilots will not be advised of being in radar contact since that continued status cannot be guaranteed and since the purpose of the radar identification is not to establish a link for the provision of radar services.

e. A few of the radar equipped towers are authorized to use the radar to ensure separation between aircraft in specific situations, while still others may function as limited radar approach controls. The various radar uses are strictly a function of FAA operational need. The facilities may be indistinguishable to pilots since they are all referred to as tower and no publication lists the degree of radar use. Therefore, WHEN IN COMMUNICATION WITH A TOWER CONTROLLER WHO MAY HAVE RADAR AVAILABLE, DO NOT ASSUME THAT CONSTANT RADAR MONITORING AND COMPLETE ATC RADAR SERVICES ARE BEING PROVIDED.

4–53. VISUAL INDICATORS AT UNCONTROLLED AIRPORTS

a. At those airports *without an operating control tower,* a segmented circle visual indicator system, if installed, is designed to provide traffic pattern information. (Reference—Traffic Advisory Practices at Airports Where a Tower is Not in Operation, paragraph 4–8). The segmented circle system consists of the following components:

1. The segmented circle—Located in a position affording maximum visibility to pilots in the air and on the ground and providing a centralized location for other elements of the system.

2. The wind direction indicator—A wind cone, wind sock, or wind tee installed near the operational runway to indicate wind direction. The large end of the wind cone/wind sock points into the wind as does the large end (cross bar) of the wind tee. In lieu of a tetrahedron and where a wind sock or wind cone is collocated with a wind tee, the wind tee may be manually aligned with the runway in use to indicate landing direction. These signaling devices may be located in the center of the segmented circle and may be lighted for night use. Pilots are cautioned against using a tetrahedron to indicate wind direction.

3. The landing direction indicator—A tetrahedron is installed when conditions at the airport warrant its use. It may be used to indicate the direction of landings and takeoffs. A tetrahedron may be located at the center of a segmented circle and may be lighted for night operations. The small end of the tetrahedron points in the direction of landing. Pilots are cautioned against using a tetrahedron for any purpose other than as an indicator of landing direction. Further, pilots should use extreme caution when making runway selection by use of a tetrahedron in very light or calm wind conditions as the tetrahedron may not be aligned with the designated calm-wind runway. At airports with control towers, the tetrahedron should only be referenced when the control tower is not in operation. Tower instructions supersede tetrahedron indications.

4. Landing strip indicators—Installed in pairs as shown in the segmented circle diagram and used to show the alignment of landing strips.

5. Traffic pattern indicators—Arranged in pairs in conjunction with landing strip indicators and used to indicate the direction of turns when there is a variation from the normal left traffic pattern. (If there is no segmented circle installed at the airport, traffic pattern indicators may be installed on or near the end of the runway.)

b. Preparatory to landing at an airport without a control tower, or when the control tower is not in operation, the pilot should concern himself with the indicator for the approach end of the runway to be used. When approaching for landing, all turns must be made to the left unless a traffic pattern indicator indicates that turns should be made to the right. If the pilot will mentally enlarge the indicator for the runway to be used, the base and final approach legs of the traffic pattern to be flown immediately become apparent. Similar treatment of the indicator at the departure end of the runway will clearly indicate the direction of turn after takeoff.

c. When two or more aircraft are approaching an airport for the purpose of landing, the aircraft at the lower altitude has the right of way, but it shall not take advantage of this rule to cut in front of another which is on final approach to land, or to overtake that aircraft (FAR 91.113(f)).

4–54. TRAFFIC PATTERNS

At most airports and military air bases, traffic pattern altitudes for propeller-driven aircraft generally extend from 600 feet to as high as 1,500 feet above the ground. Also, traffic pattern altitudes for military turbojet aircraft sometimes extend up to 2,500 feet above the ground. Therefore, pilots of en route aircraft should be constantly on the alert for other aircraft in traffic patterns and avoid these areas whenever possible. Traffic pattern altitudes should be maintained unless otherwise required by the applicable distance from cloud criteria (FAR 91.155). (See Figure 4–54[1] and Figure 4–54[2].)

Figure 4–54[1]

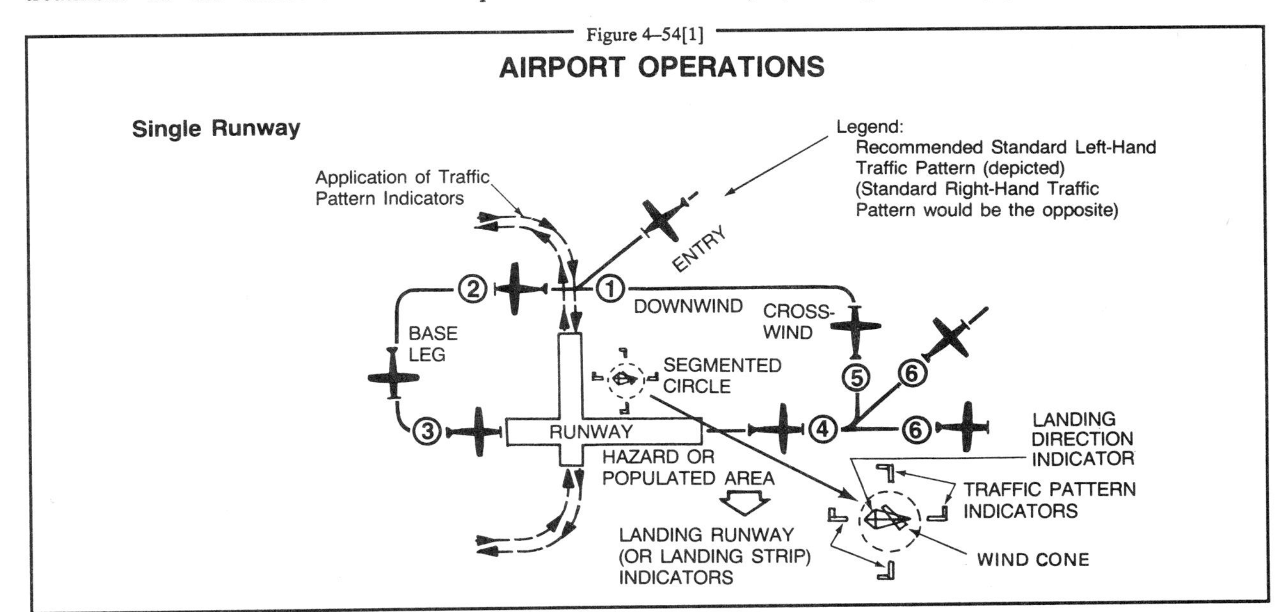

Figure 4–54[2]

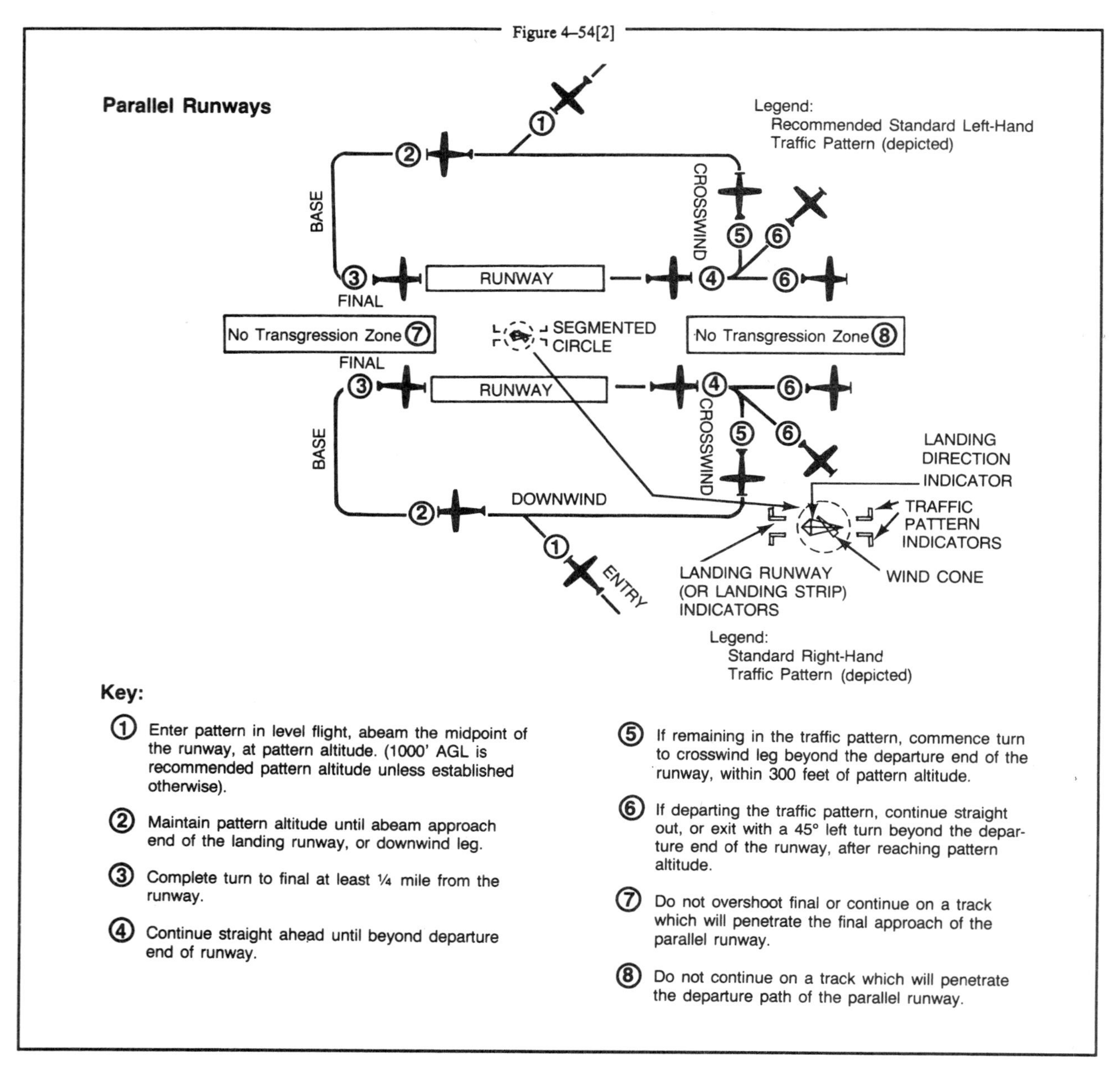

4–55. UNEXPECTED MANEUVERS IN THE AIRPORT TRAFFIC PATTERN

There have been several incidents in the vicinity of controlled airports that were caused primarily by aircraft executing unexpected maneuvers. ATC service is based upon observed or known traffic and airport conditions. Controllers establish the sequence of arriving and departing aircraft by requiring them to adjust flight as necessary to achieve proper spacing. These adjustments can only be based on observed traffic, accurate pilot reports, and anticipated aircraft maneuvers. Pilots are expected to cooperate so as to preclude disrupting traffic flows or creating conflicting patterns. The pilot-in-command of an aircraft is directly responsible for and is the final authority as to the operation of his aircraft. On occasion it may be necessary for a pilot to maneuver his aircraft to maintain spacing with the traffic he has been sequenced to follow. The controller can anticipate minor maneuvering such as shallow ''S'' turns. The controller cannot, however, anticipate a major maneuver such as a 360 degree turn. If a pilot makes a 360 degree turn after he has obtained a landing sequence, the result is usually a gap in the landing interval and, more importantly, it causes a chain reaction which may result in a conflict with following traffic and an interruption of the sequence established by the tower or approach controller. Should a pilot decide he needs to make maneuvering turns to maintain spacing behind a preceding aircraft, he should always advise the controller if at all possible. Except when requested by the controller or in emergency situations, a 360 degree turn should never be executed in the traffic pattern or when receiving radar service without first advising the controller.

4–56. USE OF RUNWAYS/DECLARED DISTANCES

a. Runways are identified by numbers which indicate the nearest 10–degree increment of the azimuth of the runway centerline. For example, where the magnetic azimuth is 183 degrees, the runway designation would be 18; for a magnetic azimuth of 87 degrees, the runway designation would be 9. For a magnetic azimuth ending in the number 5, such as 185, the runway designation could be either 18 or 19. Wind direction issued by the tower is also magnetic and wind velocity is in knots.

b. Airport proprietors are responsible for taking the lead in local aviation noise control. Accordingly, they may propose specific noise abatement plans to the FAA. If approved, these plans are applied in the form of Formal or Informal Runway Use Programs for noise abatement purposes. (Reference—Pilot/Controller Glossary, Runway Use Program).

1. At airports where no runway use program is established, ATC clearances may specify:

(a) The runway most nearly aligned with the wind when it is 5 knots or more;

(b) The "calm wind" runway when wind is less than 5 knots, or;

(c) Another runway if operationally advantageous.

4–56b1c NOTE—It is not necessary for a controller to specifically inquire if the pilot will use a specific runway or to offer him a choice of runways. If a pilot prefers to use a different runway from that specified or the one most nearly aligned with the wind, he is expected to inform ATC accordingly.

2. At airports where a runway use program is established, ATC will assign runways deemed to have the least noise impact. If in the interest of safety a runway different from that specified is preferred, the pilot is expected to advise ATC accordingly. ATC will honor such requests and advise pilots when the requested runway is noise sensitive. When use of a runway other than the one assigned is requested, pilot cooperation is encouraged to preclude disruption of traffic flows or the creation of conflicting patterns.

c. At some airports, the airport proprietor may declare that sections of a runway at one or both ends are not available for landing or takeoff. For these airports, the declared distance of runway length available for a particular operation is published in the Airport/Facility Directory. Declared distances (TORA, TODA, ASDA, and LDA) are defined in the Pilot/Controller Glossary. These distances are calculated by adding to the full length of paved runway any applicable clearway or stopway and subtracting from that sum the sections of the runway unsuitable for satisfying the required takeoff run, takeoff, accelerate/stop, or landing distance.

4–57. LOW LEVEL WIND SHEAR ALERT SYSTEM (LLWAS)

a. This computerized system detects the presence of a possible hazardous low-level wind shear by continuously comparing the winds measured by sensors installed around the periphery of an airport with the wind measured at the center field location. If the difference between the center field wind sensor and a peripheral wind sensor becomes excessive, a thunderstorm or thunderstorm gust front wind shear is probable. When this condition exists, the tower controller will provide arrival and departure aircraft with an advisory of the situation, which includes the center field wind plus the remote site location and wind.

b. Since the sensors are not all associated with specific runways, descriptions of the remote sites will be based on an eight-point compass system.

EXAMPLE:

DELTA ONE TWENTY FOUR CENTER FIELD WIND TWO SEVEN ZERO AT ONE ZERO. SOUTH BOUNDARY WIND ONE FOUR ZERO AT THREE ZERO.

c. An airport equipped with the Low Level Wind Shear Alert System is so indicated in the Airport/Facility Directory under *Weather Data Sources* for that particular airport.

4–58. BRAKING ACTION REPORTS AND ADVISORIES

a. When available, ATC furnishes pilots the quality of braking action received from pilots or airport management. The quality of braking action is described by the terms "good," "fair," "poor," and "nil," or a combination of these terms. When pilots report the quality of braking action by using the terms noted above, they should use descriptive terms that are easily understood, such as, "braking action poor the first/last half of the runway," together with the particular type of aircraft.

b. For NOTAM purposes, braking action reports are classified according to the most critical term used. Reports containing the terms "good" or "fair" are classified as NOTAM(L). Reports containing the terms "poor" or "nil" are classified as NOTAM(D).

c. When tower controllers have received runway braking action reports which include the terms *poor* or *nil,* or whenever weather conditions are conducive to deteriorating or rapidly changing runway braking conditions, the tower will include on the ATIS broadcast the statement, *"BRAKING ACTION ADVISORIES ARE IN EFFECT."*

d. During the time that braking action advisories are in effect, ATC will issue the latest braking action report for the runway in use to each arriving and departing aircraft. Pilots should be prepared for deteriorating braking conditions and should request current runway condition information if not volunteered by controllers. Pilots should also be prepared to provide a descriptive runway condition report to controllers after landing.

4–59. RUNWAY FRICTION REPORTS AND ADVISORIES

a. Friction is defined as the ratio of the tangential force needed to maintain uniform relative motion between two contacting surfaces (aircraft tires to the pavement surface) to the perpendicular force holding them in contact (distributed aircraft weight to the aircraft tire area). Simply stated, friction quantifies slipperiness of pavement surfaces.

b. The greek letter MU (pronounced "myew"), is used to designate a friction value representing runway surface conditions.

c. MU (friction) values range from 0 to 100 where zero is the lowest friction value and 100 is the maximum friction value obtainable. For frozen contaminants on runway surfaces, a MU value of 40 or less is the level when the aircraft braking performance starts to deteriorate and directional control begins to be less responsive. The lower the MU value, the less effective braking performance becomes and the more difficult directional control becomes.

d. At airports with friction measuring devices, airport management should conduct friction measurements on runways covered with compacted snow and/or ice.

1. Numerical readings may be obtained by using any FAA approved friction measuring device. It is not necessary to designate the type of friction measuring device since they provide essentially the same numerical reading when the values are 40 or less.

2. When the MU value for any one–third zone of an active runway is 40 or less, a report should be given to ATC by airport management for dissemination to pilots. The report will identify the runway, the time of measurement, MU values for each zone, and the contaminant conditions, e.g., wet snow, dry snow, slush, deicing chemicals, etc. Measurements for each one–third zone will be given in the direction of takeoff and landing on the runway. A report should also be given when MU values rise above 40 in all zones of a runway previously reporting a MU below 40.

3. Airport management should initiate a NOTAM(D) when the friction measuring device is out of service.

e. When MU reports are provided by airport management, the ATC facility providing approach control or local airport advisory will provide the report to any pilot upon request.

f. Pilots should use MU information with other knowledge including aircraft performance characteristics, type, and weight, previous experience, wind conditions, and aircraft tire type (i.e., bias ply vs. radial constructed) to determine runway suitability.

g. No correlation has been established between MU values and the descriptive terms "good," "fair," "poor," and "nil" used in braking action reports.

4–60. INTERSECTION TAKEOFFS

a. In order to enhance airport capacities, reduce taxiing distances, minimize departure delays, and provide for more efficient movement of air traffic, controllers may initiate intersection takeoffs as well as approve them when the pilot requests. If for ANY reason a pilot prefers to use a different intersection or the full length of the runway or desires to obtain the distance between the intersection and the runway end, HE IS EXPECTED TO INFORM ATC ACCORDINGLY.

b. An aircraft is expected to taxi to (but not onto) the end of the assigned runway unless prior approval for an intersection departure is received from ground control.

c. Pilots should state their position on the airport when calling the tower for takeoff from a runway intersection.

EXAMPLE:

CLEVELAND TOWER, APPACHE 3722P, AT THE INTERSECTION OF TAXIWAY OSCAR AND RUNWAY TWO THREE RIGHT, READY FOR DEPARTURE.

d. Controllers are required to separate small aircraft (12,500 pounds or less, maximum certificated takeoff weight) departing (same or opposite direction) from an intersection behind a large nonheavy aircraft on the same runway, by ensuring that at least a 3–minute interval exists between the time the preceding large aircraft has taken off and the succeeding small aircraft begins takeoff roll. To inform the pilot of the required 3–minute hold, the controller will state, "Hold for wake turbulence." If after considering wake turbulence hazards, the pilot feels that a lesser time interval is appropriate, he may request a waiver to the 3–minute interval. Pilots must initiate such a request by stating, "Request waiver to 3–minute interval," or by making a similar statement. Controllers may then issue a takeoff clearance if other traffic permits, since the pilot has accepted responsibility for his own wake turbulence separation.

e. The 3–minute interval is not required when the intersection is 500 feet or less from the departure point of the preceding aircraft and both aircraft are taking off in the same direction. Controllers may permit the small aircraft to alter course after takeoff to avoid the flight path of the preceding departure.

f. The 3–minute interval is mandatory behind a heavy aircraft in all cases.

4–61. SIMULTANEOUS OPERATIONS ON INTERSECTING RUNWAYS

a. Despite the many new and lengthened runways which have been added to the nation's airports in recent years, limited runway availability remains a major contributing factor to operational delays. Many high-density airports have gained operational experience with intersecting runways which clearly indicates that simultaneous operations are safe and feasible. Tower controllers may authorize simultaneous landings or a simultaneous landing and takeoff on intersecting runways when the following conditions are met:

1. The runways are dry and the controller has received no reports that braking action is less than good.

2. A simultaneous takeoff and landing operation may be conducted only in VFR conditions.

3. Instructions are issued to restrict one aircraft from entering the intersecting runway being used by another aircraft.

4. Traffic information issued is acknowledged by the pilots of both aircraft.

5. The measured distance from runway threshold to intersection is issued if the pilot requests it.

6. The conditions specified in 3, 4 and 5 are met at or before issuance of the landing clearance.

7. The distance from landing threshold to the intersection is adequate for the category of aircraft being held short. Controllers are provided a general table of aircraft category/minimum runway length requirements as a guide. Operators of STOL aircraft should identify their aircraft as such on initial contact with the tower, unless a Letter of Agreement concerning this fact, is in effect. WHENEVER A HOLD SHORT CLEARANCE IS RECEIVED, IT IS INCUMBENT ON THE PILOT TO DETERMINE HIS/HER ABILITY TO HOLD SHORT OF AN INTERSECTION AFTER LANDING WHEN INSTRUCTED TO DO SO. ADDITIONALLY, PILOTS SHOULD INCLUDE THE WORDS "HOLD SHORT OF (POINT)" IN THE ACKNOWLEDGEMENT OF SUCH CLEARANCES.

8. There is no tailwind for the landing aircraft restricted to hold short of the intersection.

b. THE SAFETY AND OPERATION OF AN AIRCRAFT REMAIN THE RESPONSIBILITY OF THE PILOT. IF FOR ANY REASON; e.g. DIFFICULTY IN DISCERNING LOCATION OF AN INTERSECTION AT NIGHT, INABILITY TO HOLD SHORT OF AN INTERSECTION, WIND FACTORS, ETC., A PILOT ELECTS TO USE THE FULL LENGTH OF THE RUNWAY, A DIFFERENT RUNWAY OR DESIRES TO OBTAIN THE DISTANCE FROM THE LANDING THRESHOLD TO THE INTERSECTION, HE IS EXPECTED TO PROMPTLY INFORM ATC ACCORDINGLY.

4–62. LOW APPROACH

a. A low approach (sometimes referred to as a low pass) is the go-around maneuver following an approach. Instead of landing or making a touch-and-go, a pilot may wish to go around (low approach) in order to expedite a particular operation (a series of practice instrument approaches is an example of such an operation). Unless otherwise authorized by ATC, the low approach should be made straight ahead, with no turns or climb made until the pilot has made a thorough visual check for other aircraft in the area.

b. When operating within an Airport Traffic Area, a pilot intending to make a low approach should contact the tower for approval. This request should be made prior to starting the final approach.

c. When operating to an airport, not within an airport traffic area, a pilot intending to make a low approach should, prior to leaving the final approach fix inbound (non precision approach) or the outer marker or fix used in lieu of the outer marker inbound (precision approach), so advise the FSS, UNICOM, or make a broadcast as appropriate. (Reference—Traffic Advisory Practices at

Airports Without Operating Control Towers, paragraph 4–8).

4–63. TRAFFIC CONTROL LIGHT SIGNALS

a. The following procedures are used by ATCTs in the control of aircraft, ground vehicles, equipment, and personnel not equipped with radio. These same procedures will be used to control aircraft, ground vehicles, equipment, and personnel equipped with radio if radio contact cannot be established. ATC personnel use a directive traffic control signal which emits an intense narrow light beam of a selected color (either red, white, or green) when controlling traffic by light signals.

b. Although the traffic signal light offers the advantage that some control may be exercised over nonradio equipped aircraft, pilots should be cognizant of the disadvantages which are:

1. The pilot may not be looking at the control tower at the time a signal is directed toward him.

2. The directions transmitted by a light signal are very limited since only approval or disapproval of a pilot's anticipated actions may be transmitted. No supplement or explanatory information may be transmitted except by the use of the "General Warning Signal" which advises the pilot to be on the alert.

c. Between sunset and sunrise, a pilot wishing to attract the attention of the control tower should turn on a landing light and taxi the aircraft into a position, clear of the active runway, so that light is visible to the tower. The landing light should remain on until appropriate signals are received from the tower.

d. Air Traffic Control Tower Light Gun Signals: (See Table 4–63[1].)

Table 4–63[1]

ATCT Light Gun Signals

COLOR AND TYPE OF SIGNAL	MEANING		
	MOVEMENT OF VEHICLES EQUIPMENT AND PERSONNEL	AIRCRAFT ON THE GROUND	AIRCRAFT IN FLIGHT
Steady green	Cleared to cross, proceed or go	Cleared for takeoff	Cleared to land
Flashing green	Not applicable	Cleared for taxi	Return for landing (to be followed by steady green at the proper time)
Steady red	STOP	STOP	Give way to other aircraft and continue circling
Flashing red	Clear the taxiway/runway	Taxi clear of the runway in use	Airport unsafe, do not land
Flashing white	Return to starting point on airport	Return to starting point on airport	Not applicable
Alternating red and green	Exercise extreme caution	Exercise extreme caution	Exercise extreme caution

e. During daylight hours, acknowledge tower transmissions or light signals by moving the ailerons or rudder. At night, acknowledge by blinking the landing or navigation lights. If radio malfunction occurs after departing the parking area, watch the tower for light signals or monitor tower frequency.

4–64. COMMUNICATIONS

a. Pilots of departing aircraft should communicate with the control tower on the appropriate ground control/clearance delivery frequency prior to starting engines to receive engine start time, taxi and/or clearance information. Unless otherwise advised by the tower, remain on that frequency during taxiing and runup, then change to local control frequency when ready to request takeoff clearance. (Reference—Automatic Terminal Information Service (ATIS) for continuous broadcast of terminal information paragraph 4–12).

b. The majority of ground control frequencies are in the 121.6–121.9 MHz bandwidth. Ground control frequencies are provided to eliminate frequency congestion on the tower (local control) frequency and are limited to communications between the tower and aircraft on the ground and between the tower and utility vehicles on the airport, provide a clear VHF channel for arriving and departing aircraft. They are used for issuance of taxi information, clearances, and other necessary contacts between the tower and aircraft or other vehicles operated on the airport. A pilot who has just landed should not change from the tower frequency to the ground control frequency until he is directed to do so by the controller. Normally, only one ground control frequency is assigned at an airport; however, at locations where the amount of traffic so warrants, a second ground control frequency and/or another frequency designated as a clearance delivery frequency, may be assigned.

c. A controller may omit the ground or local control frequency if the controller believes the pilot knows which frequency is in use. If the ground control frequency is in the 121 MHz bandwidth the controller may omit the numbers preceding the decimal point; e.g., 121.7, "CONTACT GROUND POINT SEVEN." However, if any doubt exists as to what frequency is in use, the pilot should promptly request the controller to provide that information.

d. Controllers will normally avoid issuing a radio frequency change to helicopters, known to be single-piloted, which are hovering, air taxiing, or flying near the ground. At times, it may be necessary for pilots to alert ATC regarding single pilot operations to minimize delay of essential ATC communications. Whenever possible, ATC instructions will be relayed through the frequency being monitored until a frequency change can be accomplished. You must promptly advise ATC if you are unable to comply with a frequency change. Also, you should advise ATC if you must land to accomplish the frequency change unless it is clear the landing; e.g., on a taxiway or in a helicopter operating area, will have no impact on other air traffic.

4–65. GATE HOLDING DUE TO DEPARTURE DELAYS

a. Pilots should contact ground control or clearance delivery prior to starting engines as gate hold procedures will be in effect whenever departure delays exceed or are anticipated to exceed 15 minutes. The sequence for departure will be maintained in accordance with initial call up unless modified by flow control restrictions. Pilots should monitor the ground control or clearance delivery frequency for engine startup advisories or new proposed start time if the delay changes.

b. The tower controller will consider that pilots of turbine powered aircraft are ready for takeoff when they reach the runway or warm-up block unless advised otherwise.

4–66. VFR FLIGHTS IN TERMINAL AREAS

Use reasonable restraint in exercising the prerogative of VFR flight, especially in terminal areas. The weather minimums and distances from clouds are minimums. Giving yourself a greater margin in specific instances is just good judgment.

a. Approach Area—Conducting a VFR operation in a Control Zone when the official visibility is 3 or 4 miles is not prohibited, but good judgment would dictate that you keep out of the approach area.

b. Reduced Visibility—It has always been recognized that precipitation reduces forward visibility. Consequently, although again it may be perfectly legal to cancel your IFR flight plan at any time you can proceed VFR, it is good practice, when precipitation is occurring, to continue IFR operation into a terminal area until you are reasonably close to your destination.

c. Simulated Instrument Flights—In conducting simulated instrument flights, be sure that the weather is good enough to compensate for the restricted visibility of the safety pilot and your greater concentration on your flight instruments. Give yourself a little greater margin when your flight plan lies in or near a busy airway or close to an airport.

4–67. VFR HELICOPTER OPERATIONS AT CONTROLLED AIRPORTS

a. General—

1. The following ATC procedures and phraseologies recognize the unique capabilities of helicopters and were developed to improve service to all users. Helicopter design characteristics and user needs often require operations from movement and nonmovement areas within the airport boundary. In order for ATC to properly apply these procedures, it is essential that pilots familiarize themselves with the local operations and make it known to controllers when additional instructions are necessary.

2. Insofar as possible, helicopter operations will be instructed to avoid the flow of fixed-wing aircraft to minimize overall delays; however, there will be many situations where faster/larger helicopters may be integrated with fixed-wing aircraft for the benefit of all concerned. Examples would include IFR flights, avoidance of noise sensitive areas, or use of runways/taxiways to minimize the hazardous effects of rotor downwash in congested areas.

3. Because helicopter pilots are intimately familiar with the effects of rotor downwash, they are best qualified to determine if a given operation can be conducted safely. Accordingly, the pilot has the final authority with respect to the specific airspeed/altitude combinations. ATC clearances are in no way intended to place the helicopter in a hazardous position. It is expected that pilots will advise ATC if a specific clearance will cause undue hazards to persons or property.

b. Controllers normally limit ATC ground service and instruction to *movement* areas; therefore, operations from *nonmovement* areas are conducted at pilot discretion and should be based on local policies, procedures, or letters of agreement. In order to maximize the flexibility of helicopter operations, it is necessary to rely heavily on sound pilot judgment. For example, hazards such as debris, obstructions, vehicles, or personnel must be recognized by the pilot, and action should be taken as necessary to avoid such hazards. Taxi, hover taxi, and air taxi operations are considered to be ground movements. Helicopters conducting such operations are expected to adhere to the same conditions, requirements, and practices as apply to other ground taxiing and ATC procedures in the AIM.

1. The phraseology *taxi* is used when it is intended or expected that the helicopter will taxi on the airport surface, either via taxiways or other prescribed routes. *Taxi* is used primarily for helicopters equipped with wheels or in response to a pilot request. Preference should be given to this procedure whenever it is necessary to minimize effects of rotor downwash.

2. Pilots may request a *hover taxi* when slow forward movement is desired or when it may be appropriate to move very short distances. Pilots should avoid this procedure if rotor downwash is likely to cause damage to parked aircraft or if blowing dust/snow could obscure visibility. If it is necessary to operate above 25 feet AGL when hover taxiing, the pilot should initiate a request to ATC.

3. *Air taxi* is the preferred method for helicopter ground movements on airports provided ground operations and conditions permit. Unless otherwise requested or instructed, pilots are expected to remain below 100 feet AGL. However, if a higher than normal airspeed or altitude is desired, the request should be made prior to lift-off. The pilot is solely responsible for selecting a safe airspeed for the altitude/operation being conducted. Use of *air taxi* enables the pilot to proceed at an optimum airspeed/altitude, minimize downwash effect, conserve fuel, and expedite movement from one point to another. Helicopters should avoid overflight of other aircraft, vehicles, and personnel during air-taxi operations. Caution must be exercised concerning active runways and pilots must be certain that air taxi instructions are understood. Special precautions may be necessary at unfamiliar airports or airports with multiple/intersecting active runways. The taxi procedures given in: Taxiing, paragraph 4–68, Taxi During Low Visibility, paragraph 4–69, and Exiting the Runway After Landing, paragraph 4–70 also apply. (Reference—Pilot/Controller Glossary, Taxi, Hover Taxi, and Air Taxi).

c. Takeoff and Landing Procedures—

1. Helicopter operations may be conducted from a runway, taxiway, portion of a landing strip, or any clear area which could be used as a landing site such as the scene of an accident, a construction site, or the roof of a building. The terms used to describe designated areas from which helicopters operate are: movement area, landing/takeoff area, apron/ramp, heliport and helipad (See Pilot/Controller Glossary). These areas may be improved or unimproved and may be separate from or located on an airport/heliport. ATC will issue takeoff clearances from *movement* areas other than active runways, or in diverse directions from active runways, with additional instructions as necessary. Whenever possible, takeoff clearance will be issued in lieu of extended hover/air taxi operations. Phraseology will be "CLEARED FOR TAKEOFF FROM (taxiway, helipad, runway number, etc.), MAKE RIGHT/LEFT TURN FOR (direction, heading, NAVAID radial) DEPARTURE/DEPARTURE ROUTE (number, name, etc.)." Unless requested by the pilot, downwind takeoffs will not be issued if the tailwind exceeds 5 knots.

2. Pilots should be alert to wind information as well as to wind indications in the vicinity of the helicopter. ATC should be advised of the intended method of departing. A pilot request to takeoff in a given direction indicates that the pilot is willing to accept the wind condition and controllers will honor the request if traffic permits. Departure points could be a significant distance from the control tower and it may be difficult or impossible for the controller to determine the helicopter's relative position to the wind.

3. If takeoff is requested from *nonmovement* areas, the phraseology "PROCEED AS REQUESTED" will be used. Additional instructions will be issued as necessary. The pilot is responsible for operating in a safe manner and should exercise due caution. When other known traffic is not a factor and takeoff is requested from an area not visible from the tower, an area not authorized for helicopter use, an unlighted area at night, or an area not on the airport, the phraseology "DEPARTURE FROM (location) WILL BE AT YOUR OWN RISK (with reason, and additional instructions as necessary)."

4. Similar phraseology is used for helicopter landing operations. Every effort will be made to permit helicopters to proceed direct and land as near as possible to their final destination on the airport. Traffic density, the need for detailed taxiing instructions, frequency congestion, or other factors may affect the extent to which service can be expedited. As with ground movement operations, a high degree of pilot/controller cooperation and communication is necessary to achieve safe and efficient operations.

4–68. TAXIING

a. General: Approval must be obtained prior to moving an aircraft or vehicle onto the movement area during the hours an Airport Traffic Control Tower is in operation.

1. Always state your position on the airport when calling the tower for taxi instructions.

2. The movement area is normally described in local bulletins issued by the airport manager or control tower. These bulletins may be found in FSSs, fixed base operators offices, air carrier offices, and operations offices.

3. The control tower also issues bulletins describing areas where they cannot provide ATC service due to nonvisibility or other reasons.

4. A clearance must be obtained prior to taxiing on a runway, taking off, or landing during the hours an Airport Traffic Control Tower is in operation.

5. When ATC clears an aircraft to "taxi to" an assigned takeoff runway, the absence of holding instructions authorizes the aircraft to "cross" all runways which the taxi route intersects except the assigned takeoff runway. It does not include authorization to "taxi onto" or "cross" the assigned takeoff runway at any point. In order to preclude misunderstandings in radio communications, ATC will not use the word "cleared" in conjunction with authorization for aircraft to taxi.

6. In the absence of holding instructions, a clearance to "taxi to" any point other than an assigned takeoff runway is a clearance to cross all runways that intersect the taxi route to that point.

7. Air traffic control will first specify the runway, issue taxi instructions, and then state any required hold short instructions, when authorizing an aircraft to taxi for departure. This does not authorize the aircraft to "enter" or "cross" the assigned departure runway at any point. AIR TRAFFIC CONTROLLERS ARE REQUIRED TO OBTAIN FROM THE PILOT A READBACK OF ALL RUNWAY HOLD SHORT INSTRUCTIONS.

b. ATC clearances or instructions pertaining to taxiing are predicated on known traffic and known physical airport conditions. Therefore, it is important that pilots clearly understand the clearance or instruction. Although an ATC clearance is issued for taxiing purposes, when operating in accordance with the FARs, it is the responsibility of the pilot to avoid collision with other aircraft. Since "the pilot-in-command of an aircraft is directly responsible for, and is the final authority as to, the operation of that aircraft" the pilot should obtain clarification of any clearance or instruction which is not understood. (Reference—General, paragraph 7–41).

1. Good operating practice dictates that pilots acknowledge all runway crossing, hold short, or takeoff clearances unless there is some misunderstanding, at which time the pilot should query the controller until the clearance is understood. AIR TRAFFIC CONTROLLERS ARE REQUIRED TO OBTAIN FROM THE PILOT A READBACK OF ALL RUNWAY HOLD SHORT INSTRUCTIONS. Pilots operating a single pilot aircraft should monitor only assigned ATC communications after being cleared onto the active runway for departure. Single pilot aircraft should not monitor other than ATC communications until flight from the Airport Traffic Area is completed. This same procedure should be practiced from after receipt of the clearance for landing until the landing and taxi activities are complete. Proper effective scanning for other aircraft, surface vehicles, or other objects should be continuously exercised in all cases.

2. If the pilot is unfamiliar with the airport or for any reason confusion exists as to the correct taxi routing, a request may be made for progressive taxi instructions which include step-by-step routing directions. Progressive instructions may also be issued if the controller deems it necessary due to traffic or field conditions; i.e., construction or closed taxiways.

c. At those airports where the U.S. Government operates the control tower and ATC has authorized noncompliance with the requirement for two-way radio communications while operating within the Airport Traffic Area, or at those airports where the U.S. Government does not operate the control tower and radio communications cannot be established, pilots shall obtain a clearance by visual light signal prior to taxiing on a runway and prior to takeoff and landing.

d. The following phraseologies and procedures are used in radio-telephone communications with aeronautical ground stations.

1. *Request for taxi instructions prior to departure:* State your aircraft identification, location, type of operation planned (VFR or IFR), and the point of first intended landing.

EXAMPLE:

Aircraft: "WASHINGTON GROUND, BEECHCRAFT ONE THREE ONE FIVE NINER AT HANGAR EIGHT, READY TO TAXI, I-F-R TO CHICAGO."

Tower: "BEECHCRAFT ONE THREE ONE FIVE NINER, WASHINGTON GROUND, TAXI TO RUNWAY THREE SIX, WIND ZERO THREE ZERO AT TWO FIVE, ALTIMETER THREE ZERO ZERO FOUR,"

or

Tower: "BEECHCRAFT ONE THREE ONE FIVE NINER, WASHINGTON GROUND, RUNWAY TWO SEVEN, TAXI VIA TAXIWAYS CHARLIE AND DELTA, HOLD SHORT OF RUNWAY THREE THREE LEFT."

Aircraft: "BEECHCRAFT ONE THREE ONE FIVE NINER, HOLD SHORT OF RUNWAY THREE THREE LEFT."

2. *Receipt of ATC clearance:* ARTCC clearances are relayed to pilots by airport traffic controllers in the following manner.

EXAMPLE:

Tower: BEECHCRAFT ONE THREE ONE FIVE NINER, CLEARED TO THE CHICAGO MIDWAY AIRPORT VIA VICTOR EIGHT, MAINTAIN EIGHT THOUSAND.

Aircraft: "BEECHCRAFT ONE THREE ONE FIVE NINER, CLEARED TO THE CHICAGO MIDWAY AIRPORT VIA VICTOR EIGHT, MAINTAIN EIGHT THOUSAND."

4–68d1 NOTE—Normally, an ATC IFR clearance is relayed to a pilot by the ground controller. At busy locations, however, pilots may be instructed by the ground controller to "CONTACT CLEARANCE DELIVERY" on a frequency designated for this purpose. No surveillance or control over the movement of traffic is exercised by this position of operation.

3. *Request for taxi instructions after landing:* State your aircraft identification, location, and that you request taxi instructions.

EXAMPLE:

Aircraft: "DULLES GROUND, BEECHCRAFT ONE FOUR TWO SIX ONE CLEARING RUNWAY ONE RIGHT ON TAXIWAY ECHO THREE, REQUEST CLEARANCE TO PAGE."

Tower: "BEECHCRAFT ONE FOUR TWO SIX ONE, DULLES GROUND, TAXI TO PAGE VIA TAXIWAYS ECHO THREE, ECHO ONE, AND ECHO NINER."

or

Aircraft: "ORLANDO GROUND, BEECHCRAFT ONE FOUR TWO SIX ONE CLEARING RUNWAY ONE EIGHT LEFT AT TAXIWAY BRAVO THREE, REQUEST CLEARANCE TO PAGE."

Tower: "BEECHCRAFT ONE FOUR TWO SIX ONE, ORLANDO GROUND, HOLD SHORT OF RUNWAY ONE EIGHT RIGHT."

Aircraft: "BEECHCRAFT ONE FOUR TWO SIX ONE, HOLD SHORT OF RUNWAY ONE EIGHT RIGHT."

4–69. TAXI DURING LOW VISIBILITY

a. Pilots and aircraft operators should be constantly aware that during certain low visibility conditions the movement of aircraft and vehicles on airports may not be visible to the tower controller. This may prevent visual confirmation of an aircraft's adherence to taxi instructions. Pilots should, therefore, exercise extreme vigilance and proceed cautiously under such conditions.

b. Of vital importance is the need for pilots to notify the controller when difficulties are encountered or at the first indication of becoming disoriented. Pilots should proceed with extreme caution when taxiing toward the sun. When vision difficulties are encountered pilots should immediately inform the controller.

4–70. EXITING THE RUNWAY AFTER LANDING

The following procedures should be followed after landing and reaching taxi speed.

a. Exit the runway without delay at the first available taxiway or on a taxiway as instructed by air traffic control (ATC).

b. Taxi clear of the runway unless otherwise directed by ATC. In the absence of ATC instructions the pilot is expected to taxi clear of the landing runway even if that requires the aircraft to protrude into or cross another taxiway, runway, or ramp area. This does not authorize an aircraft to cross a subsequent taxiway/runway/ramp after clearing the landing runway.

4–70b NOTE—The tower will issue the pilot with instructions which will normally permit the aircraft to enter another taxiway, runway, or ramp area when required to taxi clear of the runway.

c. Stop the aircraft after clearing the runway if instructions have not been received from ATC.

d. Immediately change to ground control frequency when advised by the tower and obtain a taxi clearance.

4–70d NOTE 1.—The tower will issue instructions required to resolve any potential conflictions with other ground traffic prior to advising the pilot to contact ground control.

4–70d NOTE 2.—A clearance from ATC to taxi to the ramp authorizes the aircraft to cross all runways and taxiway intersections. Pilots not familiar with the taxi route should request specific taxi instructions from ATC.

4–71. PRACTICE INSTRUMENT APPROACHES

a. Various air traffic incidents have indicated the necessity for adoption of measures to achieve more organized and controlled operations where practice instrument approaches are conducted. Practice instrument approaches are considered to be instrument approaches made by either a VFR aircraft not on an IFR flight plan or an aircraft on an IFR flight plan. To achieve this and thereby enhance air safety, it is Air Traffic Operations Service policy to provide for separation of such operations at locations where approach control facilities are located and, as resources permit, at certain other locations served by ARTCCs or parent approach control facilities. Pilot requests to practice instrument approaches may be approved by ATC subject to traffic and workload conditions. Pilots should anticipate that in some instances the controller may find it necessary to deny approval or withdraw previous approval when traffic conditions warrant. It must be clearly understood, however, that even though the controller may be providing separation, pilots on VFR flight plans are required to comply with basic visual flight rules (FAR 91.155). Application of ATC procedures or any action taken by the controller to avoid traffic conflictions does not relieve IFR and VFR pilots of their responsibility to see-and-avoid other traffic while operating in VFR conditions. (FAR 91.113) In addition to the normal IFR separation minimums (which includes visual separation) during VFR conditions, 500 feet vertical separation may be applied between VFR aircraft and between a VFR aircraft and the IFR aircraft. Pilots not on IFR flight plans desiring practice instrument approaches should always state 'practice' when making requests to ATC. Controllers will instruct VFR aircraft requesting an instrument approach to maintain VFR. This is to preclude misunderstandings between the pilot and controller as to the status of the aircraft. If the pilot wishes to proceed in accordance with instrument flight rules, he must specifically request and obtain, an IFR clearance.

b. Before practicing an instrument approach, pilots should inform the approach control facility or the tower of the type of practice approach they desire to make and how they intend to terminate it, i.e., full-stop landing, touch-and-go, or missed or low approach maneuver. This information may be furnished progressively when conducting a series of approaches. Pilots on an IFR flight plan, who have made a series of instrument approaches to full stop landings should inform ATC when they make their final landing. The controller will control flights practicing instrument approaches so as to ensure that they do not disrupt the flow of arriving and departing itinerant IFR or VFR aircraft. The priority afforded itinerant aircraft over practice instrument approaches is not intended to be so rigidly applied that it causes grossly inefficient application of services. A minimum delay to itinerant traffic may be appropriate to allow an aircraft practicing an approach to complete that approach.

4–71b NOTE—A clearance to land means that appropriate separation on the landing runway will be ensured. A landing clearance does not relieve the pilot from compliance with any previously issued restriction.

c. At airports without a tower, pilots wishing to make practice instrument approaches should notify the facility having control jurisdiction of the desired approach as indicated on the approach chart. All approach control facilities and ARTCCs are required to publish a Letter to Airmen depicting those airports where they provide standard separation to both VFR and IFR aircraft conducting practice instrument approaches.

d. The Controller will provide approved separation between both VFR and IFR aircraft when authorization is granted to make practice approaches to airports where an approach control facility is located and to certain other airports served by approach control or an ARTCC. Controller responsibility for separation of VFR aircraft begins at the point where the approach clearance becomes effective, or when the aircraft enters ARSA/TRSA/TCA airspace, whichever comes first.

e. VFR aircraft practicing instrument approaches are not automatically authorized to execute the missed approach procedure. This authorization must be specifically requested by the pilot and approved by the controller. Separation will not be provided unless the missed approach has been approved by ATC.

f. Except in an emergency, aircraft cleared to practice instrument approaches must not deviate from the approved procedure until cleared to do so by the controller.

g. At radar approach control locations when a full approach procedure (Procedure Turn, etc.,) cannot be approved, pilots should expect to be vectored to a final approach course for a practice instrument approach which is compatible with the general direction of traffic at that airport.

h. When granting approval for a practice instrument approach, the controller will usually ask the pilot to report to the tower prior to or over the final approach fix inbound (non precision approaches) or over the outer marker or fix used in lieu of the outer marker inbound (precision approaches).

i. When authorization is granted to conduct practice instrument approaches to an airport with a tower, but where approved standard separation is not provided to aircraft conducting practice instrument approaches, the tower will approve the practice approach, instruct the aircraft to maintain VFR and issue traffic information, as required.

j. When an aircraft notifies a FSS providing Airport Advisory Service to the airport concerned of the intent to conduct a practice instrument approach and whether or not separation is to be provided, the pilot will be instructed to contact the appropriate facility on a specified frequency prior to initiating the approach. At airports where separation is not provided, the FSS will acknowledge the message and issue known traffic information but will neither approve or disapprove the approach.

k. Pilots conducting practice instrument approaches should be particularly alert for other aircraft operating in the local traffic pattern or in proximity to the airport.

4–72. OPTION APPROACH

The ''Cleared for the Option'' procedure will permit an instructor, flight examiner or pilot the option to make a touch-and-go, low approach, missed approach, stop-and-go, or full stop landing. This procedure can be very beneficial in a training situation in that neither the student pilot nor examinee would know what maneuver would be accomplished. The pilot should make his request for this procedure passing the final approach fix inbound on an instrument approach or entering downwind for a VFR traffic pattern. The advantages of this procedure as a training aid are that it enables an instructor or examiner to obtain the reaction of a trainee or examinee under changing conditions, the pilot would not have to discontinue an approach in the middle of the procedure due to student error or pilot proficiency requirements, and finally it allows more flexibility and economy in training programs. This procedure will only be used at those locations with an operational control tower and will be subject to ATC approval.

4–73. USE OF AIRCRAFT LIGHTS

a. Aircraft position and anticollision lights are required to be lighted on aircraft operated from sunset to sunrise. Anticollision lights, however, need not be lighted when the pilot-in-command determines that, because of operating conditions, it would be in the interest of safety to turn off the lights (FAR 91.209). For example, strobe lights should be turned off on the ground when they adversely affect ground personnel or other pilots, and in flight when there are adverse reflection from clouds.

b. An aircraft anticollision light system can use one or more rotating beacons and/or strobe lights, be colored either red or white, and have different (higher than minimum) intensities when compared to other aircraft. Many aircraft have both a rotating beacon and a strobe light system.

c. The FAA has a voluntary pilot safety program, *Operation Lights On*, to enhance the *see-and-avoid* concept. Pilots are encouraged to turn on their anticollision lights any time the engine(s) are running, day or night. Use of these lights is especially encouraged when operating on airport surfaces during periods of reduced visibility and when snow or ice control vehicles are or may be operating. Pilots are also encouraged to turn on their landing lights during takeoff; i.e., either after takeoff clearance has been received or when beginning takeoff roll. Pilots are further encouraged to turn on their landing lights when operating below 10,000 feet, day or night, especially when operating within 10 miles of any airport, or in conditions of reduced visibility and in areas where flocks of birds may be expected, i.e., coastal areas, lake areas, around refuse dumps, etc. Although turning on aircraft lights does enhance the *see-and-avoid* concept, pilots should not become complacent about keeping a sharp lookout for other aircraft. Not all aircraft are equipped with lights and some pilots may not have their lights turned on. Aircraft manufacturer's recommendations for operation of landing lights and electrical systems should be observed.

d. Prop and jet blast forces generated by large aircraft have overturned or damaged several smaller aircraft taxiing behind them. To avoid similar results, and in the interest of preventing upsets and injuries to ground personnel from such forces, the FAA recommends that air carriers and commercial operators turn on their rotating beacons anytime their aircraft engines are in operation. General Aviation pilots using rotating beacon equipped aircraft are also encouraged to participate in this program which is designed to alert others to the potential hazard. Since this is a voluntary program, exercise caution and do not rely solely on the rotating beacon as an indication that aircraft engines are in operation.

4–74. FLIGHT INSPECTION/'FLIGHT CHECK' AIRCRAFT IN TERMINAL AREAS

a. *Flight check* is a call sign used to alert pilots and air traffic controllers when a FAA aircraft is engaged in flight inspection/certification of NAVAIDs and flight procedures. Flight Check aircraft fly preplanned high/low altitude flight patterns such as grids, orbits, DME arcs, and tracks, including low passes along the full length of the runway to verify NAVAID performance. In most instances, these flight checks are being automatically recorded and/or flown in an automated mode.

b. Pilots should be especially watchful and avoid the flight paths of any aircraft using the call sign "Flight Check" or "Flight Check Recorded." The latter call sign; e.g. "Flight Check 47 Recorded" indicates that automated flight inspections are in progress in terminal areas. These flights will normally receive special handling from ATC. Pilot patience and cooperation in allowing uninterrupted recordings can significantly help expedite flight inspections, minimize costly, repetitive runs, and reduce the burden on the U.S. taxpayer.

4–75. HAND SIGNALS

(See Figure 4–75[1][Signalman Directs Towing].)

Figure 4–75[1]

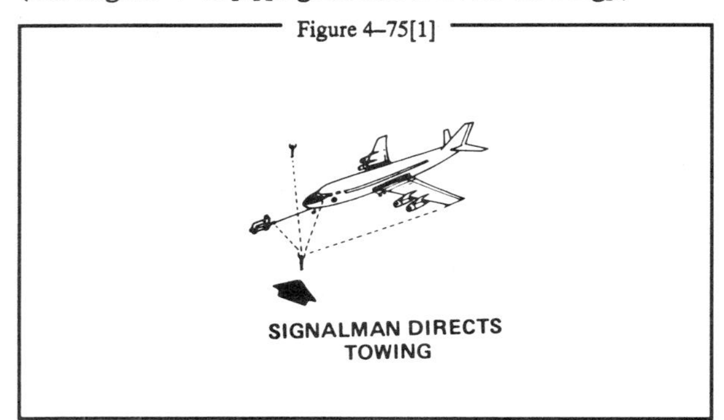

(See Figure 4–75[2][Signalman's Position].)

Figure 4–75[2]

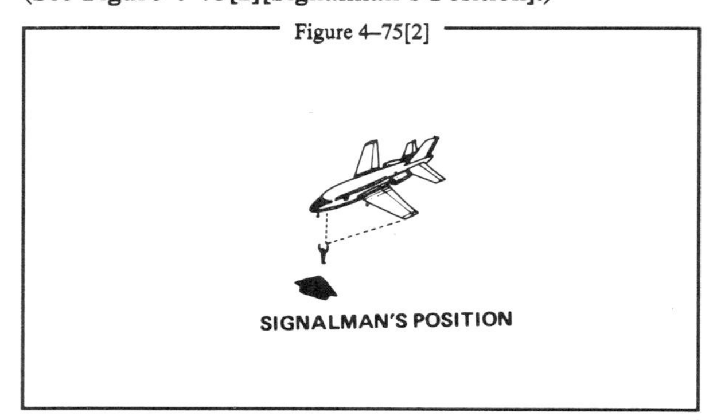

(See Figure 4–75[3][Flagman Directs Pilot].)

Figure 4–75[3]

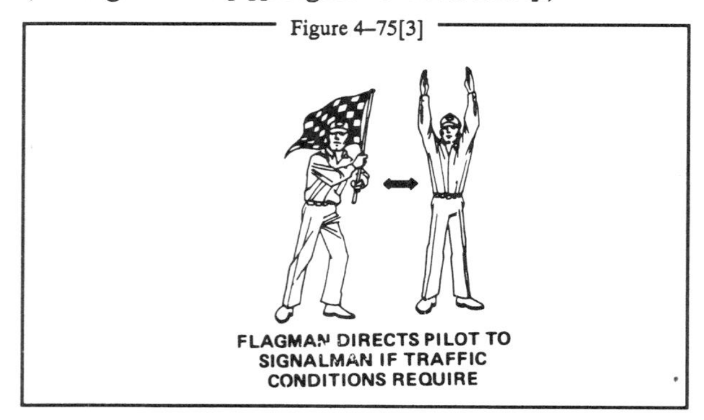

(See Figure 4–75[4][All Clear].)

Figure 4–75[4]

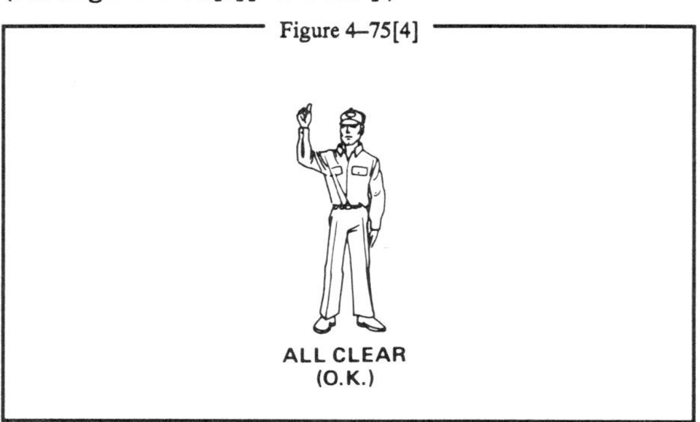

(See Figure 4–75[5][Start Engine].)

Figure 4–75[5]

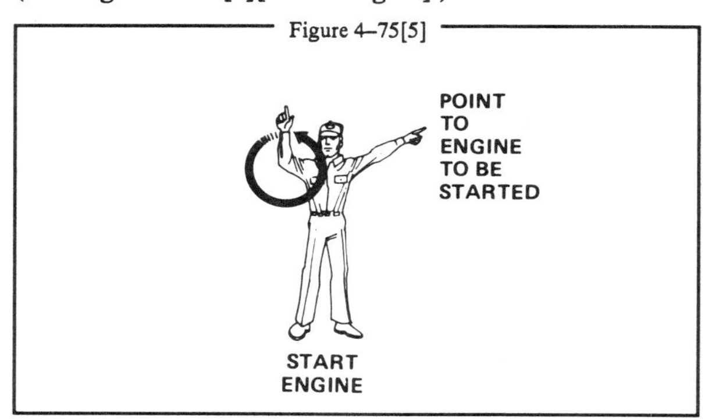

(See Figure 4–75[6][Pull Chocks].)

Figure 4–75[6]

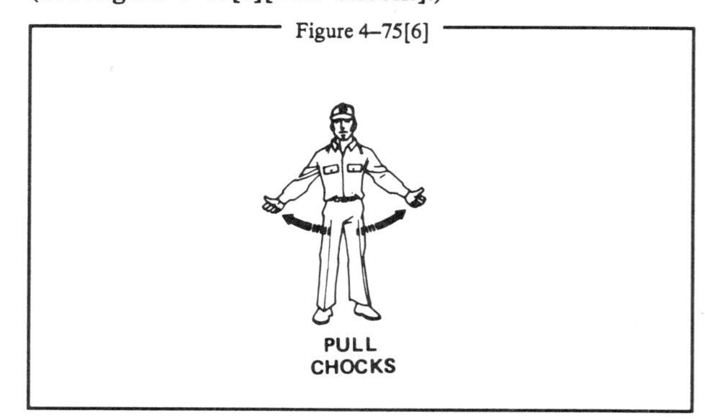

(See Figure 4–75[7][Come Ahead].)

Figure 4–75[7]

(See Figure 4–75[8][Left Turn].)

Figure 4–75[8]

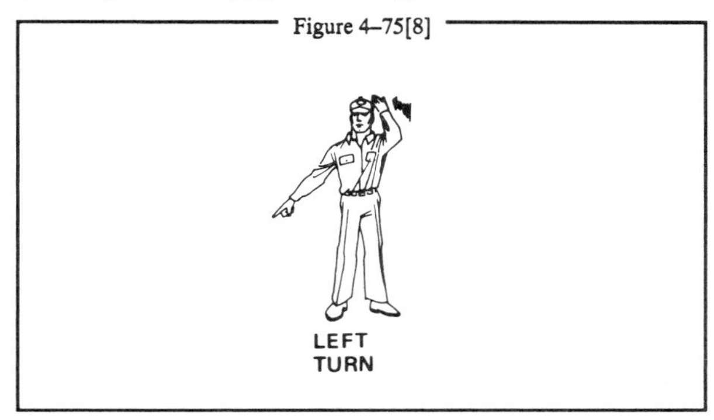

(See Figure 4–75[9][Right Turn].)

Figure 4–75[9]

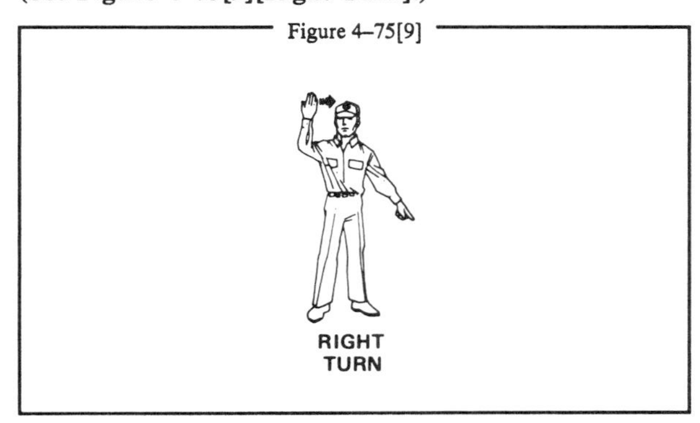

(See Figure 4–75[10][Slow Down].)

Figure 4–75[10]

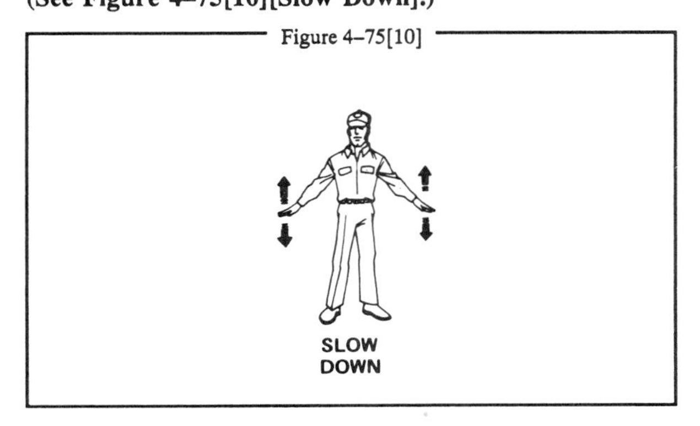

(See Figure 4–75[11][Stop].)

Figure 4–75[11]

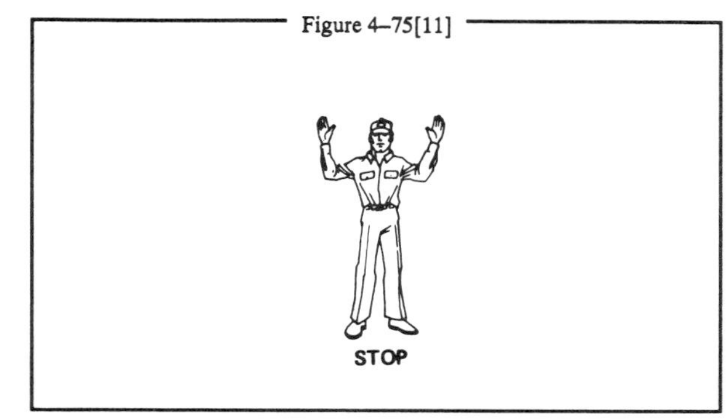

(See Figure 4–75[12][Insert Chocks].)

Figure 4–75[12]

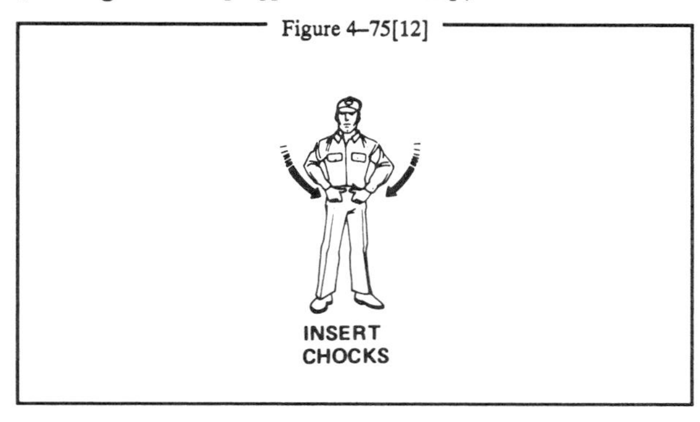

(See Figure 4–75[13][Cut Engines].)

Figure 4–75[13]

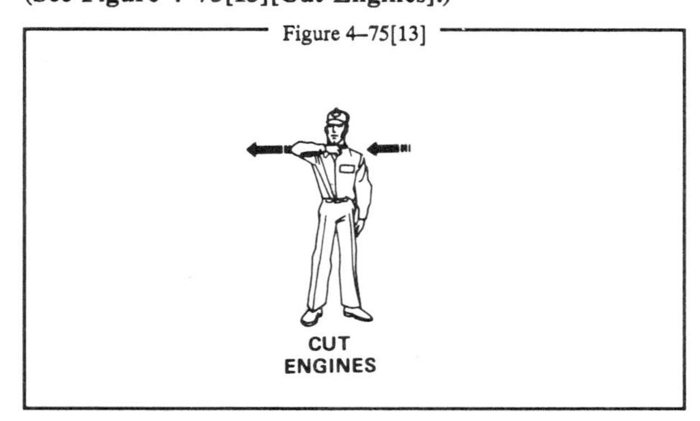

(See Figure 4–75[14][Night Operation].)

Figure 4–75[14]

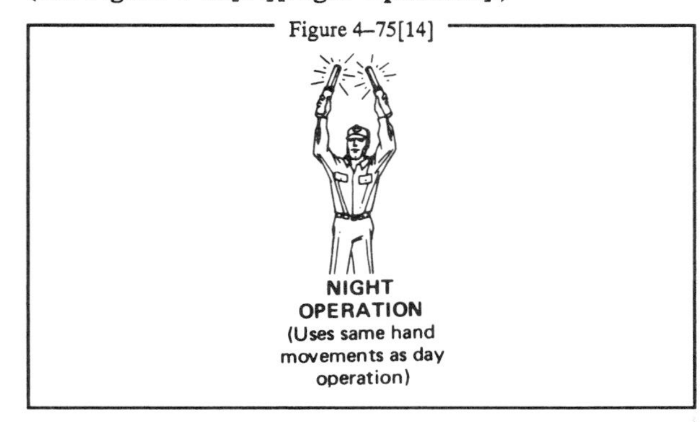

(See Figure 4–75[15][Emergency Stop].)

Figure 4–75[15]

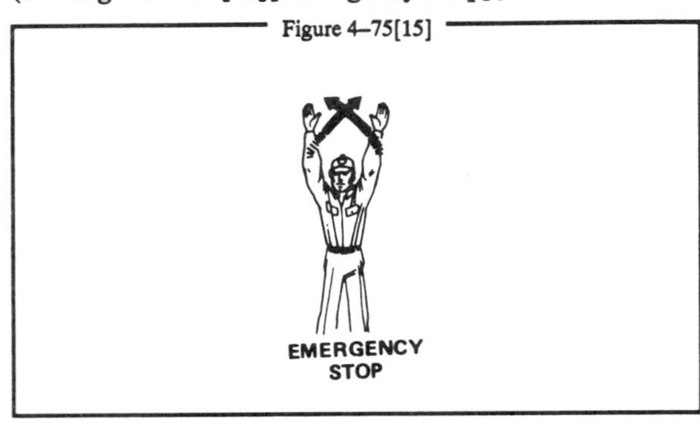

4–76 thru 4–80 RESERVED.

Section 4. ATC CLEARANCES/SEPARATIONS

4–81. CLEARANCE

a. A clearance issued by ATC is predicated on known traffic and known physical airport conditions. An ATC clearance means an authorization by ATC, for the purpose of preventing collision between known aircraft, for an aircraft to proceed under specified conditions within controlled airspace. IT IS NOT AUTHORIZATION FOR A PILOT TO DEVIATE FROM ANY RULE REGULATION OR MINIMUM ALTITUDE NOR TO CONDUCT UNSAFE OPERATION OF HIS AIRCRAFT.

b. FAR 91.3(a) states: "The pilot-in-command of an aircraft is directly responsible for, and is the final authority as to, the operation of that aircraft." If ATC issues a clearance that would cause a pilot to deviate from a rule or regulation, or in the pilot's opinion, would place the aircraft in jeopardy, IT IS THE PILOT'S RESPONSIBILITY TO REQUEST AN AMENDED CLEARANCE. Similarly, if a pilot prefers to follow a different course of action, such as make a 360 degree turn for spacing to follow traffic when established in a landing or approach sequence, land on a different runway, takeoff from a different intersection, takeoff from the threshold instead of an intersection, or delay his operation, HE IS EXPECTED TO INFORM ATC ACCORDINGLY. When he requests a different course of action, however, the pilot is expected to cooperate so as to preclude disruption of traffic flow or creation of conflicting patterns. The pilot is also expected to use the appropriate aircraft call sign to acknowledge all ATC clearances, frequency changes, or advisory information.

c. Each pilot who deviates from an ATC clearance in response to a Traffic Alert and Collision Avoidance System resolution advisory shall notify ATC of that deviation as soon as possible. (Reference—Pilot/Controller Glossary, Traffic Alert and Collision Avoidance System).

d. When weather conditions permit, during the time an IFR flight is operating, it is the direct responsibility of the pilot to avoid other aircraft since VFR flights may be operating in the same area without the knowledge of ATC. Traffic clearances provide standard separation only between IFR flights.'

4–82. CLEARANCE PREFIX

A clearance, control information, or a response to a request for information originated by an ATC facility and relayed to the pilot through an air-to-ground communication station will be prefixed by "ATC clears," "ATC advises," or "ATC requests."

4–83. CLEARANCE ITEMS

ATC clearances normally contain the following:

a. Clearance Limit.—The traffic clearance issued prior to departure will normally authorize flight to the airport of intended landing. Under certain conditions, at some locations a short-range clearance procedure is utilized whereby a clearance is issued to a fix within or just outside of the terminal area and the pilot is advised of the frequency on which he will receive the long-range clearance direct from the center controller.

b. Departure Procedure.—Headings to fly and altitude restrictions may be issued to separate a departure from other air traffic in the terminal area. (Reference—Abbreviated IFR Departure Clearance Procedures, paragraph 5–23 and Instrument Departures, paragraph 5–26). Where the volume of traffic warrants, SID's have been developed.

c. Route of Flight—

1. Clearances are normally issued for the altitude or flight level and route filed by the pilot. However, due to traffic conditions, it is frequently necessary for ATC to specify an altitude or flight level or route different from that requested by the pilot. In addition, flow patterns have been established in certain congested areas or between congested areas whereby traffic capacity is increased by routing all traffic on preferred routes. Information on these flow patterns is available in offices where preflight briefing is furnished or where flight plans are accepted.

2. When required, air traffic clearances include data to assist pilots in identifying radio reporting points. It is the responsibility of the pilot to notify ATC immediately if his radio equipment cannot receive the type of signals he must utilize to comply with his clearance.

d. Altitude Data—

1. The altitude or flight level instructions in an ATC clearance normally require that a pilot "MAINTAIN" the altitude or flight level at which the flight will operate when in controlled airspace. Altitude or flight level changes while en route should be requested prior to the time the change is desired.

2. When possible, if the altitude assigned is different from the altitude requested by the pilot, ATC will inform the pilot when to expect climb or descent clearance or to request altitude change from another facility. If this has not been received prior to crossing the boundary of the ATC facility's area and assignment at a different altitude is still desired, the pilot should reinitiate his request with the next facility.

3. The term "cruise" may be used instead of "MAINTAIN" to assign a block of airspace to a pilot from the minimum IFR altitude up to and including the altitude specified in the cruise clearance. The pilot may level off at any intermediate altitude within this block of airspace. Climb/descent within the block is to be made at the discretion of the pilot. However, once the pilot starts descent and verbally reports leaving an altitude in the block, he may not return to that altitude without additional ATC clearance.

4–83d3 NOTE—Reference—Pilot/Controller Glossary, Cruise.

e. Holding Instructions—

1. Whenever an aircraft has been cleared to a fix other than the destination airport and delay is expected, it is the responsibility of the ATC controller to issue complete holding instructions (unless the pattern is charted), an EFC time, and his best estimate of any additional en route/terminal delay.

2. If the holding pattern is charted and the controller doesn't issue complete holding instructions, the pilot is expected to hold as depicted on the appropriate chart. When the pattern is charted, the controller may omit all holding instructions except the charted holding direction and the statement *AS PUBLISHED*, e.g., *"HOLD EAST AS PUBLISHED."* Controllers shall always issue complete holding instructions when pilots request them.

4–83e2 NOTE—Only those holding patterns depicted on U.S. Government or commercially produced (meeting FAA requirements) *Low/High Altitude Enroute*, and *Area* or *STAR* charts should be used.

3. If no holding pattern is charted and holding instructions have not been issued, the pilot should ask ATC for holding instructions prior to reaching the fix. This procedure will eliminate the possibility of an aircraft entering a holding pattern other than that desired by ATC. If the pilot is unable to obtain holding instructions prior to reaching the fix (due to frequency congestion, stuck microphone, etc.), he should hold in a standard pattern on the course on which he approached the fix and request further clearance as soon as possible. In this event, the altitude/flight level of the aircraft at the clearance limit will be protected so that separation will be provided as required.

4. When an aircraft is 3 minutes or less from a clearance limit and a clearance beyond the fix has not been received, the pilot is expected to start a speed reduction so that he will cross the fix, initially, at or below the maximum holding airspeed.

5. When no delay is expected, the controller should issue a clearance beyond the fix as soon as possible and, whenever possible, at least 5 minutes before the aircraft reaches the clearance limit.

6. Pilots should report to ATC the time and altitude/flight level at which the aircraft reaches the clearance limit and report leaving the clearance limit.

4–83e6 NOTE—In the event of two-way communications failure, pilots are required to comply with FAR 91.185.

4–84. AMENDED CLEARANCES

a. Amendments to the initial clearance will be issued at any time an air traffic controller deems such action necessary to avoid possible confliction between aircraft. Clearances will require that a flight "hold" or change altitude prior to reaching the point where standard separation from other IFR traffic would no longer exist.

4–84a NOTE—Some pilots have questioned this action and requested "traffic information" and were at a loss when the reply indicated "no traffic report." In such cases the controller has taken action to prevent a traffic confliction which would have occurred at a distant point.

b. A pilot may wish an explanation of the handling of his flight at the time of occurrence; however, controllers are not able to take time from their immediate control duties nor can they afford to overload the ATC communications channels to furnish explanations. Pilots may obtain an explanation by directing a letter or telephone call to the chief controller of the facility involved.

c. The pilot has the privilege of requesting a different clearance from that which has been issued by ATC if he feels that he has information which would make another course of action more practicable or if aircraft equipment limitations or company procedures forbid compliance with the clearance issued.

4–85. SPECIAL VFR CLEARANCES

a. An ATC clearance must be obtained *prior* to operating within a Control Zone when the weather is less than that required for VFR flight. A VFR pilot may request and be given a clearance to enter, leave, or operate within most control zones in Special VFR conditions, traffic permitting, and providing such flight will not delay IFR operations. All Special VFR flights must remain clear of clouds. The visibility requirements for Special VFR aircraft (other than helicopters) are:

1. At least 1 statute mile flight visibility for operations within the control zone.

2. At least 1 statute mile ground visibility if taking off or landing. If ground visibility is not reported at that airport, the flight visibility must be at least 1 statute mile.

3. The restriction in (1) and (2) do not apply to helicopters. Helicopters must remain clear of clouds and may operate in control zones with less than 1 statute mile visibility.

b. When a control tower is located within the control zone, requests for clearances should be to the tower. If no tower is located within the control zone, a clearance may be obtained from the nearest tower, FSS, or center.

c. It is not necessary to file a complete flight plan with the request for clearance, but the pilot should state his intentions in sufficient detail to permit ATC to fit his flight into the traffic flow. The clearance will not contain a specific altitude as the pilot must remain clear of clouds. The controller may require the pilot to fly at or below a certain altitude due to other traffic, but the altitude specified will permit flight at or above the Minimum Safe Altitude. In addition, at radar locations, flights may be vectored if necessary for control purposes or on pilot request.

4–85c NOTE—The pilot is responsible for obstacle or terrain clearance (reference FAR 91.119).

d. Special VFR clearances are effective within control zones only. ATC does not provide separation after an aircraft leaves the control zone on a Special VFR clearance.

e. Special VFR operations by fixed-wing aircraft are prohibited in some control zones due to the volume of IFR traffic. A list of these control zones is contained in FAR 93.113. They are also depicted on Sectional Aeronautical Charts.

f. ATC provides separation between Special VFR flights and between these flights and other IFR flights.

g. Special VFR operations by fixed-wing aircraft are prohibited between sunset and sunrise unless the pilot is instrument rated and the aircraft is equipped for IFR flight.

4–86. PILOT RESPONSIBILITY UPON CLEARANCE ISSUANCE

a. Record ATC clearance—When conducting an IFR operation, make a written record of your clearance. The specified conditions which are a part of your air traffic clearance may be somewhat different from those included in your flight plan. Additionally, ATC may find it necessary to ADD conditions, such as particular departure route. The very fact that ATC specifies different or additional conditions means that other aircraft are involved in the traffic situation.

b. ATC Clearance/Instruction Readback—Pilots of airborne aircraft should read back *those parts* of ATC clearances and instructions containing altitude assignments or vectors as a means of mutual verification. The readback of the "numbers" serves as a double check between pilots and controllers and reduces the kinds of communications errors that occur when a number is either "misheard" or is incorrect.

1. Precede all readbacks and acknowledgements with the aircraft identification. This aids controllers in determining that the correct aircraft received the clearance or instruction. The requirement to include aircraft identification in all readbacks and acknowledgements becomes more important as frequency congestion increases and when aircraft with similar call signs are on the same frequency.

2. Read back altitudes, altitude restrictions, and vectors in the same sequence as they are given in the clearance or instruction.

3. Altitudes contained in charted procedures, such as SID's, instrument approaches, etc., should not be read back unless they are specifically stated by the controller.

c. It is the responsibility of the pilot to accept or refuse the clearance issued.

4–87. IFR CLEARANCE VFR-ON-TOP

a. A pilot on an IFR flight plan operating in VFR weather conditions, may request VFR-ON-TOP in lieu of an assigned altitude. This would permit the pilot to select an altitude or flight level of his choice (subject to any ATC restrictions.)

b. Pilots desiring to climb through a cloud, haze, smoke, or other meteorological formation and then either cancel their IFR flight plan or operate VFR-ON-TOP may request a climb to VFR-ON-TOP. The ATC authorization shall contain either a top report or a statement that no top report is available, and a request to report reaching VFR-ON-TOP. Additionally, the ATC authorization may contain a clearance limit, routing and an alternative clearance if VFR-ON-TOP is not reached by a specified altitude.

c. A pilot on an IFR flight plan, operating in VFR conditions, may request to climb/descend in VFR conditions.

d. ATC may not authorize VFR-ON-TOP/VFR CONDITIONS operations unless the pilot requests the VFR operation or a clearance to operate in VFR CONDITIONS will result in noise abatement benefits where part of the IFR departure route does not conform to an FAA approved noise abatement route or altitude.

e. When operating in VFR conditions with an ATC authorization to "MAINTAIN VFR-ON-TOP/ MAINTAIN VFR CONDITIONS" pilots on IFR flight plans must:

1. Fly at the appropriate VFR altitude as prescribed in FAR 91.159.

2. Comply with the VFR visibility and distance from cloud criteria in FAR 91.155 (BASIC VFR WEATHER MINIMUMS.)

3. Comply with instrument flight rules that are applicable to this flight; i.e., minimum IFR altitudes, position reporting, radio communications, course to be flown, adherence to ATC clearance, etc.

4–87e3 NOTE—Pilots should advise ATC prior to any altitude change to insure the exchange of accurate traffic information.

f. ATC authorization to "MAINTAIN VFR-ON-TOP" is not intended to restrict pilots so that they must operate only *above* an obscuring meteorological formation (layer). Instead, it permits operation above, below, between layers, or in areas where there is no meteorological obscuration. It is imperative, however, that pilots understand that clearance to operate "VFR-ON-TOP/VFR CONDITIONS" does not imply cancellation of the IFR flight plan.

g. Pilots operating VFR-ON-TOP/VFR CONDITIONS may receive traffic information from ATC on other pertinent IFR or VFR aircraft. However, aircraft operating in TCA's/TRSA's shall be separated as required by FAA Handbook 7110.65.

4–87g NOTE—When operating in VFR weather conditions, it is the pilot's responsibility to be vigilant so as to see-and-avoid other aircraft.

h. ATC will not authorize VFR or VFR-ON-TOP operations in PCA's. (Reference—Positive Control Area, paragraph 3–24).

4–88. VFR/IFR FLIGHTS

A pilot departing VFR, either intending to or needing to obtain an IFR clearance en route, must be aware of the position of the aircraft and the relative terrain/obstructions. When accepting a clearance below the MEA/MIA/MVA, pilots are responsible for their own terrain/obstruction clearance until reaching the MEA/MIA/MVA. If the pilot is unable to maintain terrain/obstruction clearance the controller will advise the pilot to state intentions.

4–89. ADHERENCE TO CLEARANCE

a. When air traffic clearance has been obtained under either Visual or Instrument Flight Rules, the pilot-in-command of the aircraft shall not deviate from the provisions thereof unless an amended clearance is obtained. When ATC issues a clearance or instruction, pilots are expected to execute its provisions upon receipt. ATC, in certain situations, will include the word "IMMEDIATELY" in a clearance or instruction to impress urgency of an imminent situation and expeditious compliance by the pilot is expected and necessary for safety. The addition of a VFR or other restriction; i.e., climb or descent point or time, crossing altitude, etc., does not authorize a pilot to deviate from the route of flight or any other provision of the ATC clearance.

b. When a heading is assigned or a turn is requested by ATC, pilots are expected to promptly initiate the turn, to complete the turn, and maintain the new heading unless issued additional instructions.

c. The term "AT PILOT'S DISCRETION" included in the altitude information of an ATC clearance means that ATC has offered the pilot the option to start climb or descent when he wishes. He is authorized to conduct the climb or descent at any rate he wishes and to temporarily level off at any intermediate altitude he may desire. However, once he has vacated an altitude, he may not return to that altitude.

d. When ATC has not used the term "AT PILOT'S DISCRETION" nor imposed any climb or descent restrictions, pilots should initiate climb or descent promptly on acknowledgement of the clearance. Descend or climb at an optimum rate consistent with the operating characteristics of the aircraft to 1,000 feet above or below the assigned altitude, and then attempt to descend or climb at a rate of between 500 and 1,500 fpm until the assigned altitude is reached. If at anytime the pilot is unable to climb or descend at a rate of at least 500 feet a minute, advise ATC. If it is necessary to level off at an intermediate altitude during climb or descent, advise ATC, except for level off at 10,000 feet MSL on descent or 3,000 feet above airport elevation (prior to entering an Airport Traffic Area), when required for speed reduction (FAR 91.117).

4–89d NOTE—Leveling off at 10,000 feet MSL on descent or 3,000 feet above airport elevation (prior to entering in an airport traffic area) to comply with FAR 91.117 airspeed restrictions is commonplace. Controllers anticipate this action and plan accordingly. Leveling off at any other time on climb or descent may seriously affect air traffic handling by ATC. Consequently, it is imperative that pilots make every effort to fulfill the above expected actions to aid ATC in safely handling and expediting traffic.

e. If the altitude information of an ATC DESCENT clearance includes a provision to "CROSS (fix) AT" or "AT OR ABOVE/BELOW (altitude)," the manner in which the descent is executed to comply with the crossing altitude is at the pilot's discretion. This authorization to descend at pilot's discretion is only applicable to that portion of the flight to which the crossing altitude restriction applies, and the pilot is expected to comply with the crossing altitude as a provision of the clearance. Any other clearance in which pilot execution is optional will so state "AT PILOT'S DISCRETION."

EXAMPLE:

"UNITED FOUR SEVENTEEN, DESCEND AND MAINTAIN SIX THOUSAND."

4–89e NOTE—The pilot is expected to commence descent upon receipt of the clearance and to descend at the suggested rates until reaching the assigned altitude of 6,000 feet.

EXAMPLE:

"UNITED FOUR SEVENTEEN, DESCEND AT PILOT'S DISCRETION, MAINTAIN SIX THOUSAND."

4–89e NOTE—The pilot is authorized to conduct descent within the context of the term AT PILOT'S DISCRETION as described above.

EXAMPLE:

"UNITED FOUR SEVENTEEN, CROSS LAKEVIEW V-O-R AT OR ABOVE FLIGHT LEVEL TWO ZERO ZERO, DESCEND AND MAINTAIN SIX THOUSAND."

4–89e NOTE—The pilot is authorized to conduct descent AT PILOT'S DISCRETION until reaching Lakeview VOR. He must comply with the clearance provision to cross the Lakeview VOR at or above FL 200. After passing Lakeview VOR, he is expected to descend at the suggested rates until reaching the assigned altitude of 6,000 feet.

EXAMPLE:

"UNITED FOUR SEVENTEEN, CROSS LAKEVIEW V-O-R AT SIX THOUSAND, MAINTAIN SIX THOUSAND."

4–89e NOTE—The pilot is authorized to conduct descent AT PILOT'S DISCRETION, however, he must comply with the clearance provision to cross the Lakeview VOR at 6,000 feet.

EXAMPLE

"UNITED FOUR SEVENTEEN, DESCEND NOW TO FLIGHT LEVEL TWO SEVEN ZERO, CROSS LAKEVIEW V-O-R AT OR BELOW ONE ZERO THOUSAND, DESCEND AND MAINTAIN SIX THOUSAND."

4–89e NOTE—The pilot is expected to promptly execute and complete descent to FL 270 upon receipt of the clearance. After reaching FL 270 he is authorized to descend "at pilot's discretion" until reaching Lakeview VOR. He must comply with the clearance provision to cross Lakeview VOR at or below 10,000 feet. After Lakeview VOR he is expected to descend at the suggested rates until reaching 6,000 feet.

EXAMPLE:

"UNITED THREE TEN, DESCEND NOW AND MAINTAIN FLIGHT LEVEL TWO FOUR ZERO, PILOT'S DISCRETION AFTER REACHING FLIGHT LEVEL TWO EIGHT ZERO."

4–89e NOTE—The pilot is expected to commence descent upon receipt of the clearance and to descend at the suggested rates until reaching flight level 280. At that point, the pilot is authorized to continue descent to flight level 240 within the context of the term "AT PILOT'S DISCRETION" as described above

f. In case emergency authority is used to deviate from provisions of an ATC clearance, the pilot-in-command shall notify ATC as soon as possible and obtain an amended clearance. In an emergency situation which does not result in a deviation from the rules prescribed in FAR 91 but which requires ATC to give priority to an aircraft, the pilot of such aircraft shall, when requested by ATC, make a report within 48 hours of such emergency situation to the manager of that ATC facility.

g. The guiding principle is that the last ATC clearance has precedence over the previous ATC clearance. When the route or altitude in a previously issued clearance is amended, the controller will restate applicable altitude restrictions. If altitude to maintain is changed or restated, whether prior to departure or while airborne, and previously issued altitude restrictions are omitted, those altitude restrictions are canceled, including SID altitude restrictions.

EXAMPLE:

A departure flight receives a clearance to destination airport to maintain FL 290. The clearance incorporates a SID which has certain altitude crossing restrictions. Shortly after takeoff, the flight receives a new clearance changing the maintaining FL from 290 to 250. If the altitude restrictions are still applicable, the controller restates them.

EXAMPLE:

A departing aircraft is cleared to cross Fluky intersection at or above 3,000 feet, Gordonville VOR at or above 12,000 feet, maintain FL 200. Shortly after departure, the altitude to be maintained is changed to FL 240. If the altitude restrictions are still applicable, the controller issues an amended clearance as follows: "CROSS FLUKY INTERSECTION AT OR ABOVE THREE THOUSAND, CROSS GORDONVILLE V-O-R AT OR ABOVE ONE TWO THOUSAND, MAINTAIN FLIGHT LEVEL TWO FOUR ZERO.":

EXAMPLE:

An arriving aircraft is cleared to his destination airport via V45 Delta VOR direct; he is cleared to cross Delta VOR at 10,000 feet, and then to maintain 6,000 feet. Prior to

Delta VOR, the controller issues an amended clearance as follows: "TURN RIGHT HEADING ONE EIGHT ZERO FOR VECTOR TO RUNWAY THREE SIX I-L-S APPROACH, MAINTAIN SIX THOUSAND."

4–89g NOTE—Because the altitude restriction "cross Delta V-O-R at 10,000 feet" was omitted from the amended clearance, it is no longer in effect.

h. Pilots of turbojet aircraft equipped with afterburner engines should advise ATC prior to takeoff if they intend to use afterburning during their climb to the en route altitude. Often, the controller may be able to plan his traffic to accommodate a high performance climb and allow the pilot to climb to his planned altitude without restriction.

4–90. IFR SEPARATION STANDARDS

a. ATC effects separation of aircraft vertically by assigning different altitudes; longitudinally by providing an interval expressed in time or distance between aircraft on the same, converging, or crossing courses, and laterally by assigning different flight paths.

b. Separation will be provided between all aircraft operating on IFR flight plans except during that part of the flight (outside a TCA or TRSA) being conducted on a VFR-ON-TOP/VFR CONDITIONS clearance. Under these conditions, ATC may issue traffic advisories, but it is the sole responsibility of the pilot to be vigilant so as to see and avoid other aircraft.

c. When radar is employed in the separation of aircraft at the same altitude, a minimum of 3 miles separation is provided between aircraft operating within 40 miles of the radar antenna site, and 5 miles between aircraft operating beyond 40 miles from the antenna site. These minima may be increased or decreased in certain specific situations.

4–90 NOTE—Certain separation standards are increased in the terminal environment when CENRAP is being utilized.

4–91. SPEED ADJUSTMENTS

a. ATC will issue speed adjustments to pilots of radar-controlled aircraft to achieve or maintain required or desire spacing.

b. ATC will express all speed adjustments in terms of knots based on indicated airspeed (IAS) in 10 knot increments except that at or above FL 240 speeds may be expressed in terms of Mach numbers in 0.01 increments. The use of Mach numbers is restricted to turbojet aircraft with Mach meters.

c. Pilots complying with speed adjustments are expected to maintain a speed within plus or minus 10 knots or 0.02 Mach number of the specified speed.

d. Unless pilot concurrence is obtained, ATC requests for speed adjustments will be in accordance with the following minimums:

1. To aircraft operating between FL 280 and 10,000 feet, a speed not less than 250 knots or the equivalent Mach number.

2. To turbine powered aircraft operating below 10,000 feet:

(a) A speed not less than 210 knots, except;

(b) Within 20 flying miles of the airport of intended landing, a speed not less than 170 knots.

3. Reciprocating engine or turboprop aircraft within 20 flying miles of the runway threshold of the airport of intended landing, a speed not less than 150 knots.

4. To departing aircraft:

(a) Turbine powered aircraft, a speed not less than 230 knots.

(b) Reciprocating engine aircraft, a speed not less than 150 knots.

e. When ATC combines a speed adjustment with a descent clearance, the sequence of delivery, with the word "then" between, indicates the expected order of execution;

EXAMPLE:

DESCEND AND MAINTAIN (altitude);
THEN, REDUCE SPEED TO (speed).

EXAMPLE:

REDUCE SPEED TO (speed); THEN,
DESCEND AND MAINTAIN (altitude).

4–91e NOTE—The maximum speeds below 10,000 feet as established in FAR 91.117 still apply. If there is any doubt concerning the manner in which such a clearance is to be executed, request clarification from ATC.

f. If ATC determines (before an approach clearance is issued) that it is no longer necessary to apply speed adjustment procedures, they will inform the pilot to resume normal speed. Approach clearances supersede any prior speed adjustment assignments, and pilots are expected to make their own speed adjustments, as necessary, to complete the approach. Under certain circumstances however, it may be necessary for ATC to issue further speed adjustments after approach clearance is issued to maintain separation between successive arrivals. Under such circumstances, previously issued speed adjustments will be restated if that speed is to be maintained or additional speed adjustments are requested. ATC must obtain pilot concurrence for speed adjustments after approach clearances are issued. Speed adjustments should not be assigned inside the final approach fix on final or a point 5 miles from the runway, whichever is closer to the runway.

g. The pilots retain the prerogative of rejecting the application of speed adjustment by ATC if the minimum safe airspeed for any particular operation is greater than the speed adjustment. IN SUCH CASES, PILOTS ARE EXPECTED TO ADVISE ATC OF THE SPEED THAT WILL BE USED.

h. Pilots are reminded that they are responsible for rejecting the application of speed adjustment by ATC if, in their opinion, it will cause them to exceed the maximum indicated airspeed prescribed by FAR 91.117(a). IN SUCH CASES, THE PILOT IS EXPECTED TO SO INFORM ATC. Pilots operating at or above 10,000 feet MSL who are issued speed adjustments which exceed 250 knots IAS and are subsequently cleared below 10,000 feet MSL are expected to comply with FAR 91.117(a).

i. For operations conducted below 10,000 feet MSL when outside the United States and beneath a TCA, airspeed restrictions apply to all U.S. registered aircraft. For operations conducted below 10,000 feet MSL when outside the United States within a TCA, there are no speed restrictions.

j. For operations in an Airport Traffic Area, ATC is authorized to request or approve a speed greater than the maximum indicated airspeeds prescribed for operation within that airspace (FAR 91.117(b)).

k. When in communication with the ARTCC, pilots should, as a good operating practice, state any ATC assigned speed restriction on initial radio contact associated with an ATC communications frequency change.

4–92. RUNWAY SEPARATION

Tower controllers establish the sequence of arriving and departing aircraft by requiring them to adjust flight or ground operation as necessary to achieve proper spacing. They may "HOLD" an aircraft short of the runway to achieve spacing between it and an arriving aircraft; the controller may instruct a pilot to "EXTEND DOWNWIND" in order to establish spacing from an arriving or departing aircraft. At times a clearance may include the word "IMMEDIATE." For example: "CLEARED FOR IMMEDIATE TAKEOFF." In such cases "IMMEDIATE" is used for purposes of *air traffic separation.* It is up to the pilot to refuse the clearance if, in his opinion, compliance would adversely affect his operation. (Reference—Gate Holding Due to Departure Delays, paragraph 4–65).

4–93. VISUAL SEPARATION

a. Visual separation is a means employed by ATC to separate aircraft only in terminal areas. There are two methods employed to effect this separation:

1. The tower controller sees the aircraft involved and issues instructions, as necessary, to ensure that the aircraft avoid each other.

2. A pilot sees the other aircraft involved and upon instructions from the controller provides his own separation by maneuvering his aircraft to avoid it. This may involve following in-trail behind another aircraft or keeping it in sight until it is no longer a factor.

b. A pilot's acceptance of instructions to follow another aircraft or provide visual separation from it is an acknowledgment that the pilot will maneuver his/her aircraft as necessary to avoid the other aircraft or to maintain in-trail separation. In operations conducted behind heavy jet aircraft, it is also an acknowledgment that the pilot accepts the responsibility for wake turbulence separation.

c. WHEN A PILOT HAS BEEN TOLD TO FOLLOW ANOTHER AIRCRAFT OR TO PROVIDE VISUAL SEPARATION FROM IT HE/SHE SHOULD PROMPTLY NOTIFY THE CONTROLLER IF VISUAL CONTACT WITH THE OTHER AIRCRAFT IS LOST OR CANNOT BE MAINTAINED OR IF THE PILOT CANNOT ACCEPT THE RESPONSIBILITY FOR THE SEPARATION FOR ANY REASON.

d. Pilots should remember, however, that they have a regulatory responsibility (FAR 91.113(a)) to see and avoid other aircraft when weather conditions permit.

4–94. USE OF VISUAL CLEARING PROCEDURES

a. Before Takeoff—Prior to taxiing onto a runway or landing area in preparation for takeoff, pilots should scan the approach areas for possible landing traffic, executing appropriate clearing maneuvers to provide him a clear view of the approach areas.

b. Climbs and Descents—During climbs and descents in flight conditions which permit visual detection of other traffic, pilots should execute gentle banks, left and right at a frequency which permits continuous visual scanning of the airspace about them.

c. Straight and Level—Sustained periods of straight and level flight in conditions which permit visual detection of other traffic should be broken at intervals with appropriate clearing procedures to provide effective visual scanning.

d. Traffic Pattern—Entries into traffic patterns while descending create specific collision hazards and should be avoided.

e. Traffic at VOR Sites—All operators should emphasize the need for sustained vigilance in the vicinity of VORs and airway intersections due to the convergence of traffic.

f. Training Operations—Operators of pilot training programs are urged to adopt the following practices:

1. Pilots undergoing flight instruction at all levels should be requested to verbalize clearing procedures (call out "clear" left, right, above, or below) to instill and sustain the habit of vigilance during maneuvering.

2. High-wing airplane: momentarily raise the wing in the direction of the intended turn and look.

3. Low-wing airplane: momentarily lower the wing in the direction of the intended turn and look.

4. Appropriate clearing procedures should precede the execution of all turns including chandelles, lazy eights, stalls, slow flight, climbs, straight and level, spins, and other combination maneuvers.

4–95. TRAFFIC ALERT AND COLLISION AVOIDANCE SYSTEM (TCAS I & II)

a. TCAS I provides proximity warning only, to assist the pilot in the visual acquisition of intruder aircraft. No recommended avoidance maneuvers are provided nor authorized as a direct result of a TCAS I warning. It is intended for use by smaller commuter aircraft holding 10 to 30 passenger seats, and general aviation aircraft.

b. TCAS II provides traffic advisories (TA's) and resolution advisories (RA's). Resolution advisories provide recommended maneuvers in a vertical direction (climb or descend only) to avoid conflicting traffic. Airline aircraft, and larger commuter and business aircraft holding 31 passenger seats or more, use TCAS II equipment.

1. Each pilot who deviates from an ATC clearance in response to a TCAS II RA shall notify ATC of that deviation as soon as practicable and expeditiously return to the current ATC clearance when the traffic conflict is resolved.

2. Deviations from rules, policies, or clearances should be kept to the minimum necessary to satisfy a TCAS II RA.

3. The serving IFR air traffic facility is not responsible to provide approved standard IFR separation to an aircraft after a TCAS II RA maneuver until one of the following conditions exists:

(a) The aircraft has returned to its assigned altitude and course.

(b) Alternate ATC instructions have been issued.

c. TCAS does not alter or diminish the pilot's basic authority and responsibility to ensure safe flight. Since TCAS does not respond to aircraft which are not transponder equipped or aircraft with a transponder failure, TCAS alone does not ensure safe separation in every case.

d. At this time, no air traffic service nor handling is predicated on the availability of TCAS equipment in the aircraft.

4–96 thru 4–100. RESERVED

Chapter 5. AIR TRAFFIC PROCEDURES

Section 1. PREFLIGHT

5–1. PREFLIGHT PREPARATION

a. Every pilot is urged to receive a preflight briefing and to file a flight plan. This briefing should consist of the latest or most current weather, airport, and en route NAVAID information. Briefing service may be obtained from an FSS either by telephone or interphone, by radio when airborne, or by a personal visit to the station. Pilots with a current medical certificate in the 48 contiguous States may access toll-free the Direct User Access Terminal System (DUATS) through a personal computer. DUATS will provide alpha-numeric preflight weather data and allow pilots to file domestic VFR or IFR flight plans. (Reference—FAA Weather Services, paragraph 7–2c(5) lists DUATS vendors).

5–1a NOTE—Pilots filing flight plans via "fast file" who desire to have their briefing recorded, should include a statement at the end of the recording as to the source of their weather briefing.

b. The information required by the FAA to process flight plans is contained on FAA Form 7233–1, Flight Plan. (Reference—Flight Plan—VFR Flights, paragraph 5–4 and Flight Plan—IFR Flights, paragraph 5–7). The forms are available at all flight service stations. Additional copies will be provided on request.

c. Consult an FSS or a Weather Service Office (WSO) for preflight weather briefing.

d. FSS's are required to advise of pertinent NOTAM's if a *standard* briefing is requested, but if they are overlooked, don't hesitate to remind the specialist that you have not received NOTAM information.

5–1d NOTE—NOTAM's which are known in sufficient time for publication and are of 7 days duration or longer are normally incorporated into the Notices to Airmen publication and carried there until cancellation time. FDC NOTAM's, which apply to instrument flight procedures, are also included in the Notices to Airmen publication up to and including the number indicated in the FDC NOTAM legend. Printed NOTAM's are not provided during a briefing unless specifically requested by the pilot since the FSS specialist has no way of knowing whether the pilot has already checked the Notices to Airmen publication prior to calling. Remember to *ask* for NOTAM's in the Notices to Airmen publication. This information is not normally furnished during your briefing. (Reference—Notice to Airmen (NOTAM) System, paragraph 5–3).

e. Pilots are urged to use only the latest issue of aeronautical charts in planning and conducting flight operations. Aeronautical charts are revised and reissued on a regular scheduled basis to ensure that depicted data are current and reliable. In the conterminous U.S., Sectional Charts are updated each 6 months, IFR En route Charts each 56 days, and amendments to civil IFR Approach Charts are accomplished on a 56 day cycle with a change notice volume issued on the 28 day midcycle. Charts that have been superseded by those of a more recent date may contain obsolete or incomplete flight information. (Reference—General Description of each Chart Series, paragraph 9–4).

f. When requesting a preflight briefing, identify yourself as a pilot and provide the following:

1. Type of flight planned; e.g., VFR or IFR.
2. Aircraft's number or pilot's name.
3. Aircraft type.
4. Departure Airport.
5. Route of flight.
6. Destination.
7. Flight altitude(s).
8. ETD and ETE.

g. Prior to conducting a briefing, briefers are required to have the background information listed above so that they may tailor the briefing to the needs of the proposed flight. The objective is to communicate a "picture" of meteorological and aeronautical information necessary for the conduct of a safe and efficient flight. Briefers use all available weather and aeronautical information to summarize data applicable to the proposed flight. They do not read weather reports and forecasts verbatim unless specifically requested by the pilot. Refer to paragraph 7–3 *PREFLIGHT BRIEFINGS* for those items of a weather briefing that should be expected or requested.

h. The Federal Aviation Administration (FAA) by Federal Aviation Regulation, Part 93, Subpart K, has designated High Density Traffic Airports (HDTA's) and has prescribed air traffic rules and requirements for operating aircraft (excluding helicopter operations) to and from these airports (Reference—Airport/Facility Directory, Special Notices Section, for details).

i. In addition to the filing of a flight plan, if the flight will traverse or land in one or more foreign countries, it is particularly important that pilots leave a complete itinerary with someone directly concerned, keep that person advised of the flight's progress, and inform him that, if serious doubt arises as to the safety of the flight, he should first contact the FSS. (Reference—Flights Outside the United States and U.S. Territories, paragraph 5–9).

j. Pilots operating under provisions of FAR 135 and not having an FAA assigned 3–letter designator, are urged to prefix the normal registration (N) number with the letter "T" on flight plan filing; e.g., TN1234B. (Reference—Aircraft Call Signs, paragraph 4–34).

5–2. FOLLOW IFR PROCEDURES EVEN WHEN OPERATING VFR

a. To maintain IFR proficiency, pilots are urged to practice IFR procedures whenever possible, even when operating VFR. Some suggested practices include:

1. Obtain a complete preflight and weather briefing. Check the NOTAM's.

2. File a flight plan. This is an excellent low cost insurance policy. The cost is the time it takes to fill it out. The insurance includes the knowledge that someone will be looking for you if you become overdue at your destination.

3. Use current charts.

4. Use the navigation aids. Practice maintaining a good course-keep the needle centered.

5. Maintain a constant altitude which is appropriate for the direction of flight.

6. Estimate en route position times.

7. Make accurate and frequent position reports to the FSS's along your route of flight.

b. Simulated IFR flight is recommended (under the hood); however, pilots are cautioned to review and adhere to the requirements specified in FAR 91.109 before and during such flight.

c. When flying VFR at night, in addition to the altitude appropriate for the direction of flight, pilots should maintain an altitude which is at or above the minimum en route altitude as shown on charts. This is especially true in mountainous terrain, where there is usually very little ground reference. Do not depend on your eyes alone to avoid rising unlighted terrain, or even lighted obstructions such as TV towers.

5–3. NOTICE TO AIRMEN (NOTAM) SYSTEM

a. Time-critical aeronautical information which is of either a temporary nature or not sufficiently known in advance to permit publication on aeronautical charts or in other operational publications receives immediate dissemination via the National Notice to Airmen (NOTAM) System.

5–3a NOTE—NOTAM information is that aeronautical information that could affect a pilot's decision to make a flight. It includes such information as airport or primary runway closures, changes in the status of navigational aids, ILS's, radar service availability, and other information essential to planned enroute, terminal, or landing operations.

b. NOTAM information is classified into three categories. These are NOTAM (D) or distant, NOTAM (L) or local, and Flight Data Center (FDC) NOTAM's.

1. NOTAM (D) information is disseminated for all navigational facilities that are part of the National Airspace System (NAS), all public use airports, seaplane bases, and heliports listed in the Airport/Facility Directory (A/FD). The complete file of all NOTAM (D) information is maintained in a computer data base at the National Communications Center (NATCOM), located in Kansas City. This category of information is distributed automatically, appended to the hourly weather reports, via the Service A telecommunications system. Air traffic facilities, primarily FSS's, with Service A capability have access to the entire NATCOM data base of NOTAM's. These NOTAM's remain available via Service A for the duration of their validity or until published.

2. NOTAM (L)

(a) NOTAM (L) information includes such data as taxiway closures, men and equipment near or crossing runways, airport rotating beacon outages, and airport lighting aids that do not affect instrument approach criteria, such as VASI.

(b) NOTAM (L) information is distributed locally only and is not attached to the hourly weather reports. A separate file of local NOTAM's is maintained at each FSS for facilities in their area only. NOTAM (L) information for other FSS areas must be specifically requested directly from the FSS that has responsibility for the airport concerned.

5–3a2b NOTE—DUATS vendors are not required to provide NOTAM L information.

3. FDC NOTAM's

(a) On those occasions when it becomes necessary to disseminate information which is regulatory in nature, the National Flight Data Center (NFDC), in Washington, DC, will issue an FDC NOTAM. FDC NOTAM's contain such things as amendments to published IAP's and other current aeronautical charts. They are also used to advertise temporary flight restrictions caused by such things as natural disasters or large-scale public events that may generate a congestion of air traffic over a site.

(b) FDC NOTAM's are transmitted via Service A only once and are kept on file at the FSS until published or canceled. FSS's are responsible for maintaining a file of current, unpublished FDC NOTAM's concerning conditions within 400 miles of their facilities. FDC information concerning conditions that are more than 400 miles from the FSS, or that is already published, is given to a pilot only on request.

5–3a3b NOTE 1—DUATS vendors will provide FDC NOTAM's only upon site-specific requests using a location identifier.

5–3a3b NOTE 2—NOTAM data may not always be current due to the changeable nature of National Airspace System components, delays inherent in processing information, and occasional temporary outages of the United States NOTAM System. While en route, pilots should contact FSS's and obtain updated information for their route of flight and destination.

c. An integral part of the NOTAM System is the biweekly Notices to Airmen publication (NTAP). Data is included in this publication to reduce congestion on the telecommunications circuits and, therefore, is not available via Service A. Once published, the information is not provided during pilot weather briefings unless specifically requested by the pilot. This publication contains two sections.

1. The first section consists of notices that meet the criteria for NOTAM (D) and are expected to remain in effect for an extended period and FDC NOTAM's that are current at the time of publication. Occasionally, some NOTAM (L) and other unique information is included in this section when it will contribute to flight safety.

2. The second section contains special notices that are either too long or concern a wide or unspecified geographic area and are not suitable for inclusion in the first section. The content of these notices vary widely and there are no specific criteria for their inclusion, other than their enhancement of flight safety.

3. The number of the last FDC NOTAM included in the publication is noted on the first page to aid the user in updating the listing with any FDC NOTAM's which may have been issued between the cut-off date and the date the publication is received. All information contained will be carried until the information expires, is canceled, or in the case of permanent conditions, is published in other publications, such as the A/FD.

4. All new notices entered, excluding FDC NOTAM's, will be published only if the information is expected to remain in effect for at least 7 days after the effective date of the publication.

5–4. FLIGHT PLAN—VFR FLIGHTS

a. Except for operations in or penetrating a Coastal or Domestic ADIZ or DEWIZ a flight plan is not required for VFR flight. (Reference—National Security, paragraph 5–91). However, it is strongly recommended that one be filed with an FAA FSS. This will ensure that you receive VFR Search and Rescue Protection. (Reference—Search and Rescue, paragraph 6–17g for the proper method of filing).

b. To obtain maximum benefits from the flight plan program, flight plans should be filed directly with the nearest FSS. For your convenience, FSS's provide aeronautical and meteorological briefings while accepting flight plans. Radio may be used to file if no other means are available.

5–4b NOTE— Some states operate aeronautical communications facilities which will accept and forward flight plans to the FSS for further handling.

c. When a "stopover" flight is anticipated, it is recommended that a separate flight plan be filed for each "leg" when the stop is expected to be more than 1 hour duration.

d. Pilots are encouraged to give their departure times directly to the FSS serving the departure airport or as otherwise indicated by the FSS when the flight plan is filed. This will ensure more efficient flight plan service and permit the FSS to advise you of significant changes in aeronautical facilities or meteorological conditions. When a VFR flight plan is filed, it will be held by the FSS until 1 hour after the proposed departure time unless:

1. The actual departure time is received.

2. A revised proposed departure time is received.

3. At a time of filing, the FSS is informed that the proposed departure time will be met, but actual time cannot be given because of inadequate communications (assumed departures).

e. On pilot's request, at a location having an active tower, the aircraft identification will be forwarded by the tower to the FSS for reporting the actual departure time. This procedure should be avoided at busy airports.

f. Although position reports are not required for VFR flight plans, periodic reports to FAA FSS's along the route are good practice. Such contacts permit significant information to be passed to the transiting aircraft and also serve to check the progress of the flight should it be necessary for any reason to locate the aircraft.

EXAMPLE:

BONANZA 314K, OVER KINGFISHER AT (time), VFR FLIGHT PLAN, TULSA TO AMARILLO.

EXAMPLE:

CHEROKEE 5133J, OVER OKLAHOMA CITY AT (time), SHREVEPORT TO DENVER, NO FLIGHT PLAN.

g. Pilots not operating on an IFR flight plan and when in level cruising flight, are cautioned to conform with VFR cruising altitudes appropriate to the direction of flight.

h. When filing VFR flight plans, indicate aircraft equipment capabilities by appending the appropriate suffix to aircraft type in the same manner as that prescribed for IFR flight. (Reference—Flight Plan—IFR Flights, paragraph 5–7). Under some circumstances, ATC computer tapes can be useful in constructing the radar history of a downed or crashed aircraft. In each case, knowledge of the aircraft's transponder equipment is necessary in determining whether or not such computer tapes might prove effective.

i. Flight Plan Form—(See Figure 5–4[1]).

j. Explanation of VFR Flight Plan Items—

Block 1. Check the type flight plan. Check both the VFR and IFR blocks if composite VFR/ IFR.

Block 2. Enter your complete aircraft identification including the prefix "N" if applicable.

Block 3. Enter the designator for the aircraft, or if unknown, the aircraft manufacturer's name.

Block 4. Enter your true airspeed (TAS).

Block 5. Enter the departure airport identifier code, or if unknown, the name of the airport.

Block 6. Enter the proposed departure time in Coordinated Universal Time (UTC). If airborne, specify the actual or proposed departure time as appropriate.

Block 7. Enter the appropriate VFR altitude (to assist the briefer in providing weather and wind information).

Block 8. Define the route of flight by using NAVAID identifier codes and airways.

Block 9. Enter the destination airport identifier code, or if unknown, the airport name .

Figure 5–4[1]

Form Approved: OMB No. 2120-0026

U.S. DEPARTMENT OF TRANSPORTATION
FEDERAL AVIATION ADMINISTRATION

FLIGHT PLAN

(FAA USE ONLY) ☐ PILOT BRIEFING ☐ VNR ☐ STOPOVER

TIME STARTED | SPECIALIST INITIALS

1. TYPE: ☐ VFR ☐ IFR ☐ DVFR
2. AIRCRAFT IDENTIFICATION
3. AIRCRAFT TYPE/ SPECIAL EQUIPMENT
4. TRUE AIRSPEED — KTS
5. DEPARTURE POINT
6. DEPARTURE TIME — PROPOSED (Z) — ACTUAL (Z)
7. CRUISING ALTITUDE
8. ROUTE OF FLIGHT
9. DESTINATION (Name of airport and city)
10. EST. TIME ENROUTE — HOURS — MINUTES
11. REMARKS
12. FUEL ON BOARD — HOURS — MINUTES
13. ALTERNATE AIRPORT(S)
14. PILOT'S NAME, ADDRESS & TELEPHONE NUMBER & AIRCRAFT HOME BASE
15. NUMBER ABOARD
16. COLOR OF AIRCRAFT
17. DESTINATION CONTACT/TELEPHONE (OPTIONAL)

CIVIL AIRCRAFT PILOTS. FAR Part 91 requires you file an IFR flight plan to operate under instrument flight rules in controlled airspace. Failure to file could result in a civil penalty not to exceed $1,000 for each violation (Section 901 of the Federal Aviation Act of 1958, as amended). Filing of a VFR flight plan is recommended as a good operating practice. See also Part 99 for requirements concerning DVFR flight plans.

FAA Form 7233-1 (8-82) CLOSE VFR FLIGHT PLAN WITH ______________ FSS ON ARRIVAL

5–4jBlock 6 NOTE— Include the city name (or even the state name) if needed for clarity.

Block 10. Enter your Estimated Time en Route in hours and minutes.

Block 11. Enter only those remarks pertinent to ATC or to the clarification of other flight plan information, such as the appropriate radiotelephony (call sign) associated with the designator filed in Block 2. Items of a personal nature are not accepted.

Block 12. Specify the fuel on board in hours and minutes.

Block 13. Specify an alternate airport if desired.

Block 14. Enter your complete name, address, and telephone number. Enter sufficient information to identify home base, airport, or operator.

5–4j Block 14 NOTE— This information is essential in the event of search and rescue operations.

Block 15. Enter total number of persons on board (POB) including crew.

Block 16. Enter the predominant colors.

Block 17. Record the FSS name for closing the flight plan. If the flight plan is closed with a different FSS or facility, state the recorded FSS name that would normally have closed your flight plan. **(Optional)**—Record a destination telephone number to assist search and rescue contact should you fail to report or cancel your flight plan within ½ hour after your Estimated Time of Arrival (ETA).

5–4j Block 17 NOTE— The information transmitted to the destination FSS will consist only of flight plans Blocks 2, 3, 9, and 10. Estimated time en route (ETE) will be converted to the correct estimated time of arrival (ETA).

5–5. FLIGHT PLAN—DEFENSE VFR (DVFR) FLIGHTS

VFR flights into a Coastal or Domestic ADIZ/DEWIZ are required to file DVFR flight plans for security purposes. Detailed ADIZ procedures are found in the National Security section of this chapter. (See FAR 99.)

5–6. COMPOSITE FLIGHT PLAN (VFR/IFR FLIGHTS)

a. Flight plans which specify VFR operation for one portion of a flight, and IFR for another portion, will be accepted by the FSS at the point of departure. If VFR flight is conducted for the first portion of the flight, the pilot should report his departure time to the FSS with which he filed his VFR/IFR flight plan; and, subsequently, close the VFR portion and request ATC clearance from the FSS nearest the point at which change from VFR to IFR is proposed. Regardless of the type facility you are communicating with (FSS, center, or tower), it is the pilot's responsibility to request that facility to "CLOSE VFR FLIGHT PLAN." The pilot must remain in VFR weather conditions until operating in accordance with the IFR clearance.

b. When a flight plan indicates IFR for the first portion of flight and VFR for the latter portion, the pilot will normally be cleared to the point at which the change is proposed. Once the pilot has reported over the clearance limit and does not desire further IFR clearance, he should advise ATC to cancel the IFR portion of his flight plan. Then, he should contact the nearest FSS to activate the VFR portion of his flight plan. If the pilot desires to continue his IFR flight plan beyond the clearance limit, he should contact ATC at least 5 minutes prior to the clearance limit and request further IFR clearance. If the requested clearance is not received prior to reaching the clearance limit fix, the pilot will be expected to established himself in a standard holding pattern on the radial or course to the fix unless a holding pattern for the clearance limit fix is depicted on a U.S. Government or commercially produced (meeting FAA requirements) Low or High Altitude En Route, Area or STAR Chart. In this case the pilot will hold according to the depicted pattern.

5–7. FLIGHT PLAN—IFR FLIGHTS

a. General—

1. Prior to departure from within, or prior to entering controlled airspace, a pilot must submit a complete flight plan and receive an air traffic clearance, if weather conditions are below VFR minimums. Instrument flight plans may be submitted to the nearest FSS or ATCT either in person or by telephone (or by radio if no other means are available). Pilots should file IFR flight plans at least 30 minutes prior to estimated time of departure to preclude possible delay in receiving a departure clearance from ATC. To minimize your delay in entering the Control Zone at destination when IFR weather conditions exist or are forecast at that airport, an IFR flight plan should be filed before departure. Otherwise, a 30 minute delay is not unusual in receiving an ATC clearance because of time spent in processing flight plan data. Traffic saturation frequently prevents control personnel from accepting flight plans by radio. In such cases, the pilot is advised to contact the nearest FSS for the purpose of filing the flight plan.

5–7a1 NOTE— There are several methods of obtaining IFR clearances at nontower, non-FSS, and outlying airports. The procedure may vary due to geographical features, weather conditions, and the complexity of the ATC system. To determine the most effective means of receiving an IFR clearance, pilots should ask the nearest FSS the most appropriate means of obtaining the IFR clearance.

2. When filing an IFR flight plan for a TCAS/heavy equipped aircraft, add the prefix T for TCAS, H for Heavy, or B for both TCAS and heavy to the aircraft type.

EXAMPLE:

H/DC10/U T/B727/A B/B747/R

3. When filing an IFR flight plan for flight in an aircraft equipped with a radar beacon transponder, DME equipment, TACAN-only equipment or a combination of both, identify equipment capability by adding a suffix to the AIRCRAFT TYPE preceded by a slant, as follows:

/X—no transponder.

/T—transponder with no altitude encoding capability.

/U—transponder with altitude encoding capability.

/D—DME, but no transponder.

/B—DME and transponder, but no altitude encoding capability.

/A—DME and transponder with altitude encoding capability.

/M—TACAN only, but no transponder.

/N—TACAN only and transponder, but with no altitude encoding capability.

/P—TACAN only and transponder with altitude encoding capability.

/C—RNAV and transponder, but with no altitude encoding capability.

/R—RNAV and transponder with altitude encoding capability.

/W—RNAV but no transponder.

4. It is recommended that pilots file the maximum transponder or navigation capability of their aircraft in the equipment suffix. This will provide ATC with the necessary information to utilize all facets of navigational equipment and transponder capabilities available. In the case of area navigation equipped aircraft, pilots should file the /C, /R, or /W capability of the aircraft even though an RNAV route or random RNAV route has not been requested. This will ensure ATC awareness of the pilot's ability to navigate point-to-point and may be utilized to expedite the flight.

5–7a4 NOTE— The suffix is not to be added to the aircraft identification or be transmitted by radio as part of the aircraft identification.

b. Airways and Jet Routes Depiction on Flight Plan—

1. It is vitally important that the route of flight be accurately and completely described in the flight plan. To simplify definition of the proposed route, and to facilitate ATC, pilots are requested to file via airways or jet routes established for use at the altitude or Flight Level planned.

2. If flight is to be conducted via designated airways or jet routes, describe the route by indicating the type and number designators of the airway(s) or jet route(s) requested. If more than one airway or jet route is to be used, clearly indicate points of transition. If the transition is made at an unnamed intersection, show the next succeeding NAVAID or named intersection on the intended route and the complete route from that point. Reporting points may be identified by using authorized name/code as depicted on appropriate aeronautical charts. The following two examples illustrate the need to specify the transition point when two routes share more than one transition fix.

EXAMPLE:

ALB J37 BUMPY J14 BHM

SPELLED OUT: From Albany, New York, via Jet Route 37 transitioning to Jet Route 14 at BUMPY intersection, thence via Jet Route 14 to Birmingham, Alabama.

EXAMPLE:

ALB J37 ENO J14 BHM

SPELLED OUT: From Albany, New York, via Jet Route 37 transitioning to Jet Route 14 at Kenton VORTAC (ENO) thence via Jet Route 14 to Birmingham, Alabama.

(a) The route of flight may also be described by naming the reporting points or NAVAID's over which the flight will pass, provided the points named are established for use at the altitude or flight level planned.

EXAMPLE:

BWI V44 SWANN V433 DQO

SPELLED OUT: From Baltimore-Washington International, via Victor 44 To Swann Intersection, transitioning to Victor 433 at Swann, thence via V433 to Dupont.

(b) When the route of flight is defined by named reporting points, whether alone or in combination with airways or jet routes, and the navigational aids (VOR, VORTAC, TACAN, NDB) to be used for the flight are a combination of different types of aids, enough information should be included to clearly indicate the route requested.

EXAMPLE:

LAX J5 LKV J3 GEG YXC FL 330 J500 VLR J515 YWG

SPELLED OUT: From Los Angeles International via Jet Route 5 Lakeview, Jet Route 3 Spokane, direct Cranbrook, British Columbia VOR/DME, Flight Level 330 Jet Route 500 to Langruth, Manitoba VORTAC, Jet Route 515 to Winnepeg, Manitoba.

(c) When filing IFR, it is to the pilot's advantage to file a preferred route.

5–72c NOTE— Preferred IFR routes are described and tabulated in the Airport/Facility Directory.

(d) ATC may issue a SID or a STAR, as appropriate. (Reference—Instrument Departure, paragraph 5–26 and Standard Terminal Arrival (STAR), paragraph 5–41).

5–72d NOTE— Pilots not desiring a SID or STAR should so indicate in the remarks section of the flight plan as "NO SID" or "NO STAR".

c. Direct Flights—

1. All or any portions of the route which will not be flown on the radials or courses of established airways or routes, such as direct route flights, must be defined by indicating the radio fixes over which the flight will pass. Fixes selected to define the route shall be those over which the position of the aircraft can be accurately determined. Such fixes automatically become compulsory reporting points for the flight, unless advised otherwise by ATC. Only those navigational aids established for use in a particular structure; i.e., in the low or high structures, may be used to define the en route phase of a direct flight within that altitude structure.

2. The azimuth feature of VOR aids and that azimuth and distance (DME) features of VORTAC and TACAN aids are assigned certain frequency protected areas of airspace which are intended for application to established airway and route use, and to provide guidance for planning flights outside of established airways or routes. These areas of airspace are expressed in terms of cylindrical service volumes of specified dimensions called "class limits" or "categories." (Reference—Navaid Service Volumes, paragraph 1–8). An operational service volume has been established for each class in which adequate signal coverage and frequency protection can be assured. To facilitate use of VOR, VORTAC, or TACAN aids, consistent with their operational service volume limits, pilot use of such aids for defining a direct route of flight in controlled airspace should not exceed the following:

(a) Operations above FL 450–Use aids not more than 200 NM apart. These aids are depicted on En Route High Altitude Charts.

(b) Operation off established routes from 18,000 feet MSL to FL 450–Use aids not more than 260 NM apart. These aids are depicted on En Route High Altitude Charts.

(c) Operation off established airways below 18,000 feet MSL-Use aids not more than 80 NM apart. These aids are depicted on Enroute Low Altitude Charts.

(d) Operation off established airways between 14,500 feet MSL and 17,999 feet MSL in the conterminous U.S.-(H) facilities not more than 200 NM apart may be used.

3. Increasing use of self-contained airborne navigational systems which do not rely on the VOR/VORTAC/TACAN system has resulted in pilot requests for direct routes which exceed NAVAID service volume limits. These direct route requests will be approved only in a radar environment, with approval based on pilot responsibility for navigation on the authorized direct route. Radar flight following will be provided by ATC for ATC purposes.

4. At times, ATC will initiate a direct route in a radar environment which exceeds NAVAID service volume limits. In such cases ATC will provide radar monitoring and navigational assistance as necessary.

5. Airway or jet route numbers, appropriate to the stratum in which operation will be conducted, may also be included to describe portions of the route to be flown.

EXAMPLE:

MDW V262 BDF V10 BRL STJ SLN GCK

SPELLED OUT: From Chicago Midway Airport via Victor 262 to Bradford, Victor 10 to Burlington, Iowa, direct St. Joseph, Missouri, direct Salina, Kansas, direct Garden City, Kansas.

5–7c5 NOTE— When route of flight is described by radio fixes, the pilot will be expected to fly a direct course between the points named.

6. Pilots are reminded that they are responsible for adhering to obstruction clearance requirements on those segments of direct routes that are outside of controlled airspace. The MEA's and other altitudes shown on Low Altitude IFR Enroute Charts pertain to those route segments within controlled airspace, and those altitudes may not meet obstruction clearance criteria when operating off those routes.

d. Area Navigation (RNAV)—

1. Random RNAV routes can only be approved in a radar environment. Factors that will be considered by ATC in approving random RNAV routes include the capability to provide radar monitoring and compatibility with traffic volume and flow. ATC will radar monitor each flight, however, navigation on the random RNAV route is the responsibility of the pilot.

2. To be certified for use in the National Airspace System, RNAV equipment must meet the specifications outlined in AC 90–45. The pilot is responsible for variations in equipment capability and must advise ATC if a RNAV clearance can not be accepted as specified. The controller need only be concerned that the aircraft is RNAV equipped; if the flight plan equipment suffix denotes RNAV capability, the RNAV routing can be applied.

3. Pilots of aircraft equipped with operational area navigation equipment may file for random RNAV routes throughout the National Airspace System, where radar monitoring by ATC is available, in accordance with the following procedures.

(a) File airport-to-airport flight plans prior to departure.

(b) File the appropriate RNAV capability certification suffix in the flight plan.

(c) Plan the random route portion of the flight plan to begin and end over appropriate arrival and departure transition fixes or appropriate navigation aids for the altitude stratum within which the flight will be conducted. The use of normal preferred departure and arrival routes (SID/STAR), where established, is recommended.

(d) File route structure transitions to and from the random route portion of the flight.

(e) Define the random route by waypoints. File route description waypoints by using degree-distance fixes based on navigational aids which are appropriate for the altitude stratum.

(f) File a minimum of one route description waypoint for each ARTCC through whose area the random route will be flown. These waypoints must be located within 200 NM of the preceding center's boundary.

(g) File an additional route description waypoint for each turnpoint in the route.

(h) Plan additional route description waypoints as required to ensure accurate navigation via the filed route of flight. Navigation is the pilot's responsibility unless ATC assistance is requested.

(i) Plan the route of flight so as to avoid Prohibited and Restricted Airspace by 3 NM unless permission has been obtained to operate in that airspace and the appropriate ATC facilities are advised.

4. Pilots of aircraft equipped with latitude/longitude coordinate navigation capability, independent of VOR/TACAN references, may file for random RNAV routes at and above FL 390 within the conterminous United States using the following procedures.

(a) File airport-to-airport flight plans prior to departure.

(b) File the appropriate RNAV capability certification suffix in the flight plan.

(c) Plan the random route portion of the flight to begin and end over published departure/arrival transition fixes or appropriate navigation aids for airports without published transition procedures. The use of preferred departure and arrival routes, such as SID and STAR where established, is recommended.

(d) Plan the route of flight so as to avoid prohibited and restricted airspace by 3 NM unless permission has been obtained to operate in that airspace and the appropriate ATC facility is advised.

(e) Define the route of flight after the departure fix, including each intermediate fix (turnpoint) and the arrival fix for the destination airport in terms of latitude/longitude coordinates plotted to the nearest minute. The arrival fix must be identified by both the latitude/longitude coordinates and a fix identifier.

Example:

MIA[1] SRQ[2] 3407/10615[3] 3407/11546 TNP[4] LAX[5]

[1] Departure airport.

[2] Departure fix.

[3] Intermediate fix (turning point).

[4] Arrival fix.

[5] Destination airport.

(f) Record latitude/longitude coordinates by four figures describing latitude in degrees and minutes followed by a solidus and five figures describing longitude in degrees and minutes.

(g) File at FL 390 or above for the random RNAV portion of the flight.

(h) Fly all routes/route segments on Great Circle tracks.

(i) Make any inflight requests for random RNAV clearances or route amendments to an en route ATC facility.

e. Flight Plan Form—See Figure 5-7[1].

f. Explanation of IFR Flight Plan Items—

Block 1. Check the type flight plan. Check both the VFR and IFR blocks if composite VFR/IFR.

Block 2. Enter your complete aircraft identification including the prefix "N" if applicable.

Block 3. Enter the designator for the aircraft, or if unknown, the aircraft manufacturer's name; e.g., Cessna, followed by a slant(/), and the transponder or DME equipment code letter; e.g., C–182/U. Heavy aircraft, add prefix "H" to aircraft type; example: H/DC10/U.

Block 4. Enter your computed true airspeed (TAS).

5–7f Block 4 NOTE— If the average TAS changes plus or minus 5 percent or 10 knots, whichever is greater, advise ATC.

Figure 5–7[1]

Form Approved: OMB No. 2120-0026

U.S. DEPARTMENT OF TRANSPORTATION FEDERAL AVIATION ADMINISTRATION **FLIGHT PLAN**	(FAA USE ONLY) ☐ PILOT BRIEFING ☐ VNR ☐ STOPOVER	TIME STARTED	SPECIALIST INITIALS

1. TYPE	2. AIRCRAFT IDENTIFICATION	3. AIRCRAFT TYPE/ SPECIAL EQUIPMENT	4. TRUE AIRSPEED	5. DEPARTURE POINT	6. DEPARTURE TIME		7. CRUISING ALTITUDE
☐ VFR ☐ IFR ☐ DVFR			KTS		PROPOSED (Z)	ACTUAL (Z)	

8. ROUTE OF FLIGHT

9. DESTINATION (Name of airport and city)	10. EST. TIME ENROUTE		11. REMARKS
	HOURS	MINUTES	

12. FUEL ON BOARD		13. ALTERNATE AIRPORT(S)	14. PILOT'S NAME, ADDRESS & TELEPHONE NUMBER & AIRCRAFT HOME BASE	15. NUMBER ABOARD
HOURS	MINUTES		17. DESTINATION CONTACT/TELEPHONE (OPTIONAL)	

16. COLOR OF AIRCRAFT	CIVIL AIRCRAFT PILOTS. FAR Part 91 requires you file an IFR flight plan to operate under instrument flight rules in controlled airspace. Failure to file could result in a civil penalty not to exceed $1,000 for each violation (Section 901 of the Federal Aviation Act of 1958, as amended). Filing of a VFR flight plan is recommended as a good operating practice. See also Part 99 for requirements concerning DVFR flight plans.

FAA Form 7233-1 (8-82) CLOSE VFR FLIGHT PLAN WITH ____________ FSS ON ARRIVAL

Block 5. Enter the departure airport identifier code (or the name if the identifier is unknown).

5–7 Block 5 NOTE— Use of identifier codes will expedite the processing of your flight plan.

Block 6. Enter the proposed departure time in Coordinated Universal Time (UTC) (Z). If airborne, specify the actual or proposed departure time as appropriate.

Block 7. Enter the requested en route altitude or flight level.

5–7 Block 7 NOTE— Enter only the initial requested altitude in this block. When more than one IFR altitude or flight level is desired along the route of flight, it is best to make a subsequent request direct to the controller.

Block 8. Define the route of flight by using NAVAID identifier codes (or names if the code is unknown), airways, jet routes, and waypoints (for RNAV).

5–7 Block 8 NOTE— Use NAVAID's or WAYPOINT's to define direct routes and radials/bearings to define other unpublished routes.

Block 9. Enter the destination airport identifier code (or name if the identifier is unknown).

Block 10. Enter your Estimated Time en Route based on latest forecast winds.

Block 11. Enter only those remarks pertinent to ATC or to the clarification of other flight plan information, such as the appropriate radiotelephony (call sign) associated with the designator filed in Block 2. Items of a personal nature are not accepted. Do not assume that remarks will be automatically transmitted to every controller. Specific ATC or en route requests should be made directly to the appropriate controller.

Block 12. Specify the fuel on board, computed from the departure point.

Block 13. Specify an alternate airport if desired or required, but do not include routing to the alternate airport.

Block 14. Enter the complete name, address, and telephone number of pilot-in-command, or in the case of a formation flight, the formation commander. Enter sufficient information to identify home base, airport, or operator.

5–7 Block 14 NOTE— This information would be essential in the event of search and rescue operation.

Block 15. Enter the total number of persons on board including crew.

Block 16. Enter the predominant colors.

5–7 Block 16 NOTE— Close IFR flight plans with tower, approach control, or ARTCC, or if unable, with FSS. When landing at an airport with a functioning control tower, IFR flight plans are automatically canceled.

g. The information transmitted to the ARTCC for IFR flight plans will consist of only flight plan blocks 2, 3, 4, 5, 6, 7, 8, 9, 10, and 11.

h. A description of the International Flight Plan Form is contained in the International Flight Information Manual (IFIM).

5–8. IFR OPERATIONS TO HIGH ALTITUDE DESTINATIONS

Pilots planning IFR flights to airports located in mountainous terrain are cautioned to consider the necessity for an alternate airport even when the forecast weather conditions would technically relieve them from the requirement to file one (Reference: FAR 91.167 and paragraph 4–17(b)). The FAA has identified three possible situations where the failure to plan for an alternate airport when flying IFR to such a destination airport could result in a critical situation if the weather is less than forecast and sufficient fuel is not available to proceed to a suitable airport.

a. An IFR flight to an airport where the MDA's or landing visibility minimums for *ALL INSTRUMENT APPROACHES* are higher than the forecast weather minimums specified in FAR 91.167(b). For example, there are 11 high altitude airports in the United States with approved instrument approach procedures where all of the Minimum Descent Altitudes (MDA's) are greater than 2,000 feet and/or the landing visibility minimums are greater than 3 miles (Bishop, California; South Lake Tahoe, California; Ukiah, California; Aspen-Pitkin Co./Sardy Field, Colorado; Butte, Montana; Helena, Montana; Missoula, Montana; Chadron, Nebraska; Ely, Nevada; Klamath Falls, Oregon; and Omak, Washington). In the case of these 11 airports, it is possible for a pilot to elect, on the basis of forecasts, not to carry sufficient fuel to get to an alternate when the ceiling and/or visibility is actually lower than that necessary to complete the approach.

b. A small number of other airports in mountainous terrain have MDA's which are slightly (100 to 300 feet) below 2,000 feet AGL. In situations where there is an option as to whether to plan for an alternate, pilots should bear in mind that just a slight worsening of the weather conditions from those forecast could place the airport below the published IFR landing minimums.

c. An IFR flight to an airport which requires special equipment; i.e., DME, glide slope, etc., in order to make the available approaches to the lowest minimums. Pilots should be aware that all other minimums on the approach charts may require weather conditions better than those specified in FAR 91.167(b). An inflight equipment malfunction could result in the inability to comply with the published approach procedures or, again, in the position of having the airport below the published IFR landing minimums for all remaining instrument approach alternatives.

5–9. FLIGHTS OUTSIDE THE UNITED STATES AND U.S. TERRITORIES

a. When conducting flights, particularly extended flights, outside the U.S. and its territories, full account should be taken of the amount and quality of air navigation services available in the airspace to be traversed. Every effort should be made to secure information on the location and range of navigational aids, availability of communications and meteorological services, the provision of air traffic services, including alerting service, and the existence of search and rescue services.

b. Pilots should remember that there is a need to continuously guard the VHF emergency frequency 121.5 mHz when on long over-water flights, except when communications on other VHF channels, equipment limitations, or cockpit duties prevent simultaneous guarding of two channels. Guarding of 121.5 mHz is particularly critical when operating in proximity to Flight Information Region (FIR) boundaries, for example, operations on Route R220 between Anchorage and Tokyo, since it serves to facilitate communications with regard to aircraft which may experience in-flight emergencies, communications, or navigational difficulties. (Reference ICAO Annex 10, Vol II Paras 5.2.2.1.1.1 and 5.2.2.1.1.2.)

c. The filing of a flight plan, always good practice, takes on added significance for extended flights outside U.S. airspace and is, in fact, usually required by the laws of the countries being visited or overflown. It is also particularly important in the case of such flights that pilots leave a complete itinerary and schedule of the flight with someone directly concerned, keep that person advised of the flight's progress and inform him that if serious doubt arises as to the safety of the flight he should first contact the appropriate FSS. Round Robin Flight Plans to Mexico are not accepted.

d. All pilots should review the foreign airspace and entry restrictions published in the IFIM during the flight planning process. Foreign airspace penetration without official authorization can involve both danger to the aircraft and the imposition of severe penalties and inconvenience to both passengers and crew. A flight plan on file with ATC authorities does not necessarily constitute the prior permission required by certain other authorities. The possibility of fatal consequences cannot be ignored in some areas of the world.

e. Current NOTAM's for foreign locations must also be reviewed. The publication International Notices to Airmen, published biweekly, contains considerable information pertinent to foreign flight. Current foreign NOTAM's are also available from the U.S. International NOTAM Office in Washington, D.C., through any local FSS.

f. When customs notification is required, it is the responsibility of the pilot to arrange for customs notification in a timely manner. The following guidelines are applicable:

1. When customs notification is required on flights to Canada and Mexico and a predeparture flight plan cannot be filed or an advise customs message (ADCUS) cannot be included in a predeparture flight plan, call the nearest en route domestic or International FSS as soon as radio communication can be established and file a VFR or DVFR flight plan, as required, and include as the last item the advise customs information. The station with which such a flight plan is filed will forward it to the appropriate FSS who will notify the customs office responsible for the destination airport.

2. If the pilot fails to include ADCUS in the radioed flight plan, it will be assumed that other arrangements have been made and FAA will not advise customs.

3. The FAA assumes no responsibility for any delays in advising customs if the flight plan is given too late for delivery to customs before arrival of the aircraft. It is still the pilot's responsibility to give timely notice even though a flight plan is given to FAA.

5–10. CHANGE IN FLIGHT PLAN

In addition to altitude or flight level, destination and/or route changes, increasing or decreasing the speed of an aircraft constitutes a change in a flight plan. Therefore, at any time the average true airspeed at cruising altitude between reporting points varies or is expected to vary from that given in the flight plan by *plus or minus 5 percent, or 10 knots, whichever is greater,* ATC should be advised.

5–11. CHANGE IN PROPOSED DEPARTURE TIME

a. To prevent computer saturation in the en route environment, parameters have been established to delete proposed departure flight plans which have not been activated. Most centers have this parameter set so as to delete these flight plans a minimum of 1 hour after the proposed departure time. To ensure that a flight plan remains active, pilots whose actual departure time will be delayed 1 hour or more beyond their filed departure time, are requested to notify ATC of their departure time.

b. Due to traffic saturation, control personnel frequently will be unable to accept these revisions via radio. It is recommended that you forward these revisions to the nearest FSS.

5–12. CLOSING VFR/DVFR FLIGHT PLANS

A pilot is responsible for ensuring that his VFR or DVFR flight plan is canceled (FAR 91.153 and FAR 91.169). You should close your flight plan with the nearest FSS, or if one is not available, you may request any ATC facility to relay your cancellation to the FSS. *Control towers do not automatically close VFR or DVFR flight plans* since they do not know if a particular VFR aircraft is on a flight plan. If you fail to report or cancel your flight plan within ½ hour after your ETA, search and rescue procedures are started.

5–13. CANCELING IFR FLIGHT PLAN

a. FAR 91.153 and FAR 91.169 includes the statement "When a flight plan has been filed, the pilot-in-command, upon canceling or completing the flight under the flight plan, shall notify the nearest FSS or ATC facility."

b. An IFR flight plan may be canceled at any time the flight is operating in VFR conditions outside positive controlled airspace by the pilot stating "CANCEL MY IFR FLIGHT PLAN" to the controller or air/ground station with which he is communicating. Immediately after canceling an IFR flight plan, a pilot should take necessary action to change to the appropriate air/ground frequency, VFR radar beacon code and VFR altitude or flight level.

c. ATC separation and information services will be discontinued, including radar services (where applicable). Consequently, if the canceling flight desires VFR radar advisory service, the pilot must specifically request it.

5–13c NOTE— Pilots must be aware that other procedures may be applicable to a flight that cancels an IFR flight plan within an area where a special program, such as a designated TRSA, ARSA, or TCA, has been established.

d. If a DVFR flight plan requirement exists, the pilot is responsible for filing this flight plan to replace the canceled IFR flight plan. If a subsequent IFR operation becomes necessary, a new IFR flight plan must be filed and an ATC clearance obtained before operating in IFR conditions.

e. If operating on an IFR flight plan to an airport with a functioning control tower, the flight plan is automatically closed upon landing.

f. If operating on an IFR flight plan to an airport where there is no functioning control tower, the pilot must initiate cancellation of the IFR flight plan. This can be done after landing if there is a functioning FSS or other means of direct communications with ATC. In the event there is no FSS and air/ground communications with ATC is not possible below a certain altitude, the pilot should, weather conditions permitting, cancel his IFR flight plan while still airborne and able to communicate with ATC by radio. This will not only save the time and expense of canceling the flight plan by telephone but will quickly release the airspace for use by other aircraft.

5–14 thru 5–20. RESERVED

Section 2. DEPARTURE PROCEDURES

5-21. PRE-TAXI CLEARANCE PROCEDURES

a. Certain airports have established Pre-taxi Clearance programs whereby pilots of departing IFR aircraft may elect to receive their IFR clearances before they start taxiing for takeoff. The following provisions are included in such procedures:

1. Pilot participation is not mandatory.

2. Participating pilots call clearance delivery or ground control not more than 10 minutes before proposed taxi time.

3. IFR clearance (or delay information, if clearance cannot be obtained) is issued at the time of this initial call-up.

4. When the IFR clearance is received on clearance delivery frequency, pilots call ground control when ready to taxi.

5. Normally, pilots need not inform ground control that they have received IFR clearance on clearance delivery frequency. Certain locations may, however, require that the pilot inform ground control of a portion of his routing or that he has received his IFR clearance.

6. If a pilot cannot establish contact on clearance delivery frequency or has not received his IFR clearance before he is ready to taxi, he contacts ground control and informs the controller accordingly.

b. Locations where these procedures are in effect are indicated in the Airport/Facility Directory.

5-22. TAXI CLEARANCE

Pilots on IFR flight plans should communicate with the control tower on the appropriate ground control or clearance delivery frequency, prior to starting engines, to receive engine start time, taxi and/or clearance information.

5-23. ABBREVIATED IFR DEPARTURE CLEARANCE (CLEARED...AS FILED) PROCEDURES

a. ATC facilities will issue an abbreviated IFR departure clearance based on the ROUTE of flight filed in the IFR flight plan, provided the filed route can be approved with little or no revision. These abbreviated clearance procedures are based on the following conditions:

1. The aircraft is on the ground or it has departed VFR and the pilot is requesting IFR clearance while airborne.

2. That a pilot will not accept an abbreviated clearance if the route or destination of a flight plan filed with ATC has been changed by him or the company or the operations officer before departure.

3. That it is the responsibility of the company or operations office to inform the pilot when they make a change to the filed flight plan.

4. That it is the responsibility of the pilot to inform ATC in his initial call-up (for clearance) when the filed flight plan has been either:

(a) amended, or

(b) canceled and replaced with a new filed flight plan.

5-23a3b NOTE—The facility issuing a clearance may not have received the revised route or the revised flight plan by the time a pilot requests clearance.

b. The controller will issue a detailed clearance when he knows that the original filed flight plan has been changed or when the pilot requests a full route clearance.

c. The clearance as issued will include the destination airport filed in the flight plan.

d. ATC procedures now require the controller to state the SID name, the current number and the SID Transition name after the phrase "Cleared to (destination) airport" and prior to the phrase, "then as filed," for ALL departure clearances when the SID or SID Transition is to be flown. The procedures apply whether or not the SID is filed in the flight plan.

e. STARs, when filed in a flight plan, are considered a part of the filed route of flight and will not normally be stated in an initial departure clearance. If the ARTCC's jurisdictional airspace includes both the departure airport and the fix where a STAR or STAR Transition begins, the STAR name, the current number and the STAR Transition name MAY be stated in the initial clearance.

f. "Cleared to (destination) airport as filed" does NOT include the en route altitude filed in a flight plan. An en route altitude will be stated in the clearance or the pilot will be advised to expect an assigned or filed altitude within a given time frame or at a certain point after departure. This may be done verbally in the departure instructions or stated in the SID.

g. In both radar and nonradar environments, the controller will state "Cleared to (destination) airport as filed" or:

1. If a SID or SID Transition is to be flown, specify the SID name, the current SID number, the SID Transition name, the assigned altitude/ Flight Level, and any additional instructions (departure control frequency, beacon code assignment, etc.) necessary to clear a departing aircraft via the SID or SID Transition and the route filed.

EXAMPLE:
NATIONAL SEVEN TWENTY CLEARED TO MIAMI AIRPORT INTERCONTINENTAL ONE DEPARTURE, LAKE CHARLES TRANSITION THEN AS FILED, MAINTAIN FLIGHT LEVEL TWO SEVEN ZERO.

2. When there is no SID or when the pilot cannot accept a SID, the controller will specify the assigned altitude or Flight Level, and any additional instructions necessary to clear a departing aircraft via an appropriate departure routing and the route filed.

5–23g2 NOTE—A detailed departure route description or a radar vector may be used to achieve the desired departure routing.

3. If it is necessary to make a minor revision to the filed route, the controller will specify the assigned SID or SID Transition (or departure routing), the revision to the filed route, the assigned altitude or flight level and any additional instructions necessary to clear a departing aircraft.

EXAMPLE:
JET STAR ONE FOUR TWO FOUR CLEARED TO ATLANTA AIRPORT, SOUTH BOSTON TWO DEPARTURE THEN AS FILED EXCEPT CHANGE ROUTE TO READ SOUTH BOSTON VICTOR 20 GREENSBORO, MAINTAIN ONE SEVEN THOUSAND.

4. Additionally, in a nonradar environment, the controller will specify one or more fixes, as necessary, to identify the initial route of flight.

EXAMPLE:
CESSNA THREE ONE SIX ZERO FOXTROT CLEARED TO CHARLOTTE AIRPORT AS FILED VIA BROOKE, MAINTAIN SEVEN THOUSAND.

h. To ensure success of the program, pilots should:

1. Avoid making changes to a filed flight plan just prior to departure.

2. State the following information in the initial call-up to the facility when no change has been made to the filed flight plan: Aircraft call sign, location, type operation (IFR) and the name of the airport (or fix) to which you expect clearance.

EXAMPLE:
"WASHINGTON CLEARANCE DELIVERY (or ground control if appropriate) AMERICAN SEVENTY SIX AT GATE ONE, IFR LOS ANGELES."

3. If the flight plan has been changed, state the change and request a full route clearance.

EXAMPLE:
"WASHINGTON CLEARANCE DELIVERY, AMERICAN SEVENTY SIX AT GATE ONE. IFR SAN FRANCISCO. MY FLIGHT PLAN ROUTE HAS BEEN AMENDED (or destination changed). REQUEST FULL ROUTE CLEARANCE."

4. Request verification or clarification from ATC if ANY portion of the clearance is not clearly understood.

5. When requesting clearance for the IFR portion of a VFR/IFR flight, request such clearance prior to the fix where IFR operation is proposed to commence in sufficient time to avoid delay. Use the following phraseology:

EXAMPLE:
"LOS ANGELES CENTER, APACHE SIX ONE PAPA, VFR ESTIMATING PASO ROBLES VOR AT THREE TWO, ONE THOUSAND FIVE HUNDRED, REQUEST IFR TO BAKERSFIELD."

5–24. DEPARTURE RESTRICTIONS, CLEARANCE VOID TIMES, HOLD FOR RELEASE, AND RELEASE TIMES

a. ATC may assign departure restrictions, clearance void times, hold for release, and release times, when necessary, to separate departures from other traffic or to restrict or regulate the departure flow.

1. CLEARANCE VOID TIMES—If operating from an airport not served by a control tower, the pilot may receive a clearance containing a provision that if the flight has not departed by a specific time, the clearance is void. The pilot who does not depart prior to the void time, and is still on the ground, must advise ATC as soon as possible of his intentions. Thirty minutes is the maximum amount of time allotted for the pilots to notify ATC that they were not able to depart prior to the void time, and are still on the ground (controllers will advise pilots of the actual amount of time allotted, e.g., 10, 15, 30 minutes). Failure of the pilot to notify ATC that they could not meet their clearance void time will result in the aircraft being considered overdue 30 minutes after the clearance void time. At this point search and rescue procedures are initiated and IFR traffic is suspended.

5–24a1 NOTE—Pilots who depart at or after their void time are not afforded IFR separation and may be in violation of FAR 91.173 which requires that pilots receive an appropriate ATC clearance before operating IFR in controlled airspace.

EXAMPLE:
CLEARANCE VOID IF NOT OFF BY (clearance void time) and, if required, IF NOT OFF BY (clearance void time) ADVISE (facility) NOT LATER THAN (time) OF INTENTIONS.

2. HOLD FOR RELEASE—ATC may issue "hold for release" instructions in a clearance to delay an aircraft's departure for traffic management reasons (i.e., weather, traffic volume, etc.). When ATC states in the clearance, "hold for release," the pilot may not depart until he receives a release time or is given additional instructions by ATC. In addition, ATC will include departure delay information in conjunction with "hold for release" instructions.

EXAMPLE:
(aircraft identification) CLEARED TO (destination) AIRPORT AS FILED, MAINTAIN (altitude), and, if required (additional instructions or information), HOLD FOR RELEASE, EXPECT (time in hours and/or minutes) DEPARTURE DELAY.

3. RELEASE TIMES—A "release time" is a departure restriction issued to a pilot by ATC, specifying the earliest time an aircraft may depart. ATC will use "release times" in conjunction with traffic management procedures and/or to separate a departing aircraft from other traffic.

EXAMPLE:

(aircraft identification) RELEASED FOR DEPARTURE AT (time in hours and/or minutes).

5–25. DEPARTURE CONTROL

a. Departure Control is an approach control function responsible for ensuring separation between departures. So as to expedite the handling of departures, Departure Control may suggest a take off direction other than that which may normally have been used under VFR handling. Many times it is preferred to offer the pilot a runway that will require the fewest turns after take off to place the pilot on his filed course or selected departure route as quickly as possible. At many locations particular attention is paid to the use of preferential runways for local noise abatement programs, and route departures away from congested areas.

b. Departure Control utilizing radar will normally clear aircraft out of the terminal area using SIDs via radio navigation aids. When a departure is to be vectored immediately following take off, the pilot will be advised prior to take off of the initial heading to be flown but may not be advised of the purpose of the heading. Pilots operating in a radar environment are expected to associate departure headings with vectors to their planned route or flight. When given a vector taking his aircraft off a previously assigned nonradar route, the pilot will be advised briefly what the vector is to achieve. Thereafter, radar service will be provided until the aircraft has been reestablished "on-course" using an appropriate navigation aid and the pilot has been advised of his position or a handoff is made to another radar controller with further surveillance capabilities.

c. Controllers will inform pilots of the departure control frequencies and, if appropriate, the transponder code before takeoff. Pilots should not operate their transponder until ready to start the takeoff roll or change to the departure control frequency until requested. Controllers may omit the departure control frequency if a SID has or will be assigned and the departure control frequency is published on the SID.

5–26. INSTRUMENT DEPARTURES

a. STANDARD INSTRUMENT DEPARTURES (SID)

1. A SID is an ATC coded departure procedure which has been established at certain airports to simplify clearance delivery procedures.

2. Pilots of civil aircraft operating from locations where SID procedures are effective may expect ATC clearance containing a SID. Use of a SID requires pilot possession of at least the textual description of the SID procedures. Controllers may omit the departure control frequency if a SID clearance is issued and the departure control frequency is published on the SID. If the pilot does not possess a charted SID or a preprinted SID description or, for any other reason, does not wish to use a SID, he is expected to advise ATC. Notification may be accomplished by filing "NO SID" in the remarks section of the filed flight plan or by the less desirable method of verbally advising ATC.

3. All effective SIDs are published in textual and graphic form by the National Ocean Service in Terminal Procedures Publication (TPP).

4. SID procedures will be depicted in one of two basic forms.

(a) **Pilot Navigation (Pilot NAV) SIDs:** are established where the pilot is primarily responsible for navigation on the SID route. They are established for airports when terrain and safety related factors indicate the necessity for a pilot NAV SID. Some pilot NAV SIDs may contain vector instructions which pilots are expected to comply with until instruction are received to resume normal navigation on the filed/assigned route or SID procedure.

(b) **Vector SIDs:** are established where ATC will provide radar navigational guidance to a filed/assigned route or to a fix depicted on the SID.

b. OBSTRUCTION CLEARANCE DURING DEPARTURE

1. Published instrument departure procedures and SIDs assist pilots conducting IFR flight in avoiding obstacles during climbout to Minimum en Route Altitude (MEA). These procedures are established only at locations where instrument approach procedures are published. Standard instrument takeoff minimums and departure procedures are prescribed in FAR 91.175. Airports with takeoff minimums other than standard (one statute mile for aircraft having two engines or less and one-half statute mile for aircraft having more than two engines) are described in airport listings on separate pages titled IFR TAKE-OFF MINIMUMS AND DEPARTURE PROCEDURES, at the front of each U.S. Government Terminal Procedures Publication (TPP). The approach chart and SID chart for each airport where takeoff minimums are not standard and/or departure procedures are published is annotated with a special symbol ▼. The use of this symbol indicates that the separate listing should be consulted. These minimums also apply to SIDs unless the SIDs specify different minimums.

2. Obstacle clearance is based on the aircraft climbing at least 200 feet per nautical mile, crossing the end of the runway at least 35 feet AGL, and climbing to 400 feet above airport elevation before turning, unless otherwise specified in the procedure. A slope of 152 feet per nautical mile, starting no higher than 35 feet above the departure end of the runway, is assessed for obstacles. A minimum obstacle clearance of 48 feet per nautical mile is provided in the assumed climb gradient.

(a) If no obstacles penetrate the 152 feet per nautical mile slope, IFR departure procedures are not published.

(b) If obstacles do penetrate the slope, avoidance procedures are specified. These procedures may be: a ceiling and visibility to allow the obstacles to be seen and avoided; a climb gradient greater than 200 feet per nautical mile; detailed flight maneuvers; or a combination of the above. In extreme cases, IFR takeoff may not be authorized for some runways.

EXAMPLE:

Rwy 17, 300–1 or standard with minimum climb of 220 feet per NM to 1100.

3. Climb gradients are specified when required for obstacle clearance. Crossing restrictions in the SIDs may be established for traffic separation or obstacle clearance. When no gradient is specified, the pilot is expected to climb at least 200 feet per nautical mile to MEA unless required to level off by a crossing restriction.

EXAMPLE:

"CROSS ALPHA INTERSECTION AT OR BELOW 4000; MAINTAIN 6000." The pilot climbs at least 200 feet per nautical mile to 6000. If 4000 is reached before ALPHA, the pilot levels off at 4000 until passing ALPHA; then immediately resumes at least 200 feet per nautical mile climb.

4. Climb gradients may be specified to an altitude/fix, above which the normal gradient applies.

EXAMPLE:

"MINIMUM CLIMB 340 FEET PER NM TO 2700." The pilot climbs at least 340 feet per nautical mile to 2700, then at least 200 feet per NM to MEA.

5. Some IFR departure procedures require a climb in visual conditions to cross the airport (or an on-airport NAVAID) in a specified direction, at or above a specified altitude.

EXAMPLE:

"CLIMB IN VISUAL CONDITIONS SO AS TO CROSS THE McELORY AIRPORT SOUTHBOUND, AT OR ABOVE 6000, THEN CLIMB VIA KEEMMLING R–033 TO KEEMMLING VORTAC."

(a) When climbing in visual conditions it is the pilot's responsibility to see and avoid obstacles. Specified ceiling and visibility minimums will allow visual avoidance of obstacles until the pilot enters the standard obstacle protection area. Obstacle avoidance is not guaranteed if the pilot maneuvers farther from the airport than the visibility minimum.

(b) That segment of the procedure which requires the pilot to see and avoid obstacles ends when the aircraft crosses the specified point at the required altitude. Thereafter, standard obstacle protection is provided.

6. Each pilot, prior to departing an airport on an IFR flight should consider the type of terrain and other obstacles on or in the vicinity of the departure airport and:

(a) Determine whether a departure procedure and/or SID is available for obstacle avoidance.

(b) Determine if obstacle avoidance can be maintained visually or that the departure procedure or SID should be followed.

(c) Determine what action will be necessary and take such action that will assure a safe departure.

5–26b6c NOTE—The term **Radar Contact,** when used by the controller during departure, should not be interpreted as relieving pilots of their responsibility to maintain appropriate terrain and obstruction clearance. (Reference—Pilot/Controller Glossary, Radar Contact).

Terrain/obstruction clearance is not provided by ATC until the controller begins to provide navigational guidance, i.e., Radar Vectors.

5–27 thru 5–30. RESERVED

Section 3. EN ROUTE PROCEDURES

5–31. ARTCC COMMUNICATIONS

a. Direct Communications, Controllers and Pilots—

1. ARTCC's are capable of direct communications with IFR air traffic on certain frequencies. Maximum communications coverage is possible through the use of Remote Center Air/Ground (RCAG) sites comprised of both VHF and UHF transmitters and receivers. These sites are located throughout the U.S. Although they may be several hundred miles away from the ARTCC, they are remoted to the various ARTCC's by land lines or microwave links. Since IFR operations are expedited through the use of direct communications, pilots are requested to use these frequencies strictly for communications pertinent to the control of IFR aircraft. Flight plan filing, en route weather, weather forecasts, and similar data should be requested through FSS's, company radio, or appropriate military facilities capable of performing these services.

2. An ARTCC is divided into sectors. Each sector is handled by one or a team of controllers and has its own sector discrete frequency. As a flight progresses from one sector to another, the pilot is requested to change to the appropriate sector discrete frequency.

b. ATC Frequency Change Procedures—

1. The following phraseology will be used by controllers to effect a frequency change:

EXAMPLE:

(Aircraft identification) CONTACT (facility name or location name and terminal function) (frequency) AT (time, fix, or altitude).

5–31b1 NOTE—Pilots are expected to maintain a listening watch on the transferring controller's frequency until the time, fix, or altitude specified. ATC will omit frequency change restrictions whenever pilot compliance is expected upon receipt.

2. The following phraseology should be utilized by pilots for establishing contact with the designated facility:

(a) When a position report will be made:

EXAMPLE:

(Name) CENTER, (aircraft identification), (position).

(b) When no position report will be made:

EXAMPLE:

(Name) CENTER, (aircraft identification), (position) ESTIMATING (reporting point and time) AT (altitude or Flight Level) (CLIMBING or DESCENDING) TO MAINTAIN (altitude or flight level).

(c) When operating in a radar environment and no position report is required: On initial contact, the pilot should inform the controller of the pilot's assigned altitude preceded by the words "level," or "climbing to," or "descending to," as appropriate; and the pilot's present vacating altitude, if applicable. Also, when on other than published routes, the pilot should include the presently assigned routing on initial contact with each air traffic controller.

EXAMPLES:

1. (Name) CENTER, (aircraft identification), LEVEL (altitude or flight level), HEADING (exact heading).

2. (Name) CENTER, (aircraft identification), LEAVING (exact altitude or flight level), CLIMBING TO (altitude or flight level), PROCEEDING TO (name) V-O-R VIA THE (VOR name) (number) RADIAL.

3. (Name) CENTER, (aircraft identification), LEAVING (exact altitude or flight level), DESCENDING TO (altitude or flight level), DIRECT (name) V-O-R.

5–31b2c NOTE—Exact altitude or flight level means to the nearest 100 foot increment. Exact altitude or flight level reports on initial contact provide ATC with information required prior to using MODE C altitude information for separation purposes.

3. At times controllers will ask pilots to verify that they are at a particular altitude. The phraseology used will be: "VERIFY AT (altitude)." In climbing or descending situations, controllers may ask pilots to "VERIFY ASSIGNED ALTITUDE AS (altitude)." Pilots should confirm that they are at the altitude stated by the controller or that the assigned altitude is correct as stated. If this is not the case, they should inform the controller of the actual altitude being maintained or the different assigned altitude.

CAUTION: Pilots should not take action to change their actual altitude or different assigned altitude to the altitude stated in the controllers verification request unless the controller specifically authorizes a change.

c. ARTCC Radio Frequency Outage. ARTCC's normally have at least one back-up radio receiver and transmitter system for each frequency, which can usually be placed into service quickly with little or no disruption of ATC service. Occasionally, technical problems may cause a delay but switchover seldom takes more than 60 seconds. When it appears that the outage will not be quickly remedied, the ARTCC will usually request a nearby aircraft, if there is one, to switch to the affected frequency to broadcast communications instructions. It is important, therefore, that the pilot wait at least 1 minute before deciding that the ARTCC has actually experienced a radio frequency failure. When such an outage does occur, the pilot should, if workload and equipment capability permit, maintain a listening watch on the affected frequency while attempting to comply with the following recommended communications procedures:

1. If two-way communications cannot be established with the ARTCC after changing frequencies, a pilot should attempt to recontact the transferring controller for the assignment of an alternative frequency or other instructions.

2. When an ARTCC radio frequency failure occurs after two-way communications have been established, the pilot should attempt to reestablish contact with the center on any other known ARTCC frequency, preferably that of the next responsible sector when practicable, and ask for instructions. However, when the next normal frequency change along the route is known to involve another ATC facility, the pilot should contact that facility, if feasible, for instructions. If communications cannot be reestablished by either method, the pilot is expected to request communications instructions from the FSS appropriate to the route of flight.

5–31c2 NOTE—The exchange of information between an aircraft and an ARTCC through an FSS is quicker than relay via company radio because the FSS has direct interphone lines to the responsible ARTCC sector. Accordingly, when circumstances dictate a choice between the two, during an ARTCC frequency outage, relay via FSS radio is recommended.

5–32. POSITION REPORTING

The safety and effectiveness of traffic control depends to a large extent on accurate position reporting. In order to provide the proper separation and expedite aircraft movements, ATC must be able to make accurate estimates of the progress of every aircraft operating on an IFR flight plan.

a. Position Identification—

1. When a position report is to be made passing a VOR radio facility, the time reported should be the time at which the first complete reversal of the "to/from" indicator is accomplished.

2. When a position report is made passing a facility by means of an airborne ADF, the time reported should be the time at which the indicator makes a complete reversal.

3. When an aural or a light panel indication is used to determine the time passing a reporting point, such as a fan marker, Z marker, cone of silence or intersection of range courses, the time should be noted when the signal is first received and again when it ceases. The mean of these two times should then be taken as the actual time over the fix.

4. If a position is given with respect to distance and direction from a reporting point, the distance and direction should be computed as accurately as possible.

5. Except for terminal area transition purposes, position reports or navigation with reference to aids not established for use in the structure in which flight is being conducted will not normally be required by ATC.

b. Position Reporting Points—FAR's require pilots to maintain a listening watch on the appropriate frequency and, unless operating under the provisions of (c), to furnish position reports passing certain reporting points. Reporting points are indicated by symbols on en route charts. The designated compulsory reporting point symbol is a solid triangle and the "on request" reporting point symbol is the open triangle. Reports passing an "on request" reporting point are only necessary when requested by ATC.

c. Position Reporting Requirements—

1. Flights along airways or routes—A position report is required by all flights regardless of altitude, including those operating in accordance with an ATC clearance specifying "VFR ON TOP," over each designated compulsory reporting point along the route being flown.

2. Flight Along a Direct Route—Regardless of the altitude or flight level being flown, including flights operating in accordance with an ATC clearance specifying "VFR ON TOP", pilots shall report over each reporting point used in the flight plan to define the route of flight.

3. Flights in a Radar Environment—When informed by ATC that their aircraft are in "Radar Contact," pilots should discontinue position reports over designated reporting points. They should resume normal position reporting when ATC advises "RADAR CONTACT LOST" or "RADAR SERVICE TERMINATED."

5–32c3 NOTE—ATC will inform a pilot that he is in "RADAR CONTACT" (a) when his aircraft is initially identified in the ATC system; and (b) when radar identification is reestablished after radar service has been terminated or radar contact lost. Subsequent to being advised that the controller has established radar contact, this fact will not be repeated to the pilot when handed off to another controller. At times, the aircraft identity will be *confirmed* by the receiving controller; however, this should not be construed to mean that radar contact has been lost. The identity of transponder equipped aircraft will be confirmed by asking the pilot to "IDENT," "SQUAWK STANDBY," or to change codes. Aircraft without transponders will be advised of their position to confirm identity. In this case, *the pilot is expected to advise the controller if he disagrees with the position given.* If the pilot cannot confirm the accuracy of the position given because he is not tuned to the NAVAID referenced by the controller, the pilot should ask for another radar position relative to the NAVAID to which he is tuned.

d. Position Report Items—

1. Position reports should include the following items:

(a) Identification.

(b) Position.

(c) Time.

(d) Altitude or flight level (include actual altitude or flight level when operating on a clearance specifying VFR-ON-TOP.)

(e) Type of flight plan (not required in IFR position reports made directly to ARTCC's or approach control),

(f) ETA and name of next reporting point.

(g) The name only of the next succeeding reporting point along the route of flight, and

(h) Pertinent remarks.

5-33. ADDITIONAL REPORTS

a. The following reports should be made to ATC or FSS facilities without a specific ATC request:

1. At all times:

(a) When vacating any previously assigned altitude or flight level for a newly assigned altitude or flight level.

(b) When an altitude change will be made if operating on a clearance specifying VFR ON TOP.

(c) When *unable* to climb/descend at a rate of a least 500 feet per minute.

(d) When approach has been missed. (Request clearance for specific action; i.e., to alternative airport, another approach, etc.)

(e) Change in the average true airspeed (at cruising altitude) when it varies by 5 percent or 10 knots (whichever is greater) from that filed in the flight plan.

(f) The time and altitude or flight level upon reaching a holding fix or point to which cleared.

(g) When leaving any assigned holding fix or point.

5-33a1g NOTE—The reports in subparagraphs (f) and (g) may be omitted by pilots of aircraft involved in instrument training at military terminal area facilities when radar service is being provided.

(h) Any loss, in controlled airspace, of VOR, TACAN, ADF, low frequency navigation receiver capability, complete or partial loss of ILS receiver capability or impairment of air/ground communications capability. Reports should include aircraft identification, equipment affected, degree to which the capability to operate under IFR in the ATC system is impaired, and the nature and extent of assistance desired from ATC.

5-33a1h NOTE—Other equipment installed in an aircraft may effectively impair safety and/or the ability to operate under IFR. If such equipment (e.g. airborne weather radar) malfunctions and in the pilot's judgment either safety or IFR capabilities are affected, reports should be made as above.

(i) Any information relating to the safety of flight.

2. When not in radar contact:

(a) When leaving final approach fix inbound on final approach (non precision approach) or when leaving the outer marker or fix used in lieu of the outer marker inbound on final approach (precision approach).

(b) A corrected estimate at anytime it becomes apparent that an estimate as previously submitted is in error in excess of 3 minutes.

b. Pilots encountering weather conditions which have not been forecast, or hazardous conditions which have been forecast, are expected to forward a report of such weather to ATC. (Reference—Pilot Weather Reports (PIREPs), paragraph 7-19 and FAR 91.183(b) and (c).)

5-34. AIRWAYS AND ROUTE SYSTEMS

a. Two fixed route systems are established for air navigation purposes. They are the VOR and L/MF system, and the jet route system. To the extent possible, these route systems are aligned in an overlying manner to facilitate transition between each.

1. The VOR and L/MF Airway System consists of airways designated from 1,200 feet above the surface (or in some instances higher) up to but not including 18,000 feet MSL. These airways are depicted on En Route Low Altitude Charts.

5-34a1 NOTE—The altitude limits of a Victor airway should not be exceeded except to effect transition within or between route structures.

(a) Except in Alaska and coastal North Carolina, the VOR airways are predicated solely on VOR or VORTAC navigation aids; are depicted in blue on aeronautical charts; and are identified by a "V" (Victor) followed by the airway number (e.g., V12).

5-34a1a NOTE—Segments of VOR airways in Alaska and North Carolina (V56, V290) are based on L/MF navigation aids and charted in brown instead of blue on en route charts.

(1) A segment of an airway which is common to two or more routes carries the numbers of all the airways which coincide for that segment. When such is the case, a pilot filing a flight plan needs to indicate only that airway number of the route which he is using.

(2) *Alternate Airways* are identified by their location with respect to the associated main airway. "Victor 9 West" indicates an alternate airway associated with, and lying to the west of Victor 9.

5-34a1a2 NOTE—A pilot who intends to make an airway flight, using VOR facilities, will simply specify the appropriate "Victor" airways(s) in his flight plan. For example, if a flight is to be made from Chicago to New Orleans at 8,000 feet, using omniranges only, the route may be indicated as "Departing from Chicago-Midway, cruising 8,000 feet via Victor 9 to Moisant International." If flight is to be conducted in part by means of L/MF navigation aids and in part on omniranges, specifications of the appropriate airways in the flight plan will indicate which types of facilities will be used along the described routes, and, for IFR flight, permit ATC to issue a traffic clearance accordingly. A route may also be described by specifying the station over which the flight will pass, but in this case since many VOR's and L/MF aids have the same name, the pilot must be careful to indicate which aid will be used at a particular location. This will be indicated in the route of flight portion of the flight plan by specifying the type of facility to be used after the location name in the following manner: Newark L/MF, Allentown VOR.

(3) With respect to position reporting, reporting points are designated for VOR Airway Systems. Flights using Victor Airways will report over these points unless advised otherwise by ATC.

(b) The L/MF airways (colored airways) are predicated solely on L/MF navigation aids and are depicted in brown on aeronautical charts and are identified by color name and number (e.g., Amber One). Green and Red airways are plotted east and west. Amber and Blue airways are plotted north and south.

5–34a1ab NOTE—Except for G13 in North Carolina, the colored airway system exists only in the State of Alaska. All other such airways formerly so designated in the conterminous U.S. have been rescinded.

2. The Jet Route system consists of jet routes established from 18,000 feet MSL to FL 450 inclusive.

(a) These routes are depicted on En Route High Altitude Charts. Jet routes are depicted in blue on aeronautical charts and are identified by a "J" (Jet) followed by the airway number (e.g., J 12). Jet routes, as VOR airways, are predicated solely on VOR or VORTAC navigation facilities (except in Alaska).

5–34a2a NOTE—Segments of jet routes in Alaska are based on L/MF navigation aids and are charted in brown color instead of blue on en route charts.

(b) With respect to position reporting, reporting points are designated for Jet Route systems. Flights using Jet Routes will report over these points unless otherwise advised by ATC.

3. Area Navigation (RNAV) Routes—

(a) RNAV is a method of navigation that permits aircraft operations on any desired course within the coverage of station referenced navigation signals or within the limits of a self-contained system capability or combination of these.

(b) Fixed RNAV routes are permanent, published routes which can be flight planned for use by aircraft with RNAV capability. A previously established fixed RNAV route system has been terminated except for a few high altitude routes in Alaska.

(c) Random RNAV routes are direct routes, based on area navigation capability, between waypoints defined in terms of latitude/longitude coordinates, degree-distance fixes, or offsets from established routes/airways at a specified distance and direction. Radar monitoring by ATC is required on all random RNAV routes.

b. Operation above FL 450 may be conducted on a point-to-point basis. Navigational guidance is provided on an area basis utilizing those facilities depicted on the En Route High Altitude Charts.

c. Radar Vectors. Controllers may vector aircraft within controlled airspace for separation purposes, noise abatement considerations, when an operational advantage will be realized by the pilot or the controller, or when requested by the pilot. Vectors outside of controlled airspace will be provided only on pilot request. Pilots will be advised as to what the vector is to achieve when the vector is controller initiated and will take the aircraft off a previously assigned nonradar route. To the extent possible, aircraft operating on RNAV routes will be allowed to remain on their own navigation.

d. When flying in Canadian airspace, pilots are cautioned to review Canadian Air Regulations.

1. Special attention should be given to the parts which differ from U.S. FAR's.

(a) The Canadian Airways CLASS B airspace restriction is an example. CLASS B airspace is all controlled low level airspace above 12,500 feet MSL or the MEA, whichever is higher, within which only IFR and controlled VFR flights are permitted. (Low level airspace means an airspace designated and defined as such in the *Designated Airspace Handbook.)*

(b) Regardless of the weather conditions or the height of the terrain, no person shall operate an aircraft under VFR conditions within CLASS B airspace except in accordance with a clearance for VFR flight issued by ATC.

(c) Unless otherwise authorized by the Director General, Civil Aeronautics of the Department of Transportation, no person holding a commercial or private pilot license without an instrument rating shall operate an aircraft in VFR flight within CLASS B airspace unless successful completion of a written examination demonstrating knowledge of radio navigation and of ATC procedures applicable to IFR flight including clearances and position reports is accomplished. The pilot license must be endorsed to that effect.

2. Segments of VOR airways and high level routes in Canada are based on L/MF navigation aids and are charted in brown color instead of blue on en route charts.

5–35. AIRWAY OR ROUTE COURSE CHANGES

a. Pilots of aircraft are required to adhere to airways or routes being flown. Special attention must be given to this requirement during course changes. Each course change consists of variables that make the technique applicable in each case a matter only the pilot can resolve. Some variables which must be considered are turn radius, wind effect, airspeed, degree of turn, and cockpit instrumentation. An early turn, as illustrated below, is one method of adhering to airways or routes. The use of any available cockpit instrumentation, such as Distance Measuring Equipment, may be used by the pilot to lead his turn when making course changes. This *is consistent* with the intent of FAR 91.181, which requires pilots to operate along the centerline of an airway and along the direct course between navigational aids or fixes.

Figure 5–35[1]

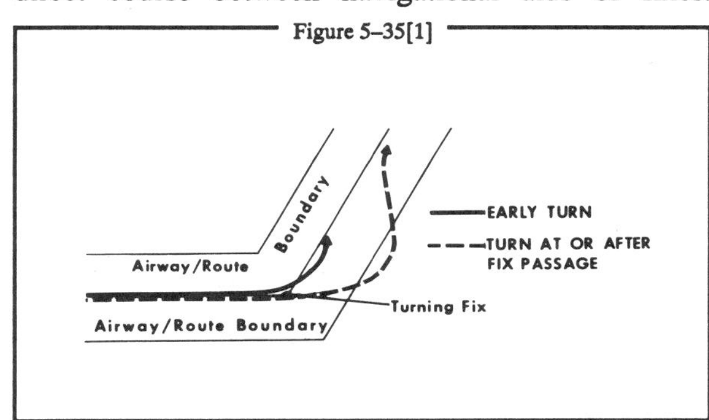

b. Turns which begin at or after fix passage may exceed airway or route boundaries. Figure 5–35[1].) contains an example flight track depicting this, together with an example of an early turn.

c. Without such actions as leading a turn, aircraft operating in excess of 290 knots TAS can exceed the normal airway or route boundaries depending on the amount of course change required, wind direction and velocity, the character of the turn fix (DME, overhead navigation aid, or intersection), and the pilot's technique in making a course change. For example, a flight operating at 17,000 feet MSL with a TAS of 400 knots, a 25 degree bank, and a course change of more than 40 degrees would exceed the width of the airway or route; i.e., 4 nautical miles each side of centerline. However, in the airspace below 18,000 feet MSL, operations in excess of 290 knots TAS are not prevalent and the provision of additional IFR separation in all course change situations for the occasional aircraft making a turn in excess of 290 knots TAS creates an unacceptable waste of airspace and imposes a penalty upon the preponderance of traffic which operate at low speeds. Consequently, the FAA expects pilots to lead turns and take other actions they consider necessary during course changes to adhere as closely as possible to the airways or route being flown.

d. Due to the high airspeeds used at 18,000 feet MSL and above, FAA provides additional IFR separation protection for course changes made at such altitude levels.

5–36. CHANGEOVER POINTS (COP'S)

a. COP's are prescribed for Federal Airways, jet routes, Area Navigation routes, or other direct routes for which an MEA is designated under FAR 95. The COP is a point along the route or airway segment between two adjacent navigation facilities or way points where changeover in navigation guidance should occur. At this point, the pilot should change navigation receiver frequency from the station behind the aircraft to the station ahead.

b. The COP is located midway between the navigation facilities for straight route segments, or at the intersection of radials or courses forming a dogleg in the case of dogleg route segments. When the COP is NOT located at the midway point, aeronautical charts will depict the COP location and give the mileage to the radio aids.

c. COP's are established for the purpose of preventing loss of navigation guidance, to prevent frequency interference from other facilities, and to prevent use of different facilities by different aircraft in the same airspace. Pilots are urged to observe COP's to the fullest extent.

5–37. HOLDING

a. Whenever an aircraft is cleared to a fix other than the destination airport and delay is expected, it is the responsibility of the ATC controller to issue complete holding instructions (unless the pattern is charted), an EFC time and his best estimate of any additional en route/terminal delay.

5–37a NOTE—Only those holding patterns depicted on U.S. Government or commercially produced (meeting FAA requirements) *Low/High Altitude Enroute,* and *Area* or *STAR* charts should be used.

b. If the holding pattern is charted and the controller doesn't issue complete holding instructions, the pilot is expected to hold as depicted on the appropriate chart. When the pattern is charted, the controller may omit all holding instructions except the charted holding direction and the statement *AS PUBLISHED;* e.g., *HOLD EAST AS PUBLISHED.* Controllers shall always issue complete holding instructions when pilots request them.

c. If no holding pattern is charted and holding instructions have not been issued, the pilot should ask ATC for holding instructions prior to reaching the fix. This procedure will eliminate the possibility of an aircraft entering a holding pattern other than that desired by ATC. If the pilot is unable to obtain holding instructions prior to reaching the fix (due to frequency congestion, stuck microphone, etc.), he should hold in a standard pattern on the course on which he approached the fix and request further clearance as soon as possible. In this event, the altitude/flight level of the aircraft at the clearance limit will be protected so that separation will be provided as required.

d. When an aircraft is 3 minutes or less from a clearance limit and a clearance beyond the fix has not been received, the pilot is expected to start a speed reduction so that he will cross the fix, initially, at or below the maximum holding airspeed.

e. When no delay is expected, the controller should issue a clearance beyond the fix as soon as possible and, whenever possible, at least 5 minutes before the aircraft reaches the clearance limit.

f. Pilots should report to ATC the time and altitude/flight level at which the aircraft reaches the clearance limit and report leaving the clearance limit.

5–37f NOTE—In the event of two-way communications failure, pilots are required to comply with FAR 91.185.

g. When holding at a VOR station, pilots should begin the turn to the outbound leg at the time of the first complete reversal of the to/from indicator.

h. Patterns at the most generally used holding fixes are depicted (charted) on U.S. Government or commercially produced (meeting FAA requirements) Low or High Altitude En route, Area and STAR Charts. Pilots are expected to hold in the pattern depicted unless specifically advised otherwise by ATC.

i. An ATC clearance requiring an aircraft to hold at a fix where the pattern is not charted will include the following information: (See Figure 5–37[1].)

1. Direction of holding from the fix in terms of the eight cardinal compass points (i.e., N, NE, E, SE, etc.).

2. Holding fix (the fix may be omitted if included at the beginning of the transmission as the clearance limit).

3. Radial, course, bearing, airway or route on which the aircraft is to hold.

4. Leg length in miles if DME or RNAV is to be used (leg length will be specified in minutes on pilot request or if the controller considers it necessary).

5. Direction of turn if left turns are to be made, the pilot requests, or the controller considers it necessary.

Figure 5–37[1]

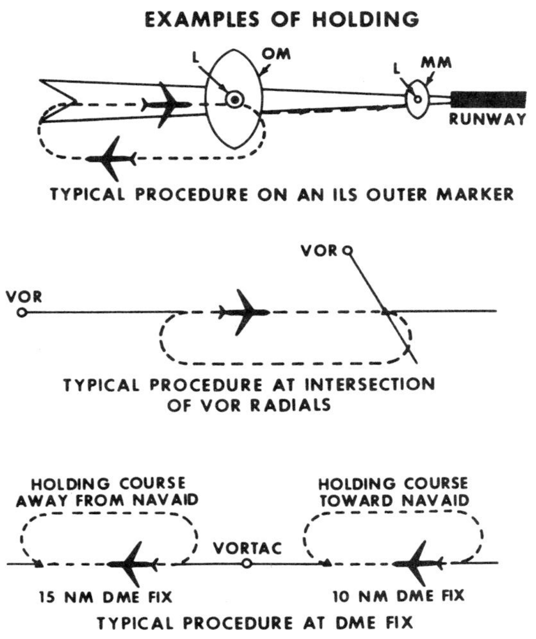

EXAMPLES OF HOLDING

L OM L MM RUNWAY

TYPICAL PROCEDURE ON AN ILS OUTER MARKER

VOR VOR

TYPICAL PROCEDURE AT INTERSECTION OF VOR RADIALS

HOLDING COURSE AWAY FROM NAVAID HOLDING COURSE TOWARD NAVAID VORTAC

15 NM DME FIX 10 NM DME FIX

TYPICAL PROCEDURE AT DME FIX

6. Time to expect further clearance and any pertinent additional delay information.

j. Holding pattern airspace protection is based on the following procedures. They are the only procedures for entry and holding recommended by FAA.

1. Descriptive Terms—

Figure 5–37[2]

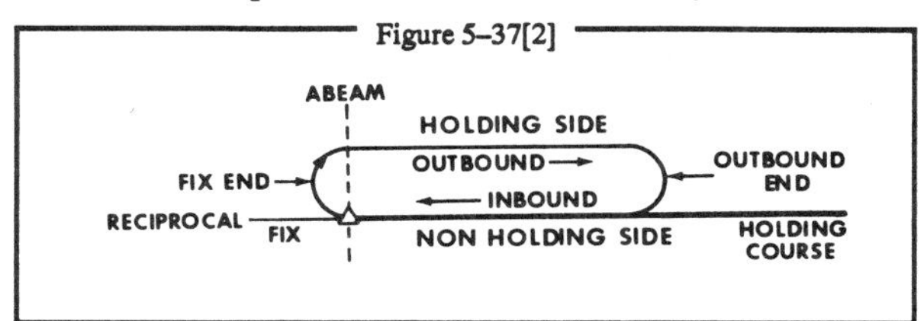

(a) Standard Pattern: Right turns (See Figure 5–37[2].)

(b) Nonstandard Pattern: Left turns

2. Airspeeds (maximum)—(See Table 5–37[1][Propeller–Driven], Table 5–37[2][Civil Turbojet], and Table 5–37[3][Military Turbojet]).

Table 5–37[1]

PROPELLER-DRIVEN	MAXIMUM AIRSPEED
1. All (including turboprop)	175K IAS

Table 5–37[2]

CIVIL TURBOJET	MAXIMUM AIRSPEED
1. MHA through 14,000 feet	230K IAS
2. Above 14,000 feet	265K IAS

Table 5–37[3]

MILITARY TURBOJET	MAXIMUM AIRSPEED
1. All—Except aircraft listed in 2, 3, and 4	230K IAS
2. USAF F–4 aircraft	280K IAS
3. B–1, F–111, and F–5	310K IAS
4. T–37	175K IAS

5–37j2 NOTE 1—Additional military exceptions may be added to Table 5–37[3].

5–37j2 NOTE 2—Holding speed depends upon weight and drag configuration.

5–37j2 NOTE 3—Civil aircraft holding at military or joint civil/military use airports should expect to operate at a maximum holding pattern airspeed of 230 knots.

3. Entry (See Figure 5–37[3])—

(a) Parallel Procedure: Parallel holding course, turn left, and return to holding fix or intercept holding course.

Figure 5–37[3]

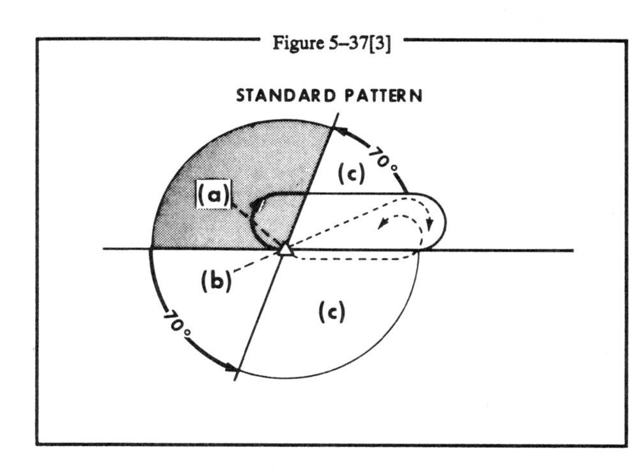

(b) Teardrop Procedure: Proceed on outbound track of 30 degrees (or less) to holding course, turn right to intercept holding course.

(c) Direct Entry Procedure: Turn right and fly the pattern.

4. Timing—

(a) Inbound Leg:

(1) At or below 14,000 Ft. MSL–1 minute.

(2) Above 14,000 Ft. MSL–1½ minutes.

5–37j4a2 NOTE—The *initial* outbound leg should be flown for 1 minute or 1 and ½ min. (appropriate to altitude). Timing for subsequent outbound legs should be adjusted, as necessary, to achieve proper inbound leg time. Pilots may use any navigational means available; i.e. DME, RNAV, etc., to insure the appropriate inbound leg times.

(b) Outbound leg timing begins *over/abeam* the fix, whichever occurs later. If the abeam position cannot be determined, start timing when turn to outbound is completed.

Figure 5–37[4]

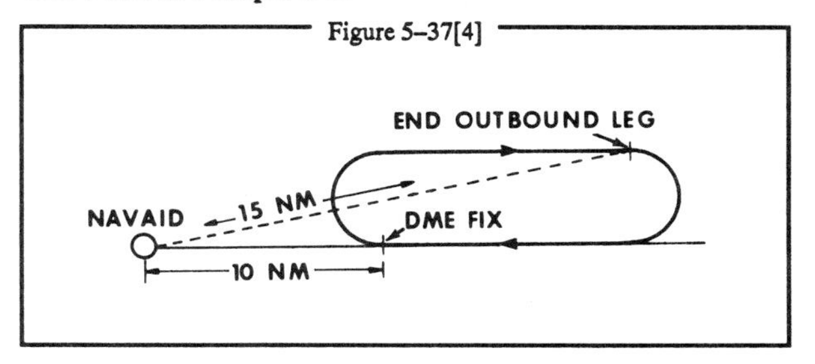

NOTE—When the inbound course is *toward* the NAVAID and the fix distance is 10 NM, and the leg length is 5 NM, then the end of the outbound leg will be reached when the DME reads 15 NM.

Figure 5–37[5]

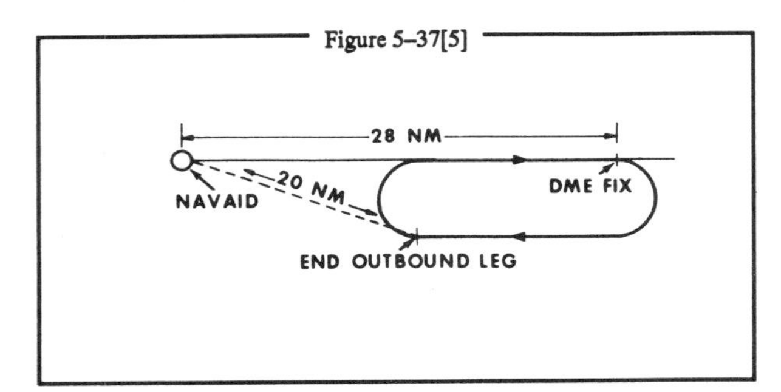

NOTE—When the inbound course is *away* from the NAVAID and the fix distance is 28 NM, and the leg length is 8 NM, then the end of the outbound leg will be reached when the DME reads 20 NM.

5. Distance Measuring Equipment (DME)—DME holding is subject to the same entry and holding procedures except that distances (nautical miles) are used in lieu of time values. The *outbound course* of a DME holding pattern is called the outbound leg of the pattern. The length of the outbound leg will be specified by the controller. The end of the outbound leg is determined by the odometer reading. (See Figure 5–37[4] and Figure 5–37[5].)

6. Pilot Action—

(a) Start speed reduction when 3 minutes or less from the holding fix. Cross the holding fix, initially, at or below the maximum holding airspeed.

(b) Make all turns during entry and while holding at:

(1) 3 degrees per second, or

(2) 30 degree bank angle, or

(3) 25 degree bank provided a flight director system is used.

5–37j6b3 NOTE—Use whichever requires the least bank angle.

(c) Compensate for wind effect primarily by drift correction on the inbound and outbound legs. When outbound, double the inbound drift correction to avoid major turning adjustments; e.g., if correcting left by 8 degrees when inbound, correct right by 16 degrees when outbound.

(d) Determine entry turn from aircraft heading upon arrival at the holding fix; +/–5 degrees in heading is considered to be within allowable good operating limits for determining entry.

(e) Advise ATC immediately if any increased airspeed is necessary due to turbulence, icing, etc., or if unable to accomplish any part of the holding procedures. When such higher speeds become no longer necessary, operate according to the appropriate published holding speed and notify ATC.

5–37j6e NOTE—Airspace protection for turbulent air holding is based on a maximum of 280K IAS/Mach 0.8, whichever is lower. Considerable impact on traffic flow will result when turbulent air holding patterns are used; thus, pilot discretion will ensure their use is limited to bona fide conditions or requirements.

7. Nonstandard Holding Pattern—Fix end and outbound end turns are made to the left. Entry procedures to a nonstandard pattern are oriented in relation to the 70 degree line on the holding side just as in the standard pattern.

k. When holding at a fix and instructions are received specifying the time of departure from the fix, the pilot should adjust his flight path within the limits of the established holding pattern in order to leave the fix at the exact time specified. After departing the holding fix, normal speed is to be resumed with respect to other governing speed requirements, such as terminal area speed limits, specific ATC requests, etc. Where the fix is associated with an instrument approach and timed approaches are in effect, a procedure turn shall not be executed unless the pilot advises ATC, since aircraft holding are expected to proceed inbound on final approach directly from the holding pattern when approach clearance is received.

l. Radar surveillance of outer fix holding pattern airspace areas.

1. Whenever aircraft are holding at an outer fix, ATC will usually provide radar surveillance of the outer fix holding pattern airspace area, or any portion of it, if it is shown on the controller's radar scope.

2. The controller will attempt to detect any holding aircraft that stray outside the holding pattern airspace area and will assist any detected aircraft to return to the assigned airspace area.

5–37l2 NOTE—Many factors could prevent ATC from providing this additional service, such as workload, number of targets, precipitation, ground clutter, and radar system capability. These circumstances may make it unfeasible to maintain radar identification of aircraft to detect aircraft straying from the holding pattern. The provision of this service depends entirely upon whether the controller believes he is in a position to provide it and does not relieve a pilot of his responsibility to adhere to an accepted ATC clearance.

3. If an aircraft is established in a published holding pattern at an assigned altitude above the published minimum holding altitude and subsequently cleared for the approach, the pilot may descend to the published minimum holding altitude. The holding pattern would only be a segment of the IAP *if* it is published on the instrument procedure chart and is used in lieu of a procedure turn.

m. For those holding patterns where there are no published minimum holding altitudes, the pilot, upon receiving an approach clearance, must maintain his last assigned altitude until leaving the holding pattern and established on the inbound course. Thereafter, the published minimum altitude of the route segment being flown will apply. It is expected that the pilot will be assigned a holding altitude that will permit a normal descent on the inbound course.

5–38 thru 5–40. RESERVED

Section 4. ARRIVAL PROCEDURES

5-41. STANDARD TERMINAL ARRIVAL (STAR)

a. A STAR is an ATC coded IFR arrival route established for application to arriving IFR aircraft destined for certain airports. Its purpose is to simplify clearance delivery procedures.

b. Pilots of IFR civil aircraft destined to locations for which STARs have been published may be issued a clearance containing a STAR whenever ATC deems it appropriate. Until military STAR publications and distribution is accomplished, STARs will be issued to military pilots only when requested in the flight plan or verbally by the pilot.

c. Use of STARs requires pilot possession of at least the approved textual description. As with any ATC clearance or portion thereof, it is the responsibility of each pilot to accept or refuse an issued STAR. A pilot should notify ATC if he does not wish to use a STAR by placing "NO STAR" in the remarks section of the flight plan or by the less desirable method of verbally stating the same to ATC.

d. STAR charts are published in the *Terminal Procedures Publication (TPP)* and are available on subscription from the National Ocean Service.

5-42. LOCAL FLOW TRAFFIC MANAGEMENT PROGRAM

a. This program is a continuing effort by the FAA to enhance safety, minimize the impact of aircraft noise and conserve aviation fuel. The enhancement of safety and reduction of noise is achieved in this program by minimizing low altitude maneuvering of arriving turbojet and turboprop aircraft weighing more than 12,500 pounds and, by permitting departure aircraft to climb to higher altitudes sooner, as arrivals are operating at higher altitudes at the points where their flight paths cross. The application of these procedures also reduces exposure time between controlled aircraft and uncontrolled aircraft at the lower altitudes in and around the terminal environment. Fuel conservation is accomplished by absorbing any necessary arrival delays for aircraft included in this program operating at the higher and more fuel efficient altitudes.

b. A fuel efficient descent is basically an uninterrupted descent (except where level flight is required for speed adjustment) from cruising altitude to the point when level flight is necessary for the pilot to stabilize his final approach. The procedure for a fuel efficient descent is based on an altitude loss which is most efficient for the majority of aircraft being served. This will generally result in a descent gradient window of 250-350 feet per nautical mile.

c. When crossing altitudes and speed restrictions are issued verbally or are depicted on a chart, ATC will expect the pilot to descend first to the crossing altitude and then reduce speed. Verbal clearances for descent will normally permit an uninterrupted descent in accordance with the procedure as described in paragraph b. above. Acceptance of a charted fuel efficient descent (Runway Profile Descent) clearance requires the pilot to adhere to the altitudes, speeds, and headings depicted on the charts unless otherwise instructed by ATC. PILOTS RECEIVING A CLEARANCE FOR A FUEL EFFICIENT DESCENT ARE EXPECTED TO ADVISE ATC IF THEY DO NOT HAVE RUNWAY PROFILE DESCENT CHARTS PUBLISHED FOR THAT AIRPORT OR ARE UNABLE TO COMPLY WITH THE CLEARANCE.

5-43. APPROACH CONTROL

a. Approach control is responsible for controlling all instrument flight operating within its area of responsibility. Approach control may serve one or more airfields, and control is exercised primarily by direct pilot and controller communications. Prior to arriving at the destination radio facility, instructions will be received from ARTCC to contact approach control on a specified frequency.

b. Radar Approach Control

1. Where radar is approved for approach control service, it is used not only for radar approaches (ASR and PAR) but is also used to provide vectors in conjunction with published nonradar approaches based on radio NAVAIDs (ILS, MLS, VOR, NDB, TACAN). Radar vectors can provide course guidance and expedite traffic to the final approach course of any established IAP or to the traffic pattern for a visual approach. Approach control facilities that provide this radar service will operate in the following manner:

(a) Arriving aircraft are either cleared to an outer fix most appropriate to the route being flown with vertical separation and, if required, given holding information or, when radar handoffs are effected between the ARTCC and approach control, or between two approach control facilities, aircraft are cleared to the airport or to a fix so located that the handoff will be completed prior to the time the aircraft reaches the fix. When radar handoffs are utilized, successive arriving flights may be handed off to approach control with radar separation in lieu of vertical separation.

(b) After release to approach control, aircraft are vectored to the final approach course (ILS, MLS, VOR, ADF, etc.). Radar vectors and altitude or Flight Levels will be issued as required for spacing and separating aircraft. *Therefore, pilots must not deviate from the headings issued by approach control.* Aircraft will normally be informed when it is necessary to vector across the final approach course for spacing or other reasons. If approach course crossing is imminent and the pilot has not been informed that he will be vectored across the final approach course, he should query the controller.

(c) The pilot is not expected to turn inbound on the final approach course unless an approach clearance has been issued. This clearance will normally be issued with the final vector for interception of the final approach course, and the vector will be such as to enable the pilot to establish his aircraft on the final approach course prior to reaching the final approach fix.

(d) In the case of aircraft already inbound on the final approach course, approach clearance will be issued prior to the aircraft reaching the final approach fix. When established inbound on the final approach course, radar separation will be maintained and the pilot will be expected to complete the approach utilizing the approach aid designated in the clearance (ILS, MLS, VOR, radio beacons, etc.) as the primary means of navigation. Therefore, once established on the final approach course, pilots must not deviate from it unless a clearance to do so is received from ATC.

(e) After passing the final approach fix on final approach, aircraft are expected to continue inbound on the final approach course and complete the approach or effect the missed approach procedure published for that airport.

2. Whether aircraft are vectored to the appropriate final approach course or provide their own navigation on published routes to it, radar service is automatically terminated when the landing is completed or when instructed to change to advisory frequency at uncontrolled airports, whichever occurs first.

5–44. ADVANCE INFORMATION ON INSTRUMENT APPROACH

a. When landing at airports with approach control services and where two or more IAPs are published, pilots will be provided in advance of their arrival with the type of approach to expect or that they may be vectored for a visual approach. This information will be broadcast either by a controller or on ATIS. It will not be furnished when the visibility is three miles or better and the ceiling is at or above the highest initial approach altitude established for any low altitude IAP for the airport.

b. The purpose of this information is to aid the pilot in planning arrival actions; however, it is not an ATC clearance or commitment and is subject to change. Pilots should bear in mind that fluctuating weather, shifting winds, blocked runway, etc., are conditions which may result in changes to approach information previously received. It is important that the pilot advise ATC immediately if he is unable to execute the approach ATC advised will be used, or if he prefers another type of approach.

c. When making an IFR approach to an airport not served by a tower or FSS, after the ATC controller advises "CHANGE TO ADVISORY FREQUENCY APPROVED" you should broadcast your intentions, including the type of approach being executed, your position, and when over the final approach fix inbound (non precision approach) or when over the outer marker or fix used in lieu of the outer marker inbound (precision approach). Continue to monitor the appropriate frequency (UNICOM, etc.) for reports from other pilots.

5–45. INSTRUMENT APPROACH PROCEDURE CHARTS

a. FAR 91.175a (Instrument Approaches to Civil Airports) requires the use of SIAPs prescribed for the airport in Part 97 unless otherwise authorized by the Administrator (including ATC). FAR 91.175g (Military Airports) requires civil pilots flying into or out of military airports to comply with the IAPs and takeoff and landing minimums prescribed by the authority having jurisdiction at those airports.

1. All IAPs (standard and special, civil and military) are based on joint civil and military criteria contained in the U.S. Standard for TERPs. The design of IAPs based on criteria contained in TERPs, takes into account the interrelationship between airports, facilities, and the surrounding environment, terrain, obstacles, noise sensitivity, etc. Appropriate altitudes, courses, headings, distances, and other limitations are specified and, once approved, the procedures are published and distributed by government and commercial cartographers as instrument approach charts.

2. Not all IAPs are published in chart form. Radar IAPs are established where requirements and facilities exist but they are printed in tabular form in appropriate U.S. Government Flight Information Publications.

3. A pilot adhering to the altitudes, flight paths, and weather minimums depicted on the IAP chart or vectors and altitudes issued by the radar controller, is assured of terrain and obstruction clearance and runway or airport alignment during approach for landing.

4. IAPs are designed to provide an IFR descent from the en route environment to a point where a safe landing can be made. They are prescribed and approved by appropriate civil or military authority to ensure a safe descent during instrument flight conditions at a specific airport. It is important that pilots understand these procedures and their use prior to attempting to fly instrument approaches.

5. TERPs criteria are provided for the following type of instrument approach procedures:

(a) Precision approaches where an electronic glide slope is provided (PAR and ILS) and,

(b) Nonprecision approaches where glide slope information is not provided (all approaches except PAR and ILS).

b. The method used to depict prescribed altitudes on instrument approach charts differs according to techniques employed by different chart publishers. Prescribed altitudes may be depicted in three different configurations: Minimum, maximum, and mandatory. The U.S. Government distributes charts produced by Defense Mapping Agency (DMA) and NOS. Altitudes are depicted on these charts in the profile view with underscore, overscore, or both to identify them as minimum, maximum, or mandatory.

1. Minimum Altitude will be depicted with the altitude value underscored. Aircraft are required to maintain altitude at or above the depicted value.

2. Maximum Altitude will be depicted with the altitude value overscored. Aircraft are required to maintain altitude at or below the depicted value.

3. Mandatory Altitude will be depicted with the altitude value both underscored and overscored. Aircraft are required to maintain altitude at the depicted value.

5–45b3 NOTE— The underscore and overscore to identify mandatory altitudes and the overscore to identify maximum altitudes are used almost exclusively by DMA for military charts. With very few exceptions, civil approach charts produced by NOS utilize only the underscore to identify minimum altitudes. Pilots are cautioned to adhere to altitudes as prescribed because, in certain instances, they may be used as the basis for vertical separation of aircraft by ATC. When a depicted altitude is specified in the ATC clearance, that altitude becomes mandatory as defined above.

c. Minimum Safe Altitudes (MSA) are published for emergency use on approach procedure charts utilizing NDB or VOR type facilities. The altitude shown provides at least 1,000 feet of clearance above the highest obstacle in the defined sector to a distance of 25 NM from the facility. As many as four sectors may be depicted with different altitudes for each sector displayed in rectangular boxes in the plan view of the chart. A single altitude for the entire area may be shown in the lower right portion of the plan view. Navigational course guidance is not assured at the MSA within these sectors.

Figure 5–45[1]

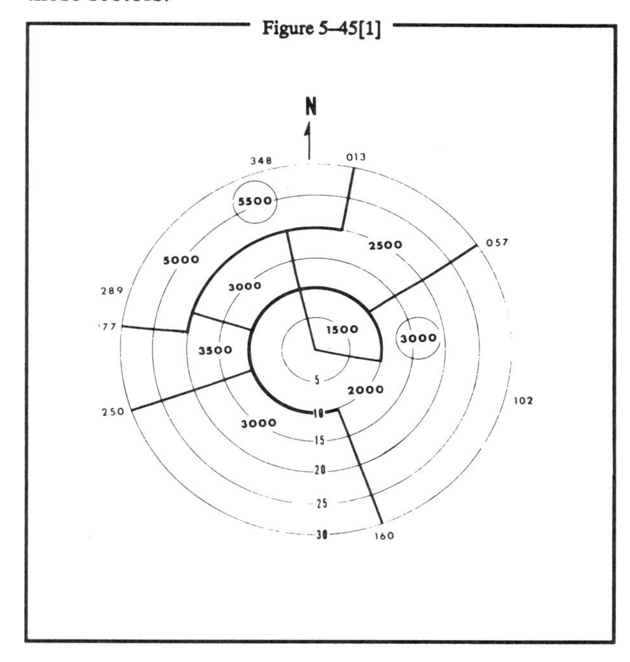

d. Minimum Vectoring Altitudes (MVA) are established for use by ATC when radar ATC is exercised. MVA charts are prepared by air traffic facilities at locations where there are numerous different minimum IFR altitudes. Each MVA chart has sectors large enough to accommodate vectoring of aircraft within the sector at the MVA. Each sector boundary is at least 3 miles from the obstruction determining the MVA. To avoid a large sector with an excessively high MVA due to an isolated prominent obstruction, the obstruction may be enclosed in a buffer area whose boundaries are at least 3 miles from the obstruction. This is done to facilitate vectoring around the obstruction. (See Figure 5–45[1].)

1. The minimum vectoring altitude in each sector provides 1,000 feet above the highest obstacle in nonmountainous areas and 2,000 feet above the highest obstacle in designated mountainous areas. Where lower MVAs are required in designated mountainous areas to achieve compatibility with terminal routes or to permit vectoring to an IAP, 1,000 feet of obstacle clearance may be authorized with the use of Airport Surveillance Radar (ASR). The minimum vectoring altitude will provide at least 300 feet above the floor of controlled airspace.

2. Because of differences in the areas considered for MVA, and those applied to other minimum altitudes, and the ability to isolate specific obstacles, some MVAs may be lower than the nonradar Minimum en Route Altitudes (MEA), Minimum Obstruction Clearance Altitudes (MOCA) or other minimum altitudes depicted on charts for a given location. While being radar vectored, IFR altitude assignments by ATC will be at or above MVA.

e. Visual Descent Points (VDP) are incorporated in selected nonprecision approach procedures. The VDP is a defined point on the final approach course of a nonprecision straight-in approach procedure from which normal descent from the MDA to the runway touchdown point may be commenced, provided visual reference required by FAR 91.175(c)(3) is established. The VDP will normally be identified by DME on VOR and LOC procedures. The VDP is identified on the profile view of the approach chart by the symbol: V .

1. VDPs are intended to provide additional guidance where they are implemented. No special technique is required to fly a procedure with a VDP. The pilot should not descend below the MDA prior to reaching the VDP and acquiring the necessary visual reference.

2. Pilots not equipped to receive the VDP should fly the approach procedure as though no VDP had been provided.

5–46. APPROACH CLEARANCE

a. An aircraft which has been cleared to a holding fix and subsequently "cleared . . . approach" has not received new routing. Even though clearance for the approach may have been issued prior to the aircraft reaching the holding fix, ATC would expect the pilot to proceed via the holding fix (his last assigned route), and the feeder route associated with that fix (if a feeder route is published on the approach chart) to the initial approach fix (IAF) to commence the approach. *When cleared for the approach, the published off airway (feeder) routes that lead from the en route structure to the IAF are part of the approach clearance.*

b. If a feeder route to an IAF begins at a fix located along the route of flight prior to reaching the holding fix, and clearance for an approach is issued, a pilot should commence his approach via the published feeder route; i.e., the aircraft would not be expected to overfly the feeder route and return to it. The pilot is expected to commence his approach in a similar manner at the IAF, if the IAF for the procedure is located along the route of flight to the holding fix.

c. If a route of flight directly to the initial approach fix is desired, it should be so stated by the controller with phraseology to include the words "direct . . .," "proceed direct" or a similar phrase which the pilot can interpret without question. If the pilot is uncertain of his clearance, he should immediately query ATC as to what route of flight is desired.

5–47. INSTRUMENT APPROACH PROCEDURES

a. Minimums are specified for various aircraft approach categories based upon a value 1.3 times the stalling speed of the aircraft in the landing configuration at maximum certificated gross landing weight. (See FAR 97.3(b)). If it is necessary, while circling-to-land, to maneuver at speeds in excess of the upper limit of the speed range for each category, due to the possibility of extending the circling maneuver beyond the area for which obstruction clearance is provided, the circling minimum for the next higher approach category should be used. For example, an aircraft which falls in Category C, but is circling to land at a speed of 141 knots or higher should use the approach category "D" minimum when circling to land.

b. When operating on an unpublished route or while being radar vectored, the pilot, when an approach clearance is received, shall, in addition to complying with the minimum altitudes for IFR operations (FAR 91.177), maintain his last assigned altitude unless a different altitude is assigned by ATC, or until the aircraft is established on a segment of a published route or IAP. After the aircraft is so established, published altitudes apply to descent within each succeeding route or approach segment unless a different altitude is assigned by ATC. Notwithstanding this pilot responsibility, for aircraft operating on unpublished routes or while being radar vectored, ATC will, except when conducting a radar approach, issue an IFR approach clearance only after the aircraft is established on a segment of a published route or IAP, or assign an altitude to maintain until the aircraft is established on a segment of a published route or instrument approach procedure. For this purpose, the Procedure Turn of a published IAP shall not be considered a segment of that IAP until the aircraft reaches the initial fix or navigation facility upon which the procedure turn is predicated.

EXAMPLE:

CROSS REDDING VOR AT OR ABOVE FIVE THOUSAND, CLEARED VOR RUNWAY THREE FOUR APPROACH. *or*

EXAMPLE:

FIVE MILES FROM OUTER MARKER, TURN RIGHT HEADING THREE THREE ZERO, MAINTAIN TWO THOUSAND UNTIL ESTABLISHED ON THE LOCALIZER, CLEARED ILS RUNWAY THREE SIX APPROACH.

5–47b NOTE— The altitude assigned will assure IFR obstruction clearance from the point at which the approach clearance is issued until established on a segment of a published route or IAP. If a pilot is uncertain of the meaning of his clearance, he shall immediately request clarification from ATC.

c. Several IAPs, using various navigation and approach aids may be authorized for an airport. ATC may advise that a particular approach procedure is being used, primarily to expedite traffic. (Reference—Advance Information on Instrument Approach, paragraph 5–44). If a pilot is issued a clearance that specifies a particular approach procedure, he is expected to notify ATC immediately if he desires a different one. In this event it may be necessary for ATC to withhold clearance for the different approach until such time as traffic conditions permit. However, if the pilot is involved in an emergency situation he will be given priority. If the pilot is not familiar with the specific approach procedure, ATC should be advised and they will provide detailed information on the execution of the procedure.

d. At times ATC may not specify a particular approach procedure in the clearance, but will state "CLEARED FOR APPROACH." Such clearance indicates that the pilot may execute any one of the authorized IAPs for that airport. This clearance does not constitute approval for the pilot to execute a contact approach or a visual approach.

e. When cleared for a specifically prescribed IAP; i.e., "cleared ILS runway one niner approach" or when "cleared for approach" i.e., execution of any procedure prescribed for the airport, pilots shall execute the entire procedure as described on the IAP Chart unless an appropriate new or revised ATC clearance is received, or the IFR flight plan is canceled.

f. Pilots planning flights to locations served by special IAPs should obtain advance approval from the owner of the procedure. Approval by the owner is necessary because special procedures are for the exclusive use of the single interest unless otherwise authorized by the owner. Additionally, some special approach procedures require certain crew qualifications training, or other special considerations in order to execute the approach. Also, some of these approach procedures are based on privately owned navigational aids. Owners of aids that are not for public use may elect to turn off the aid for whatever reason they may have; i.e., maintenance, conservation, etc. Air traffic controllers are not required to question pilots to determine if they have permission to use the procedure. Controllers presume a pilot has obtained approval and is aware of any details of the procedure if he files an IFR flight plan to that airport.

g. When executing an instrument approach and in radio contact with an FAA facility, unless in "radar contact," report passing the final approach fix inbound (non precision approach) or the outer marker or fix used in lieu of the outer marker inbound (precision approach).

h. If a missed approach is required, advise ATC and include the reason (unless initiated by ATC). Comply with the missed approach instructions for the instrument approach procedure being executed, unless otherwise directed by ATC. (Reference—Missed Approach, paragraph 5–57 and Missed Approach, paragraph 5–75).

5–48. PROCEDURE TURN

a. A procedure turn is the maneuver prescribed when it is necessary to reverse direction to establish the aircraft inbound on an intermediate or final approach course. It is a required maneuver except when the symbol NoPT is shown, when RADAR VECTORING is provided, when a holding pattern is published in lieu of procedure turn, when conducting a timed approach, or when the procedure turn is not authorized. The altitude prescribed for the procedure turn is a *minimum* altitude until the aircraft is established on the inbound course. The maneuver must be completed within the distance specified in the profile view.

1. On U.S. Government charts, a barbed arrow indicates the direction or side of the outbound course on which the procedure turn is made. Headings are provided for course reversal using the 45 degree type procedure turn. However, the point at which the turn may be commenced and the type and rate of turn is left to the discretion of the pilot. Some of the options are the 45 degree procedure turn, the racetrack pattern, the tear-drop procedure turn, or the 80 degree - 260 degree course reversal. Some procedure turns are specified by procedural track. These turns must be flown exactly as depicted.

2. When the approach procedure involves a procedure turn, a maximum speed of not greater than 250 knots (IAS) should be observed and the turn should be executed within the distance specified in the profile view. The normal procedure turn distance is 10 miles. This may be reduced to a minimum of 5 miles where only Category A or helicopter aircraft are to be operated or increased to as much as 15 miles to accommodate high performance aircraft.

3. A teardrop procedure or penetration turn may be specified in some procedures for a required course reversal. The teardrop procedure consists of departure from an initial approach fix on an outbound course followed by a turn toward and intercepting the inbound course at or prior to the intermediate fix or point. Its purpose is to permit an aircraft to reverse direction and lose considerable altitude within reasonably limited airspace. Where no fix is available to mark the beginning of the intermediate segment, it shall be assumed to commence at a point 10 miles prior to the final approach fix. When the facility is located on the airport, an aircraft is considered to be on final approach upon completion of the penetration turn. However, the final approach segment begins on the final approach course 10 miles from the facility.

4. A procedure turn need not be established when an approach can be made from a properly aligned holding pattern. In such cases, the holding pattern is established over an intermediate fix or a final approach fix. The holding pattern maneuver is completed when the aircraft is established on the inbound course after executing the appropriate entry. If cleared for the approach prior to returning to the holding fix, and the aircraft is at the prescribed altitude, additional circuits of the holding pattern are not necessary nor expected by ATC. If the pilot elects to make additional circuits to lose excessive altitude or to become better established on course, it is his responsibility to so advise ATC when he receives his approach clearance.

5. A procedure turn is not required when an approach can be made directly from a specified intermediate fix to the final approach fix. In such cases, the term ''NoPT'' is used with the appropriate course and altitude to denote that the procedure turn is not required. If a procedure turn is desired, and when cleared to do so by ATC, descent below the procedure turn altitude should not be made until the aircraft is established on the inbound course, since some NoPT altitudes may be lower than the procedure turn altitudes.

b. Limitations on Procedure Turns.

1. In the case of a radar initial approach to a final approach fix or position, or a timed approach from a holding fix, or where the procedure specifies ''NoPT'', no pilot may make a procedure turn unless, when he receives his final approach clearance, he so advises ATC and a clearance is received.

2. When a teardrop procedure turn is depicted and a course reversal is required, this type turn must be executed.

3. When holding pattern replaces the procedure turn, the standard entry and the holding pattern must be followed except when RADAR VECTORING is provided or when NoPT is shown on the approach course. As in the procedure turn, the descent from the minimum holding pattern altitude to the final approach fix altitude (when lower) may not commence until the aircraft is established on the inbound course.

4. The absence of the procedure turn barb in the Plan View indicates that a procedure turn is not authorized for that procedure.

5–49. TIMED APPROACHES FROM A HOLDING FIX

a. TIMED APPROACHES may be conducted when the following conditions are met:

1. A control tower is in operation at the airport where the approaches are conducted.

2. Direct communications are maintained between the pilot and the center or approach controller until the pilot is instructed to contact the tower.

3. If more than one missed approach procedure is available, none require a course reversal.

4. If only one missed approach procedure is available, the following conditions are met:

(a) Course reversal is not required; and,

(b) Reported ceiling and visibility are equal to or greater than the highest prescribed circling minimums for the IAP.

5. When cleared for the approach, pilots shall not execute a procedure turn. (FAR 91.175.)

b. Although the controller will not specifically state that ''timed approaches are in progress'', his assigning a time to depart the final approach fix inbound (non precision approach) or the outer marker or fix used in lieu of the outer marker inbound (precision approach) is indicative that timed approach procedures are being utilized, or in lieu of holding, he may use radar vectors to the Final Approach Course to establish a mileage interval between aircraft that will insure the appropriate time sequence between the final approach fix/outer marker or fix used in lieu of the outer marker and the airport.

c. Each pilot in an approach sequence will be given advance notice as to the time he should leave the holding point on approach to the airport. When a time to leave the holding point has been received, the pilot should adjust his flight path to leave the fix as closely as possible to the designated time. (See Figure 5–49[1].)

5–50. RADAR APPROACHES

a. The only airborne radio equipment required for radar approaches is a functioning radio transmitter and receiver. The radar controller vectors the aircraft to align it with the runway centerline. The controller continues the vectors to keep the aircraft on course until the pilot can complete the approach and landing by visual reference to the surface. There

Figure 5–49[1]

Time Approach Example — a final approach procedure from a holding pattern at a final approach fix (FAF).

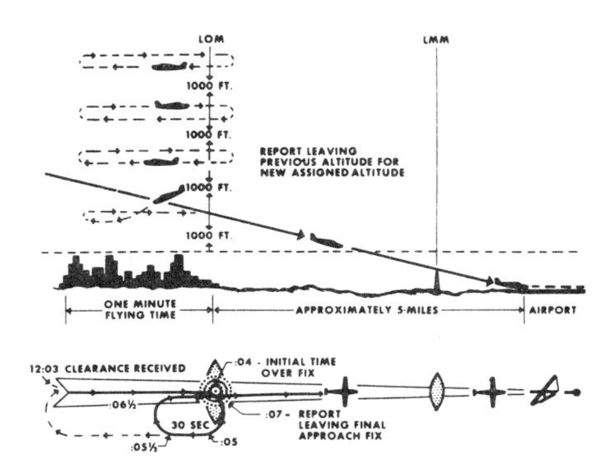

At 12:03 local time, in the example shown, a pilot holding, receives instructions to leave the fix inbound at 12:07. These instructions are received just as the pilot has completed turn at the outbound end of the holding pattern and is proceeding inbound towards the fix. Arriving back over the fix, the pilot notes that the time is 12:04 and that he has 3 minutes to lose in order to leave the fix at the assigned time. Since the time remaining is more than 2 minutes, the pilot plans to fly a race track pattern rather than a 360 degrees turn, which would use up 2 minutes. The turns at the ends of the race track pattern will consume approximately 2 minutes. Three minutes to go, minus 2 minutes required for turns, leaves 1 minute for level flight. Since two portions of level flight will be required to get back to the fix inbound, the pilot halves the 1 minute remaining and plans to fly level for 30 seconds outbound before starting his turn back toward the fix on final approach. If the winds were negligible at flight altitude, this procedure would bring the pilot inbound across the fix precisely at the specified time of 12:07. However, if the pilot expected a headwind on final approach, he should shorten his 30 seconds outbound course somewhat, knowing that the wind will carry him away from the fix faster while outbound and decrease his ground speed while returning to the fix. On the other hand, if the pilot knew he would have a tailwind on final approach, he should lengthen his calculated 30-second outbound heading somewhat, knowing that the wind would tend to hold him closer to the fix while outbound and increase his ground speed while returning to the fix.

are two types of radar approaches: Precision (PAR) and Surveillance (ASR).

b. A radar approach may be given to any aircraft upon request and may be offered to pilots of aircraft in distress or to expedite traffic, however, an ASR might not be approved unless there is an ATC operational requirement, or in an unusual or emergency situation. Acceptance of a PAR or ASR by a pilot does not waive the prescribed weather minimums for the airport or for the particular aircraft operator concerned. The decision to make a radar approach when the reported weather is below the established minimums rests with the pilot.

c. PAR and ASR minimums are published on separate pages in the NOS Terminal Procedures Publication (TPP).

1. A PRECISION APPROACH (PAR) is one in which a controller provides highly accurate navigational guidance in azimuth and elevation to a pilot. Pilots are given headings to fly, to direct them to, and keep their aircraft aligned with the extended centerline of the landing runway. They are told to anticipate glide path interception approximately 10 to 30 seconds before it occurs and when to start descent. The published Decision Height will be given only if the pilot requests it. If the aircraft is observed to deviate above or below the glide path, the pilot is given the relative amount of deviation by use of terms "slightly" or "well" and is expected to adjust his rate of descent to return to the glide path. Trend information is also issued with respect to the elevation of the aircraft and may be modified by the terms "rapidly" and "slowly"; e.g., "well above glide path, coming down rapidly." Range from touchdown is given at least once each mile. If an aircraft is observed by the controller to proceed outside of specified safety zone limits in azimuth and/or elevation and continue to operate outside these prescribed limits, the pilot will be directed to execute a missed approach or to fly a specified course unless he has the runway environment (runway, approach lights, etc.) in sight. Navigational guidance in azimuth and elevation is provided the pilot until the aircraft reaches the published Decision Height (DH). Advisory course and glidepath information is furnished by the controller until the aircraft passes over the landing threshold, at which point the pilot is advised of any deviation from the runway centerline. Radar service is automatically terminated upon completion of the approach.

2. A SURVEILLANCE APPROACH (ASR) is one in which a controller provides navigational guidance in azimuth only. The pilot is furnished headings to fly to align his aircraft with the extended centerline of the landing runway. Since the radar information used for a surveillance approach is considerably less precise than that used for a precision approach, the accuracy of the approach will not be as great and higher minimums will apply. Guidance in elevation is not possible but the pilot will be advised when to commence descent to the Minimum Descent Altitude (MDA) or, if appropriate, to an intermediate step-down fix Minimum Crossing Altitude and subsequently to the prescribed MDA. In addition, the pilot will be advised of the location of the Missed Approach Point (MAP) prescribed for the procedure and his position each mile on final from the runway, airport or heliport or MAP, as appropriate. If requested by the pilot, recommended altitudes will be issued at each mile, based on the descent gradient established for the procedure, down to the last mile that is at or above the MDA. Normally, navigational guidance will be provided until the aircraft reaches the MAP. Controllers will terminate guidance and instruct the pilot to execute a missed approach unless at the MAP the pilot has the runway, airport or heliport in sight or, for a helicopter point-in-space approach, the prescribed visual reference with the surface is established. Also, if, at any time during the approach the controller considers that safe guidance for the remainder of the approach can not be provided, he will terminate guidance and instruct the pilot to execute a missed approach. Similarly, guidance termination and missed approach will be effected upon pilot request and, for civil aircraft only, controllers may terminate guidance when the pilot reports the runway, airport/heliport or visual surface route (point-in-space approach) in sight or otherwise indicates that continued guidance is not required. Radar service is automatically terminated at the completion of a radar approach.

5–50c2 NOTE— The published MDA for straight-in approaches will be issued to the pilot before beginning descent. When a surveillance approach will terminate in a circle-to-land maneuver, the pilot must furnish the aircraft approach category to the controller. The controller will then provide the pilot with the appropriate MDA.

5–50c2 NOTE— ASR approaches are not available when an ATC facility is using CENRAP.

3. A NO-GYRO APPROACH is available to a pilot under radar control who experiences circumstances wherein his directional gyro or other stabilized compass is inoperative or inaccurate. When this occurs, he should so advise ATC and request a No-Gyro vector or approach. Pilots of aircraft not equipped with a directional gyro or other stabilized compass who desire radar handling may also request a No-Gyro vector or approach. The pilot should make all turns at standard rate and should execute the turn immediately upon receipt of instructions. For example, "TURN RIGHT," "STOP TURN." When a surveillance or precision approach is made, the pilot will be advised after his aircraft has been turned onto final approach to make turns at half standard rate.

5–51. RADAR MONITORING OF INSTRUMENT APPROACHES

a. PAR facilities operated by the FAA and the military services at some joint-use (civil and military) and military installations monitor aircraft on instrument approaches and issue radar advisories to the pilot when weather is below VFR minimums (1,000 and 3), at night, or when requested by a pilot. This service is provided only when the PAR Final Approach Course coincides with the final approach of the navigational aid and only during the operational hours of the PAR. The radar advisories serve only as a secondary aid since the pilot has selected the navigational aid as the primary aid for the approach.

b. Prior to starting final approach, the pilot will be advised of the frequency on which the advisories will be transmitted. If, for any reason, radar advisories cannot be furnished, the pilot will be so advised.

c. Advisory information, derived from radar observations, includes information on:

1. Passing the final approach fix inbound (non precision approach) or passing the outer marker or fix used in lieu of the outer marker inbound (precision approach).

5–51c1 NOTE— At this point, the pilot may be requested to report sighting the approach lights or the runway.

2. Trend advisories with respect to elevation and/or azimuth radar position and movement will be provided.

5–51c2 NOTE— Whenever the aircraft nears the PAR safety limit, the pilot will be advised that he is well above or below the glidepath or well left or right of course. Glidepath information is given only to those aircraft executing a precision approach, such as ILS or MLS. Altitude information is not transmitted to aircraft executing other than precision approaches because the descent portions of these approaches generally do not coincide with the depicted PAR glidepath. At locations where the MLS glidepath and PAR glidepath are not coincidental, only azimuth monitoring will be provided.

3. If, after repeated advisories, the aircraft proceeds outside the PAR safety limit or if a radical deviation is observed, the pilot will be advised to execute a missed approach unless the prescribed visual reference with the surface is established.

d. Radar service is automatically terminated upon completion of the approach.

5–52. SIMULTANEOUS ILS/MLS APPROACHES

(See Figure 5–52[1].)

Figure 5–52[1]

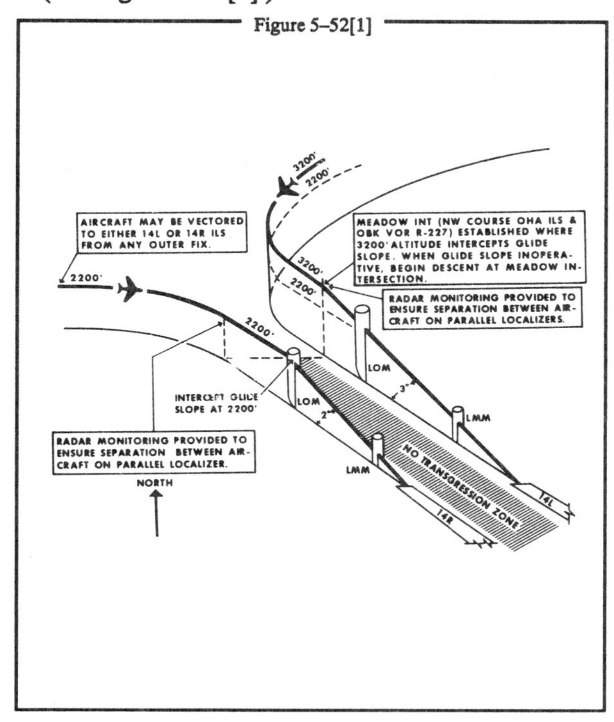

a. System: An approach system permitting simultaneous ILS/MLS, or ILS and MLS approaches to airports having parallel runways separated by at least 4,300 feet between centerlines. Integral parts of a total system are ILS or MLS, radar, communications, ATC procedures, and appropriate airborne equipment. The Approach Procedure Chart permitting simultaneous approaches will contain the note "simultaneous approaches authorized Rwys 14L and 14R" identifying the appropriate runways as the case may be. When advised that simultaneous ILS approaches are in progress, pilots shall advise approach control immediately of malfunctioning or inoperative receivers or if simultaneous approach is not desired.

5–52a NOTE— Simultaneous ILS/MLS Approaches are not available when CENRAP is in use.

b. Radar Monitoring: This service is provided for each ILS/MLS approach to insure prescribed lateral separation during simultaneous ILS/MLS approaches. Radar Monitoring includes instructions when an aircraft nears or exceeds the prescribed no transgression zone (an area at least 2,000 feet wide). This service will be provided as follows:

1. The monitor controller will have the capability of overriding the tower controller on the tower frequency.

2. Pilots will be advised to monitor the tower frequency to receive advisories and instructions.

3. Aircraft deviating from either final approach course to the point where the no transgression zone (an area at least 2.000 feet wide) may be penetrated will be instructed to take corrective action. If an aircraft fails to respond to such instruction, the aircraft on the adjacent final approach course may be instructed to alter course.

4. The monitor will automatically be terminated no more than one mile from the runway threshold.

5. The monitor controller will *not* advise when the monitor is terminated.

5–53. PARALLEL ILS/MLS APPROACHES

a. Parallel approaches are an ATC procedure permitting parallel ILS, MLS, or ILS and MLS approaches to airports having parallel runways separated by at least 2,500 feet between centerlines. Integral parts of a total system are ILS or MLS, radar, communications, ATC procedures, and appropriate airborne equipment.

b. A parallel approach differs from a simultaneous approach in that the minimum distance between parallel runway centerlines is reduced; there is no requirement for radar monitoring or advisories; and a staggered separation of aircraft on the adjacent localizer course is required.

c. Aircraft are afforded a minimum of two miles radar separation between successive aircraft on the adjacent localizer course and a minimum of three miles radar separation from aircraft on the same final approach course. In addition, a minimum of 1,000 feet vertical or a minimum of three miles radar separation is provided between aircraft during turn on.

d. Whenever parallel approaches are in progress, aircraft are informed that approaches to both runways are in use. In addition, the radar controller will have the interphone capability of communicating directly with the tower controller where the responsibility for radar separation is not performed by the tower controller.

5–54. SIMULTANEOUS CONVERGING INSTRUMENT APPROACHES

a. ATC may conduct instrument approaches simultaneously to converging runways; i.e., runways having an included angle from 15 to 100 degrees, at airports where a program has been specifically approved to do so.

b. The basic concept requires that dedicated, separate standard instrument approach procedures be developed for each converging runway included. Missed Approach Points must be as least 3 miles apart and missed approach procedures ensure that missed approach protected airspace does not overlap.

c. Other requirements are: radar availability, nonintersecting final approach courses, precision (ILS/MLS) approach systems on each runway and, if runways intersect, controllers must be able to apply visual separation as well as intersecting runway separation criteria. Intersecting runways also require minimums of at least 700 and 2. Straight in approaches and landings must be made.

d. Whenever simultaneous converging approaches are in progress, aircraft will be informed by the controller as soon as feasible after initial contact or via ATIS. Additionally, the radar controller will have direct communications capability with the tower controller where separation responsibility has not been delegated to the tower.

5–55. SIDE-STEP MANEUVER

a. ATC may authorize an approach procedure which serves either one of parallel runways that are separated by 1,200 feet or less followed by a straight-in landing on the adjacent runway.

b. Aircraft that will execute a side-step maneuver will be cleared for a specified approach and landing on the adjacent parallel runway. Example, "cleared ILS runway 7 left approach, side-step to runway 7 right." Pilots are expected to commence the side-step maneuver as soon as possible after the runway or runway environment is in sight.

c. Landing minimums to the adjacent runway will be higher than the minimums to the primary runway, but will normally be lower than the published circling minimums.

5–56. APPROACH AND LANDING MINIMUMS

a. Landing Minimums. The rules applicable to landing minimums are contained in FAR 91.175.

b. Published Approach Minimums. Approach minimums are published for different aircraft categories and consist of a minimum altitude (DH, MDA) and required visibility. These minimums are determined by applying the appropriate TERPS criteria. When a fix is incorporated in a nonprecision final segment, two sets of minimums may be published: one, for the pilot that is able to identify the fix, and a second for the pilot that cannot. Two sets of minimums may also be published when a second altimeter source is used in the procedure.

c. Obstacle Clearance. Final approach obstacle clearance is provided from the start of the final segment to the runway or Missed Approach Point, whichever occurs last. Side-step obstacle protection is provided by increasing the width of the final approach obstacle clearance area. Circling approach protected areas are defined by the tangential connection of arcs drawn from each runway end. The arc radii distance differs by aircraft approach category. Because of obstacles near the airport, a portion of the circling area may be restricted by a procedural note: e.g., "Circling NA E of RWY 17–35." Obstacle clearance is provided at the published minimums for the pilot that makes a straight-in approach, side-steps, circles, or executes the missed approach. Missed approach obstacle clearance requirements may dictate the published minimums for the approach. (See Figure 5–56[1].)

Figure 5–56[1]

CIRCLING APPROACH AREA RADII

Approach Category	Radius (Miles)
A	1.3
B	1.5
C	1.7
D	2.3
E	4.5

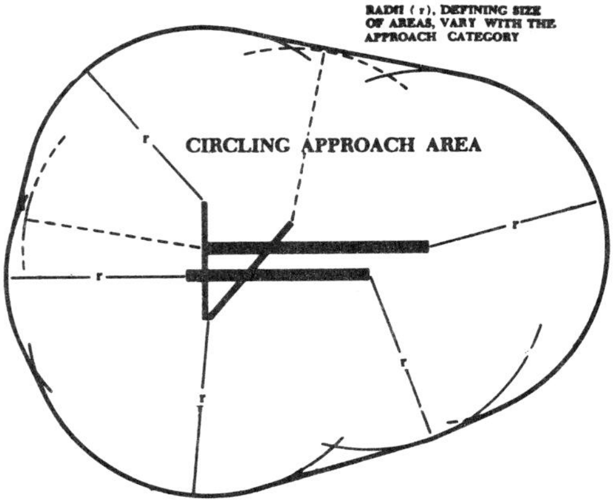

CONSTRUCTION OF CIRCLING APPROACH AREA.

d. Straight-in Minimums. Are shown on the IAP when the final approach course is within 30 degrees of the runway alignment and a normal descent can be made from the IFR altitude shown on the IAP to the runway surface. When either the normal rate of descent or the runway alignment factor of 30 degrees is exceeded, a straight-in minimum is not published and a circling minimum applies. The fact that a straight-in minimum is not published does not preclude pilots from landing straight-in if they have the active runway in sight and have sufficient time to make a normal approach for landing. Under such conditions and when ATC has cleared them for landing on that runway, pilots are not expected to circle even though only circling minimums are published. If they desire to circle, they should advise ATC.

e. Side-Step Maneuver Minimums. Landing minimums for a side-step maneuver to the adjacent runway will normally be higher than the minimums to the primary runway.

f. Circling Minimums. In some busy terminal areas, ATC may not allow circling and circling minimums will not be published. Published circling minimums provide obstacle clearance when pilots remain within the appropriate area of protection. Pilots should remain at or above the circling altitude until the aircraft is continuously in a position from which a descent to a landing on the intended runway can be made at a normal rate of descent using normal maneuvers. Circling may require maneuvers at low altitude, at low airspeed, and in marginal weather conditions. Pilots must use sound judgment, have an indepth knowledge of their capabilities, and fully understand the aircraft performance to determine the exact circling maneuver since weather, unique airport design, and the aircraft position, altitude, and airspeed must all be considered. The following basic rules apply:

1. Maneuver the shortest path to the base or downwind leg, as appropriate, considering existing weather conditions. There is no restriction from passing over the airport or other runways.

2. It should be recognized that circling maneuvers may be made while VFR or other flying is in progress at the airport. Standard left turns or specific instruction from the controller for maneuvering must be considered when circling to land.

3. At airports without a control tower, it may be desirable to fly over the airport to observe wind and turn indicators and other traffic which may be on the runway or flying in the vicinity of the airport.

g. Instrument Approach At a Military Field. When instrument approaches are conducted by civil aircraft at military airports, they shall be conducted in accordance with the procedures and minimums approved by the military agency having jurisdiction over the airport.

5–57. MISSED APPROACH

a. When a landing cannot be accomplished, advise ATC and, upon reaching the Missed Approach Point defined on the approach procedure chart, the pilot must comply with the missed approach instructions for the procedure being used or with an alternate missed approach procedure specified by ATC.

b. Protected obstacle clearance areas for missed approach are predicated on the assumption that the abort is initiated at the missed approach point not lower than the MDA or DH. Reasonable buffers are provided for normal maneuvers. However, no consideration is given to an abnormally early turn. Therefore, when an early missed approach is executed, pilots should, unless otherwise cleared by ATC, fly the IAP as specified on the approach plate to the missed approach point at or above the MDA or DH before executing a turning maneuver.

Figure 5–57[1]

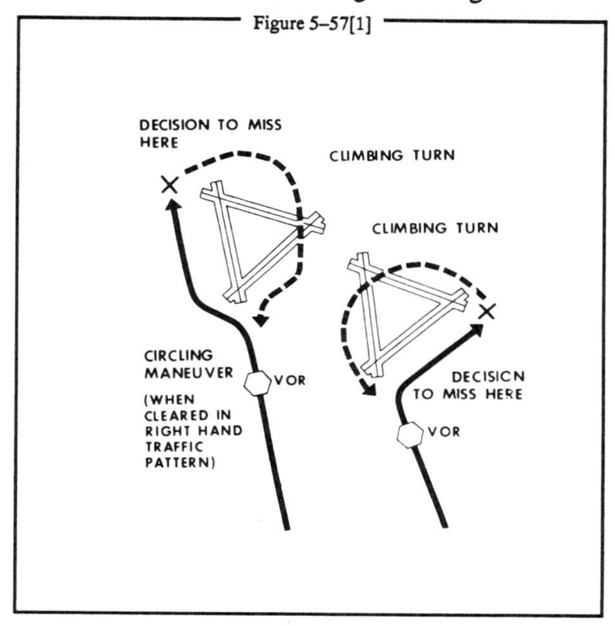

c. If visual reference is lost while circling-to-land from an instrument approach, the missed approach specified for that particular procedure must be followed (unless an alternate missed approach procedure is specified by ATC). To become established on the prescribed missed approach course, the pilot should make an initial climbing turn toward the landing runway and continue the turn until he is established on the missed approach course. Inasmuch as the circling maneuver may be accomplished in more than one direction, different patterns will be required to become established on the prescribed missed approach course, depending on the aircraft position at the time visual reference is lost. Adherence to the procedure will assure that an aircraft will remain within the circling and missed approach obstruction clearance areas. (See Figure 5–57[1].)

d. At locations where ATC Radar Service is provided, the pilot should conform to radar vectors

Figure 5–57[2]

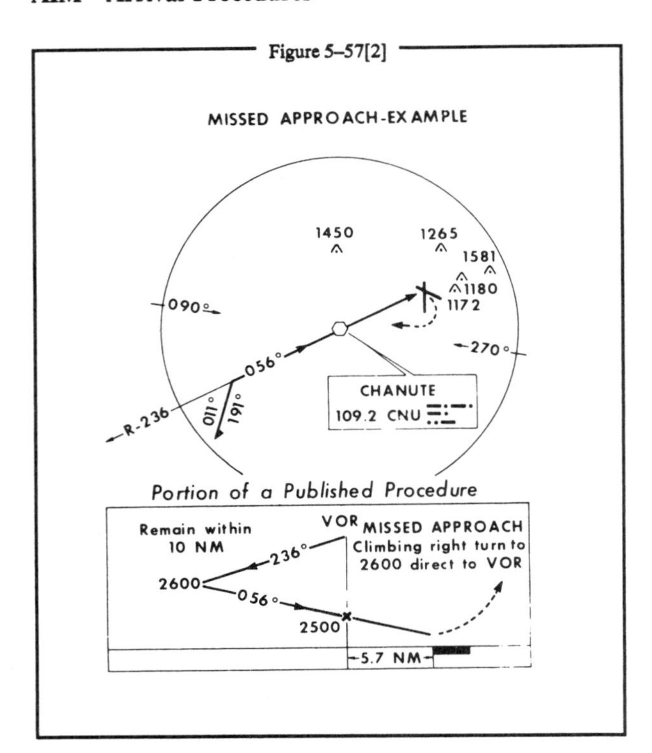

when provided by ATC in lieu of the published missed approach procedure. (See Figure 5–57[2].)

e. When approach has been missed, request clearance for specific action; i.e., to alternative airport, another approach, etc.

5–58. VISUAL APPROACH

a. When it will be operationally beneficial, ATC may authorize an aircraft to conduct a visual approach to an airport or to follow another aircraft when flight to, and landing at, the airport can be accomplished in VFR weather. The aircraft must have the airport or the identified preceding aircraft in sight before the clearance is issued. If the pilot has the airport in sight but cannot see the aircraft he is following, ATC may still clear the aircraft for a visual approach; however, ATC retains both separation and wake vortex separation responsibility. When visually following a preceding aircraft, acceptance of the visual approach clearance, constitutes acceptance of pilot responsibility for maintaining a safe approach interval and adequate wake turbulence separation.

b. When operating to an airport with an operating control tower, aircraft may be authorized to conduct a visual approach to one runway while other aircraft are conducting IFR or VFR approaches to another parallel, intersecting, or converging runway. When operating to airports with parallel runways separated by less than 2,500 feet, the succeeding aircraft must report sighting the preceding aircraft unless standard separation is being provided by ATC. When operating to parallel runways separated by at least 2,500 feet but less than 4,300 feet, controllers will clear/vector aircraft to the final at an angle not greater than 30 degrees unless radar, vertical, or visual separation is provided during the turn-on. The purpose of the 30 degree intercept angle is to reduce the potential for overshoots of the final and to preclude side-by-side operations with one or both aircraft in a *belly-up* configuration during the turn-on. Once the aircraft are established within 30 degrees of final, or on the final, these operations may be conducted simultaneously. When the parallel runways are separated by 4,300 feet or more, or intersecting/converging runways are in use, ATC may authorize a visual approach after advising all aircraft involved that other aircraft are conducting operations to the other runway. This may be accomplished through use of the ATIS.

c. When operating to an airport without weather reporting service, ATC may initiate a visual approach provided area weather reports indicate that VFR conditions exist at the airport and there is reasonable assurance that descent and flight to the airport can be made in VFR conditions. Pilot acceptance of a visual approach clearance or a pilot's request for a visual approach will indicate to ATC that the pilot can comply with FAR 91.155. ATC will advise the pilot when weather is not available at the destination airport.

d. Authorization to conduct a visual approach is an IFR authorization and does not alter IFR flight plan cancellation responsibility. (Reference—Canceling IFR Flight Plan, paragraph 5–13).

e. A visual approach is not an IAP and therefore has no missed approach segment. If a go around is necessary for any reason, aircraft operating at controlled airports will be issued an appropriate advisory/clearance/instruction by the tower. At uncontrolled airports, aircraft are expected to remain in VFR conditions and complete a landing as soon as possible. If a landing cannot be accomplished, the aircraft is expected to remain in VFR conditions and contact ATC as soon as possible for further clearance. Separation from other IFR aircraft will be maintained under these circumstances.

f. Visual approaches are initiated by ATC to reduce pilot/controller workload and expedite traffic by shortening flight paths to the airport. It is the pilot's responsibility to advise ATC as soon as possible if a visual approach is not desired.

g. Radar service is automatically terminated, without advising the pilot, when the aircraft is instructed to change to advisory frequency.

5–59. CHARTED VISUAL FLIGHT PROCEDURES (CVFP)

a. CVFPs are charted visual approaches established at locations with jet operations for noise abatement purposes. The approach charts depict prominent landmarks, courses, and recommended altitudes to specific runways.

b. These procedures will be used only in a radar environment at airports with an operating control tower.

c. Most approach charts will depict some NAVID information which is for supplemental navigational guidance only.

d. Unless indicating a TCA floor, all depicted altitudes are for noise abatement purposes and are recommended only. Pilots are not prohibited from flying other than recommended altitudes if operational requirements dictate.

e. When landmarks used for navigation are not visible at night, the approach will be annotated "PROCEDURE NOT AUTHORIZED AT NIGHT."

f. CVFPs usually begin within 15 flying miles from the airport.

g. Published weather minimums for CVFPs are based on minimum vectoring altitudes rather than the recommended altitudes depicted on charts.

h. CVFPs are not instrument approaches and do not have missed approach segments.

i. ATC will not issue clearances for CVFPs when the weather is less than the published minimum.

j. ATC will clear aircraft for a CVFP after the pilot reports siting a charted landmark or a preceding aircraft. If instructed to follow a preceding aircraft, pilots are responsible for maintaining a safe approach interval and wake turbulence separation.

k. Pilots should advise ATC if at any point they are unable to continue an approach or lose sight of a preceding aircraft. Missed approaches will be handled as a go-around.

5–60. CONTACT APPROACH

a. Pilots operating in accordance with an IFR flight plan, provided they are clear of clouds and have at least 1 mile flight visibility and can reasonably expect to continue to the destination airport in those conditions, may request ATC authorization for a contact approach.

b. Controllers may authorize a contact approach provided:

1. The Contact Approach is specifically requested by the pilot. ATC cannot initiate this approach.

EXAMPLE:

REQUEST CONTACT APPROACH.

2. The reported ground visibility at the destination airport is at least 1 statute mile.

3. The contact approach will be made to an airport having a standard or special instrument approach procedure.

4. Approved separation is applied between aircraft so cleared and between these aircraft and other IFR or special VFR aircraft.

EXAMPLE:

CLEARED CONTACT APPROACH (and, if required) AT OR BELOW (altitude) (Routing) IF NOT POSSIBLE (alternative procedures) AND ADVISE.

c. A Contact Approach is an approach procedure that may be used by a pilot (with prior authorization from ATC) in lieu of conducting a standard or special IAP to an airport. It is not intended for use by a pilot on an IFR flight clearance to operate to an airport not having an authorized IAP. Nor is it intended for an aircraft to conduct an instrument approach to one airport and then, when "in the clear", to discontinue that approach and proceed to another airport. In the execution of a contact approach, the pilot assumes the responsibility for obstruction clearance. If radar service is being received, it will automatically terminate when the pilot is told to contact the tower.

5–61. LANDING PRIORITY

A clearance for a specific type of approach (ILS, MLS, ADF, VOR or Straight-in Approach) to an aircraft operating on an IFR flight plan does not mean that landing priority will be given over other traffic. ATCTs handle all aircraft, regardless of the type of flight plan, on a "first-come, first-served" basis. Therefore, because of local traffic or runway in use, it may be necessary for the controller in the interest of safety, to provide a different landing sequence. In any case, a landing sequence will be issued to each aircraft as soon as possible to enable the pilot to properly adjust his flight path.

5–62 thru 5–70. RESERVED

Section 5. PILOT/CONTROLLER ROLES AND RESPONSIBILITIES

5–71. GENERAL

a. The roles and responsibilities of the pilot and controller for effective participation in the ATC system are contained in several documents. Pilot responsibilities are in the FARs and the air traffic controller's are in the Air Traffic Control Handbook (7110.65) and supplemental FAA directives. Additional and supplemental information for pilots can be found in the current Airman's Information Manual (AIM), Notices to Airmen, Advisory Circulars and aeronautical charts. Since there are many other excellent publications produced by non-government organizations, as well as other government organizations, with various updating cycles, questions concerning the latest or most current material can be resolved by cross-checking with the above mentioned documents.

b. The pilot in command of an aircraft is directly responsible for, and is the final authority as to the safe operation of that aircraft. In an emergency requiring immediate action, the pilot in command may deviate from any rule in the General Subpart A and Flight Rules Subpart B in accordance with FAR 91.3.

c. The air traffic controller is responsible to give first priority to the separation of aircraft and to the issuance of radar safety alerts, second priority to other services that are required, but do not involve separation of aircraft and third priority to additional services to the extent possible.

d. In order to maintain a safe and efficient air traffic system, it is necessary that each party fulfill his responsibilities to the fullest.

e. The responsibilities of the pilot and the controller intentionally overlap in many areas providing a degree of redundancy. Should one or the other fail in any manner, this overlapping responsibility is expected to compensate, in many cases, for failures that may affect safety.

f. The following, while not intended to be all inclusive, is a brief listing of pilot and controller responsibilities for some commonly used procedures or phases of flight. More detailed explanations are contained in other portions of this publication, the appropriate FARs, ACs and similar publications. The information provided is an overview of the principles involved and is not meant as an interpretation of the rules nor is it intended to extend or diminish responsibilities.

5–72. AIR TRAFFIC CLEARANCE

a. Pilot—

1. Acknowledges receipt and understanding of an ATC clearance.

2. Readbacks any hold short of runway instructions issued by ATC.

3. Requests clarification or amendment, as appropriate, any time a clearance is not fully understood or considered unacceptable from a safety standpoint.

4. Promptly complies with an air traffic clearance upon receipt except as necessary to cope with an emergency. Advises ATC as soon as possible and obtains an amended clearance, if deviation is necessary.

5–72a Note—A clearance to land means that appropriate separation on the landing runway will be ensured. A landing clearance does not relieve the pilot from compliance with any previously issued altitude crossing restriction.

b. Controller—

1. Issues appropriate clearances for the operation to be conducted, or being conducted, in accordance with established criteria.

2. Assigns altitudes in IFR clearances that are at or above the Minimum IFR Altitudes in controlled airspace.

3. Ensures acknowledgement by the pilot for issued information, clearances, or instructions.

4. Ensures that readbacks by the pilot of altitude, heading, or other items are correct. If incorrect, distorted, or incomplete, makes corrections as appropriate.

5–73. CONTACT APPROACH

a. Pilot—

1. Must request a contact approach and makes it in lieu of a standard or special instrument approach.

2. By requesting the contact approach, indicates that the flight is operating clear of clouds, has at least one mile flight visibility, and reasonably expects to continue to the destination airport in those conditions.

3. Assumes responsibility for obstruction clearance while conducting a contact approach.

4. Advises ATC immediately if unable to continue the contact approach or if encounters less than 1 mile flight visibility.

5. Is aware that if radar service is being received, it may automatically terminated when told to contact the tower. (Reference—Pilot/Controller Glossary, Radar Service Terminated).

b. Controller—

1. Issues clearance for a contact approach only when requested by the pilot. Does not solicit the use of this procedure.

2. Before issuing the clearance, ascertains that reported ground visibility at destination airport is at least 1 mile.

3. Provides approved separation between the aircraft cleared for a contact approach and other IFR or special VFR aircraft. When using vertical separation, does not assign a fixed altitude, but clears the aircraft at or below an altitude which is at least 1,000 feet below any IFR traffic but not below Minimum Safe Altitudes prescribed in FAR 91.119.

4. Issues alternative instructions if, in his judgment, weather conditions may make completion of the approach impracticable.

5–74. INSTRUMENT APPROACH

a. Pilot—

1. Be aware that the controller issues clearance for approach based only on known traffic.

2. Follows the procedure as shown on the IAP, including all restrictive notations, such as:

(a) Procedure not authorized at night;

(b) Approach not authorized when local area altimeter not available;

(c) Procedure not authorized when control tower not in operation;

(d) Procedure not authorized when glide slope not used;

(e) Straight-in minimums not authorized at night; etc.

(f) Radar required; or

(g) The circling minimums published on the instrument approach chart provide adequate obstruction clearance and the pilot should not descend below the circling altitude until the aircraft is in a position to make final descent for landing. Sound judgment and knowledge of his and the aircraft's capabilities are the criteria for a pilot to determine the exact maneuver in each instance since airport design and the aircraft position, altitude and airspeed must all be considered. (Reference—Approach and Landing Minimums, paragraph 5–56f).

3. Upon receipt of an approach clearance while on an unpublished route or being radar vectored:

(a) Complies with the minimum altitude for IFR, and

(b) Maintains the last assigned altitude until established on a segment of a published route or IAP, at which time published altitudes apply.

b. Controller—

1. Issues an approach clearance based on known traffic.

2. Issues an IFR approach clearance only after the aircraft is established on a segment of published route or IAP, or assigns an appropriate altitude for the aircraft to maintain until so established.

5–75. MISSED APPROACH

a. Pilot—

1. Executes a missed approach when one of the following conditions exist:

(a) Arrival at the Missed Approach Point (MAP) or the Decision Height (DH) and visual reference to the runway environment is insufficient to complete the landing.

(b) Determined that a safe landing is not possible.

(c) Instructed to do so by ATC.

2. Advises ATC that a missed approach will be made. Include the reason for the missed approach unless the missed approach is initiated by ATC.

3. Complies with the missed approach instructions for the IAP being executed unless other *missed approach* instructions are specified by ATC.

4. If executing a missed approach prior to reaching the MAP or DH, flies the instrument procedure to the MAP at an altitude at or above the Minimum Descent Altitude (MDA) or DH before executing a turning maneuver.

5. Radar vectors issued by ATC when informed that a missed approach is being executed supersedes the previous missed approach procedure.

6. If making a missed approach from a radar approach, executes the missed approach procedure previously given or climbs to the altitude and flies the heading specified by the controller.

7. Following a missed approach, requests clearance for specific action; i.e., another approach, hold for improved conditions, proceed to an alternate airport, etc.

b. Controller—

1. Issues an approved alternate missed approach procedure if it is desired that the pilot execute a procedure other than as depicted on the instrument approach chart.

2. May vector a radar identified aircraft executing a missed approach when operationally advantageous to the pilot or the controller.

3. In response to the pilot's stated intentions, issues a clearance to an alternate airport, to a holding fix, or for reentry into the approach sequence, as traffic conditions permit.

5–76. RADAR VECTORS

a. Pilot—

1. Promptly complies with headings and altitudes assigned to you by the controller.

2. Questions any assigned heading or altitude believed to be incorrect.

3. If operating VFR and compliance with any radar vector or altitude would cause a violation of any FAR, advises ATC and obtains a revised clearance or instructions.

b. Controller—

1. Vectors aircraft in controlled airspace:

(a) For separation.

(b) For noise abatement.

(c) To obtain an operational advantage for the pilot or controller.

2. Vectors aircraft in controlled and uncontrolled airspace when requested by the pilot.

3. Vectors IFR aircraft at or above Minimum Vectoring Altitudes.

4. May vector VFR aircraft, not at an ATC assigned altitude, at any altitude. In these cases, terrain separation is the pilot's responsibility.

5–77. SAFETY ALERT

a. Pilot—

1. Initiates appropriate action if a safety alert is received from ATC.

2. Be aware that this service is not always available and that many factors affect the ability of the controller to be aware of a situation in which unsafe proximity to terrain, obstructions, or another aircraft may be developing.

b. Controller—

1. Issues a safety alert if he is aware an aircraft under his control is at an altitude which, in the controller's judgment, places the aircraft in unsafe proximity to terrain, obstructions or another aircraft. Types of safety alerts are:

(a) Terrain or Obstruction Alert—Immediately issued to an aircraft under his control if he is aware the aircraft is at an altitude believed to place the aircraft in unsafe proximity to terrain or obstructions.

(b) Aircraft Conflict Alert—Immediately issued to an aircraft under his control if he is aware of an aircraft not under his control at an altitude believed to place the aircraft in unsafe proximity to each other. With the alert, he offers the pilot an alternative, if feasible.

2. Discontinues further alerts if informed by the pilot that he is taking action to correct the situation or that he has the other aircraft in sight.

5–78. SEE AND AVOID

a. Pilot—When meteorological conditions permit, regardless of type of flight plan or whether or not under control of a radar facility, the pilot is responsible to see and avoid other traffic, terrain, or obstacles.

b. Controller—

1. Provides radar traffic information to radar identified aircraft operating outside positive control airspace on a workload permitting basis.

2. Issues a safety alert to an aircraft under his control if he is aware the aircraft is at an altitude believed to place the aircraft in unsafe proximity to terrain, obstructions, or other aircraft.

5–79. SPEED ADJUSTMENTS

a. Pilot—

1. Advises ATC any time cruising airspeed varies plus or minus 5 percent or 10 knots, whichever is greater, from that given in the flight plan.

2. Complies with speed adjustments from ATC unless:

(a) The minimum or maximum safe airspeed for any particular operation is greater or less than the requested airspeed. In such cases, advises ATC.

(b) Operating at or above 10,000 feet MSL on an ATC assigned SPEED ADJUSTMENT of more than 250 knots IAS and subsequent clearance is received for descent below 10,000 feet MSL. In such cases, pilots are expected to comply with FAR 91.117(a).

3. When complying with speed adjustment assignments, maintains an indicated airspeed within plus or minus 10 knots or 0.02 mach number of the specified speed.

b. Controller—

1. Assigns speed adjustments to aircraft when necessary but not as a substitute for good vectoring technique.

2. Adheres to the restrictions published in the Controllers Handbook (7110.65) as to when speed adjustment procedures may be applied.

3. Avoids speed adjustments requiring alternate decreases and increases.

4. Assigns speed adjustments to a specified IAS (KNOTS)/mach number or to increase or decrease speed using increments of 10 knots or multiples thereof.

5. Advises pilots to resume normal speed when speed adjustments are no longer required.

6. Gives due consideration to aircraft capabilities to reduce speed while descending.

5-80. TRAFFIC ADVISORIES (Traffic Information)

a. Pilot—

1. Acknowledges receipt of traffic advisories.

2. Informs controller if traffic in sight.

3. Advises ATC if a vector to avoid traffic is desired.

4. Do not expect to receive radar traffic advisories on all traffic. Some aircraft may not appear on the radar display. Be aware that the controller may be occupied with higher priority duties and unable to issue traffic information for a variety of reasons.

5. Advises controller if service not desired.

b. Controller—

1. Issues radar traffic to the maximum extent consistent with higher priority duties except in positive controlled airspace.

2. Provides vectors to assist aircraft to avoid observed traffic when requested by the pilot.

3. Issues traffic information to aircraft in the Airport Traffic Area for sequencing purposes.

5-81. VISUAL APPROACH

a. Pilot—

1. If a visual approach is not desired, advises ATC.

2. Complies with controller's instructions for vectors toward the airport of intended landing or to a visual position behind a preceding aircraft.

3. After being cleared for a visual approach, proceeds to the airport in a normal manner or follows the designated traffic and/or charted flight procedures, as appropriate, remaining in VFR at all times.

4. Acceptance of a visual approach clearance to visually follow a preceding aircraft is pilot acknowledgement that he will establish a safe landing interval behind the preceding aircraft if so cleared, and that he accepts responsibility for his own wake turbulence separation.

5. Advise ATC immediately if you are unable to continue following a designated aircraft or encounter less than basic VFR weather conditions.

6. Be aware that radar service is automatically terminated, without advising the pilot, when the aircraft is instructed to change to advisory frequency.

7. Be aware that there may be other traffic in the traffic pattern and the landing sequence may differ from the traffic sequence assigned by approach control or ARTCC.

b. Controller—

1. Does not vector an aircraft for a visual approach to an airport with weather reporting service unless the reported ceiling at the airport is 500 feet or more above the MVA and visibility is 3 miles or more.

2. Informs the pilot when weather is not available for the destination airport, and does not vector for a visual approach to those airports unless there is reasonable assurance that descent and flight to the airport can be made in VFR conditions.

3. Does not clear an aircraft for a visual approach unless the aircraft is and can remain in VFR conditions.

4. Issues visual approach clearance when the pilot reports sighting the airport or a preceding aircraft which is to be followed.

5. Provides separation except when visual separation is being applied by the pilot of the aircraft executing the visual approach.

6. Continues flight following and traffic information until the aircraft has landed or has been instructed to change to advisory frequency.

7. Informs the pilot conducting the visual approach of the aircraft class when pertinent traffic is known to be a heavy aircraft.

5-82. VISUAL SEPARATION

a. Pilot—

1. Acceptance of instructions to follow another aircraft or to provide visual separation from it is an acknowledgment that the pilot will maneuver his/her aircraft as necessary to avoid the other aircraft or to maintain in-trail separation.

2. If instructed by ATC to follow another aircraft or to provide visual separation from it, promptly notify the controller if you lose sight of that aircraft, are unable to maintain continued visual contact with it, or cannot accept the responsibility for your own separation for any reason.

3. The pilot also accepts responsibility for wake turbulence separation under these conditions.

b. Controller—Applies visual separation only:

1. In conjunction with visual approaches.

2. Within the terminal area when a controller has both aircraft in sight or by instructing a pilot who sees the other aircraft to maintain visual separation from it.

3. Within en route airspace when aircraft are on opposite courses and one pilot reports having seen the other aircraft and that the aircraft have passed each other.

5-83. VFR-ON-TOP

a. Pilot—

1. This clearance must be requested by the pilot on an IFR flight plan, and if approved, permits the pilot to select an altitude or Flight Level of his choice (subject to any ATC restrictions) in lieu of an assigned altitude.

5–83a1 Note 1—VFR-ON-TOP is not permitted in certain airspace areas, such as positive control airspace, certain restricted areas, etc. Consequently, IFR flights operating VFR-ON-TOP will avoid such airspace.

5–83a1 Note 2—Reference—IFR Altitudes and Flight Levels, paragraph 3–29, IFR Clearance VFR-ON-TOP, paragraph 4–87, IFR Separation Standards, paragraph 4–90, Position Reporting, paragraph 5–32, and Additional Reports, paragraph 5–33.

2. By requesting a VFR-ON-TOP clearance, the pilot indicates that he is assuming the sole responsibility to be vigilant so as to see and avoid other aircraft and that he will:

(a) Fly at the appropriate VFR altitude as prescribed in FAR 91.159.

(b) Comply with the VFR visibility and distance from criteria in FAR 91.155 (Basic VFR Weather Minimums).

(c) Comply with instrument flight rules that are applicable to this flight; i.e., minimum IFR altitudes, position reporting, radio communications, course to be flown, adherence to ATC clearance, etc.

3. Should advise ATC prior to any altitude change to ensure the exchange of accurate traffic information.

b. Controller—

1. May clear an aircraft to maintain VFR-ON-TOP if the pilot of an aircraft on an IFR flight plan requests the clearance.

2. Informs the pilot of an aircraft cleared to climb to VFR-ON-TOP the reported height of the tops or that no top report is available; issues an alternate clearance if necessary; and once the aircraft reports reaching VFR-ON-TOP, reclears the aircraft to maintain VFR-ON-TOP.

3. Before issuing clearance, ascertains that the aircraft is not in or will not enter positive control airspace.

5–84. INSTRUMENT DEPARTURES

a. Pilot—

1. Prior to departure considers the type of terrain and other obstructions on or in the vicinity of the departure airport.

2. Determines if obstruction avoidance can be maintained visually or that the departure procedure should be followed.

3. Determines whether a departure procedure and/or SID is available for obstruction avoidance.

4. At airports where IAP's have not been published, hence no published departure procedure, determines what action will be necessary and takes such action that will assure a safe departure.

b. Controller—

1. At locations with airport traffic control service, when necessary, specifies direction of takeoff, turn, or initial heading to be flown after takeoff.

2. At locations without airport traffic control service but within a Control Zone when necessary to specify direction of takeoff, turn, or initial heading to be flown, obtains pilot's concurrence that the procedure will allow him to comply with local traffic patterns, terrain, and obstruction avoidance.

3. Includes established departure procedures as part of the ATC clearance when pilot compliance is necessary to ensure separation.

5–85. MINIMUM FUEL ADVISORY

a. Pilot—

1. Advise ATC of your minimum fuel status when your fuel supply has reached a state where, upon reaching destination, you cannot accept any undue delay.

2. Be aware this is not an emergency situation, but merely an advisory that indicates an emergency situation is possible should any undue delay occur.

3. Be aware a minimum fuel advisory does not imply a need for traffic priority.

4. If the remaining usable fuel supply suggests the need for traffic priority to ensure a safe landing, you should declare an emergency account low fuel and report fuel remaining in minutes. (Reference—Pilot/Controller Glossary, Fuel Remaining).

b. Controller—

1. When an aircraft declares a state of minimum fuel, relay this information to the facility to whom control jurisdiction is transferred.

2. Be alert for any occurrence which might delay the aircraft.

5–86 thru 5–90. RESERVED

Section 6. NATIONAL SECURITY AND INTERCEPTION PROCEDURES

5–91. NATIONAL SECURITY

a. National security in the control of air traffic is governed by (FAR 99).

b. All aircraft entering domestic U.S. airspace from points outside must provide for identification prior to entry. To facilitate early aircraft identification of all aircraft in the vicinity of U.S. and international airspace boundaries, AIR DEFENSE IDENTIFICATION ZONES (ADIZ) have been established. (Reference—ADIZ Boundaries and Designated Mountainous Areas, paragraph 5–94).

c. Operational requirements for aircraft operations associated with an ADIZ are as follows:

1. Flight Plan—Except as specified in subparagraphs **d** and **e** below, an IFR or DVFR flight plan must be filed with an appropriate aeronautical facility as follows:

(a) Generally, for all operations that enter an ADIZ.

(b) For operations that will enter or exit the United States and which will operate into, within or across the Contiguous U.S. ADIZ regardless of true airspeed.

(c) The flight plan must be filed before departure except for operations associated with the Alaskan ADIZ when the airport of departure has no facility for filing a flight plan, in which case the flight plan may be filed immediately after takeoff or when within range of the aeronautical facility.

2. Two-way Radio—For the majority of operations associated with an ADIZ, an operating two-way radio is required. See FAR 99.1 for exceptions.

3. Transponder Requirements—Unless otherwise authorized by ATC, each aircraft conducting operations into, within, or across the Contiguous U.S. ADIZ must be equipped with an operable radar beacon transponder having altitude reporting capability (Mode C), and that transponder must be turned on and set to reply on the appropriate code or as assigned by ATC.

4. Position Reporting:

(a) For IFR flight—Normal IFR position reporting.

(b) For DVFR flights—The estimated time of ADIZ penetration must be filed with the aeronautical facility at least 15 minutes prior to penetration except for flight in the Alaskan ADIZ, in which case report prior to penetration.

(c) For inbound aircraft of foreign registry—The pilot must report to the aeronautical facility at least one hour prior to ADIZ penetration.

5. Aircraft Position Tolerances:

(a) Over land, the tolerance is within plus or minus five minutes from the estimated time over a reporting point or point of penetration and within 10 NM from the centerline of an intended track over an estimated reporting point or penetration point.

(b) Over water, the tolerance is plus or minus five minutes from the estimated time over a reporting point or point of penetration and within 20 NM from the centerline of the intended track over an estimated reporting point or point of penetration (to include the Aleutian Islands).

d. Except when applicable under FAR 99.7, FAR 99 does not apply to aircraft operations:

1. Within the 48 contiguous states and the District of Columbia, or within the State of Alaska, and remains within 10 miles of the point of departure;

2. Over any island, or within three nautical miles of the coastline of any island, in the Hawaii ADIZ; or

3. Associated with any ADIZ other than the Contiguous U.S. ADIZ, when the aircraft true airspeed is less than 180 knots.

e. Authorizations to deviate from the requirements of Part 99 may also be granted by the ARTCC, on a local basis, for some operations associated with an ADIZ.

f. An Airfiled VFR Flight Plan makes an aircraft subject to interception for positive identification when entering an ADIZ. Pilots are therefore urged to file the required DVFR flight plan either in person or by telephone prior to departure.

g. Special Security Instructions

1. During defense emergency or air defense emergency conditions, additional special security instructions may be issued in accordance with the Security Control of Air Traffic and Air Navigation Aids (SCATANA) Plan.

2. Under the provisions of the SCATANA Plan, the military will direct the action to be taken-in regard to landing, grounding, diversion, or dispersal of aircraft and the control of air navigation aids in the defense of the U.S. during emergency conditions.

3. At the time a portion or all of SCATANA is implemented, ATC facilities will broadcast appropriate instructions received from the military over available ATC frequencies. Depending on instructions received from the military, VFR flights may be directed to land at the nearest available airport, and IFR flights will be expected to proceed as directed by ATC.

4. Pilots on the ground may be required to file a flight plan and obtain an approval (through FAA) prior to conducting flight operation.

5. In view of the above, all pilots should guard an ATC or FSS frequency at all times while conducting flight operations.

5–92. INTERCEPTION PROCEDURES

a. General

1. Identification intercepts during peacetime operations are vastly different than those conducted under increased states of readiness. Unless otherwise directed by the control agency, intercepted aircraft will be identified by type only. When specific information is required (i.e. markings, serial numbers, etc.) the interceptor aircrew will respond only if the request can be conducted in a safe manner. During hours of darkness or Instrument Meteorological Conditions (IMC), identification of unknown aircraft will be by type only. The interception pattern described below is the typical peacetime method used by air interceptor aircrews. In all situations, the interceptor aircrew will use caution to avoid startling the intercepted aircrew and/or passengers.

b. Intercept phases (See Figure 5–92[1].)

1. Phase One—Approach Phase: During peacetime, intercepted aircraft will be approached from the stern. Generally two interceptor aircraft will be employed to accomplish the identification. The flight leader and his wingman will coordinate their individual positions in conjunction with the ground controlling agency. Their relationship will resemble a line abreast formation. At night or in IMC, a comfortable radar trail tactic will be used. Safe vertical separation between interceptor aircraft and unknown aircraft will be maintained at all times.

2. Phase Two—Identification Phase: The intercepted aircraft should expect to visually acquire the lead interceptor and possibly the wingman during this phase in visual meteorological conditions (VMC). The wingman will assume a surveillance position while the flight leader approaches the unknown aircraft. Intercepted aircraft personnel may observe the use of different drag devices to allow for speed and position stabilization during this phase. The flight leader will then initiate a gentle closure toward the intercepted aircraft, stopping

Figure 5–92[1]

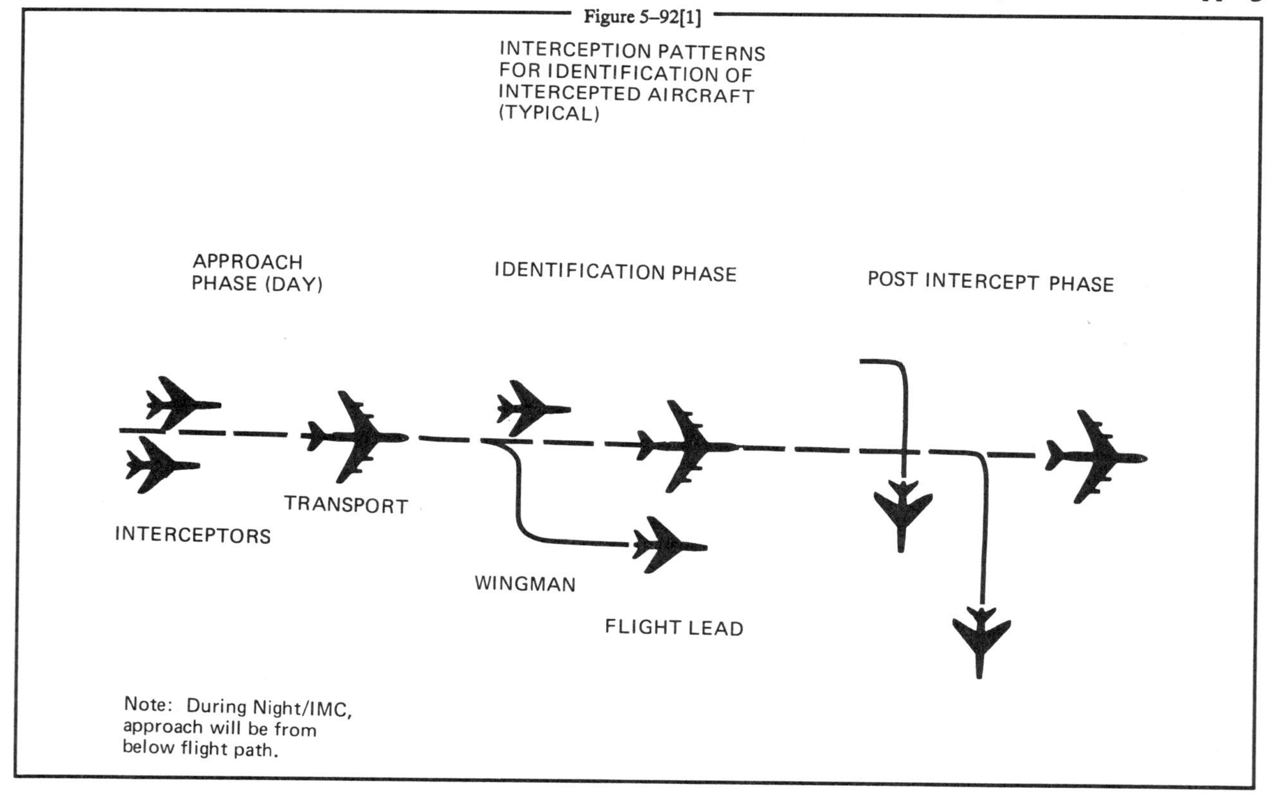

at a distance no closer than absolutely necessary to obtain the information needed. The interceptor aircraft will use every possible precaution to avoid startling intercepted aircrew or passengers. Additionally, the interceptor aircrews will constantly keep in mind that maneuvers considered normal to a fighter aircraft may be considered hazardous to passengers and crews of nonfighter aircraft. When interceptor aircrews know or believe that an unsafe condition exists, the identification phase will be terminated. As previously stated, during darkness or IMC identification of unknown aircraft will be by type only. Positive vertical separation will be maintained by interceptor aircraft throughout this phase.

3. Phase Three—Post Intercept Phase: Upon identification phase completion, the flight leader will turn away from the intercepted aircraft. The wingman will remain well clear and accomplish a rejoin with his leader.

c. Communication interface between interceptor aircrews and the ground controlling agency is essential to ensure successful intercept completion. Flight Safety is paramount. An aircraft which is intercepted by another aircraft shall immediately:

1. Follow the instructions given by the intercepting aircraft, interpreting and responding to the visual signals.

2. Notify, if possible, the appropriate air traffic services unit.

3. Attempt to establish radio communication with the intercepting aircraft or with the appropriate intercept control unit, by making a general call on the emergency frequency 243.0 MHz and repeating this call on the emergency frequency 121.5 MHz, if practicable, giving the identity and position of the aircraft and the nature of the flight.

4. If equipped with SSR transponder, select MODE 3/A Code 7700, unless otherwise instructed by the appropriate air traffic services unit. If any instructions received by radio from any sources conflict with those given by the intercepting aircraft by visual or radio signals, the intercepted aircraft shall request immediate clarification while continuing to comply with the instructions given by the intercepting aircraft.

5–93. INTERCEPTION SIGNALS

(See Table 5–93[1] and Table 5–93[2].)

Table 5–93[1]

INTERCEPTION SIGNALS

Signals initiated by intercepting aircraft and responses by intercepted aircraft
(as set forth in ICAO Annex 2—Appendix A, 2.1)

Series	INTERCEPTING Aircraft Signals	Meaning	INTERCEPTED Aircraft Responds	Meaning
1	DAY—Rocking wings from a position slightly above and ahead of, and normally to the left of, the intercepted aircraft and, after acknowledgement, a slow level turn, normally to the left, on to the desired heading. NIGHT—Same and, in addition, flashing navigational lights at irregular intervals. *Note 1.—Meteorological conditions or terrain may require the intercepting aircraft to take up a position slightly above and ahead of, and to the right of, the intercepted aircraft and to make the subsequent turn to the right.* *Note 2.—If the intercepted aircraft is not able to keep pace with the intercepting aircraft, the latter is expected to fly a series of race-track patterns and to rock its wings each time it passes the intercepted aircraft.*	You have been intercepted. Follow me.	AEROPLANES: DAY—Rocking wings and following. NIGHT—Same and, in addition, flashing navigational lights at irregular intervals. HELICOPTERS: DAY or NIGHT—Rocking aircraft, flashing navigational lights at irregular intervals and following.	Understood, will comply.
2	DAY or NIGHT—An abrupt break-away maneuver from the intercepted aircraft consisting of a climbing turn of 90 degrees or more without crossing the line of flight of the intercepted aircraft.	You may proceed.	AEROPLANES: DAY or NIGHT—Rocking wings. HELICOPTERS: DAY or NIGHT—Rocking aircraft.	Understood, will comply.
3	DAY—Circling aerodrome, lowering landing gear and overflying runway in direction of landing or, if the intercepted aircraft is a helicopter, overflying the helicopter landing area. NIGHT—Same and, in addition, showing steady landing lights.	Land at this aerodrome.	AEROPLANES: DAY—Lowering landing gear, following the intercepting aircraft and, if after overflying the runway landing is considered safe, proceeding to land. NIGHT—Same and, in addition, showing steady landing lights (if carried). HELICOPTERS: DAY or NIGHT—Following the intercepting aircraft and proceeding to land, showing a steady landing light (if carried).	Understood, will comply.

Table 5–93[2]

INTERCEPTION SIGNALS

Signals and Responses During Aircraft Intercept
Signals initiated by intercepted aircraft and responses by intercepting aircraft
(as set forth in ICAO Annex 2—Appendix A, 2.2)

Series	INTERCEPTED Aircraft Signals	Meaning	INTERCEPTING Aircraft Responds	Meaning
4	AEROPLANES: DAY—Raising landing gear while passing over landing runway at a height exceeding 300m (1,000 ft) but not exceeding 600m (2,000 ft) above the aerodrome level, and continuing to circle the aerodrome. NIGHT—Flashing landing lights while passing over landing runway at a height exceeding 300m (1,000 ft) but not exceeding 600m (2,000 ft) above the aerodrome level, and continuing to circle the aerodrome. If unable to flash landing lights, flash any other lights available.	Aerodrome you have designated is inadequate.	DAY or NIGHT—If it is desired that the intercepted aircraft follow the intercepting aircraft to an alternate aerodrome, the intercepting aircraft raises its landing gear and uses the Series 1 signals prescribed for intercepting aircraft. If it is decided to release the intercepted aircraft, the intercepting aircraft uses the Series 2 signals prescribed for intercepting aircraft.	Understood, follow me. Understood, you may proceed.
5	AEROPLANES: DAY or NIGHT—Regular switching on and off of all available lights but in such a manner as to be distinct from flashing lights.	Cannot comply.	DAY or NIGHT—Use Series 2 signals prescribed for intercepting aircraft.	Understood.
6	AEROPLANES: DAY or NIGHT—Irregular flashing of all available lights. HELICOPTERS: DAY or NIGHT—Irregular flashing of all available lights.	In distress.	DAY or NIGHT—Use Series 2 signals prescribed for intercepting aircraft.	Understood.

5–94. ADIZ BOUNDARIES AND DESIGNATED MOUNTAINOUS AREAS

(See Figure 5–94[1].)

Figure 5–94[1]

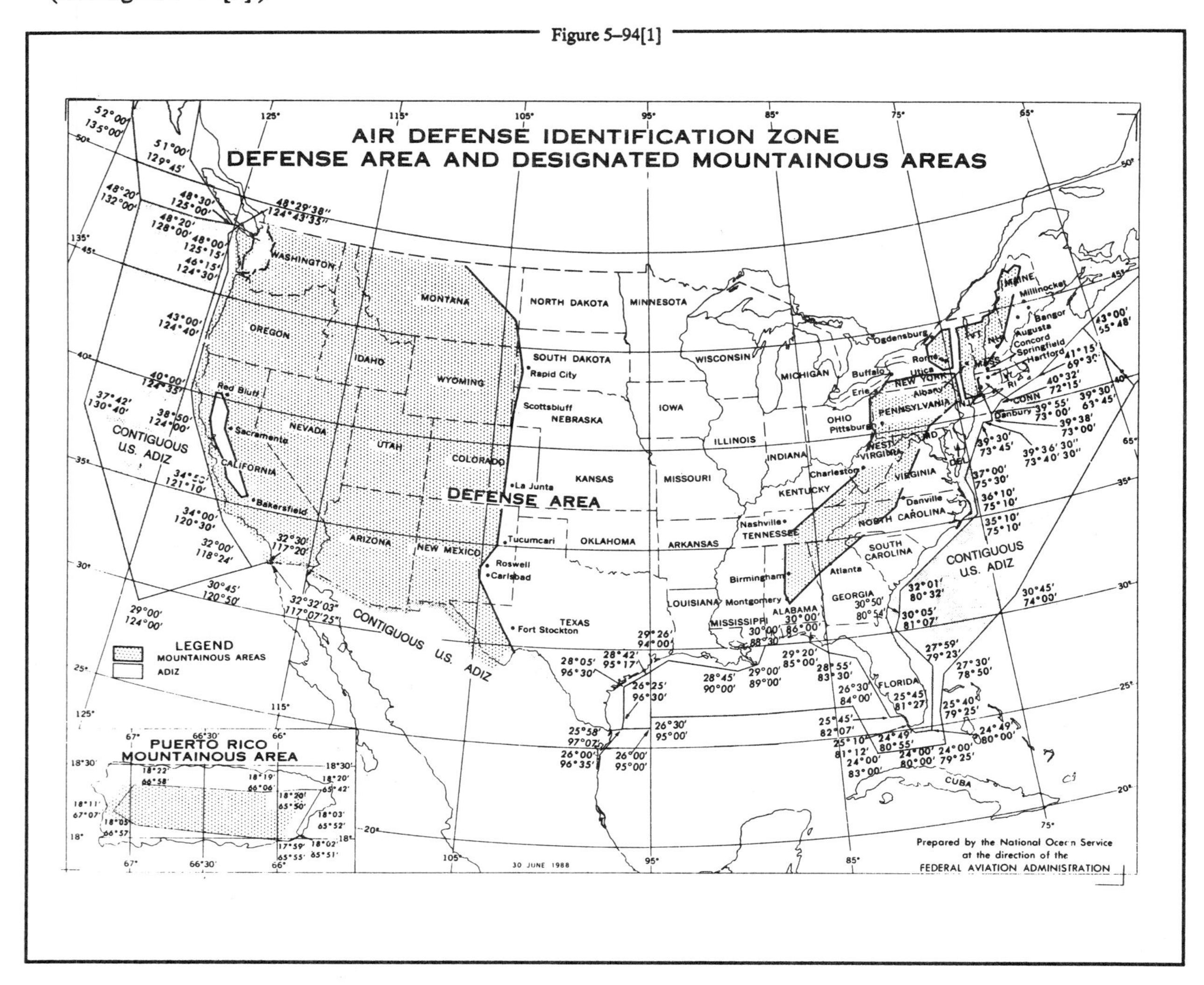

5–95 thru 5–100. RESERVED

Chapter 6. EMERGENCY PROCEDURES

Section 1. GENERAL

6–1. PILOT RESPONSIBILITY AND AUTHORITY

a. The pilot in command of an aircraft is directly responsible for and is the final authority as to the operation of that aircraft. In an emergency requiring immediate action, the pilot in command may deviate from any rule in the FAR, Subpart A, General, and Subpart B, Flight Rules, to the extent required to meet that emergency. (FAR 91.3(b).)

b. If the emergency authority of FAR 91.3.(b) is used to deviate from the provisions of an ATC clearance, the pilot in command must notify ATC as soon as possible and obtain an amended clearance.

c. Unless deviation is necessary under the emergency authority of FAR 91.3, pilots of IFR flights experiencing two-way radio communications failure are expected to adhere to the procedures prescribed under "IFR operations, two-way radio communications failure." (FAR 91.185)

6–2. EMERGENCY CONDITION—REQUEST ASSISTANCE IMMEDIATELY

a. An emergency can be either a *distress* or *urgency* condition as defined in the Pilot/Controller Glossary. Pilots do not hesitate to declare an emergency when they are faced with *distress* conditions such as fire, mechanical failure, or structural damage. However, some are reluctant to report an *urgency* condition when they encounter situations which may not be immediately perilous, but are potentially catastrophic. An aircraft is in at least an *urgency* condition the moment the pilot becomes doubtful about position, fuel endurance, weather, or any other condition that could adversely affect flight safety. This is the time to ask for help, not after the situation has developed into a *distress* condition.

b. Pilots who become apprehensive for their safety for *any* reason should *request assistance immediately.* Ready and willing help is available in the form of radio, radar, direction finding stations and other aircraft. Delay has caused accidents and cost lives. *Safety is not a luxury! Take action!*

6–3 thru 6–10. RESERVED

Section 2. EMERGENCY SERVICES AVAILABLE TO PILOTS

6-11. RADAR SERVICE FOR VFR AIRCRAFT IN DIFFICULTY

a. Radar equipped ATC facilities can provide radar assistance and navigation service (vectors) to VFR aircraft in difficulty when the pilot can talk with the controller, and the aircraft is within radar coverage. Pilots should clearly understand that authorization to proceed in accordance with such radar navigational assistance does not constitute authorization for the pilot to violate FARs. In effect, assistance is provided on the basis that navigational guidance information is advisory in nature, and the responsibility for flying the aircraft safely remains with the pilot.

b. Experience has shown that many pilots who are not qualified for instrument flight cannot maintain control of their aircraft when they encounter clouds or other reduced visibility conditions. In many cases, the controller will not know whether flight into instrument conditions will result from his instructions. To avoid possible hazards resulting from being vectored into IFR conditions, a pilot in difficulty should keep the controller advised of the weather conditions in which he is operating and the weather along the course ahead and observe the following:

1. If a course of action is available which will permit flight and a safe landing in VFR weather conditions, noninstrument rated pilots should choose the VFR condition rather than requesting a vector or approach that will take them into IFR weather conditions; or

2. If continued flight in VFR conditions is not possible, the noninstrument rated pilot should so advise the controller and indicating the lack of an instrument rating, declare a *distress* condition, or

3. If the pilot is instrument rated and current, and the aircraft is instrument equipped, the pilot should so indicate by requesting an IFR flight clearance. Assistance will then be provided on the basis that the aircraft can operate safely in IFR weather conditions.

6-12. TRANSPONDER EMERGENCY OPERATION

a. When a *distress* or *urgency* condition is encountered, the pilot of an aircraft with a coded radar beacon transponder, who desires to alert a ground radar facility, should squawk MODE 3/A, Code 7700/Emergency and MODE C altitude reporting and then immediately establish communications with the ATC facility.

b. Radar facilities are equipped so that Code 7700 normally triggers an alarm or special indicator at all control positions. Pilots should understand that they might not be within a radar coverage area. Therefore, they should continue squawking Code 7700 and establish radio communications as soon as possible.

6-13. DIRECTION FINDING INSTRUMENT APPROACH PROCEDURE

a. DF equipment has long been used to locate lost aircraft and to guide aircraft to areas of good weather or to airports. Now at most DF equipped airports, DF instrument approaches may be given to aircraft in a *distress* or *urgency* condition.

b. Experience has shown that most emergencies requiring DF assistance involve pilots with little flight experience. With this in mind, DF approach procedures provide maximum flight stability in the approach by using small turns, and wings-level descents. The DF specialist will give the pilot headings to fly and tell the pilot when to begin descent.

c. DF IAPs are for emergency use only and will not be used in IFR weather conditions unless the pilot has declared a *distress* or *urgency* condition.

d. To become familiar with the procedures and other benefits of DF, pilots are urged to request practice DF guidance and approaches in VFR weather conditions. DF specialists welcome the practice and will honor such requests, workload permitting.

6-14. INTERCEPT AND ESCORT

a. The concept of airborne intercept and escort is based on the Search and Rescue (SAR) aircraft establishing visual and/or electronic contact with an aircraft in difficulty, providing in-flight assistance, and escorting it to a safe landing. If bailout, crash landing or ditching becomes necessary, SAR operations can be conducted without delay. For most incidents, particularly those occurring at night and/or during instrument flight conditions, the availability of intercept and escort services will depend on the proximity of SAR units with suitable aircraft on alert for immediate dispatch. In limited circumstances, other aircraft flying in the vicinity of an aircraft in difficulty can provide these services.

b. If specifically requested by a pilot in difficulty or if a *distress* condition is declared, SAR coordinators *will* take steps to intercept and escort an aircraft.

Steps *may* be initiated for intercept and escort if an *urgency* condition is declared and unusual circumstances make such action advisable.

c. It is the pilot's prerogative to refuse intercept and escort services. Escort services will normally be provided to the nearest adequate airport. Should the pilot receiving escort services continue onto another location after reaching a safe airport, or decide not to divert to the nearest safe airport, the escort aircraft is not obligated to continue and further escort is discretionary. The decision will depend on the circumstances of the individual incident.

6–15. EMERGENCY LOCATOR TRANSMITTERS

a. GENERAL. Emergency Locator Transmitters (ELTS) are required for most General Aviation airplanes (FAR 91.207). ELTs of various types have been developed as a means of locating downed aircraft. These electronic, battery operated transmitters emit a distinctive downward swept audio tone on 121.5 MHz and 243.0 MHz. If "armed" and when subject to crash generated forces they are designed to automatically activate and continuously emit these signals. The transmitters will operate continuously for at least 48 hours over a wide temperature range. A properly installed and maintained ELT can expedite search and rescue operations and save lives.

b. TESTING. ELTs should be tested in accordance with the manufacturer's instructions, preferably in a shielded or screened room to prevent the broadcast of signals which could trigger a false alert. When this cannot be done, aircraft operational testing is authorized on 121.5 MHz and 243.0 MHz as follows:

1. Tests should be conducted only during the first 5 minutes after any hour. If operational tests must be made outside of this time-frame, they should be coordinated with the nearest FAA Control Tower or FSS.

2. Tests should be no longer than three audible sweeps.

3. If the antenna is removable, a dummy load should be substituted during test procedures.

4. Airborne tests are not authorized.

c. FALSE ALARMS. Caution should be exercised to prevent the inadvertent activation of ELTs in the air or while they are being handled on the ground. Accidental or unauthorized activation will generate an emergency signal that cannot be distinguished from the real thing, leading to expensive and frustrating searches. A false ELT signal could also interfere with genuine emergency transmissions and hinder or prevent the timely location of crash sites. Frequent false alarms could also result in complacency and decrease the vigorous reaction that must be attached to all ELT signals. Numerous cases of inadvertent activation have occurred as a result of aerobatics, hard landings, movement by ground crews and aircraft maintenance. These false alarms can be minimized by monitoring 121.5 MHz and/or 243.0 MHz as follows:

1. In flight when a receiver is available.

2. Prior to engine shut down at the end of each flight.

3. When the ELT is handled during installation or maintenance.

4. When maintenance is being performed in the vicinity of the ELT.

5. When the aircraft is moved by a ground crew.

6. If an ELT signal is heard, turn off the ELT to determine if it is transmitting. If it has been activated, maintenance might be required before the unit is returned to the "ARMED" position.

d. IN-FLIGHT MONITORING AND REPORTING. Pilots are encouraged to monitor 121.5 MHz and/or 243.0 MHz while in-flight to assist in identifying possible emergency ELT transmissions. On receiving a signal, report the following information to the nearest air traffic facility:

1. Your position at the time the signal was first heard.

2. Your position at the time the signal was last heard.

3. Your position at maximum signal strength.

4. Your flight altitudes and frequency on which the emergency signal was heard—121.5 MHz or 243.0 MHz. If possible, positions should be given relative to a navigation aid. If the aircraft has homing equipment, provide the bearing to the emergency signal with each reported position.

6–16. FAA SPONSORED EXPLOSIVES DETECTION (DOG/HANDLER TEAM) LOCATIONS

a. At many of our major airports a program has been established by the FAA to make available explosives detection dog/handler teams. The dogs are trained by the Air Force and the overall program is run by FAA's Office of Civil Aviation Security. Local police departments are the caretakers of the dogs and are allowed to use the dogs in their normal police patrol functions. The local airport, however, has first call on the teams' services. The explosives detection teams were established so that no aircraft in flight is more than 1 hour from an airport at which it can be searched if a bomb threat is received. The following list contains those locations that presently have a team in existence. This list will be updated as more teams are established. If you desire this service, notify your company or an FAA facility.

b. Team Locations: (See Table 6–16[1].)

Table 6–16[1]

Airport Symbol	Location
ATL	Atlanta, Georgia
BWI	Baltimore, Maryland
BHM	Birmingham, Alabama
BIS	Bismarck, North Dakota
BOS	Boston, Massachusetts
BUF	Buffalo, New York
ORD	Chicago, Illinois
CLE	Cleveland, Ohio
DFW	Dallas, Texas
IAH	Houston, Texas
JAX	Jacksonville, Florida
MCI	Kansas City, Missouri
LAX	Los Angeles, California
MEM	Memphis, Tennessee
MIA	Miami, Florida
MSY	New Orleans, Louisiana
PHX	Phoenix, Arizona
PIT	Pittsburgh, Pennsylvania
PDX	Portland, Oregon
SLC	Salt Lake City, Utah
SAN	San Diego, California
SFO	San Francisco, California
SJU	San Juan, Puerto Rico
SEA	Seattle, Washington
GEG	Spokane, Washington
STL	St. Louis, Missouri
TUS	Tucson, Arizona
TUL	Tulsa, Oklahoma

c. If due to weather or other considerations an aircraft with a suspected hidden explosive problem were to land or intended to land at an airport other than those listed in b. above, it is recommended that they call the FAA's Washington Operations Center (telephone 202–426–3333, if appropriate) or have an air traffic facility with which you can communicate contact the above center requesting assistance.

6–17. SEARCH AND RESCUE

a. GENERAL. SAR is a lifesaving service provided through the combined efforts of the federal agencies signatory to the National SAR Plan, and the agencies responsible for SAR within each state. Operational resources are provided by the U.S. Coast Guard, DOD components, the Civil Air Patrol, the Coast Guard Auxiliary, state, county and local law enforcement and other public safety agencies, and private volunteer organizations. Services include search for missing aircraft, survival aid, rescue, and emergency medical help for the occupants after an accident site is located.

b. NATIONAL SEARCH AND RESCUE PLAN. By federal interagency agreement, the National Search and Rescue Plan provides for the effective use of all available facilities in all types of SAR missions. These facilities include aircraft, vessels, pararescue and ground rescue teams, and emergency radio fixing. Under the Plan, the U.S. Coast Guard is responsible for the coordination of SAR in the Maritime Region, and the USAF is responsible in the Inland Region. To carry out these responsibilities, the Coast Guard and the Air Force have established Rescue Coordination Centers (RCCs) to direct SAR activities within their Regions. For aircraft emergencies, distress, and urgency, information normally will be passed to the appropriate RCC through an ARTCC or FSS.

c. COAST GUARD RESCUE COORDINATION CENTERS.

(See Table 6–17[1]).

Table 6–17[1]
Coast Guard Rescue Coordination Centers

Boston, MA 617–223–8555	Long Beach, CA 213–590–2225 310–499–5380
New York, NY 212–668–7055	San Francisco, CA 415–437–3700
Portsmouth, VA 804–398–6231	Seattle, WA 206–553–5886
Miami, FL 305–536–5611	Juneau, AK 907–463–2000
New Orleans, LA 504–589–6225	Honolulu, HI 808–541–2500
Cleveland, OH 216–522–3984	San Juan, Puerto Rico 809–729–6770
St. Louis, MO 314–262–3706	

d. AIR FORCE RESCUE COORDINATION CENTERS.

Air Force Rescue Coordination Center—48 Contiguous States.

(See Table 6–17[2]).

Table 6–17[2]
Air Force Rescue Coordination Centers

Scott AFB, Illinois	Phone
Commercial	618–256–4815
WATS	800–851–3051
DSN	576–4815

Alaskan Air Command Rescue Coordination Center—Alaska.

(See Table 6–17[3]).

Table 6–17[3]
Alaskan Air Command Rescue Coordination Center

Elmendorf AFB, Alaska	Phone
Commercial	907–552–5375
DSN	317–552–2426

e. JOINT RESCUE COORDINATION CENTERS

Honolulu Joint Rescue Coordination Center—Hawaii.

(See Table 6–17[4]).

Table 6–17[4]
Honolulu Joint Rescue Coordination Center

HQ 14th CG District Honolulu	Phone
Commercial	808–541–2500
DSN	448–0301

f. OVERDUE AIRCRAFT.

1. ARTCCs and FSSs will alert the SAR system when information is received from any source that an aircraft is in difficulty, overdue, or missing. *A filed flight plan is the most timely and effective indicator that an aircraft is overdue.* Flight plan information is invaluable to SAR forces for search planning and executing search efforts.

2. Prior to departure on every flight, local or otherwise, someone at the departure point should be advised of your destination and route of flight if other than direct. Search efforts are often wasted and rescue is often delayed because of pilots who thoughtlessly takeoff without telling anyone where they are going. File a flight plan for *your* safety.

3. According to the National Search and Rescue Plan, "The life expectancy of an injured survivor decreases as much as 80 percent during the first 24 hours, while the chances of survival of uninjured survivors rapidly diminishes after the first 3 days."

4. An Air Force Review of 325 SAR missions conducted during a 23–month period revealed that "Time works against people who experience a *distress* but are not on a flight plan, since 36 hours normally pass before family concern initiates an (alert)."

g. VFR SEARCH AND RESCUE PROTECTION.

1. To receive this valuable protection, *file a VFR or DVFR Flight Plan* with an FAA FSS. For maximum protection, file only to the point of first intended landing, and refile for each leg to final destination. When a lengthy flight plan is filed, with several stops en route and an ETE to final destination, a mishap could occur on any leg, and unless other information is received, it is probably that no one would start looking for you until 30 minutes after your ETA at your final destination.

2. If you land at a location other than the intended destination, report the landing to the nearest FAA FSS and advise them of your original destination.

3. If you land en route and are delayed more than 30 minutes, report this information to the nearest FSS and give them your original destination.

4. If your ETE changes by 30 minutes or more, report a new ETA to the nearest FSS and give them your original destination. Remember that if you fail to respond within one-half hour after your ETA at final destination, a search will be started to locate you.

5. It is important that you *close your flight plan IMMEDIATELY AFTER ARRIVAL AT YOUR FINAL DESTINATION WITH THE FSS DESIGNATED WHEN YOUR FLIGHT PLAN WAS FILED. The pilot is responsible for closure of a VFR or DVFR flight plan; they are not closed automatically.* This will prevent needless search efforts.

6. The rapidity of rescue on land or water will depend on how accurately your position may be determined. If a flight plan has been followed and your position is on course, rescue will be expedited.

h. SURVIVAL EQUIPMENT.

1. For flight over uninhabited land areas, it is wise to take and know how to use survival equipment for the type of climate and terrain.

2. If a forced landing occurs at sea, chances for survival are governed by the degree of crew proficiency in emergency procedures and by the availability and effectiveness of water survival equipment.

HOW TO USE THEM

If you are forced down and are able to attract the attention of the pilot of a rescue airplane, the body signals illustrated on these pages can be used to transmit messages to him as he circles over your location. Stand in the open when you make the signals. Be sure the background, as seen from the air, is not confusing. Go through the motions slowly and repeat each signal until you are positive that the pilot understands you.

(See Figure 6–17[1][Ground–Air Visual Code for Use by Survivors].)

Figure 6–17[1]

GROUND-AIR VISUAL CODE FOR USE BY SURVIVORS

NO.	MESSAGE	CODE SYMBOL
1	Require assistance	V
2	Require medical assistance	X
3	No or Negative	N
4	Yes or Affirmative	Y
5	Proceeding in this direction	↑

IF IN DOUBT, USE INTERNATIONAL SYMBOL SOS

INSTRUCTIONS

1. Lay out symbols by using strips of fabric or parachutes, pieces of wood, stones, or any available material.
2. Provide as much color contrast as possible between material used for symbols and background against which symbols are exposed.
3. Symbols should be at least 10 feet high or larger. Care should be taken to lay out symbols exactly as shown.
4. In addition to using symbols, every effort is to be made to attract attention by means of radio, flares, smoke, or other available means.
5. On snow covered ground, signals can be made by dragging, shoveling or tramping. Depressed areas forming symbols will appear black from the air.
6. Pilot should acknowledge message by rocking wings from side to side.

(See Figure 6–17[2][Ground–Air Visual Code for use by Ground Search Parties].)

Figure 6–17[2]

GROUND-AIR VISUAL CODE FOR USE BY GROUND SEARCH PARTIES		
NO.	MESSAGE	CODE SYMBOL
1	Operation completed.	LLL
2	We have found all personnel.	LL
3	We have found only some personnel.	++
4	We are not able to continue. Returning to base.	XX
5	Have divided into two groups. Each proceeding in direction indicated.	⇄
6	Information received that aircraft is in this direction.	→ →
7	Nothing found. Will continue search.	NN

"Note: These visual signals have been accepted for international use and appear in Annex 12 to the Convention on International Civil Aviation."

Figure 6–17[3]

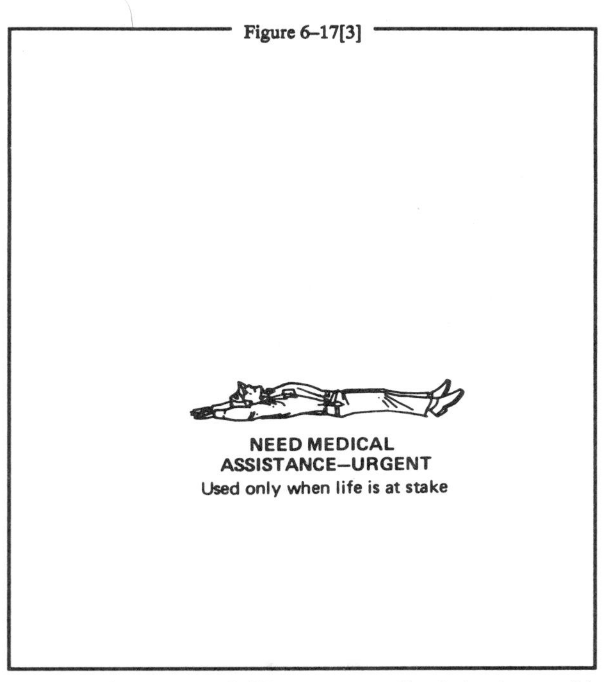

(See Figure 6–17[3][Urgent Medical Assistance].)

Figure 6–17[4]

(See Figure 6–17[4][All Ok].)

Figure 6–17[5]

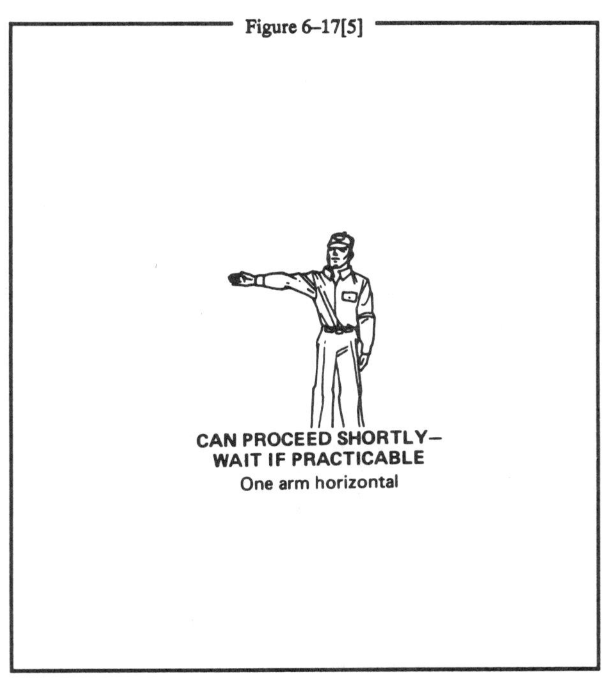

(See Figure 6–17[5][Short Delay].)

Figure 6–17[6]

(See Figure 6–17[6][Long Delay].)

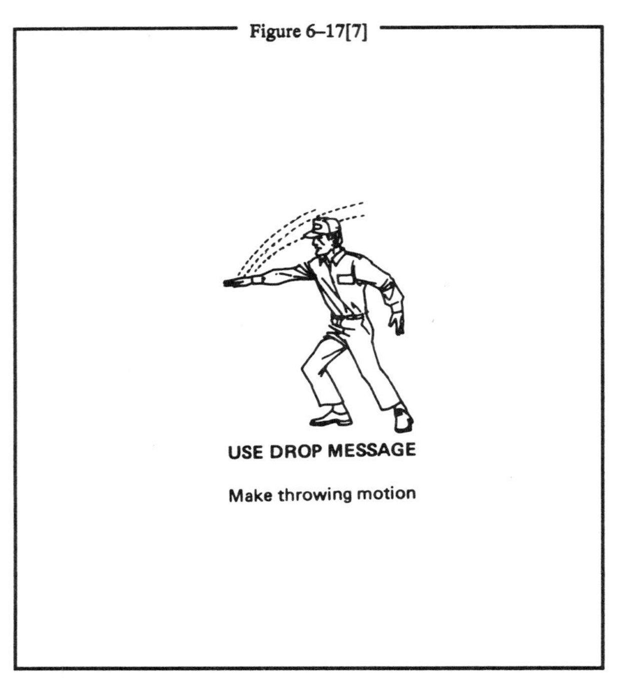

(See Figure 6–17[7][Drop Message].)

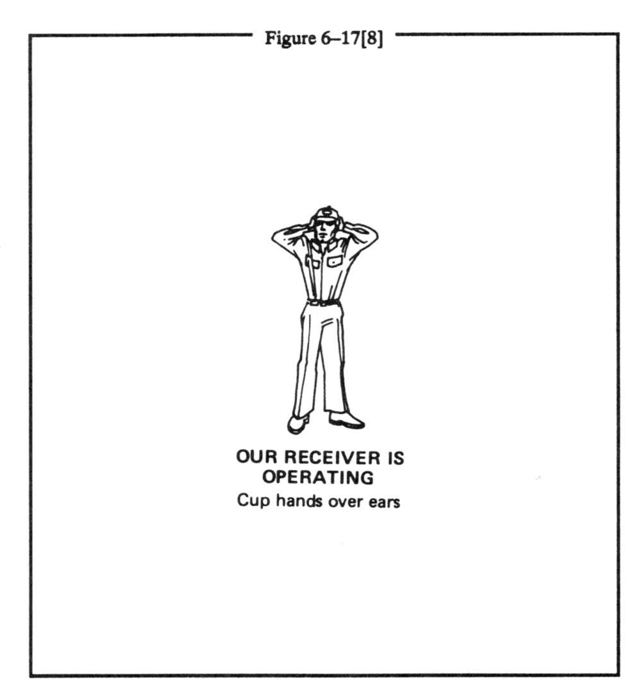

(See Figure 6–17[8][Receiver Operates].)

(See Figure 6–17[9][Do Not Land Here].)

(See Figure 6–17[10][Land Here].)

Figure 6–17[11]

(See Figure 6–17[11][Negative (Ground)].)

Figure 6–17[12]

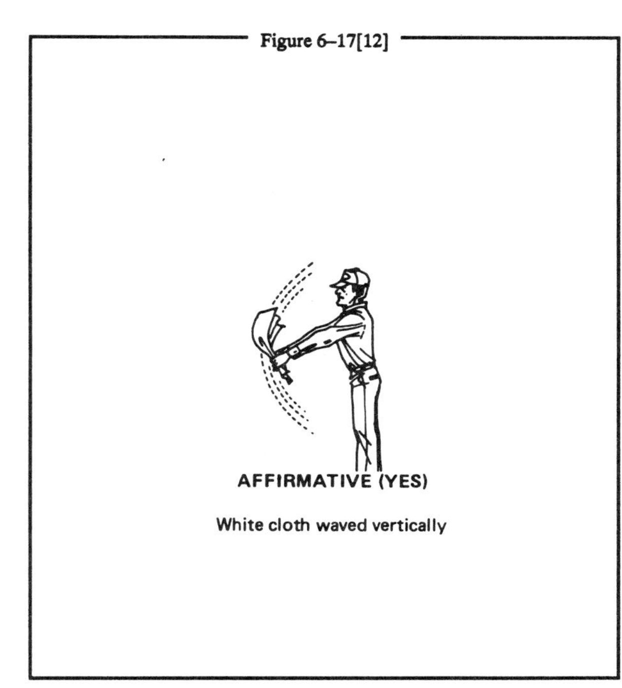

(See Figure 6–17[12][Affirmative (Ground)].)

Figure 6–17[13]

(See Figure 6–17[13][Pick Us Up].)

Figure 6–17[14]

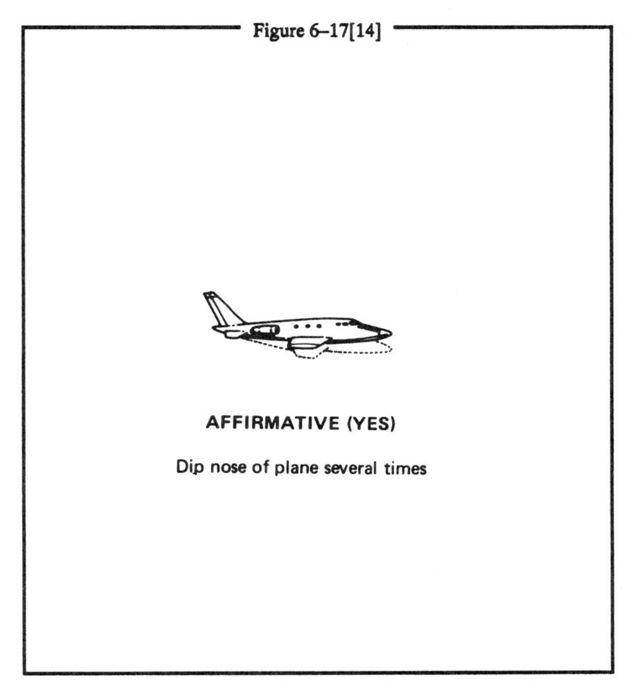

(See Figure 6–17[14][Affirmative (Aircraft)].)

(See Figure 6–17[15][Negative (Aircraft)].)

Figure 6–17[15]

i. OBSERVANCE OF DOWNED AIRCRAFT.

1. Determine if crash is marked with a yellow cross; if so, the crash has already been reported and identified.

2. If possible, determine type and number of aircraft and whether there is evidence of survivors.

3. Fix the position of the crash as accurately as possible with reference to a navigational aid. If possible provide geographic or physical description of the area to aid ground search parties.

4. Transmit the information to the nearest FAA or other appropriate radio facility.

5. If circumstances permit, orbit the scene to guide in other assisting units until their arrival or until you are relieved by another aircraft.

6. Immediately after landing, make a complete report to the nearest FAA facility, or Air Force or Coast Guard Rescue Coordination Center. The report can be made by long distance collect telephone.

6–18 thru 6–20. RESERVED

Section 3. DISTRESS AND URGENCY PROCEDURES

6–21. DISTRESS AND URGENCY COMMUNICATIONS

a. A pilot who encounters a *distress* or *urgency* condition can obtain assistance simply by contacting the air traffic facility or other agency in whose area of responsibility the aircraft is operating, stating the nature of the difficulty, pilot's intentions and assistance desires. *Distress* and *urgency* communications procedures are prescribed by the International Civil Aviation Organization (ICAO), however, and have decided advantages over the informal procedure described above.

b. *Distress* and *urgency* communications procedures discussed in the following paragraphs relate to the use of air ground voice communications.

c. The initial communication, and if considered necessary, any subsequent transmissions by an aircraft in *distress* should begin with the signal MAYDAY, preferably repeated three times. The signal PAN-PAN should be used in the same manner for an *urgency* condition.

d. *Distress* communications have absolute priority over all other communications, and the word MAYDAY commands radio silence on the frequency in use. *Urgency* communications have priority over all other communications except *distress,* and the word PAN-PAN warns other stations not to interfere with *urgency* transmissions.

e. Normally, the station addressed will be the air traffic facility or other agency providing air traffic services, on the frequency in use at the time. If the pilot is not communicating and receiving services, the station to be called will normally be the air traffic facility or other agency in whose area of responsibility the aircraft is operating, on the appropriate assigned frequency. If the station addressed does not respond, or if time or the situation dictates, the *distress* or *urgency* message may be broadcast, or a collect call may be used, addressing "Any Station (Tower)(Radio)(Radar)."

f. The station addressed should immediately acknowledge a *distress* or *urgency* message, provide assistance, coordinate and direct the activities of assisting facilities, and alert the appropriate search and rescue coordinator if warranted. Responsibility will be transferred to another station only if better handling will result.

g. All other stations, aircraft and ground, will continue to listen until it is evident that assistance is being provided. If any station becomes aware that the station being called either has not received a *distress* or *urgency* message, or cannot communicate with the aircraft in difficulty, it will attempt to contact the aircraft and provide assistance.

h. Although the frequency in use or other frequencies assigned by ATC are preferable, the following emergency frequencies can be used for distress or urgency communications, if necessary or desirable:

1. 121.5 MHz and 243.0 MHz—Both have a range generally limited to line of sight. 121.5 MHz is guarded by direction finding stations and some military and civil aircraft. 243.0 MHz is guarded by military aircraft. Both 121.5 MHz and 243.0 MHz are guarded by military towers, most civil towers, FSSs, and radar facilities. Normally ARTCC emergency frequency capability does not extend to radar coverage limits. If an ARTCC does not respond when called on 121.5 MHz or 243.0 MHz, call the nearest tower or FSS.

2. 2182 kHz—The range is generally less than 300 miles for the average aircraft installation. It can be used to request assistance from stations in the maritime service. 2182 kHz is guarded by major radio stations serving Coast Guard Rescue Coordination Centers, and Coast Guard units along the sea coasts of the U.S. and shores of the Great Lakes. The call "Coast Guard" will alert all Coast Guard Radio Stations within range. 2182 kHz is also guarded by most commercial coast stations and some ships and boats.

6–22. OBTAINING EMERGENCY ASSISTANCE

a. A pilot in any *distress* or *urgency* condition should *immediately* take the following action, not necessarily in the order listed, to obtain assistance:

1. Climb, if possible, for improved communications, and better radar and direction finding detection. However, it must be understood that unauthorized climb or descent under IFR conditions within controlled airspace is prohibited, except as permitted by FAR 91.3(b).

2. If equipped with a radar beacon transponder (civil) or IFF/SIF (military):

(a) Continue squawking assigned MODE A/3 discrete code/VFR code and MODE C altitude encoding when in radio contact with an air traffic facility or other agency providing air traffic services, unless instructed to do otherwise.

(b) If unable to immediately establish communications with an air traffic facility/agency, squawk MODE A/3, Code 7700/Emergency and MODE C.

3. Transmit a *distress* or *urgency* message consisting of *as many as necessary* of the following elements, preferably in the order listed:

(a) If distress, MAYDAY, MAYDAY, MAYDAY; if *urgency,* PAN-PAN, PAN-PAN, PAN-PAN.

(b) Name of station addressed.

(c) Aircraft identification and type.

(d) Nature of *distress* or *urgency.*

(e) Weather.

(f) Pilots intentions and request.

(g) Present position, and heading; or if *lost,* last known position, time, and heading since that position.

(h) altitude or Flight Level.

(i) Fuel remaining in minutes.

(j) Number of people on board.

(k) Any other useful information.

Reference—Pilot/Controller Glossary, Fuel Remaining.

b. After establishing radio contact, comply with advice and instructions received. Cooperate. Do not hesitate to ask questions or clarify instructions when you do not understand or if you cannot comply with clearance. Assist the ground station to control communications on the frequency in use. Silence interfering radio stations. Do not change frequency or change to another ground station unless absolutely necessary. If you do, advise the ground station of the new frequency and station name prior to the change, transmitting in the blind if necessary. If two-way communications cannot be established on the new frequency, return immediately to the frequency or station where two-way communications last existed.

c. When in a distress condition with bailout, crash landing or ditching imminent, take the following additional actions to assist search and rescue units:

1. Time and circumstances permitting, transmit as many as necessary of the message elements in subpargraph a(3) and any of the following that you think might be helpful

(a) ELT status.

(b) Visible landmarks.

(c) Aircraft color.

(d) Number of persons on board.

(e) Emergency equipment on board.

2. Actuate your ELT if the installation permits.

3. For bailout, and for crash landing or ditching if risk of fire is not a consideration, set your radio for continuous transmission.

4. If it becomes necessary to ditch, make every effort to ditch near a surface vessel. If time permits, an FAA facility should be able to get the position of the nearest commercial or Coast Guard vessel from a Coast Guard Rescue Coordination Center.

5. After a crash landing unless you have good reason to believe that you will not be located by search aircraft or ground teams, it is best to remain with your aircraft and prepare means for signalling search aircraft.

6–23. DITCHING PROCEDURES

(See Figure 6–23[1][Wind–Swell–Ditch Heading].)

Figure 6–23[1]

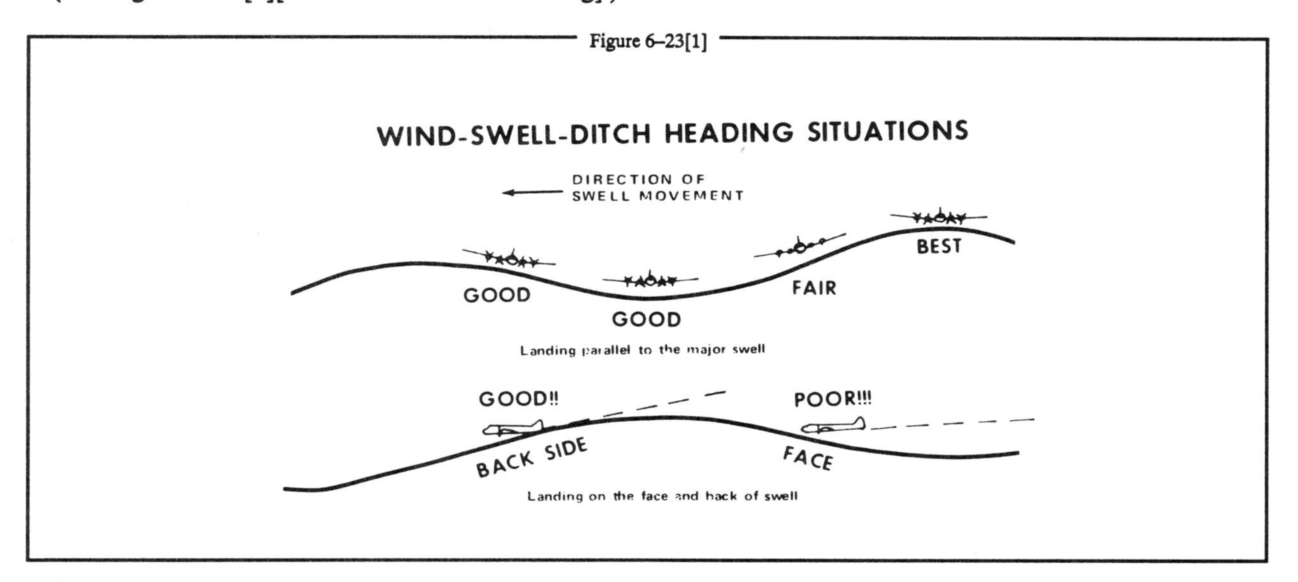

(See Figure 6–23[2][Single Swell (15 knot wind)].)

Figure 6–23[2]

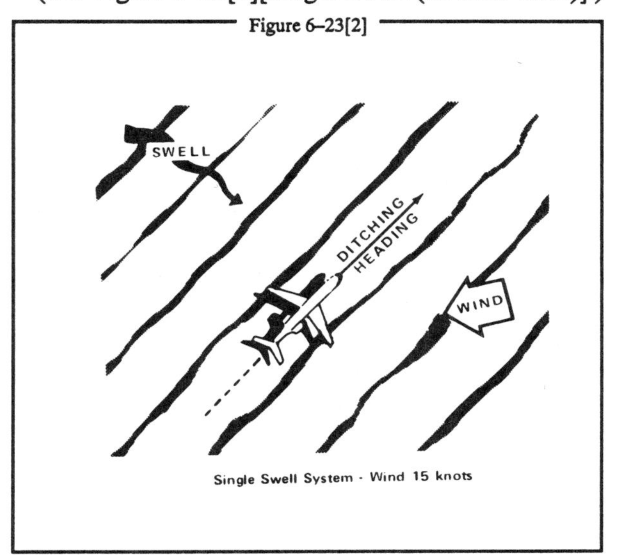

Single Swell System - Wind 15 knots

(See Figure 6–23[3][Double Swell (15 knot wind)].)

Figure 6–23[3]

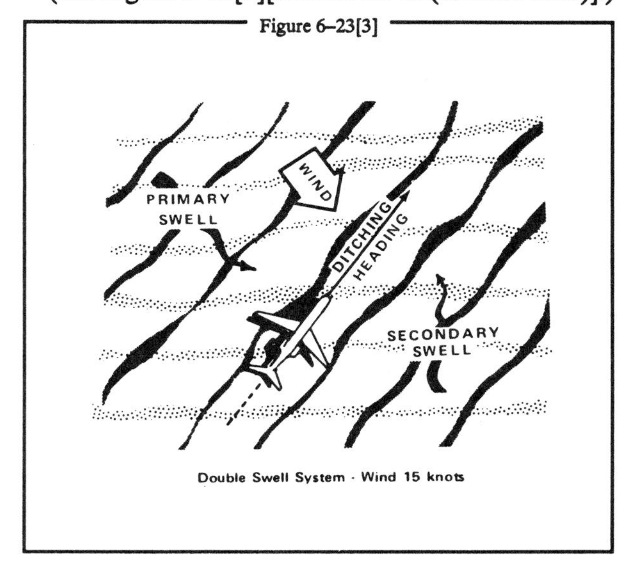

Double Swell System - Wind 15 knots

(See Figure 6–23[4][Double Swell (30 knot wind).])

Figure 6–23[4]

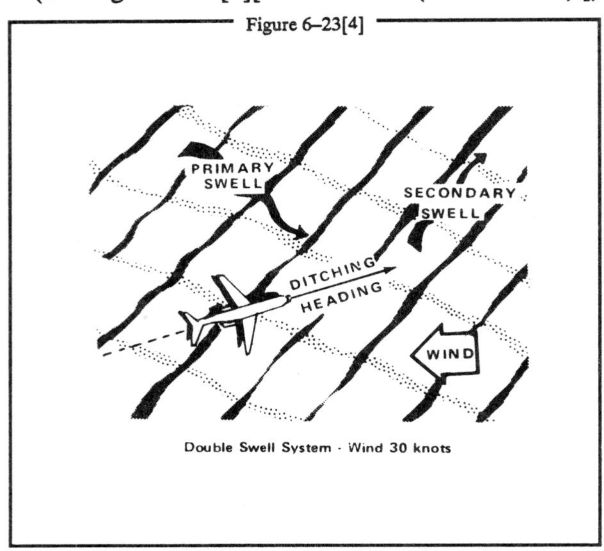

Double Swell System - Wind 30 knots

(See Figure 6–23[5][(50 knot wind)].)

Figure 6–23[5]

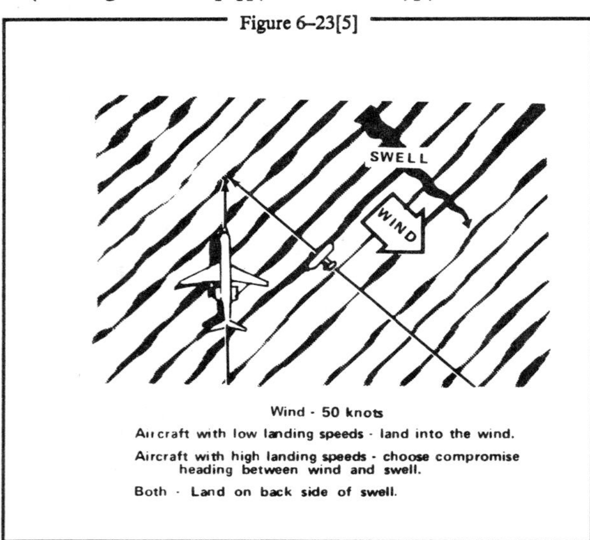

Wind - 50 knots
Aircraft with low landing speeds - land into the wind.
Aircraft with high landing speeds - choose compromise heading between wind and swell.
Both - Land on back side of swell.

a. A successful aircraft ditching is dependent on three primary factors. In order of importance they are:

1. Sea conditions and wind.

2. Type of aircraft.

3. Skill and technique of pilot.

b. Common oceanographic terminology:

1. Sea. The condition of the surface that is the result of both waves and swells.

2. Wave (or Chop). The condition of the surface caused by the local winds.

3. Swell. The condition of the surface which has been caused by a distance disturbance.

4. Swell Face. The side of the swell toward the observer. The backside is the side away from the observer. These definitions apply regardless of the direction of swell movement.

5. Primary Swell. The swell system having the greatest height from trough to crest.

6. Secondary Swells. Those swell systems of less height than the primary swell.

7. Fetch. The distance the waves have been driven by a wind blowing in a constant direction, without obstruction.

8. Swell Period. The time interval between the passage of two successive crests at the same spot in the water, measured in seconds.

9. Swell Velocity. The speed and direction of the swell with relation to a fixed reference point, measured in knots. There is little movement of water in the horizontal direction. Swells move primarily in a vertical motion, similar to the motion observed when shaking out a carpet.

10. Swell Direction. The direction *from* which a swell is moving. This direction is not necessarily the result of the wind present at the scene. The swell may be moving into or across the local wind. Swells, once set in motion, tend to maintain their original direction for as long as they continue in deep water, regardless of changes in wind direction.

11. Swell Height. The height between crest and trough, measured in feet. The vast majority of ocean swells are lower than 12 to 15 feet, and swells over 25 feet are not common at any spot on the oceans. Successive swells may differ considerably in height.

c. In order to select a good heading when ditching an aircraft, a basic evaluation of the sea is required. Selection of a good ditching heading may well minimize damage and could save your life. It can be extremely dangerous to land into the wind without regard to sea conditions; the swell system, or systems, must be taken into consideration. Remember one axiom—AVOID THE FACE OF A SWELL.

1. In ditching parallel to the swell, it makes little difference whether touchdown is on the top of the crest or in the trough. It is preferable, however, to land on the top or back side of the swell, if possible. After determining which heading (and its reciprocal) will parallel the swell, select the heading with the most into the wind component.

2. If only one swell system exists, the problem is relatively simple—even with a high, fast system. Unfortunately, most cases involve two or more swell systems running in different directions. With more than one system present, the sea presents a confused appearance. One of the most difficult situations occurs when two swell systems are at right angles. For example, if one system is eight feet high, and the other three feet, plan to land parallel to the primary system, and on the down swell of the secondary system. If both systems are of equal height, a compromise may be advisable—select an intermediate heading at 45 degrees down swell to both systems. When landing down a secondary swell, attempt to touch down on the back side, not on the face of the swell.

3. *If the swell system is formidable, it is considered advisable, in landplanes, to accept more crosswind in order to avoid landing directly into the swell.*

4. The secondary swell system is often from the same direction as the wind. Here, the landing may be made parallel to the primary system, with the wind and secondary system at an angle. There is a choice to two directions paralleling the primary system. One direction is downwind and down the secondary swell, and the other is into the wind and into the secondary swell, the choice will depend on the velocity of the wind versus the velocity and height of the secondary swell.

d. The simplest method of estimating the wind direction and velocity is to examine the windstreaks on the water. These appear as long streaks up and down wind. Some persons may have difficulty determining wind direction after seeing the streaks on the water. Whitecaps fall forward with the wind but are overrun by the waves thus producing the illusion that the foam is sliding backward. Knowing this, and by observing the direction of the streaks, the wind direction is easily determined. Wind velocity can be estimated by noting the appearance of the whitecaps, foam and wind streaks.

1. The behavior of the aircraft on making contact with the water will vary within wide limits according to the state of the sea. If landed parallel to a single swell system, the behavior of the aircraft may approximate that to be expected on a smooth sea. If landed into a heavy swell or into a confused sea, the deceleration forces may be extremely great—resulting in breaking up of the aircraft. Within certain limits, the pilot is able to minimize these forces by proper sea evaluation and selection of ditching heading.

2. When on final approach the pilot should look ahead and observe the surface of the sea. There may be shadows and whitecaps—signs of large seas. Shadows and whitecaps close together indicate short and rough seas. Touchdown in these areas is to be avoided. Select and touchdown in any area (only about 500 feet is needed) where the shadows and whitecaps are not so numerous.

3. Touchdown should be at the *lowest* speed and rate of descent which permit safe handling and optimum nose up attitude on impact. Once first impact has been made, there is often little the pilot can do to control a landplane.

e. Once preditching preparations are completed, the pilot should turn to the ditching heading and commence let-down. The aircraft should be flown low over the water, and slowed down until ten knots or so above stall. At this point, additional power should be used to overcome the increased drag caused by the nose up attitude. When a smooth stretch of water appears ahead, cut power, and touchdown at the best recommended speed as fully stalled as possible. By cutting power when approaching a relatively smooth area, the pilot will prevent overshooting and will touchdown with less chance of planing off into a second uncontrolled landing. Most experienced seaplane pilots prefer to make contact with the water in a semi-stalled attitude, cutting power as the tail makes contact. This technique eliminates the chance of misjudging altitude with a resultant heavy drop in a fully stalled condition. Care must be taken not to drop the aircraft from too high altitude or to balloon due to excessive speed. The altitude above water depends on the aircraft. Over glassy smooth water, or at night without sufficient light, it is very easy, for even the most experienced pilots to misjudge altitude by 50 feet or more. Under such conditions, carry enough power to maintain nine to twelve degrees nose up attitude, and 10 to 20 percent over stalling speed until contact is made with the water. The proper use of power on the approach is of great importance. If power is available on one side only, a little power should be used to flatten the approach; however, the engine should not be used to such an extent that the aircraft cannot be turned against the good engines right down to the stall with a margin of rudder movement available. When near the stall, sudden application of excessive unbalanced power may result in loss of directional control. If power is available on one side only, a slightly higher than normal glide approach speed should be used. This will insure good control and some margin of speed after leveling off without excessive use of power. The use of power in ditching is so important that when it is certain that the coast cannot be reached, the pilot should, if possible, ditch before fuel is exhausted. The use of power in a night or instrument ditching is far more essential than under daylight contact conditions.

1. If no power is available, a greater than normal approach speed should be used down to the flare-out. This speed margin will allow the glide to be broken early and more gradually, thereby giving the pilot time and distance to feel for the surface—decreasing the possibility of stalling high or flying into the water. When landing parallel to a swell system, little difference is noted between landing on top of a crest or in the trough. If the wings of aircraft are trimmed to the surface of the sea rather than the horizon, there is little need to worry about a wing hitting a swell crest. The actual slope of a swell is very gradual. If forced to land into a swell, touchdown should be made just after passage of the crest. If contact is made on the face of the swell, the aircraft may be swamped or thrown violently into the air, dropping heavily into the next swell. If control surfaces remain intact, the pilot should attempt to maintain the proper nose above the horizon attitude by rapid and positive use of the controls.

f. After Touchdown: In most cases drift, caused by crosswind can be ignored; the forces acting on the aircraft after touchdown are of such magnitude that drift will be only a secondary consideration. If the aircraft is under good control, the "crab" may be kicked out with rudder just prior to touchdown. This is more important with high wing aircraft, for they are laterally unstable on the water in a crosswind and may roll to the side in ditching.

6–23f NOTE—This information has been extracted from Appendix H of the "National Search And Rescue Manual."

6–24. SPECIAL EMERGENCY (AIR PIRACY)

a. A special emergency is a condition of air piracy, or other hostile act by a person(s) aboard an aircraft, which threatens the safety of the aircraft or its passengers.

b. The pilot of an aircraft reporting a special emergency condition should:

1. If circumstances permit, apply *distress or urgency* radio-telephony procedures (Reference—Distress and Urgency Communications, paragraph 6–21). Include the details of the special emergency.

2. If circumstances do not permit the use of prescribed *distress* or *urgency* procedures, transmit:

(a) On the air/ground frequency in use at the time.

(b) As many as possible of the following elements spoken distinctly and in the following order:

(1) Name of the station addressed (time and circumstances permitting).

(2) The identification of the aircraft and present position.

(3) The nature of the special emergency condition and pilot intentions (circumstances permitting).

(4) If unable to provide this information, use code words and/or transponder as follows: state "TRANSPONDER SEVEN FIVE ZERO ZERO". Meaning: "I am being hijacked/forced to a new destination"; and/or use Transponder Setting MODE 3/A, Code 7500.

6–24b2b4 NOTE—Code 7500 will never be assigned by ATC without prior notification from the pilot that his aircraft is being subjected to unlawful interference. The pilot should refuse the assignment of Code 7500 in any other situation and inform the controller accordingly. Code 7500 will trigger the special emergency indicator in all radar ATC facilities.

c. Air traffic controllers will acknowledge and confirm receipt of transponder Code 7500 by asking the pilot to verify it. If the aircraft is not being subjected to unlawful interference, the pilot should respond to the query by broadcasting in the clear that he is not being subjected to unlawful interference. Upon receipt of this information, the controller will request the pilot to verify the code selection depicted in the code selector windows in the transponder control panel and change the code to the appropriate setting. If the pilot replies in the affirmative or does not reply, the controller will not ask further questions but will flight follow, respond to pilot requests and notify appropriate authorities.

d. If it is possible to do so without jeopardizing the safety of the flight, the pilot of a hijacked passenger aircraft, after departing from the cleared routing over which the aircraft was operating, will attempt to do one or more of the following things, insofar as circumstances may permit:

1. Maintain a true airspeed of no more than 400 knots, and preferably an altitude of between 10,000 and 25,000 feet.

2. Fly a course toward the destination which the hijacker has announced.

e. If these procedures result in either radio contact or air intercept, the pilot will attempt to comply with any instructions received which may direct him to an appropriate landing field.

6–25. FUEL DUMPING

a. Should it become necessary to dump fuel, the pilot should immediately advise ATC. Upon receipt of information that an aircraft will dump fuel, ATC will broadcast or cause to be broadcast immediately and every 3 minutes thereafter the following on appropriate ATC and FSS radio frequencies:

EXAMPLE:

ATTENTION ALL AIRCRAFT—FUEL DUMPING IN PROGRESS OVER—(location) AT (altitude) BY (type aircraft) (flight direction).

b. Upon receipt of such a broadcast, pilots of aircraft affected, which are not on IFR flight plans or special VFR clearances, should clear the area specified in the advisory. Aircraft on IFR flight plans or special VFR clearances will be provided specific separation by ATC. At the termination of the fuel dumping operation, pilots should advise ATC. Upon receipt of such information, ATC will issue, on the appropriate frequencies, the following:

EXAMPLE:

ATTENTION ALL AIRCRAFT—FUEL DUMPING BY—(type aircraft)—TERMINATED.

6–26 thru 6–30. RESERVED

Section 4. TWO-WAY RADIO COMMUNICATIONS FAILURE

6–31. TWO-WAY RADIO COMMUNICATIONS FAILURE

a. It is virtually impossible to provide regulations and procedures applicable to all possible situations associated with two-way radio communications failure. During two-way radio communications failure, when confronted by a situation not covered in the regulation, pilots are expected to exercise good judgment in whatever action they elect to take. Should the situation so dictate they should not be reluctant to use the emergency action contained in FAR 91.3(b).

b. Whether two-way communications failure constitutes an emergency depends on the circumstances, and in any event, it is a determination made by the pilot. FAR 91.3(b) authorizes a pilot to deviate from any rule in Subparts A and B to the extent required to meet an emergency.

c. In the event of two-way radio communications failure, ATC service will be provided on the basis that the pilot is operating in accordance with FAR 91.185. A pilot experiencing two-way communications failure should (unless emergency authority is exercised) comply with FAR 91.185 quoted below:

6–31c NOTE—Capitalization and examples added for emphasis.

1. General. Unless otherwise authorized by ATC, each pilot who has two-way radio communications failure when operating under IFR shall comply with the rules of this section.

2. VFR conditions. If the failure occurs in VFR conditions, or if VFR conditions are encountered after the failure, each pilot shall continue the flight under VFR and land as soon as practicable.

6–31c2 NOTE—This procedure also applies when two-way radio failure occurs while operating in Positive Control Airspace (PCA). The primary objective of this provision in FAR 91.185 is to preclude extended IFR operation in the ATC system in VFR weather conditions. Pilots should recognize that operation under these conditions may unnecessarily as well as adversely affect other users of the airspace, since ATC may be required to reroute or delay other users in order to protect the failure aircraft. However, it is not intended that the requirement to "land as soon as practicable" be construed to mean "as soon as possible." The pilot retains his prerogative of exercising his best judgment and is not required to land at an unauthorized airport, at an airport unsuitable for the type of aircraft flown, or to land only minutes short of his destination.

3. IFR conditions. If the failure occurs in IFR conditions, or if paragraph (2) of this section cannot be complied with, each pilot shall continue the flight according to the following:

(a) Route.

(1) By the route assigned in the last ATC clearance received;

(2) If being radar vectored, by the direct route from the point of radio failure to the fix, route, or airway specified in the vector clearance;

(3) In the absence of an assigned route, by the route that ATC has advised may be expected in a further clearance; or

(4) In the absence of an assigned route or a route that ATC has advised may be expected in a further clearance by the route filed in the flight plan.

(b) Altitude. At the HIGHEST of the following altitudes or Flight Levels FOR THE ROUTE SEGMENT BEING FLOWN:

(1) The altitude or flight level assigned in the last ATC clearance received;

(2) The minimum altitude (converted, if appropriate, to minimum flight level as prescribed in FAR 91.121. (c)) for IFR operations; or

(3) The altitude or flight level ATC has advised may be expected in a further clearance.

6–31cb2 NOTE—The intent of the rule is that a pilot who has experienced two-way radio failure should select the appropriate altitude for the particular route segment being flown and make the necessary altitude adjustments for subsequent route segments. If the pilot received an "expect further clearance" containing a higher altitude to expect at a specified time or fix, he/she should maintain the highest of the following altitudes until that time/fix: (1) His/her last assigned altitude, or (2) The minimum altitude/flight level for IFR operations.

Upon reaching the time/fix specified, the pilot should commence his/her climb to the altitude he/she was advised to expect. If the radio failure occurs after the time/fix specified, the altitude to be expected is not applicable and the pilot should maintain an altitude consistent with 1 or 2 above.

If the pilot receives an "expect further clearance" containing a lower altitude, the pilot should maintain the highest of 1 or 2 above until that time/fix specified in paragraph 6–31c(3)(c).

EXAMPLE:

A pilot experiencing two-way radio failure at an assigned altitude of 7,000 feet is cleared along a direct route which will require a climb to a minimum IFR altitude of 9,000 feet, should climb to reach 9,000 feet at the time or place where it becomes necessary (see FAR 91.177(b)). Later while proceeding along an airway with an MEA of 5,000 feet, the pilot would descend to *7,000 feet* (the last assigned altitude), because that altitude is *higher* than the MEA.

EXAMPLE:

A pilot experiencing two-way radio failure while being progressively descended to lower altitudes to begin an approach is assigned 2,700 feet until crossing the VOR and then cleared for the approach. The MOCA along the airway is 2,700 feet and MEA is 4,000 feet. The aircraft is within 22 NM of the VOR. The pilot should remain at 2,700 feet until crossing the VOR because that altitude is the minimum IFR altitude for the route segment being flown.

EXAMPLE:

The MEA between **A** and **B** –5,000 feet. The MEA between **B** and **C** –5,000 feet. The MEA between **C** and **D** –11,000 feet. The MEA between **D** and **E** –7,000 feet. A pilot had been cleared via **A, B, C, D,** to **E.** While flying between A and B his assigned altitude was 6,000 feet and he was told to expect a clearance to 8,000 feet at **B.** Prior to receiving the higher altitude assignment, he experienced two-way failure. The pilot would maintain 6,000 to **B,** then climb to 8,000 feet (the altitude he was advised to expect.) He would maintain 8,000 feet, then climb to 11,000 at **C,** or prior to **C** if necessary to comply with an MCA at **C.** (FAR 91.177(b)). Upon reaching **D,** the pilot would descend to *8,000 feet* (even though the MEA was 7,000 feet), as 8,000 was the highest of the altitude situations stated in the rule (FAR 91.185).

(c) Leave clearance limit.

(1) When the clearance limit is a fix from which an approach begins, commence descent or descent and approach as close as possible to the expect further clearance time if one has been received, or if one has not been received, as close as possible to the Estimated Time of Arrival as calculated from the filed or amended (with ATC) Estimated Time en Route.

(2) If the clearance limit is not a fix from which an approach begins, leave the clearance limit at the expect further clearance time if one has been received, or if none has been received, upon arrival over the clearance limit, and proceed to a fix from which an approach begins and commence descent or descent and approach as close as possible to the estimated time of arrival as calculated from the filed or amended (with ATC) estimated time en route.

6–32. TRANSPONDER OPERATION DURING TWO-WAY COMMUNICATIONS FAILURE

a. If a pilot of an aircraft with a coded radar beacon transponder experiences a loss of two-way radio capability he should:

1. Adjust his transponder to reply on MODE A/3, Code 7700 for a period of 1 minute.

2. Then change to Code 7600 and remain on 7600 for a period of 15 minutes or the remainder of the flight, whichever comes first.

3. Repeat steps (1) and (2) as practicable.

b. The pilot should understand that he may not be in an area of radar coverage.

6–33. REESTABLISHING RADIO CONTACT

a. In addition to monitoring the NAVAID voice feature, the pilot should attempt to reestablish communications by attempting contact:

1. on the previously assigned frequency, or

2. with an FSS or *ARINC.

b. If communcations are established with an FSS or ARINC, the pilot should advise that radio communications on the previously assigned frequency has been lost giving the aircraft's position, altitude, last assigned frequency and then request further clearance from the controlling facility. The preceding does not preclude the use of 121.5 MHz. There is no priority on which action should be attempted first. If the capability exists, do all at the same time.

6–33b NOTE—*AERONAUTICAL RADIO/INCORPORATED (ARINC)—is a commercial communications corporation which designs, constructs, operates, leases or otherwise engages in radio activities serving the aviation community. ARINC has the capability of relaying information to/from ATC facilities throughout the country.

6–34 thru 6–40. RESERVED

Chapter 7. SAFETY OF FLIGHT

Section 1. METEOROLOGY

7–1. NATIONAL WEATHER SERVICE AVIATION PRODUCTS

a. Weather service to aviation is a joint effort of the National Weather Service (NWS), the Federal Aviation Administration (FAA), the military weather services, and other aviation oriented groups and individuals. The NWS maintains an extensive surface, upper air, and radar weather observing program; a nationwide aviation weather forecasting service; and also provides pilot briefing service. The majority of pilot weather briefings are provided by FAA personnel at Flight Service Stations (FSS's). Surface weather observations are taken by NWS, by NWS-certified FAA, contract, and supplemental observers, and by automated observing systems. (Reference—Weather Observing Programs, paragraph 7–10).

b. Aviation forecasts are prepared by 52 Weather Service Forecast Offices (WSFO's). These offices prepare and distribute approximately 500 terminal forecasts 3 times daily for specific airports in the 50 States and the Caribbean (4 times daily in Alaska and Hawaii). These forecasts, which are amended as required, are valid for 24 hours. The last 6 hours are given in categorical outlook terms as described in Categorical Outlooks, paragraph 7–6. WSFO's also prepare a total of over 300 route forecasts and 39 synopses for Pilots Automatic Telephone Weather Answering Service (PATWAS), Transcribed Weather Broadcast (TWEB), and briefing purposes. The route forecasts that are issued during the morning and mid-day are valid for 12 hours while the evening issuance is valid for 18 hours. A centralized aviation forecast program originating from the National Aviation Weather Advisory Unit (NAWAU) in Kansas City was implemented in November 1982. In the conterminous U.S., all In-flight Advisories (SIGMETs, Convective SIGMETs, and AIRMETs) and all Area Forecasts (6 areas) are now issued by NAWAU. Area Forecasts are prepared 3 times a day in the conterminous States (4 times in Hawaii), and amended as required, while In-flight Advisories are issued only when conditions warrant. (Reference—In-Flight Weather Advisories, paragraph 7–5). Winds aloft forecasts are provided for 176 locations in the 48 contiguous States and 21 in Alaska for flight planning purposes. (Winds aloft forecasts for Hawaii are prepared locally). All the aviation weather forecasts are given wide distribution through the Weather Message Switching Center in Kansas City (WMSC).

c. Weather element values may be expressed by using different measurement systems depending on several factors, such as whether the weather products will be used by the general public, aviation interests, international services, or a combination of these users. Figure 7–1[1] provides conversion tables for the most used weather elements that will be encountered by pilots.

7–2. FAA WEATHER SERVICES

a. The FAA maintains a nationwide network of Flight Service Stations (FSS's) to serve the weather needs of pilots. In addition, NWS meteorologists are assigned to most Air Route Traffic Control Centers (ARTCC's) as part of the Center Weather Service Unit (CWSU). They provide advisory service and short-term forecasts (nowcasts) to support the needs of the FAA and other users of the system.

b. The primary source of preflight weather briefings is an individual briefing obtained from a briefer at the FSS or NWS. These briefings, which are tailored to your specific flight, are available 24 hours a day through the local FSS or through the use of toll free lines (INWATS). Numbers for these services can be found in the Airport/Facility Directory under "FAA and NWS Telephone Numbers" section. They are also listed in the U.S. Government section of your local telephone directory under Department of Transportation, Federal Aviation Administration, or Department of Commerce, National Weather Service. (Reference—Preflight Briefing, paragraph 7–3 explains the types of preflight briefings available and the information contained in each). NWS pilot briefers do not provide aeronautical information (NOTAM's, flow control advisories, etc.) nor do they accept flight plans.

Figure 7–1[1]

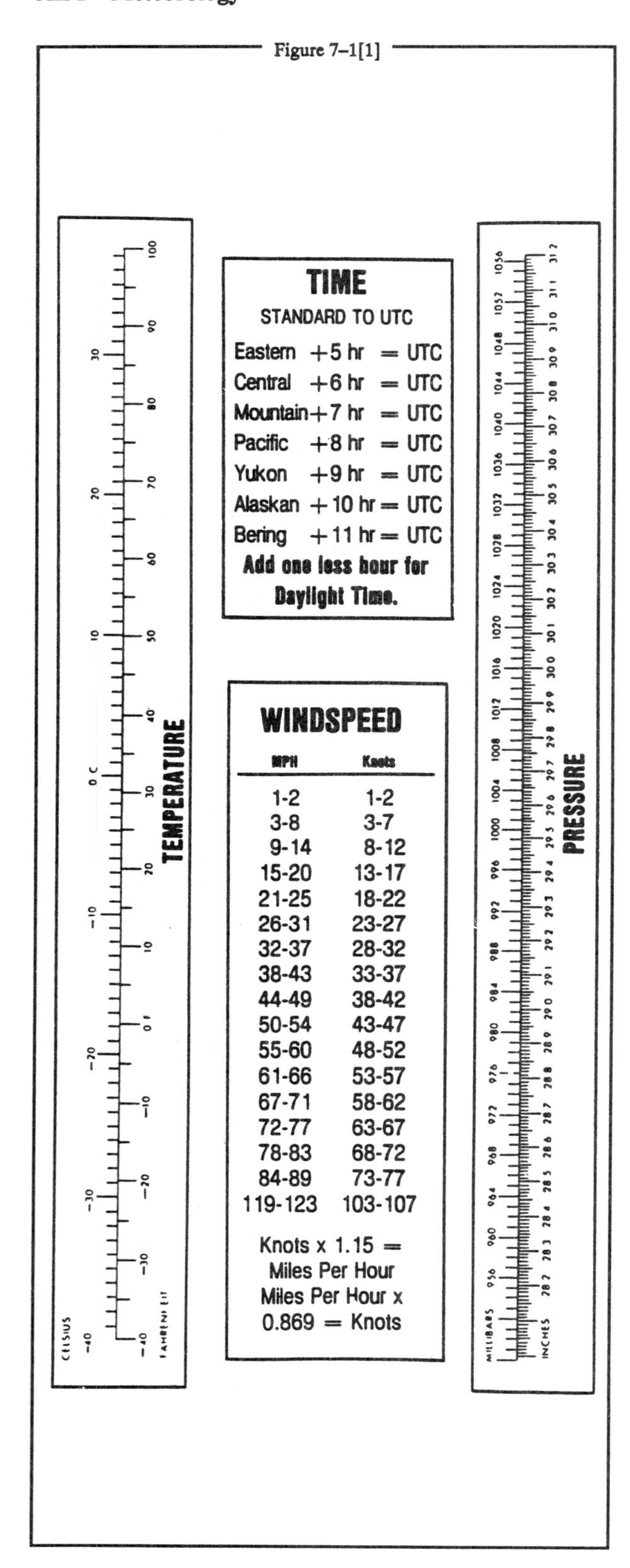

MPH	Knots
1-2	1-2
3-8	3-7
9-14	8-12
15-20	13-17
21-25	18-22
26-31	23-27
32-37	28-32
38-43	33-37
44-49	38-42
50-54	43-47
55-60	48-52
61-66	53-57
67-71	58-62
72-77	63-67
78-83	68-72
84-89	73-77
119-123	103-107

c. Other sources of weather information are as follows:

1. The A.M. Weather telecast on the PBS television network is a jointly sponsored 15–minute weather program designed for pilots. It is broadcast Monday through Friday mornings. Check TV listings in your area for station and exact times.

2. The Transcribed Weather Broadcast (TWEB), telephone access to the TWEB (TEL-TWEB), Telephone Information Briefing Service (TIBS) (AFSS) and Pilots Automatic Telephone Weather Answering Service (PATWAS) (FSS) provide continuously updated recorded weather information for short or local flights. Separate paragraphs in this section give additional information regarding these services.

3. Weather and aeronautical information is also available from numerous private industry sources on an individual or contract pay basis. Information on how to obtain this service should be available from local pilot organizations.

4. The Direct User Access System (DUATS) can be accessed by pilots with a current medical certificate toll-free in the 48 contiguous States via personal computer. Pilots can receive alpha-numeric preflight weather data and file domestic VFR and IFR flight plans. The following are the contract DUATS vendors:

CONTEL Federal Systems
15000 Conference Center Drive
Chantilly, VA 22021–3808

Telephone—*For filing flight plans and obtaining weather briefings:* (800) 767–9989

For customer service: (800) 345–3828

Data Transformation Corporation
559 Greentree Road
Turnerville, NJ 08012

Telephone—*For filing flight plans and obtaining weather briefings:* (800) 245–3828

For customer service: (800) 243–3828

d. In-flight weather information is available from any FSS within radio range. (Reference—Pilots Automatic Telephone Weather Answering Service (PATWAS) and Telephone Information Briefing Service (TIBS), paragraph 7–7, Transcribed Weather Broadcast (TWEB), paragraph 7–8, and In-Flight Weather Broadcasts, paragraph 7–9 for information on broadcasts.) En route Flight Advisory Service (EFAS) is provided to serve the nonroutine weather needs of pilots in flight. (Reference—En Route Flight Advisory Service (EFAS), paragraph 7–4 gives details on this service).

7–3. PREFLIGHT BRIEFING

a. Flight Service Stations (FSS's) are the primary source for obtaining preflight briefings and in-flight weather information. In some locations, the Weather Service Office (WSO) provides preflight briefings on a limited basis. Flight Service Specialists are qualified and certificated by the NWS as Pilot Weather Briefers. They are not authorized to make original forecasts, but are authorized to translate and interpret available forecasts and reports directly into terms describing the weather conditions which you can expect along your flight route and at your destination. Available aviation weather reports and forecasts are displayed at each FSS and WSO. Some of the larger FSS's provide a separate display for pilot use. Pilots should feel free to use these self briefing displays where available, or to ask for a briefing or assistance from the specialist on duty. Three basic types of preflight briefings are available to serve your specific needs. These are: Standard Briefing, Abbreviated Briefing, and Outlook Briefing. You should specify to the briefer the type of briefing you want, along with appropriate background information. (Reference—Preflight Preparation, paragraph 5–1 for items that are required). This will enable the briefer to tailor the information to your intended flight. The following paragraphs describe the types of briefings available and the information provided in each.

b. Standard Briefing—You should request a Standard Briefing any time you are planning a flight and you have not received a previous briefing or have not received preliminary information through mass dissemination media; e.g., TWEB, PATWAS, VRS, etc. The briefer will automatically provide the following information in the sequence listed, except as noted, when it is applicable to your proposed flight.

1. Adverse Conditions—Significant meteorological and aeronautical information that might influence the pilot to alter the proposed flight; e.g., hazardous weather conditions, runway closures, NAVAID outages, etc.

2. VFR Flight Not Recommended—When VFR flight is proposed and sky conditions or visibilities are present or forecast, surface or aloft, that in the briefer's judgment would make flight under visual flight rules doubtful, the briefer will describe the conditions, affected locations, and use the phrase *"VFR flight is not recommended."* This recommendation is advisory in nature. The final decision as to whether the flight can be conducted safely rests solely with the pilot.

3. Synopsis—A brief statement describing the type, location and movement of weather systems and/or air masses which might affect the proposed flight.

7–3b3 NOTE—These first 3 elements of a briefing may be combined in any order when the briefer believes it will help to more clearly describe conditions.

4. Current Conditions—Reported weather conditions applicable to the flight will be summarized from all available sources; e.g., SA's, PIREP's, RAREP's. This element will be omitted if the proposed time of departure is beyond 2 hours, unless the information is specifically requested by the pilot.

5. En Route Forecast—Forecast en route conditions for the proposed route are summarized in logical order; i.e., departure/climbout, en route, and descent.

6. Destination Forecast—The destination forecast for the planned ETA. Any significant changes within 1 hour before and after the planned arrival are included.

7. Winds Aloft—Forecast winds aloft will be summarized for the proposed route. The briefer will interpolate wind directions and speeds between levels and stations as necessary to provide expected conditions at planned altitudes.

8. Notices to Airmen (NOTAM's)—

(a) Available NOTAM (D) information pertinent to the proposed flight.

(b) NOTAM (L) information pertinent to the departure and/or local area, if available, and pertinent FDC NOTAM's within approximately 400 miles of the FSS providing the briefing.

7–3b8b NOTE 1—NOTAM information may be combined with current conditions when the briefer believes it is logical to do so.

7–3b8b NOTE 2—NOTAM (D) information and FDC NOTAM's which have been published in the Notices to Airmen publication are not included in pilot briefings unless a review of this publication is specifically requested by the pilot. For complete flight information you are urged to review Class II NOTAM's and the Airport/Facility Directory in addition to obtaining a briefing.

9. ATC Delays—Any known ATC delays and flow control advisories which might affect the proposed flight.

10. Pilots may obtain the following from FSS briefers upon request:

(a) Information on military training routes (MTR) and military operations area (MOA) activity within the flight plan area and a 100 NM extension around the flight plan area.

7–3b10a NOTE—Pilots are encouraged to request updated information from en route FSS's.

(b) A review of the Notices to Airmen publication for pertinent NOTAM's and Special Notices.

(c) Approximate density altitude data.

(d) Information regarding such items as air traffic services and rules, customs/immigration procedures, ADIZ rules, search and rescue, etc.

(e) LORAN-C NOTAM's.

(f) Other assistance as required.

c. Abbreviated Briefing—Request an Abbreviated Briefing when you need information to supplement mass disseminated data, update a previous briefing, or when you need only one or two specific items. Provide the briefer with appropriate background information, the time you received the previous information, and/or the specific items needed. You should indicate the source of the information already received so that the briefer can limit the briefing to the information that you have not received, and/or appreciable changes in meteorological conditions since your previous briefing. To the extent possible, the briefer will provide the information in the sequence shown for a Standard Briefing. If you request only one or two specific items, the briefer will advise you if adverse conditions are present or forecast. Details on these conditions will be provided at your request.

d. Outlook Briefing—You should request an Outlook Briefing whenever your proposed time of departure is six or more hours from the time of the briefing. The briefer will provide available forecast data applicable to the proposed flight. This type of briefing is provided for planning purposes only. You should obtain a Standard or Abbreviated Briefing prior to departure in order to obtain such items as current conditions, updated forecasts, winds aloft and NOTAM's.

e. Inflight Briefing—You are encouraged to obtain your preflight briefing by telephone or in person before departure. In those cases where you need to obtain a preflight briefing or an update to a previous briefing by radio, you should contact the nearest FSS to obtain this information. After communications have been established, advise the specialist of the type briefing you require and provide appropriate background information. You will be provided information as specified in the above paragraphs, depending on the type briefing requested. In addition, the specialist will recommend shifting to the Flight Watch frequency when conditions along the intended route indicate that it would be advantageous to do so.

f. Following any briefing, feel free to ask for any information that you or the briefer may have missed. It helps to save your questions until the briefing has been completed. This way, the briefer is able to present the information in a logical sequence, and lessens the chance of important items being overlooked.

7–4. EN ROUTE FLIGHT ADVISORY SERVICE (EFAS)

a. EFAS is a service specifically designed to provide en route aircraft with timely and meaningful weather advisories pertinent to the type of flight intended, route of flight, and altitude. In conjunction with this service, EFAS is also a central collection and distribution point for pilot reported weather information. EFAS is provided by specially trained specialists in selected AFSS's/FSS's controlling multiple Remote Communications Outlets covering a large geographical area and is normally available throughout the conterminous U.S. and Puerto Rico from 6 a.m. to 10 p.m. EFAS provides communications capabilities for aircraft flying at 5,000 feet above ground level to 17,500 feet MSL on a common frequency of 122.0 mHz. Discrete EFAS frequencies have been established to ensure communications coverage from 18,000 through 45,000 MSL serving in each specific ARTCC area. These discrete frequencies may be used below 18,000 feet when coverage permits reliable communication.

7–4a NOTE—When an EFAS outlet is located in a time zone different from the zone in which the Flight Watch control station is located, the availability of service may be plus or minus one hour from the normal operating hours.

b. Contact flight watch by using the name of the Air Route Traffic Control Center facility identification serving the area of your location, followed by your aircraft identification, name of the nearest VOR to your position. The specialist needs to know this approximate location to select the most appropriate transmitter/receiver outlet for communications coverage.

EXAMPLE:

CLEVELAND FLIGHT WATCH, CESSNA ONE TWO THREE FOUR KILO, MANSFIELD V-O-R, OVER.

c. Charts depicting the location of the flight watch control stations (parent facility) and the outlets they use are contained in the Airport Facility Directories (A/FD). If you do not know in which flight watch area you are flying, initiate contact by using the words "Flight Watch," your aircraft identification, and the name of the nearest VOR. The facility will respond using the name of the flight watch facility.

EXAMPLE:

FLIGHT WATCH, CESSNA ONE TWO THREE FOUR KILO, MANSFIELD V-O-R, OVER.

d. FSS's that provide En route Flight Advisory Service are listed regionally in the Airport/Facilities Directories.

e. EFAS is not intended to be used for filing or closing flight plans, position reporting, getting complete preflight briefings, or obtaining random weather reports and forecasts. En route flight advisories are tailored to the phase of flight that begins after climb-out and ends with descent to land. Immediate destination weather and terminal forecast will be provided on request. Pilots requesting information not within the scope of flight watch will be advised of the appropriate FSS frequency to contact to obtain the information. Pilot participation is essential to the success of EFAS by providing a continuous exchange of information on weather, winds, turbulence, flight visibility, icing, etc., between pilots and flight watch specialists. Pilots are encouraged to report good weather as well as bad, and to confirm expected conditions as well as unexpected to EFAS facilities.

7–5. IN-FLIGHT WEATHER ADVISORIES

a. The NWS issues in-flight weather advisories designated as Severe Weather Forecasts Alerts (AWW), Convective SIGMETs (WST), SIGMETs (WS), Center Weather Advisories (CWA), or AIRMET's (WA). These advisories are issued individually; however, the information contained in them is also included in relevant portions of the Area Forecast (FA). When these advisories are issued subsequent to the FA, they automatically amend appropriate portions of the FA until the FA itself has been amended. In-flight advisories serve to notify en route pilots of the possibility of encountering hazardous flying conditions which may not have been forecast at the time of the preflight briefing. Whether or not the condition described is potentially hazardous to a particular flight is for the pilot to evaluate on the basis of experience and the operational limits of the aircraft.

b. Severe Weather Forecast Alerts (AWW) are preliminary messages issued in order to alert users that a Severe Weather Bulletin (WW) is being issued. These messages define areas of possible severe thunderstorms or tornado activity. The messages are unscheduled and issued as required by the National Severe Storm Forecast Center at Kansas City, Missouri.

c. Convective SIGMETs are issued for the following phenomena:

1. Tornadoes.

2. Lines of thunderstorms.

3. Embedded thunderstorms.

4. Thunderstorm areas greater than or equal to thunderstorm intensity level 4 with an area coverage of 4/10 (40 percent) or more. (Reference—Pilot/Controller Glossary, Radar Weather Echo Intensity Levels).

5. Hail greater than or equal to 3/4 inch diameter.

7–5c5 NOTE—Since thunderstorms are the reason for issuing the Convective SIGMET, severe or greater turbulence, severe icing, and low-level wind shear (gust fronts, etc.) are implied and will not be specified in the advisory.

d. Convective Sigmet Bulletins—

1. Three Convective SIGMET bulletins, each covering a specified geographic area, are issued. These areas are the Eastern (E), Central (C), and Western (W) U.S. The boundaries that separate the Eastern from the Central and the Central from the Western U.S. are 87 and 107 degrees West, respectively. These bulletins are issued on a scheduled basis, hourly at 55 minutes past the hour (H+55), and as special bulletins on an unscheduled basis.

2. Each of the Convective SIGMET bulletins will be:

(a) Made up of one or more individually numbered Convective SIGMET's,

(b) Valid for 1 hour, and

(c) Removed from system automatically at 40 minutes past the hour (H+40).

3. On an hourly basis, an outlook is made for each of the three Convective SIGMET regions. The outlook for a particular region is appended to the Convective SIGMET bulletin for the same region. However, it is not appended to special Convective SIGMET's. The outlook is reviewed each hour and revised when necessary. The outlook is a forecast and meteorological discussion for thunderstorm systems that are expected to require Convective SIGMET issuances during a time period 2–6 hours into the future. Furthermore, an outlook will always be made for each of the three regions, even if it is a negative statement.

e. SIGMET's within the conterminous U.S. are issued by the National Aviation Weather Advisory Unit (NAWAU) for the following hazardous weather phenomena:

1. Severe and extreme turbulence.

2. Severe icing.

3. Widespread duststorms, sandstorms, or volcanic ash lowering visibilities to below three miles.

f. Center Weather Advisory: The CWA is an unscheduled in-flight, flow control, air traffic, and air crew advisory. By nature of its short lead time, the CWA is not a flight planning product. It is generally a Nowcast for conditions beginning within the next 2 hours. CWA's will be issued:

1. As a supplement to an existing SIGMET, Convective SIGMET, AIRMET, or Area Forecast (FA).

2. When an In-flight Advisory has not been issued but observed or expected weather conditions meet SIGMET/AIRMET criteria based on current pilot reports and reinforced by other sources of information about existing meteorological conditions.

3. When observed or developing weather conditions do not meet SIGMET, Convective SIGMET, or AIRMET criteria; e.g., in terms of intensity or area coverage, but current pilot reports or other weather information sources indicate that existing or anticipated meteorological phenomena will adversely affect the safe flow of air traffic within the Air Route Traffic Control Center (ARTCC) area of responsibility.

g. AIRMET's within the conterminous U.S. are issued by the NAWAU for the following weather phenomena which are potentially hazardous to aircraft:

1. Moderate icing.

2. Moderate turbulence.

3. Sustained winds of 30 knots or more at the surface.

4. Widespread area of ceilings less than 1,000 feet and/or visibility less than three miles.

5. Extensive mountain obscurement.

7–5g5 NOTE—If the above phenomena are adequately forecast in the FA, an AIRMET will not be issued.

h. SIGMET's and AIRMET's are identified by an alphanumeric designator which consists of an alphabetic identifier and issuance number. The first time an advisory is issued for a phenomenon associated with a particular weather system, it will be given the next alphabetic designator in the series, and will be numbered as the first for that designator. Subsequent advisories will retain the same alphabetic designator until the phenomenon ends. In the conterminous U.S., this means that a phenomenon that is assigned an alphabetic designator in one area will retain that designator as it moves within the area or into one or more other areas. Issuances for the same phenomenon will be sequentially numbered, using the same alphabetic designator until the phenomenon no longer exist. Alphabetic designators NOVEMBER-YANKEE, except SIERRA, TANGO and ZULU are used only for SIGMET's, while designators SIERRA, TANGO, and ZULU are used for AIRMET's.

i. Each CWA will have a phenomenon number (1–6) immediately following the ARTCC identifier. This number will be assigned to each meteorologically distinct condition or conditions; e.g., jet stream clear air turbulence or low IFR and icing northwest of a low pressure center, meeting CWA issuance criteria. Following the product type (CWA) a two digit issuance number will be entered starting at midnight local each day. In addition, those CWA's based on existing nonconvective SIGMET's/AIRMET's will include the associated alphanumeric designator; e.g., ALPHA 4.

EXAMPLE:

ZKC1 CWA01/ALPHA 4

j. Each AWW is numbered sequentially beginning January 1 of each year.

EXAMPLE:

MKC AWW 161755

WW 279 SEVERE TSTM NY PA NJ 161830Z–170000Z.

AXIS..70 STATUTE MILES EITHER SIDE OF LINE..10W MSS TO 20E ABE.

HAIL SURFACE AND ALOFT..2 INCHES. SURFACE WIND GUSTS..65 KNOTS.

MAX TOPS TO 540. MEAN WIND VECTOR 19020.

REPLACES WW 278..OH PA NY

Status reports are issued as needed on Severe Weather Watch Bulletins to show progress of storms and to delineate areas no longer under the threat of severe storm activity. Cancellation bulletins are issued when it becomes evident that no severe weather will develop or that storms have subsided and are no longer severe.

7–6. CATEGORICAL OUTLOOKS

a. Categorical outlook terms, describing general ceiling and visibility conditions for advanced planning purposes, are defined as follows:

1. LIFR (Low IFR)—Ceiling less than 500 feet and/or visibility less than 1 mile.

2. IFR—Ceiling 500 to less than 1,000 feet and/or visibility 1 to less than 3 miles.

3. MVFR (Marginal VFR)—Ceiling 1,000 to 3,000 feet and/or visibility 3 to 5 miles inclusive.

4. VFR—Ceiling greater than 3,000 feet and visibility greater than 5 miles; includes sky clear.

b. The cause of LIFR, IFR, or MVFR is indicated by either ceiling or visibility restrictions or both. The contraction "CIG" and/or weather and obstruction to vision symbols are used. If winds or gusts of 25 knots or greater are forecast for the outlook period, the word "WIND" is also included for all categories including VFR.

EXAMPLE:

LIFR CIG—Low IFR due to low ceiling.

EXAMPLE:

IFR F—IFR due to visibility restricted by fog.

EXAMPLE:

MVFR CIG HK—Marginal VFR due to both ceiling and visibility restricted by haze and smoke.

EXAMPLE:

IFR CIG R WIND—IFR due to both low ceiling and visibility restricted by rain; wind expected to be 25 knots or greater.

7–7. PILOTS AUTOMATIC TELEPHONE WEATHER ANSWERING SERVICE (PATWAS) AND TELEPHONE INFORMATION BRIEFING SERVICE (TIBS)

a. PATWAS is provided by nonautomated flight service stations. PATWAS is a continuous recording of meteorological and aeronautical information, which is available at selected locations by telephone. Normally, the recording contains a summary of data for an area within 50NM of the parent station; however, at some locations route information similar to that available on the Transcribed Weather Broadcast (TWEB) is included.

b. PATWAS is not intended to substitute for specialist-provided preflight briefings. It is, however, recommended for use as a preliminary briefing, and often will be valuable in helping you to make a "go or no go" decision.

c. TIBS is provided by Automated Flight Service Stations (AFSS) and provides continuous telephone recordings of meteorological and/or aeronautical information. Specifically, TIBS provides area and/or route briefings, airspace procedures, and special announcements (if applicable) concerning aviation interests.

d. Depending on user demand, other items may be provided; i.e., surface observations, terminal forecasts, winds/temperatures aloft forecasts, etc. A TOUCH-TONE telephone is necessary to fully utilize the TIBS program.

e. Pilots are encouraged to avail themselves of this service. PATWAS and TIBS locations are found in the Airport/Facility Directory under the FSS and National Weather Service Telephone Numbers section.

7–8. TRANSCRIBED WEATHER BROADCAST (TWEB)

Equipment is provided at selected FSS's by which meteorological and aeronautical data are recorded on tapes and broadcast continuously over selected low-frequency (190–535 kHz) navigational aids (L/MF ranges or H facilities) and/or VOR's. Broadcasts are made from a series of individual tape recordings, and changes, as they occur, are transcribed onto the tapes. The information provided varies depending on the type equipment available. Generally, the broadcast contains route-oriented data with specially prepared NWS forecasts, In-flight Advisories, and winds aloft plus preselected current information, such as weather reports, NOTAM's, and special notices. In some locations, the information is broadcast over the local VOR only and is limited to such items as the hourly weather for the parent station and up to 5 immediately adjacent stations, local NOTAM information, terminal forecast (FT) for the parent station, adverse conditions extracted from In-flight Advisories, and other potentially hazardous conditions. At selected locations, telephone access to the TWEB has been provided (TEL-TWEB). Telephone numbers for this service are found in the FSS and National Weather Service Telephone Numbers section of the Airport/Facility Directory. These broadcasts are made available primarily for preflight and in-flight planning, and as such, should not be considered as a substitute for specialist-provided preflight briefings.

7–9. IN-FLIGHT WEATHER BROADCASTS

a. Weather Advisory Broadcasts—FAA FSSs broadcast Severe Weather Forecast Alerts (AWW), Convective SIGMETs, SIGMETs, CWAs, and AIRMET's during their valid period when they pertain to the area within 150 NM of the FSS or a broadcast facility controlled by the FSS as follows:

1. Severe Weather Forecast Alerts (AWW) and Convective SIGMET—Upon receipt and at 15–minute intervals H+00, H+15, H+30, and H+45 for the first hour after issuance.

EXAMPLE:

AVIATION BROADCAST, WEATHER ADVISORY, (Severe Weather Forecast Alert or Convective SIGMET identification) (text of advisory).

2. SIGMET's, CWA's, and AIRMET's—Upon receipt and at 30–minute intervals at H+15 and H+45 for the first hour after issuance.

EXAMPLE:

AVIATION BROADCAST, WEATHER ADVISORY, (area or ARTCC identification) (SIGMET, CWA, or AIRMET identification) (text of advisory).

3. Thereafter, a summarized alert notice will be broadcast at H+15 and H+45 during the valid period of the advisories.

EXAMPLE:

AVIATION BROADCAST, WEATHER ADVISORY, A (Severe Weather Forecast Alert, Convective SIGMET, SIGMET, CWA, or AIRMET) IS CURRENT FOR (description of weather) (area affected).

4. Pilots, upon hearing the alert notice, if they have not received the advisory or are in doubt, should contact the nearest FSS and ascertain whether the advisory is pertinent to their flights.

b. ARTCC's broadcast a Severe Weather Forecast Alert (AWW), Convective SIGMET, SIGMET, or CWA alert once on all frequencies, except emergency, when any part of the area described is within 150 miles of the airspace under their jurisdiction. These broadcasts contain SIGMET or CWA (identification) and a brief description of the weather activity and general area affected.

EXAMPLE:

ATTENTION ALL AIRCRAFT, SIGMET DELTA THREE, FROM MYTON TO TUBA CITY TO MILFORD, SEVERE TURBULENCE AND SEVERE CLEAR ICING BELOW ONE ZERO THOUSAND FEET. EXPECTED TO CONTINUE BEYOND ZERO THREE ZERO ZERO ZULU.

EXAMPLE:

ATTENTION ALL AIRCRAFT, CONVECTIVE SIGMET TWO SEVEN EASTERN. FROM THE VICINITY OF ELMIRA TO PHILLIPSBURG. SCATTERED EMBEDDED THUNDERSTORMS MOVING EAST AT ONE ZERO KNOTS. A FEW INTENSE LEVEL FIVE CELLS, MAXIMUM TOPS FOUR FIVE ZERO.

EXAMPLE:

ATTENTION ALL AIRCRAFT, KANSAS CITY CENTER WEATHER ADVISORY ONE ZERO THREE. NUMEROUS REPORTS OF MODERATE TO SEVERE ICING FROM EIGHT TO NINER THOUSAND FEET IN A THREE ZERO MILE RADIUS OF ST. LOUIS. LIGHT OR NEGATIVE ICING REPORTED FROM FOUR THOUSAND TO ONE TWO THOUSAND FEET REMAINDER OF KANSAS CITY CENTER AREA.

7–9b NOTE—Terminal control facilities have the option to limit the AWW, Convective SIGMET, SIGMET, or CWA broadcast as follows: local control and approach control positions may opt to broadcast SIGMET or CWA alerts only when any part of the area described is within 50 miles of the airspace under their jurisdiction.

c. Hazardous In-Flight Weather Advisory Service (HIWAS)—This is a continuous broadcast of in-flight weather advisories including summarized AWW, SIGMET's, Convective SIGMET's, CWA's, AIRMET's, and urgent PIREP's. HIWAS has been adopted as a national program and will be implemented throughout the conterminous U.S. as resources permit. In those areas where HIWAS is commissioned, ARTCC, Terminal ATC, and FSS facilities have discontinued the broadcast of in-flight advisories as described in the preceding paragraph. HIWAS is an additional source of hazardous weather information which makes these data available on a continuous basis. It is not, however, a replacement for preflight or in-flight briefings or real-time weather updates from Flight Watch (EFAS). As HIWAS is implemented in individual center areas, the commissioning will be advertised in the Notices to Airmen publication.

1. Where HIWAS has been implemented, a HIWAS alert will be broadcast on all except emergency frequencies once upon receipt by ARTCC and terminal facilities, which will include an alert announcement, frequency instruction, number, and type of advisory updated; e.g., AWW, SIGMET, Convective SIGMET, or CWA.

EXAMPLE:

ATTENTION ALL AIRCRAFT, MONITOR HIWAS OR CONTACT A FLIGHT SERVICE STATION ON FREQUENCY ONE TWO TWO POINT ZERO OR ONE TWO TWO POINT TWO FOR NEW CONVECTIVE SIGMET (identification) INFORMATION.

2. In HIWAS ARTCC areas, FSS's will broadcast a HIWAS update announcement once on all except emergency frequencies upon completion of recording an update to the HIWAS broadcast. Included in the broadcast will be the type of advisory updated; e.g. AWW, SIGMET, Convective SIGMET, CWA, ect.

EXAMPLE:

ATTENTION ALL AIRCRAFT, MONITOR HIWAS OR CONTACT FLIGHT WATCH OR FLIGHT SERVICE FOR NEW CONVECTIVE SIGMET INFORMATION.

d. Unscheduled Broadcasts—These broadcasts are made by FSS's on VOR and selected VHF frequencies upon receipt of special weather reports, PIREP's, NOTAM's and other information considered necessary to enhance safety and efficiency of flight. These broadcasts will be made at random times and will begin with the announcement "Aviation Broadcast" followed by identification of the data.

EXAMPLE:

AVIATION BROADCAST, SPECIAL WEATHER REPORT, (Notice to Airmen, Pilot Report, etc.) (location name twice) THREE SEVEN (past the hour) OBSERVATION...(etc.).

e. Alaskan Scheduled Broadcasts—Selected FSS's in Alaska having voice capability on radio ranges (VOR) or radio beacons (NDB) broadcast weather reports and Notice to Airmen information at 15 minutes past each hour from reporting points within approximately 150 miles from the broadcast station.

7–10. WEATHER OBSERVING PROGRAMS

a. Manual Observations—Surface weather observations are taken at more than 600 locations in the United States. With only a few exceptions, these stations are located at airport sites and most are manned by FAA or NWS personnel who manually observe, perform calculations, and enter the observation into the distribution system. The format and coding of these observations are contained in Key to Aviation Weather Observations and Forecasts, paragraph 7–27.

b. Automatic Meteorological Observing Stations (AMOS)—

1. Full parameter AMOS facilities provide data for the basic weather program at remote, unstaffed, or part-time staffed locations at approximately 90 locations in the United States. They report temperature, dew point, wind, pressure, and precipitation (liquid) amount. At staffed AMOS locations, an observer may manually add visually observed and manually calculated elements to the automatic reports. The elements manually added are sky condition, visibility, weather, obstructions to vision, and sea level pressure. The content and format of AMOS reports is the same as the manually observed reports, except the acronym "AMOS" or "RAMOS" (for Remote Automatic Meteorological Observing Station) will be the first item of the report.

2. Partial parameter AMOS stations only report some of the elements contained in the full parameter locations, normally wind. These observations are not normally disseminated through aviation weather circuits.

Table 7–10[1]

DECODING OBSERVATIONS FROM AUTOB STATIONS

EXAMPLE: **ENV AUTOB E25 BKN BV7 P 33/29/3606/975 PK WND 08 001**

ENCODE	DECODE	EXPLANATION
ENV	STATION IDENTIFICATION:	(Wendover, UT) Identifies report using FAA identifiers.
AUTOB	AUTOMATIC STATION IDENTIFIER	
E25 BKN	SKY & CEILING:	(Estimated 2500 ft. broken) Figures are height in 100s of feet above ground. Contraction after height is amount of sky cover. Letter preceding height indicates ceiling. WX reported if visibility is less than 2 miles and no clouds are detected. *NO CLOUDS REPORTED ABOVE 6000 FEET.*
BV7	BACKSCATTER VISIBILITY AVERAGED IN PAST MINUTE:	Reported in whole miles from 1 to 7.
P	PRECIPITATION OCCURRENCE:	(P=precipitation in past 10 minutes).
33	TEMPERATURE:	(33 degrees F.) Minus sign indicates sub-zero temperatures.
/29	DEW POINT:	(29 degrees F.) Minus sign indicates sub-zero temperatures.
/3606	WIND:	(360 degrees true at 6 knots) Direction is first two digits and is reported in tens of degrees. To decode, add a zero to first two digits. The last digits are speed; e.g., 2524 = 250 degrees at 24 knots.
/975	ALTIMETER SETTING:	(29.75 inches) The tens digit and decimal are omitted from report. To decode, prefix a 2 to code if it begins with 8 or 9. Otherwise, prefix a 3; e.g., 982 = 29.82, 017 = 30.17.
PK WND 08	PEAK WIND SPEED:	(8 knots) Reported speed is highest detected since last hourly observation.
001	PRECIPITATION ACCUMULATION:	(0.01 inches) Amount of precipitation since last synoptic time (00, 06, 12, 1800 UTC).

NOTE: If no clouds are detected below 6,000 feet and the visibility is greater than 2 miles, the reported sky condition will be *CLR BLO 60.*

c. Automatic Observing Stations (AUTOB)—There are four AUTOB's in operation. They are located at Winslow, Arizona (INW); Sandberg, California (SDB); Del Rio, Texas (DRT); and Wendover, Utah (ENV). These stations report all normal surface aviation weather elements, but cloud height and visibility are reported in a manner different from the conventional weather report. (See Table 7–10[1] for a description of these reports.)

d. Automated Weather Observing System (AWOS)—

1. Automated weather reporting systems are increasingly being installed at airports. These systems consist of various sensors, a processor, a computer-generated voice subsystem, and a transmitter to broadcast local, minute-by-minute weather data directly to the pilot.

7–10d1 NOTE—When the barometric pressure exceeds 31.00 inches Hg., see Procedures, paragraph 7–32a2 for the altimeter setting procedures.

2. The AWOS observations will include the prefix "AWOS" to indicate that the data are derived from an automated system. Some AWOS locations will be augmented by certified observers who will provide weather and obstruction to vision information in the remarks of the report when the reported visibility is less than 3 miles. These sites, along with the hours of augmentation, are to be published in the Airport/Facility Directory. Augmentation is identified in the observation as "OBSERVER WEATHER." The AWOS wind speed, direction and gusts, temperature, dew point, and altimeter setting are exactly the same as for manual observations. The AWOS will also report density altitude when it exceeds the field elevation by more than 1,000 feet. The reported visibility is derived from a sensor near the touchdown of the primary instrument runway. The visibility sensor output is converted to a runway visibility value (RVV) equation, using a 10–minute harmonic average. The AWOS sensors have been calibrated against the FAA transmissometer standards used for runway visual range values. Since the AWOS visibility is an extrapolation of a measurement at the touchdown point of the runway, it may differ from the standard prevailing visibility. The reported sky condition/ceiling is derived from the ceilometer located next to the visibility sensor. The AWOS algorithm integrates the last 30 minutes of ceilometer data to derive cloud layers and heights. This output may also differ from the *observer* sky condition in that the AWOS is totally dependent upon the cloud advection over the sensor site.

3. These real-time systems are operationally classified into four basic levels: AWOS-A, AWOS-l, AWOS–2, and AWOS–3. AWOS-A only reports altimeter setting. AWOS-l usually reports altimeter setting, wind data, temperature, dewpoint, and density altitude. AWOS–2 provides the information provided by AWOS-l plus visibility. AWOS–3 provides the information provided by AWOS–2 plus cloud/ceiling data.

4. The information is transmitted over a discrete radio frequency or the voice portion of a local NAVAID. AWOS transmissions are receivable within 25 NM of the AWOS site, at or above 3,000 feet AGL. In many cases, AWOS signals may be received on the surface of the airport. The system transmits a 20 to 30 second weather message updated each minute. Pilots should monitor the designated frequency for the automated weather broadcast. A description of the broadcast is contained in subparagraph e. There is no two-way communication capability. Most AWOS sites also have a dial-up capability so that the minute-by-minute weather messages can be accessed via telephone.

5. AWOS information (system level, frequency, phone number, etc.) concerning specific locations is published, as the systems become operational, in the Airport/Facility Directory, and where applicable, on published Instrument Approach Procedures. Selected individual systems may be incorporated into nationwide data collection and dissemination networks in the future.

e. Automated Weather Observing System (AWOS) Broadcasts-Computer-generated voice is used in Automated Weather Observing Systems (AWOS) to automate the broadcast of the minute-by-minute weather observations. In addition, some systems are configured to permit the addition of an operator-generated voice message; e.g., weather remarks following the automated parameters. The phraseology used generally follows that used for other weather broadcasts. Following are explanations and examples of the exceptions.

1. *Location and Time*—The location/name and the phrase "AUTOMATED WEATHER OBSERVATION," followed by the time are announced.

(a) If the airport's specific location is included in the airport's name, the airport's name is announced.

EXAMPLES:

"BREMERTON NATIONAL AIRPORT AUTOMATED WEATHER OBSERVATION, ONE FOUR FIVE SIX ZULU;"

"RAVENSWOOD JACKSON COUNTY AIRPORT AUTOMATED WEATHER OBSERVATION, ONE FOUR FIVE SIX ZULU."

(b) If the airport's specific location *is not* included in the airport's name, the location is announced followed by the airport's name.

EXAMPLES:

"SAULT STE MARIE, CHIPPEWA COUNTY INTERNATIONAL AIRPORT AUTOMATED WEATHER OBSERVATION;"

"SANDUSKY, COWLEY FIELD AUTOMATED WEATHER OBSERVATION."

(c) The word "TEST" is added following "OBSERVATION" when the system is not in commissioned status.

EXAMPLE:

"BREMERTON NATIONAL AIRPORT AUTOMATED WEATHER OBSERVATION TEST, ONE FOUR FIVE SIX ZULU."

(d) The phrase "TEMPORARILY INOPERATIVE" is added when the system is inoperative.

EXAMPLE:

"BREMERTON NATIONAL AIRPORT AUTOMATED WEATHER OBSERVING SYSTEM TEMPORARILY INOPERATIVE."

2. *Ceiling and Sky Cover*—

(a) Ceiling is announced as either "CEILING" or "INDEFINITE CEILING." The phrases "MEASURED CEILING" and "ESTIMATED CEILING" are not used. With the exception of indefinite ceilings, all automated ceiling heights are measured.

EXAMPLES:

"BREMERTON NATIONAL AIRPORT AUTOMATED WEATHER OBSERVATION, ONE FOUR FIVE SIX ZULU. CEILING TWO THOUSAND OVERCAST;"

"BREMERTON NATIONAL AIRPORT AUTOMATED WEATHER OBSERVATION, ONE FOUR FIVE SIX ZULU. INDEFINITE CEILING TWO HUNDRED, SKY OBSCURED."

(b) The word "Clear" is not used in AWOS due to limitations in the height ranges of the sensors. No clouds detected is announced as "NO CLOUDS BELOW XXX" or, in newer systems as "CLEAR BELOW XXX" (where XXX is the range limit of the sensor).

EXAMPLES:

"NO CLOUDS BELOW ONE TWO THOUSAND."

"CLEAR BELOW ONE TWO THOUSAND."

(c) A sensor for determining ceiling and sky cover is not included in some AWOS. In these systems, ceiling and sky cover are not announced. "SKY CONDITION MISSING" is announced only if the system is configured with a ceilometer and the ceiling and sky cover information is not available.

3. *Visibility—*

(a) The lowest reportable visibility value in AWOS is "less than ¼." It is announced as "VISIBILITY LESS THAN ONE QUARTER."

(b) A sensor for determining visibility is not included in some AWOS. In these systems, visibility is not announced. "VISIBILITY MISSING" is announced only if the system is configured with a visibility sensor and visibility information is not available.

4. *Weather—*In the future, some AWOS's are to be configured to determine the occurrence of precipitation. However, the type and intensity may not always be determined. In these systems, the word "PRECIPITATION" will be announced if precipitation is occurring, but the type and intensity are not determined.

5. *Remarks—*If remarks are included in the observation, the word "REMARKS" is announced following the altimeter setting. Remarks are announced in the following order of priority:

(a) Automated "Remarks"

(1) Density Altitude.

(2) Variable Visibility.

(3) Variable Wind Direction.

(b) Manual Input Remarks.—Manual input remarks are prefaced with the phrase "OBSERVER WEATHER." As a general rule the manual remarks are limited to:

(1) Type and intensity of precipitation,

(2) Thunderstorms, intensity (if applicable) and direction, and

(3) Obstructions to vision when the visiblity is 3 miles or less.

EXAMPLE:

"REMARKS ... DENSITY ALTITUDE, TWO THOUSAND FIVE HUNDRED ... VISIBILITY VARIABLE BETWEEN ONE AND TWO ... WIND DIRECTION VARIABLE BETWEEN TWO FOUR ZERO AND THREE ONE ZERO ...OBSERVED WEATHER ... THUNDERSTORM MODERATE RAIN SHOWERS AND FOG ... THUNDERSTORM OVERHEAD."

(c) If an automated parameter is "missing" and no manual input for that parameter is available, the parameter is announced as "MISSING." For example, a report with the dew point "missing" and no manual input available, would be announced as follows:

EXAMPLE:

"CEILING ONE THOUSAND OVERCAST ... VISIBILITY THREE ... PRECIPITATION ... TEMPERATURE THREE ZERO, DEW POINT MISSING ... WIND CALM ... ALTIMETER THREE ZERO ZERO ONE."

(d) "REMARKS" are announced in the following order of priority:

(1) Automated "REMARKS."

I Density Altitude

II Variable Visibility

III Variable Wind Direction

(2) Manual Input "Remarks." As a general rule, the remarks are announced in the same order as the parameters appear in the basic text of the observation; i.e., Sky Condition, Visibility, Weather and Obstructions to Vision, Temperature, Dew Point, Wind, and Altimeter Setting.

EXAMPLE:

"REMARKS ... DENSITY ALTITUDE, TWO THOUSAND FIVE HUNDRED ... VISIBILITY VARIABLE BETWEEN ONE AND TWO ... WIND DIRECTION VARIABLE BETWEEN TWO FOUR ZERO AND THREE ONE ZERO ... OBSERVER CEILING ESTIMATED TWO THOUSAND BROKEN ... OBSERVER TEMPERATURE TWO, DEW POINT MINUS FIVE."

7–11. WEATHER RADAR SERVICES

(See Figure 7–11[1][NWS Radar Network].)

Figure 7–11[1]

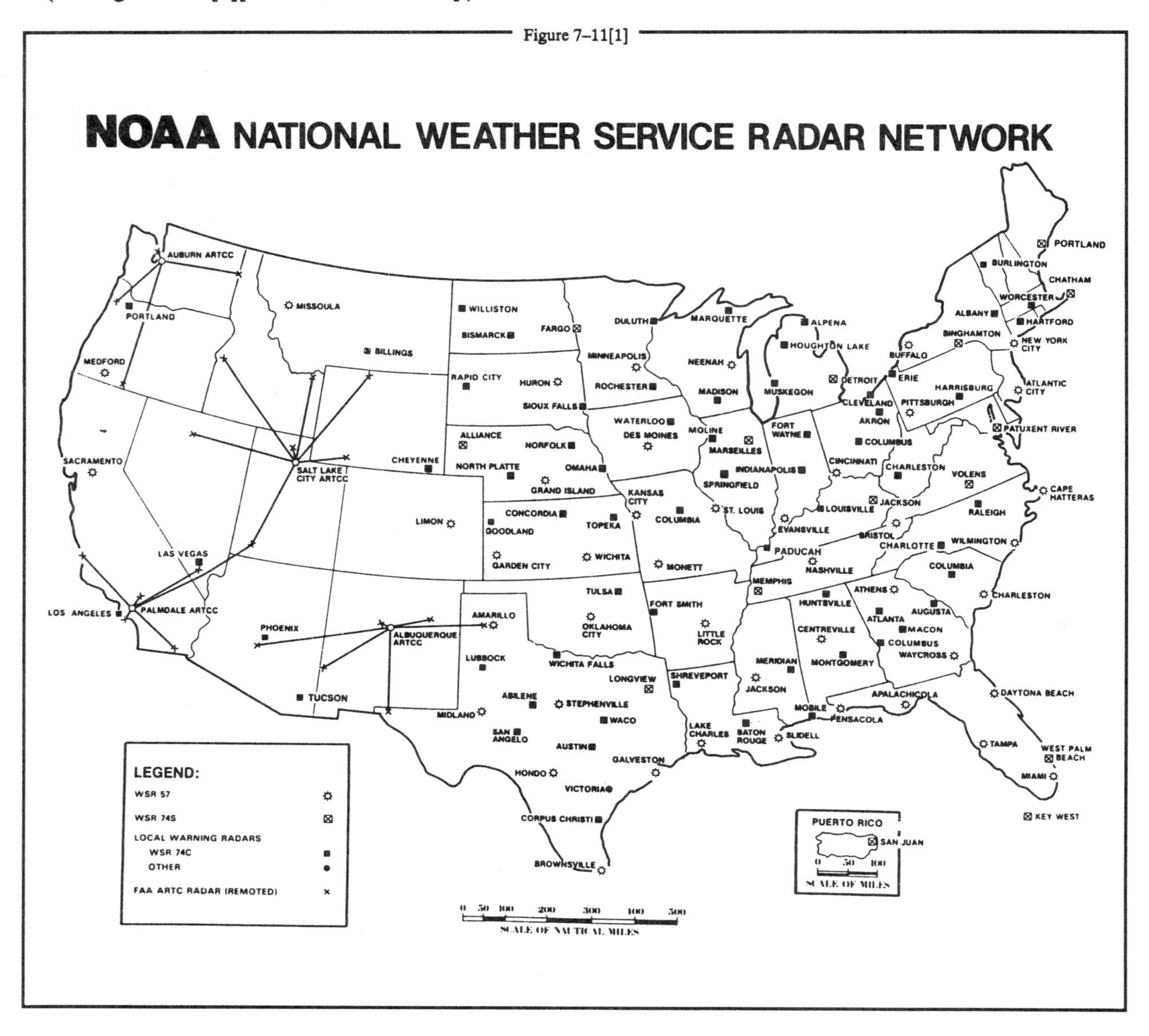

a. The National Weather Service operates a network of 56 radar sites for detecting coverage, intensity, and movement of precipitation. The network is supplemented by FAA and DOD radar sites in the western sections of the country. Another 72 local warning radar sites augment the network by operating on an as needed basis to support warning and forecast programs.

b. Scheduled radar observations are taken hourly and transmitted in alpha-numeric format on weather telecommunications circuits for flight planning purposes. Under certain conditions, special radar reports are issued in addition to the hourly transmittals. Data contained in the reports are also collected by the National Meteorological Center and used to prepare hourly national radar summary charts for dissemination on facsimile circuits.

c. All En route Flight Advisory Service facilities and many FSS's have equipment to directly access the radar displays from the individual weather radar sites. Specialists at these locations are trained to interpret the display for pilot briefing and in-flight advisory services. The Center Weather Service Units located in ARTCC's also have access to weather radar displays and provide support to all air traffic facilities within their center's area.

d. A clear radar display (no echoes) does not mean that there is no significant weather within the coverage of the radar site. Clouds and fog are not detected by the radar. However, when echoes are present, turbulence can be implied by the intensity of the precipitation, and icing is implied by the presence of the precipitation at temperatures at or below zero degrees Celsius. Used in conjunction with other weather products, radar provides invaluable information for weather avoidance and flight planning.

e. Additional information on weather radar products and services can be found in Advisory Circular 00–45, AVIATION WEATHER SERVICES. (Reference—Pilot/Controller Glossary, Radar Weather Echo Intensity Levels and Thunderstorms, paragraph 7–25). (See A/FD charts, NWS Upper Air Observing Stations and Weather Network for the location of specific radar sites.)

7–12. ATC IN-FLIGHT WEATHER AVOIDANCE ASSISTANCE

a. ATC Radar Weather Display—

1. Areas of radar weather clutter result from rain or moisture. Radars cannot detect turbulence. The determination of the intensity of the weather displayed is based on its precipitation density. Generally, the turbulence associated with a very heavy rate of rainfall will normally be more severe than any associated with a very light rainfall rate.

2. ARTCC's are phasing in computer generated digitized radar displays to replace broadband radar display. This new system, known as Narrowband Radar, provides the controller with two distinct levels of weather intensity by assigning radar display symbols for specific precipitation densities measured by the narrowband system.

b. Weather Avoidance Assistance—

1. To the extent possible, controllers will issue pertinent information on weather or chaff areas and assist pilots in avoiding such areas when requested. Pilots should respond to a weather advisory by either acknowledging the advisory or by acknowledging the advisory and requesting an alternative course of action as follows:

(a) Request to deviate off course by stating the number of miles and the direction of the requested deviation. In this case, when the requested deviation is approved, the pilot is expected to provide his own navigation, maintain the altitude assigned by ATC and to remain within the specified mileage of his original course.

(b) Request a new route to avoid the affected area.

(c) Request a change of altitude.

(d) Request radar vectors around the affected areas.

2. For obvious reasons of safety, an IFR pilot must not deviate from the course or altitude or Flight Level without a proper ATC clearance. When weather conditions encountered are so severe that an immediate deviation is determined to be necessary and time will not permit approval by ATC, the pilot's emergency authority may be exercised.

3. When the pilot requests clearance for a route deviation or for an ATC radar vector, the controller must evaluate the air traffic picture in the affected area, and coordinate with other controllers (if ATC jurisdictional boundaries may be crossed) before replying to the request.

4. It should be remembered that the controller's primary function is to provide safe separation between aircraft. Any additional service, such as weather avoidance assistance, can only be provided to the extent that it does not derogate the primary function. It's also worth noting that the separation workload is generally greater than normal when weather disrupts the usual flow of traffic. ATC radar limitations and frequency congestion may also be a factor in limiting the controller's capability to provide additional service.

5. It is very important, therefore, that the request for deviation or radar vector be forwarded to ATC as far in advance as possible. Delay in submitting it may delay or even preclude ATC approval or require that additional restrictions be placed on the clearance. Insofar as possible the following information should be furnished to ATC when requesting clearance to detour around weather activity:

(a) Proposed point where detour will commence.

(b) Proposed route and extent of detour (direction and distance).

(c) Point where original route will be resumed.

(d) Flight conditions (IFR or VFR).

(e) Any further deviation that may become necessary as the flight progresses.

(f) Advise if the aircraft is equipped with functioning airborne radar.

6. To a large degree, the assistance that might be rendered by ATC will depend upon the weather information available to controllers. Due to the extremely transitory nature of severe weather situations, the controller's weather information may be of only limited value if based on weather observed on radar only. Frequent updates by pilots giving specific information as to the area affected, altitudes intensity and nature of the severe weather can be of considerable value. Such reports are relayed by radio or phone to other pilots and controllers and also receive widespread teletypewriter dissemination.

7. Obtaining IFR clearance or an ATC radar vector to circumnavigate severe weather can often be accommodated more readily in the en route areas away from terminals because there is usually less congestion and, therefore, offer greater freedom of action. In terminal areas, the problem is more acute because of traffic density, ATC coordination requirements, complex departure and arrival routes, adjacent airports, etc. As a consequence, controllers are less likely to be able to accommodate all requests for weather detours in a terminal area or be in a position to volunteer such routing to the pilot. Nevertheless, pilots should not hesitate to advise controllers of any observed severe weather and should specifically advise controllers if they desire circumnavigation of observed weather.

7–13. RUNWAY VISUAL RANGE (RVR)

a. RVR visibility values are measured by transmissometers mounted on towers along the runway. A full RVR system consists of:

1. Transmissometer projector and related items.

2. Transmissometer receiver (detector) and related items.

3. Analogue recorder.

4. Signal data converter and related items.

5. Remote digital or remote display programmer.

b. The transmissometer projector and receiver are mounted on towers either 250 or 500 feet apart. A known intensity of light is emitted from the projector and is measured by the receiver. Any obscuring matter such as rain, snow, dust, fog, haze or smoke reduces the light intensity arriving at the receiver. The resultant intensity measurement is then converted to an RVR value by the signal data converter. These values are displayed by readout equipment in the associated air traffic facility and updated approximately once every minute for controller issuance to pilots.

c. The signal data converter receives information on the high intensity runway edge light setting in use (step 3, 4, or 5); transmission values from the transmissometer, and the sensing of day or night conditions. From the three data sources, the system will compute appropriate RVR values. Due to variable conditions, the reported RVR values may deviate somewhat from the true observed visual range due to the slant range consideration, brief time delays between the observed RVR conditions and the time they are transmitted to the pilot, and rapidly changing visibility conditions.

d. An RVR transmissometer established on a 500 foot baseline provides digital readouts to a minimum of 1,000 feet. A system established on a 250 foot baseline provides digital readouts to a minimum of 600 feet, which are displayed in 200 foot increments to 3,000 feet and in 500 foot increments from 3,000 feet to a maximum value of 6,000 feet.

e. RVR values for Category IIIa operations extend down to 700 feet RVR; however, only 600 and 800 feet are reportable RVR increments. The 800 RVR reportable value covers a range of 701 feet to 900 feet and is therefore a valid minimum indication of Category IIIa operations.

f. Approach categories with the corresponding minimum RVR values: (See Table 7–13[1].)

Table 7–13[1]

Category	*Visibility (RVR)*
Nonprecision	2,400 feet
Category I	1,800 feet
Category II	1,200 feet
Category IIIa	700 feet
Category IIIb	150 feet
Category IIIc	0 feet

g. Ten minute maximum and minimum RVR values for the designated RVR runway are reported in the remarks section of the aviation weather report when the prevailing visibility is less than one mile and/or the RVR is 6,000 feet or less. ATCT's report RVR when the prevailing visibility is 1 mile or less and/or the RVR is 6,000 feet or less.

h. Details on the requirements for the operational use of RVR are contained in FAA ADVISORY CIRCULAR 97–1, "RUNWAY VISUAL RANGE." Pilots are responsible for compliance with minimums prescribed for their class of operations in the appropriate FAR's and/or operations specifications.

7–14. REPORTING OF CLOUD HEIGHTS

a. Ceiling, by definition in the FAR's and as used in Aviation Weather Reports and Forecasts, is the height *above ground (or water) level* of the lowest layer of clouds or obscuring phenomenon that is reported as "broken," "overcast," or "obscuration" and not classified as "thin" or "partial." For example, a forecast which reads "CIGS WILL BE GENLY 1 TO 2 THSD FEET" refers to heights *above ground level* (AGL). A forecast which reads "BRKN TO OVC LYRS AT 8 TO 12 THSD MSL" states that the height is *above mean sea level* (MSL).

b. Pilots usually report height values above MSL, since they determine heights by the altimeter. This is taken in account when disseminating and otherwise applying information received from pilots. ("Ceiling" heights are always above ground level.) In reports disseminated as PIREP's, height references are given the same as received from pilots, that is, above MSL. In the following example of an hourly observation, a pilot report of the heights of the bases and tops of an overcast layer in the terminal area is converted by the reporting station to reflect AGL to report the base of the overcast layer (E12). The pilot's reported top of the overcast layer (23) is not converted to AGL and is shown in the remarks section (last item) of the weather reporting (top ovc 23)

EXAMPLE:

E12 OVC 2FK 132/49/47/0000/002/TOP OVC 23

c. In aviation forecasts (Terminal, Area, or In-flight Advisories) ceilings are denoted by the prefix "C" when used with sky cover symbols as in "LWRG to C5 OVC–1TRW," or by the contraction "CIG" before, or the contraction "AGL" after, the forecast cloud height value. When the cloud base is given in height above MSL, it is so indicated by the contraction "MSL" or "ASL" following the height value. The heights of clouds tops, freezing level, icing, and turbulence are always given in heights above ASL or MSL.

7–15. REPORTING PREVAILING VISIBILITY

a. Surface (horizontal) visibility is reported in weather observations in terms of statute miles and increments thereof; e.g., 1/16, 1/8, 1/4, 1/2, 3/4, 1, 1 1/4, etc. Visibility is determined through the ability to see and identify preselected and prominent objects at a known distance from the usual point of observation. Visibilities which are determined to be less than 7 miles, identify the obscuring atmospheric condition; e.g., fog, haze, smoke, etc., or combinations thereof.

b. Prevailing visibility is the greatest visibility equalled or exceeded throughout at least half of the horizon circle, which need not be continuous. Segments of the horizon circle which may have a significantly lower visibility may be reported in the remarks section of the weather report; i.e., the southeastern quadrant of the horizon circle may be determined to be 2 miles in fog while the remaining quadrants are determined to be 3 miles in fog.

c. When the prevailing visibility at the usual point of observation, or at the tower level, is less than 4 miles, certificated tower personnel will take visibility observations in addition to those taken at the usual point of observation. The lower of these two values will be used as the prevailing visibility for aircraft operations.

7–16. ESTIMATING INTENSITY OF PRECIPITATION

a. Light-Scattered drops or flakes that do not completely wet or cover an exposed surface, regardless of duration, to 0.10 inch per hour; maximum 0.01 inch in 6 minutes.

b. Moderate–0.11 inch to 0.30 inch per hour; more than 0.01 inch to 0.03 inch in 6 minutes.

c. Heavy—More than 0.30 inch per hour; more than 0.03 inch in 6 minutes.

7–17. ESTIMATING INTENSITY OF DRIZZLE

a. Light—Scattered drops that do not completely wet surface, regardless of duration, to 0.01 inch per hour.

b. Moderate—More than 0.01 inch to 0.02 inch per hour.

c. Heavy—More than 0.02 inch per hour.

7–18. ESTIMATING INTENSITY OF SNOW

a. Light—Visibility 5/8 statute mile or more.

b. Moderate—Visibility less than 5/8 statute mile but not less than 5/16 statute mile.

c. Heavy—Visibility less than 5/16 statute mile.

7–19. PILOT WEATHER REPORTS (PIREP'S)

a. FAA air traffic facilities are required to solicit PIREP's when the following conditions are reported or forecast: Ceilings at or below 5,000 feet; Visibility at or below 5 miles; Thunderstorms and related phenomena; Icing of light degree or greater; Turbulence of moderate degree or greater; and Windshear.

b. Pilots are urged to cooperate and promptly volunteer reports of these conditions and other atmospheric data such as: Cloud bases, tops and layers; Flight visibility; Precipitation; Visibility restrictions such as haze, smoke and dust; Wind at altitude; and Temperature aloft.

c. PIREP's should be given to the ground facility with which communications are established; i.e., EFAS, FSS, ARTCC, or terminal ATC. One of the primary duties of EFAS facilities, radio call "FLIGHT WATCH," is to serve as a collection point for the exchange of PIREP's with en route aircraft.

d. If pilots are not able to make PIREP's by radio, reporting upon landing of the in-flight conditions encountered to the nearest FSS or Weather Service Office will be helpful. Some of the uses made of the reports are:

1. The ATCT uses the reports to expedite the flow of air traffic in the vicinity of the field and for hazardous weather avoidance procedures.

2. The FSS uses the reports to brief other pilots, to provide in-flight advisories, and weather avoidance information to en route aircraft.

3. The ARTCC uses the reports to expedite the flow of en route traffic, to determine most favorable altitudes, and to issue hazardous weather information within the center's area.

4. The NWS uses the reports to verify or amend conditions contained in aviation forecast and advisories. In some cases, pilot reports of hazardous conditions are the triggering mechanism for the issuance of advisories. They also use the reports for pilot weather briefings.

5. The NWS, other government organizations, the military, and private industry groups use PIREP's for research activities in the study of meteorological phenomena.

6. All air traffic facilities and the NWS forward the reports received from pilots into the weather distribution system to assure the information is made available to all pilots and other interested parties.

Table 7–19[1]

	PIREP ELEMENTS	PIREP CODE	CONTENTS
1.	3–letter station identifier	XXX	Nearest weather reporting location to the reported phenomenon
2.	Report type	UA or UUA	Routine or Urgent PIREP
3.	Location	/OV	In relation to a VOR
4.	Time	/TM	Coordinated Universal Time
5.	Altitude	/FL	Essential for turbulence and icing reports
6.	Type Aircraft	/TP	Essential for turbulence and icing reports
7.	Sky cover	/SK	Cloud height and coverage (scattered, broken, or overcast)
8.	Weather	/WX	Flight visibility, precipitation, restrictions to visibility, etc.
9.	Temperature	/TA	Degrees Celsius
10.	Wind	/WV	Direction in degrees true and speed in knots
11.	Turbulence	/TB	See paragraph 7–21
12.	Icing	/IC	See paragraph 7–20
13.	Remarks	/RM	For reporting elements not included or to clarify previously reported items

e. The FAA, NWS, and other organizations that enter PIREP's into the weather reporting system use the format listed in Table 7–19[1]. Items 1 through 6 are included in all transmitted PIREP's along with one or more of items 7 through 13. Although the PIREP should be as complete and concise as possible, pilots should not be overly concerned with strict format or phraseology. The important thing is that the information is relayed so other pilots may benefit from your observation. If a portion of the report needs clarification, the ground station will request the information. Completed PIREP's will be transmitted to weather circuits as in the following examples:

CMH UA /OV APE 230010/TM 1516/FL085/TP BE80/SK BKN 065/WX FV03 H K/TA 20/TB LGT

Translation: One zero miles southwest of Appleton VOR; Time 1516 UTC; altitude eight thousand five hundred; aircraft type BE80; top of the broken cloud layer is six thousand five hundred; flight visibility 3 miles with haze and smoke; air Temperature 20 degrees Celsius; light turbulence.

CRW UV /OV BKW 360015–CRW/TM 1815/FL120//TP BE99/SK OVC/WX R/TA –08 /WV 290030/TB LGT-MDT/IC LGT RIME/RM MDT MXD ICG DURGC ROA NWBND FL080–100 1750

Translation: From 15 miles north of Beckley VOR to Charleston VOR; time 1815 UTC; altitude 12,000 feet; type aircraft, BE–99; in clouds; rain; temperature –08 Celsius; wind 290 degrees magnetic at 30 knots; light to moderate turbulence; light rime icing; encountered moderate mixed icing during climb northwestbound from Roanoke, VA between 8000 and 10,000 feet at 1750 UTC.

7–20. PIREP'S RELATING TO AIRFRAME ICING

a. The effects of ice on aircraft are cumulative-thrust is reduced, drag increases, lift lessens, and weight increases. The results are an increase in stall speed and a deterioration of aircraft performance. In extreme cases, 2 to 3 inches of ice can form on the leading edge of the airfoil in less than 5 minutes. It takes but ½ inch of ice to reduce the lifting power of some aircraft by 50 percent and increases the frictional drag by an equal percentage.

b. A pilot can expect icing when flying in visible precipitation, such as rain or cloud droplets, and the temperature is 0 degrees Celsius or colder. When icing is detected, a pilot should do one of two things (particularly if the aircraft is not equipped with deicing equipment), he should get out of the area of precipitation or go to an altitude where the temperature is above freezing. This "warmer" altitude may not always be a lower altitude. Proper preflight action includes obtaining information on the freezing level and the above-freezing levels in precipitation areas. Report icing to ATC/FSS, and if operating IFR, request new routing or altitude if icing will be a hazard. Be sure to give the type of aircraft to ATC when reporting icing. The following describes how to report icing conditions.

1. Trace-Ice becomes perceptible. Rate of accumulation is slightly greater than the rate of sublimation. It is not hazardous even though deicing/anti-icing equipment is not utilized unless encountered for an extended period of time (over 1 hour).

2. Light—The rate of accumulation may create a problem if flight is prolonged in this environment (over 1 hour). Occasional use of deicing/anti-icing equipment removes/prevents accumulation. It does not present a problem if the deicing/anti-icing equipment is used.

3. Moderate—The rate of accumulation is such that even short encounters become potentially hazardous and use of deicing/anti-icing equipment or flight diversion is necessary.

4. Severe—The rate of accumulation is such that deicing/anti-icing equipment fails to reduce or control the hazard. Immediate flight diversion is necessary.

EXAMPLE:

Pilot Report: Give Aircraft Identification, Location, Time (UTC), Intensity of Type, Altitude/FL, Aircraft Type, IAS, and Outside Air Temperature.

7–20b4 NOTE 1—Rime Ice: rough, milky, opaque ice formed by the instantaneous freezing of small supercooled water droplets

7–20b4 NOTE 2—Clear Ice: a glossy, clear, or translucent ice formed by the relatively slow freezing of large supercooled water droplets.

7–20b4 NOTE 3—The Outside Air Temperature (OAT) should be requested by the FSS/ATC if not included in the PIREP.

7–21. PIREP'S RELATING TO TURBULENCE

a. When encountering turbulence, pilots are urgently requested to report such conditions to ATC as soon as practicable. PIREP's relating to turbulence should state:

1. Aircraft location.

2. Time of occurrence in UTC.

3. Turbulence intensity.

4. Whether the turbulence occurred in or near clouds.

5. Aircraft altitude or Flight Level.

6. Type of Aircraft.

7. Duration of turbulence.

EXAMPLE:

OVER OMAHA, 1232Z, MODERATE TURBULENCE IN CLOUDS AT FLIGHT LEVEL THREE ONE ZERO, BOEING 707.

EXAMPLE:

FROM FIVE ZERO MILES SOUTH OF ALBUQUERQUE TO THREE ZERO MILES NORTH OF PHOENIX, 1250Z, OCCASIONAL MODERATE CHOP AT FLIGHT LEVEL THREE THREE ZERO, DC8.

b. Duration and classification of intensity should be made using Table 7–21[1].

Table 7–21[1]

TURBULENCE REPORTING CRITERIA TABLE

Intensity	Aircraft Reaction	Reaction inside Aircraft	Reporting Term-Definition
Light	Turbulence that momentarily causes slight, erratic changes in altitude and/or attitude (pitch, roll, yaw). Report as **Light Turbulence;** [1] *or* Turbulence that causes slight, rapid and somewhat rhythmic bumpiness without appreciable changes in altitude or attitude. Report as **Light Chop.**	Occupants may feel a slight strain against seat belts or shoulder straps. Unsecured objects may be displaced slightly. Food service may be conducted and little or no difficulty is encountered in walking.	Occasional—Less that ⅓ of the time. Intermittent—⅓ to ⅔. Continuous—More than ⅔.
Moderate	Turbulence that is similar to Light Turbulence but of greater intensity. Changes in altitude and/or attitude occur but the aircraft remains in positive control at all times. It usually causes variations in indicated airspeed. Report as **Moderate Turbulence;** [1] *or* Turbulence that is similar to Light Chop but of greater intensity. It causes rapid bumps or jolts without appreciable changes in aircraft altitude or attitude. Report as **Moderate Chop.** [1]	Occupants feel definite strains against seat belts or shoulder straps. Unsecured objects are dislodged. Food service and walking are difficult.	**NOTE** 1. Pilots should report location(s), time (UTC), intensity, whether in or near clouds, altitude, type of aircraft and, when applicable, duration of turbulence. 2. Duration may be based on time between two locations or over a single location. All locations should be readily identifiable.
Severe	Turbulence that causes large, abrupt changes in altitude and/or attitude. It usually causes large variations in indicated airspeed. Aircraft may be momentarily out of control. Report as **Severe Turbulence.** [1]	Occupants are forced violently against seat belts or shoulder straps. Unsecured objects are tossed about. Food service and walking are impossible.	**EXAMPLES:** a. Over Omaha. 1232Z, Moderate Turbulence, in cloud, Flight Level 310, B707.
Extreme	Turbulence in which the aircraft is violently tossed about and is practically impossible to control. It may cause structural damage. Report as **Extreme Turbulence.** [1]		b. From 50 miles south of Albuquerque to 30 miles north of Phoenix, 1210Z to 1250Z, occasional Moderate Chop, Flight Level 330, DC8.

[1] High level turbulence (normally above 15,000 feet ASL) not associated with cumuliform cloudiness, including thunderstorms, should be reported as CAT (clear air turbulence) preceded by the appropriate intensity, or light or moderate chop.

7–22. WIND SHEAR PIREP'S

a. Because unexpected changes in wind speed and direction can be hazardous to aircraft operations at low altitudes on approach to and departing from airports, pilots are urged to promptly volunteer reports to controllers of wind shear conditions they encounter. An advance warning of this information will assist other pilots in avoiding or coping with a wind shear on approach or departure.

b. When describing conditions, use of the terms "negative" or "positive" wind shear should be avoided. PIREP's of "*negative* wind shear on final," intended to describe loss of airspeed and lift, have been interpreted to mean that *no* wind shear was encountered. The recommended method for wind shear reporting is to state the loss or gain of airspeed and the altitudes at which it was encountered.

EXAMPLE:
DENVER TOWER, CESSNA 1234 ENCOUNTERED WIND SHEAR, LOSS OF 20 KNOTS AT 400 FEET.

EXAMPLE:
TULSA TOWER, AMERICAN 721 ENCOUNTERED WIND SHEAR ON FINAL, GAINED 25 KNOTS BETWEEN 600 AND 400 FEET FOLLOWED BY LOSS OF 40 KNOTS BETWEEN 400 FEET AND SURFACE.

1. Pilots who are not able to report wind shear in these specific terms are encouraged to make reports in terms of the effect upon their aircraft.

EXAMPLE:
MIAMI TOWER, GULFSTREAM 403 CHARLIE ENCOUNTERED AN ABRUPT WIND SHEAR AT 800 FEET ON FINAL, MAX THRUST REQUIRED.

2. Pilots using Inertial Navigation Systems (INS) should report the wind and altitude both above and below the shear level.

7–23. CLEAR AIR TURBULENCE (CAT) PIREP'S

CAT has become a very serious operational factor to flight operations at all levels and especially to jet traffic flying in excess of 15,000 feet. The best available information on this phenomenon must come from pilots via the PIREP reporting procedures. All pilots encountering CAT conditions are urgently requested to report *time, location,* and *intensity* (light, moderate, severe, or extreme) of the element to the FAA facility with which they are maintaining radio contact. If time and conditions permit, elements should be reported according to the standards for other PIREP's and position reports. (See Pireps Relating to Turbulence, paragraph 7–21).

7–24. MICROBURSTS

a. Relatively recent meteorological studies have confirmed the existence of microburst phenomenon. Microbursts are small-scale intense downdrafts which, on reaching the surface, spread outward in all directions from the downdraft center. This causes the presence of both vertical and horizontal wind shears that can be extremely hazardous to all types and categories of aircraft, especially at low altitudes. Due to their small size, short life-span, and the fact that they can occur over areas without surface precipitation, microbursts are not easily detectable using conventional weather radar or wind shear alert systems.

b. Parent clouds producing microburst activity can be any of the low or middle layer convective cloud types. Note, however, that microbursts commonly occur within the heavy rain portion of thunderstorms, and in much weaker, benign appearing convective cells that have little or no precipitation reaching the ground.

Figure 7–24[1]

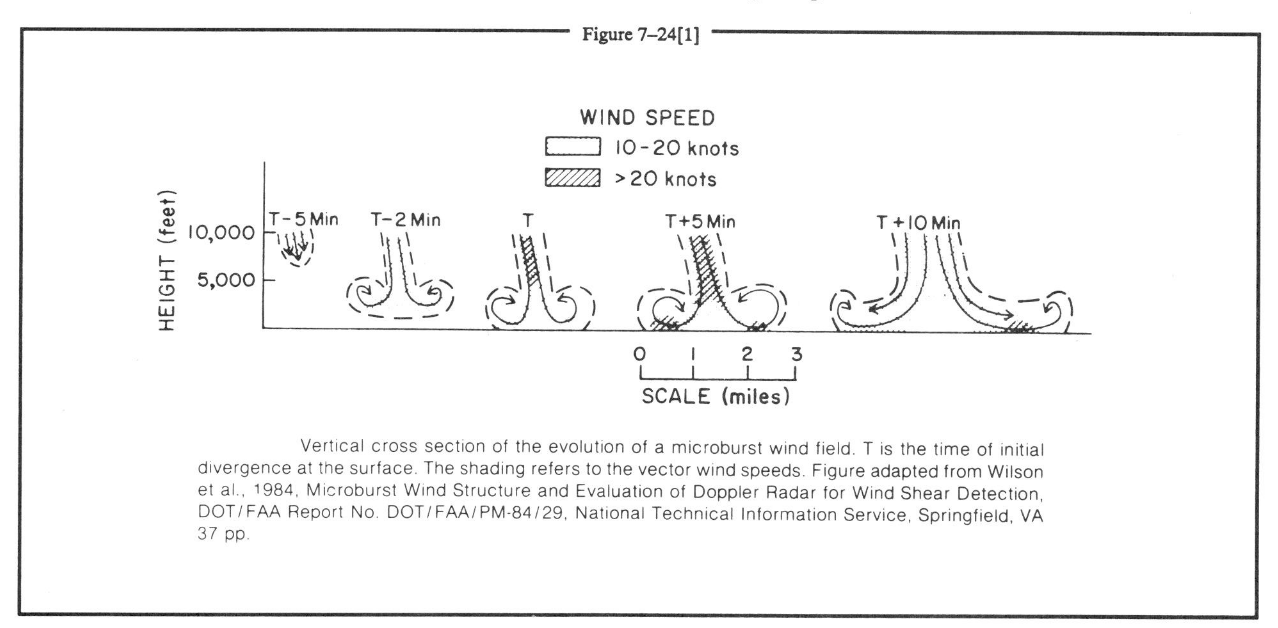

Vertical cross section of the evolution of a microburst wind field. T is the time of initial divergence at the surface. The shading refers to the vector wind speeds. Figure adapted from Wilson et al., 1984, Microburst Wind Structure and Evaluation of Doppler Radar for Wind Shear Detection, DOT/FAA Report No. DOT/FAA/PM-84/29, National Technical Information Service, Springfield, VA 37 pp.

c. The life cycle of a microburst as it descends in a convective rain shaft is seen in Figure 7–24[1]. An important consideration for pilots is the fact that the microburst intensifies for about 5 minutes after it strikes the ground.

d. Characteristics of microbursts include:

1. Size—The microburst downdraft is typically less than 1 mile in diameter as it descends from the cloud base to about 1,000–3,000 feet above the ground. In the transition zone near the ground, the downdraft changes to a horizontal outflow that can extend to approximately 2 ½ miles in diameter.

2. Intensity—The downdrafts can be as strong as 6,000 feet per minute. Horizontal winds near the surface can be as strong as 45 knots resulting in a 90 knot shear (headwind to tailwind change for a traversing aircraft) across the microburst. These strong horizontal winds occur within a few hundred feet of the ground.

3. Visual Signs—Microbursts can be found almost anywhere that there is convective activity.

They may be embedded in heavy rain associated with a thunderstorm or in light rain in benign appearing virga. When there is little or no precipitation at the surface accompanying the microburst, a ring of blowing dust may be the only visual clue of its existence.

4. Duration—An individual microburst will seldom last longer than 15 minutes from the time it strikes the ground until dissipation. The horizontal winds continue to increase during the first 5 minutes with the maximum intensity winds lasting approximately 2–4 minutes. Sometimes microbursts are concentrated into a line structure, and under these conditions, activity may continue for as long as an hour. Once microburst activity starts, multiple microbursts in the same general area are not uncommon and should be expected.

Figure 7–24[2]

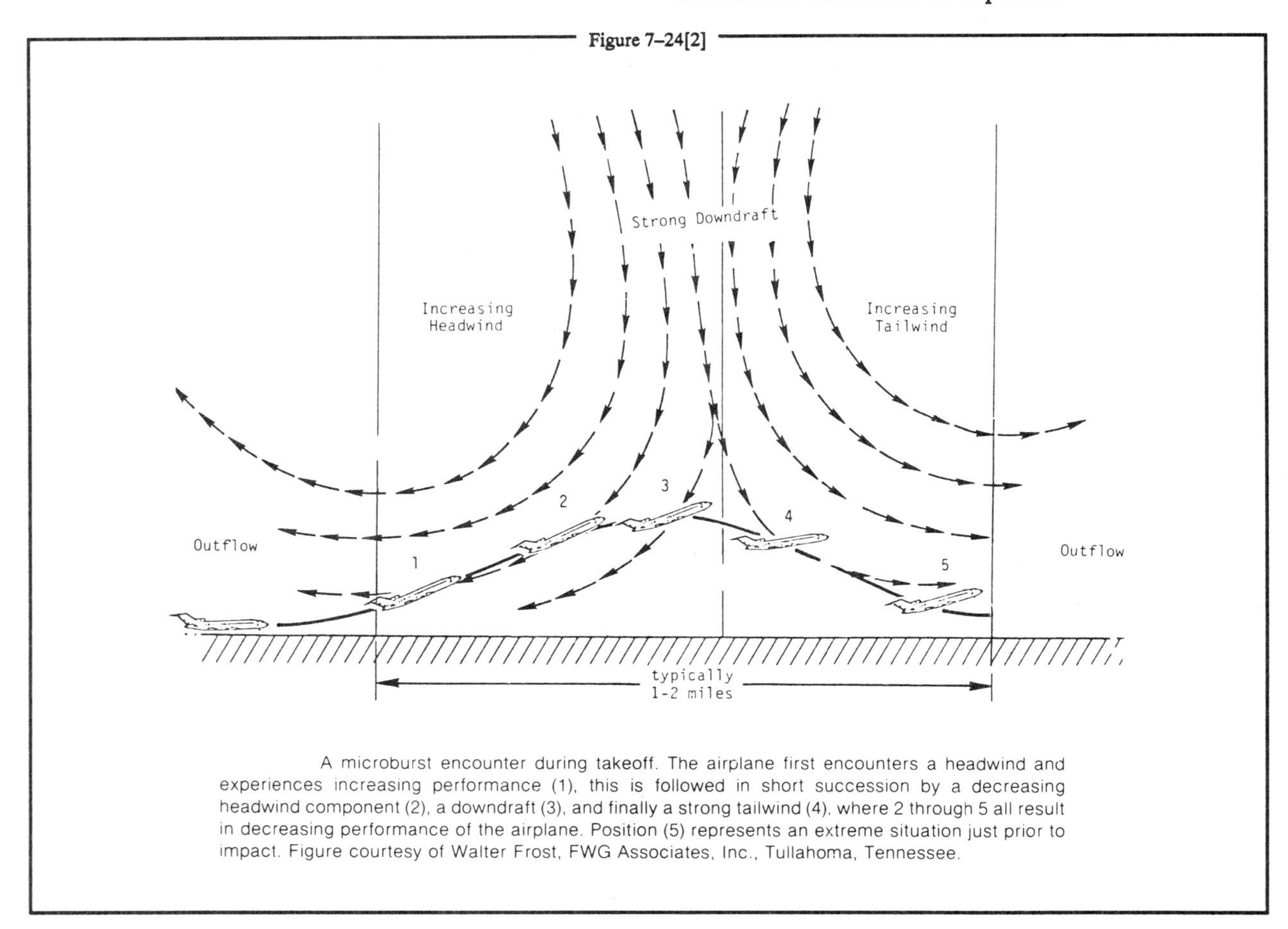

A microburst encounter during takeoff. The airplane first encounters a headwind and experiences increasing performance (1), this is followed in short succession by a decreasing headwind component (2), a downdraft (3), and finally a strong tailwind (4), where 2 through 5 all result in decreasing performance of the airplane. Position (5) represents an extreme situation just prior to impact. Figure courtesy of Walter Frost, FWG Associates, Inc., Tullahoma, Tennessee.

e. Microburst wind shear may create a severe hazard for aircraft within 1,000 feet of the ground, particularly during the approach to landing and landing and take-off phases. The impact of a microburst on aircraft which have the unfortunate experience of penetrating one is characterized in Figure 7–24[2]. The aircraft may encounter a headwind (performance increasing) followed by a downdraft and tailwind (both performance decreasing), possibly resulting in terrain impact.

f. Pilots should heed wind shear PIREP's, as a previous pilot's encounter with a microburst may be the only indication received. However, since the wind shear intensifies rapidly in its early stages, a PIREP may not indicate the current severity of a microburst. Flight in the vicinity of suspected or reported microburst activity should always be avoided. Should a pilot encounter one, a wind shear PIREP should be made at once.

7–25. THUNDERSTORMS

a. Turbulence, hail, rain, snow, lightning, sustained updrafts and downdrafts, icing conditions-all are present in thunderstorms. While there is some evidence that maximum turbulence exists at the middle level of a thunderstorm, recent studies show little variation of turbulence intensity with altitude.

b. There is no useful correlation between the external visual appearance of thunderstorms and the severity or amount of turbulence or hail within them. The visible thunderstorm cloud is only a portion of a turbulent system whose updrafts and downdrafts often extend far beyond the visible storm cloud. Severe turbulence can be expected up to 20 miles from severe thunderstorms. This distance decreases to about 10 miles in less severe storms.

c. Weather radar, airborne or ground based, will normally reflect the areas of moderate to heavy precipitation (radar does not detect turbulence). The frequency and severity of turbulence generally increases with the radar reflectivity which is closely associated with the areas of highest liquid water content of the storm. NO FLIGHT PATH THROUGH AN AREA OF STRONG OR VERY STRONG RADAR ECHOES SEPARATED BY 20–30 MILES OR LESS MAY BE CONSIDERED FREE OF SEVERE TURBULENCE.

d. Turbulence beneath a thunderstorm should not be minimized. This is especially true when the relative humidity is low in any layer between the surface and 15,000 feet. Then the lower altitudes may be characterized by strong out-flowing winds and severe turbulence.

e. The probability of lightning strikes occurring to aircraft is greatest when operating at altitudes where temperatures are between minus 5 degrees Celsius and plus 5 degrees Celsius. Lightning can strike aircraft flying in the clear in the vicinity of a thunderstorm.

f. The NWS recognizes only two classes of intensities of thunderstorms as applied to aviation surface weather observations:

1. T—Thunderstorm; and

2. T+—Severe thunderstorm.

g. NWS radar systems are able to objectively determine radar weather echo intensity levels by use of Video Integrator Processor (VIP) equipment. These thunderstorm intensity levels are on a scale of one to six. (Reference—Pilot/Controller Glossary, Radar Weather Echo Intensity Levels).

EXAMPLE:

Alert provided by an ATC facility to an aircraft:

(Aircraft identification) LEVEL FIVE INTENSE WEATHER ECHO BETWEEN TEN O'CLOCK AND TWO O'CLOCK, ONE ZERO MILES, MOVING EAST AT TWO ZERO KNOTS, TOPS FLIGHT LEVEL THREE NINE ZERO.

EXAMPLE:

Alert provided by a FSS:

(Aircraft identification) LEVEL FIVE INTENSE WEATHER ECHO, TWO ZERO MILES WEST OF ATLANTA V-O-R, TWO FIVE MILES WIDE, MOVING EAST AT TWO ZERO KNOTS, TOPS FLIGHT LEVEL THREE NINE ZERO.

7–26. THUNDERSTORM FLYING

a. Above all, remember this: never regard any thunderstorm "lightly" even when radar observers report the echoes are of light intensity. Avoiding thunderstorms is the best policy. Following are some Do's and Don'ts of thunderstorm avoidance:

1. Don't land or takeoff in the face of an approaching thunderstorm. A sudden gust front of low level turbulence could cause loss of control.

2. Don't attempt to fly under a thunderstorm even if you can see through to the other side. Turbulence and wind shear under the storm could be disastrous.

3. Don't fly without airborne radar into a cloud mass containing scattered embedded thunderstorms. Scattered thunderstorms not embedded usually can be visually circumnavigated.

4. Don't trust the visual appearance to be a reliable indicator of the turbulence inside a thunderstorm.

5. Do avoid by at least 20 miles any thunderstorm identified as severe or giving an intense radar echo. This is especially true under the anvil of a large cumulonimbus.

6. Do clear the top of a known or suspected severe thunderstorm by at least 1,000 feet altitude for each 10 knots of wind speed at the cloud top. This should exceed the altitude capability of most aircraft.

7. Do circumnavigate the entire area if the area has 6/10 thunderstorm coverage.

8. Do remember that vivid and frequent lightning indicates the probability of a severe thunderstorm.

9. Do regard as extremely hazardous any thunderstorm with tops 35,000 feet or higher whether the top is visually sighted or determined by radar.

b. If you cannot avoid penetrating a thunderstorm, following are some Do's *before* entering the storm:

1. Tighten your safety belt, put on your shoulder harness if you have one and secure all loose objects.

2. Plan and hold your course to take you through the storm in a minimum time.

3. To avoid the most critical icing, establish a penetration altitude below the freezing level or above the level of minus 15 degrees Celsius.

4. Verify that pitot heat is on and turn on carburetor heat or jet engine anti-ice. Icing can be rapid at any altitude and cause almost instantaneous power failure and/or loss of airspeed indication.

5. Establish power settings for turbulence penetration airspeed recommended in your aircraft manual.

6. Turn up cockpit lights to highest intensity to lessen temporary blindness from lightning.

7. If using automatic pilot, disengage altitude hold mode and speed hold mode. The automatic altitude and speed controls will increase maneuvers of the aircraft thus increasing structural stress.

8. If using airborne radar, tilt the antenna up and down occasionally. This will permit you to detect other thunderstorm activity at altitudes other than the one being flown.

c. Following are some Do's and Don'ts *during* the thunderstorm penetration:

1. Do keep your eyes on your instruments. Looking outside the cockpit can increase danger of temporary blindness from lightning.

2. Don't change power settings; maintain settings for the recommended turbulence penetration airspeed.

3. Do maintain constant attitude; let the aircraft "ride the waves." Maneuvers in trying to maintain constant altitude increases stress on the aircraft.

4. Don't turn back once you are in the thunderstorm. A straight course through the storm most likely will get you out of the hazards most quickly. In addition, turning maneuvers increase stress on the aircraft.

7–27. KEY TO AVIATION WEATHER OBSERVATIONS AND FORECASTS

Figure 7–27[1]

KEY TO AVIATION WEATHER OBSERVATIONS

LOCATION IDENTIFIER TYPE AND TIME OF REPORT •	SKY AND CEILING	VISIBILITY WEATHER AND OBSTRUCTION TO VISION	SEA-LEVEL PRESSURE	TEMPERATURE AND DEW POINT	WIND	ALTIMETER SETTING	REMARKS AND CODED DATA
MCI SA Ø758	15 SCT M25 OVC	1R-F	132	/58/56	/18Ø7	/993/	RØ1VR2ØV4Ø

SKY AND CEILING

Sky cover contractions are for each layer in ascending order. Figures preceding contractions are base heights in hundreds of feet above station elevation. Sky cover contractions used are:

CLR = Clear: Less than Ø.1 sky cover
SCT = Scattered: Ø.1 to Ø.5 sky cover
BKN = Broken: Ø.6 to Ø.9 sky cover
OVC = Overcast: More than Ø.9 sky cover

— = Thin (When prefixed to SCT, BKN, OVC)
—X = Partly obscured: Ø.9 or less of sky hidden by precipitation or obstruction to vision (bases at surface)
X = Obscured: 1.Ø sky hidden by precipitation or obstruction to vision (bases at surface)

A letter preceding the height of a base identifies a ceiling layer and indicates how ceiling height was determined. Thus:

E = Estimated
M = Measured
W = Vertical visibility into obscured sky
V = Immediately following the height of a base indicates a variable ceiling.

VISIBILITY

Reported in statute miles and fractions (V = Variable)

WEATHER AND OBSTRUCTION TO VISION SYMBOLS

A	Hail	IC	Ice crystals	S	Snow
BD	Blowing dust	IF	Ice fog	SG	Snow grains
BN	Blowing sand	IP	Ice pellets	SP	Snow pellets
BS	Blowing snow	IPW	Ice pellet showers	SW	Snow showers
D	Dust	K	Smoke	T	Thunderstorms
F	Fog	L	Drizzle	T+	Severe thunderstorm
GF	Ground fog	R	Rain	ZL	Freezing drizzle
H	Haze	RW	Rain showers	ZR	Freezing rain

Precipitation intensities are indicated thus: - Light; (no sign) Moderate; + Heavy

WIND Direction in tens of degrees from true north, speed in knots. ØØØØ indicates calm. G indicates gusty. Q indicates Squalls. Peak wind speed in the past 1Ø minutes follows G or Q when gusts or squalls are reported. The contraction WSHFT, followed by GMT time group in remarks, indicates windshift and its time of occurrence. (Knots x 1.15 = statute mi/hr).

EXAMPLES: 3627 = wind from 36Ø Degrees at 27 knots.
3627G4Ø = wind from 36Ø Degrees at 27 knots, peak speed in gusts 4Ø knots

ALTIMETER SETTING

The first figure of the actual altimeter setting is always omitted from the report.

RUNWAY VISUAL RANGE (RVR)

RVR is reported from some stations. For planning purposes, the value range during 1Ø minutes prior to observations and based on runway light setting 5 are reported in hundreds of feet. Runway identification precedes RVR report.

PILOT REPORTS (PIREPs)

When available, PIREPs in fixed-format may be appended to weather observations. PIREPs are designated by UA or UUA for urgent PIREPs.

DECODED REPORT

Kansas City International: Record observation completed at Ø758 UTC. 15ØØ feet scattered clouds, measured ceiling 25ØØ feet overcast, visibility 1 mile, light rain, fog, sea-level pressure 1Ø13.2 millibars, temperature 58°F, dewpoint 56°F, wind from 18Ø°, at 7 knots, altimeter setting 29.93 inches. Runway Ø1, visual range 2ØØØ feet lowest 4ØØØ feet highest in the past 1Ø minutes.

• TYPE OF REPORT

SA = a scheduled record observation
SP = an unscheduled special observation indicating a significant change in one or more elements
RS = a scheduled record observation that also qualifies as a special observation

The designator for all three types of observations (SA, SP, RS) is followed by a 24-hour-clock-time-group in Coordinated Universal Time (UTC or Z).

U.S. DEPARTMENT OF COMMERCE—NATIONAL OCEANIC AND ATMOSPHERIC ADMINISTRATION—NATIONAL WEATHER SERVICE

KEY TO AVIATION WEATHER FORECASTS

TERMINAL FORECASTS contain information for specific airports on expected ceiling, cloud heights, cloud amounts, visibility, weather and obstructions to vision, and surface wind. They are issued 3 times/day, amended as needed, and are valid for up to 24 hours. The last six hours of each forecast period are covered by a categorical statement indicating whether VFR, MVFR, IFR or LIFR conditions are expected (L in LIFR and M in MVFR indicate "low" and "marginal"). Terminal forecasts are written in the following form:

CEILING: Identified by the letter "C" (for lowest layer with cumulative sky cover greater than 5/1Ø)
CLOUD HEIGHTS: In hundreds of feet above the station (ground)
SKY COVER AMOUNT (including any obscuration)
CLOUD LAYERS: Stated in ascending order of height
VISIBILITY: In statute miles (omitted if over 6 miles)
WEATHER AND OBSTRUCTION TO VISION: Standard weather and obstruction to vision symbols are used
SURFACE WIND: In tens of degrees and knots (omitted when less than 6 knots)

EXAMPLE OF TERMINAL FORECAST

DCA 221Ø1Ø: DCA Forecast 22nd day of month - valid time 10Z-10Z.
1ØSCT C18 BKN 5SW—3415G25 OCNL C8 X 1/2SW: Scattered clouds at 1ØØØ feet, ceiling 18ØØ feet broken, visibility 5 miles, light snow showers, surface wind from 34Ø degrees at 15 knots. Gusts to 25 knots, occasional ceiling 8 hundred feet sky totally obscured, visibility 1/2 mile in moderate snow showers.
12Z C5Ø BKN 3312G22: By 12Z becoming ceiling 5ØØØ feet broken, surface wind 33Ø degrees at 12 knots. Gusts to 22.
Ø4Z MVFR CIG: Last 6 hours of FT after Ø4Z marginal VFR due to ceiling.

AREA FORECASTS are 12-hour aviation forecasts plus a 6-hour categorical outlook prepared 3 times/day, with each section amended as needed, giving general descriptions of potential hazards, airmass and frontal conditions, icing and freezing level, turbulence and low-level windshear and significant clouds and weather for an area the size of several states. Heights of cloud bases and tops, turbulence and icing are referenced ABOVE MEAN SEA LEVEL (MSL); unless indicated by Ceiling (CIG) or ABOVE GROUND LEVEL (AGL). Each SIGMET OR AIRMET affecting an FA area will also serve to amend the Area Forecast.

SIGMET, AIRMET and CWA messages (In-flight advisories) broadcast by FAA on NAVAID voice channels warn pilots of potentially hazardous weather. SIGMET's concern severe and extreme conditions of importance to all aircraft (i.e., icing, turbulence and dust storms/sandstorms or volcanic ash). Convective SIGMET's are issued for thunderstorms if they are sufficiently strong, wide spread or embedded. AIRMET's concern less severe conditions which may be hazardous to aircraft particularly smaller aircraft and less experienced or VFR only pilots. CWA's (Center Weather Advisories) concern both SIGMET and AIRMET type conditions described in greater detail and relating to a specific ARTCC area.

WINDS AND TEMPERATURES ALOFT (FD) FORECASTS are 6, 12, and 24-hour forecasts of wind direction (nearest 1Ø° true N) and speed (knots) for selected flight levels. Forecast Temperatures Aloft (°C) are included for all but the 3ØØØ-foot level.

EXAMPLES OF WINDS AND TEMPERATURES ALOFT (FD) FORECASTS
FD WBC 121645
BASED ON 1212ØØZ DATA
VALID 13ØØØØZ FOR USE 21ØØ-Ø6ØØZ. TEMPS NEG ABV 24ØØØ FT

	3ØØØ	6ØØØ	9ØØØ	12ØØØ	18ØØØ	24ØØØ	3ØØØØ	34ØØØ	39ØØØ
BOS	3127	3425-Ø7	342Ø-11	3421-16	3516-27	3512-38	311649	292451	283451
JFK	3Ø26	3327-Ø8	3324-12	3322-16	312Ø-27	2923-38	284248	28515Ø	285749

At 6ØØØ feet MSL over JFK wind from 330° at 27 knots and temperature minus 8°C.

TWEB (CONTINUOUS TRANSCRIBED WEATHER BROADCAST)-Individual route forecasts covering a 25-nautical-mile zone either side of the route. By requesting a specific route number, detailed en route weather for a 15-hour period plus a synopsis can be obtained.

PILOTS.... report in-flight weather to nearest FSS. The latest surface weather reports are available by phone at the nearest pilot weather briefing office by calling at H + 10.

U.S. DEPARTMENT OF COMMERCE—NATIONAL OCEANIC AND ATMOSPHERIC ADMINISTRATION—NATIONAL WEATHER SERVICE —REVISED JANUARY 1987

(See Figure 7–27[1].)

7–28. AUTOB DECODING KEY

(See Table 7–28[1].)

Table 7–28[1]

DECODING OBSERVATIONS FROM AUTOB STATIONS

EXAMPLE: **ENV AUTOB E25 BKN BV7 P 33/29/3606/975 PK WND 08 001**

ENCODE	DECODE	EXPLANATION
ENV	STATION IDENTIFICATION:	(Wendover, UT) Identifies report using FAA identifiers.
AUTOB	AUTOMATIC STATION IDENTIFIER	
E25 BKN	SKY & CEILING:	(Estimated 2500 ft. broken) Figures are height in 100s of feet above ground. Contraction after height is amount of sky cover. Letter preceding height indicates ceiling. WX reported if visibility is less than 2 miles and no clouds are detected. *NO CLOUDS REPORTED ABOVE 6000 FEET.*
BV7	BACKSCATTER VISIBILITY AVERAGED IN PAST MINUTE:	Reported in whole miles from 1 to 7.
P	PRECIPITATION OCCURRENCE:	(P=precipitation in past 10 minutes).
33	TEMPERATURE:	(33 degrees F.) Minus sign indicates sub-zero temperatures.
/29	DEW POINT:	(29 degrees F.) Minus sign indicates sub-zero temperatures.
/3606	WIND:	(360 degrees true at 6 knots) Direction is first two digits and is reported in tens of degrees. To decode, add a zero to first two digits. The last digits are speed; e.g., 2524 = 250 degrees at 24 knots.
/975	ALTIMETER SETTING:	(29.75 inches) The tens digit and decimal are omitted from report. To decode, prefix a 2 to code if it begins with 8 or 9. Otherwise, prefix a 3; e.g., 982 = 29.82, 017 = 30.17.
PK WND 08	PEAK WIND SPEED:	(8 knots) Reported speed is highest detected since last hourly observation.
001	PRECIPITATION ACCUMULATION:	(0.01 inches) Amount of precipitation since last synoptic time (00, 06, 12, 1800 UTC).

NOTE: If no clouds are detected below 6,000 feet and the visibility is greater than 2 miles, the reported sky condition will be *CLR BLO 60.*

7–29. INTERNATIONAL CIVIL AVIATION ORGANIZATION (ICAO) TERMINAL FORECAST (TAF)

Terminal forecasts for international locations and domestic military locations are available to the Flight Service Station specialist, via their weather computer, and to the pilot, via the Direct User Access Terminal (DUAT), but are in an international alphanumeric code. They are scheduled four times daily, for 24–hour periods, beginning at 0000Z, 0600Z, 1200Z and 1800Z.

a. Format. The TAF is a series of groups made up of digits and letters. An individual group is identified by its position in the sequence, by its alphanumeric coding or by a numerical indicator. Listed below are a few contractions used in the TAF. Some of the contractions are followed by time entries indicated by "tt" or "tttt" or by probability, "pp".

b. Significant weather change indicators.

GRADU tttt— A gradual change occurring during a period in excess of one–half hour. "tttt" are the beginning and ending times of the expected change to the nearest hour; i.e., "GRADU 1213" means the transition will occur between 1200Z and 1300Z.

RAPID tt— A rapid change occurring in one–half hour or less. "tt" is the time to the nearest hour of the change; i.e., "RAPID 23" means the change will occur about 2300Z.

Variability terms— indicate that short time period variations from prevailing conditions are expected with the total occurrence of these variations less than ½ of the time period during which they are called for.

TEMPO tttt— Temporary changes from prevailing conditions of less than one hour duration in each instance. There may be more than one (1) instance for a specified time period. "tttt" are the earliest and latest times during which the temporary changes are expected; i.e., "TEMPO 0107" means the temporary changes may occur between 0100Z and 0700Z.

INTER tttt— Changes from prevailing conditions are expected to occur frequently and briefly. "tttt" are the earliest and latest times the brief changes are expected; i.e., "INTER 1518" means that the brief, but frequent, changes may occur between 1500Z and 1800Z. INTER has shorter and more frequent changes than TEMPO.

c. Probability.

PROB pp— Probability of conditions occurring. "pp" is the probability in per cent; i.e., "PROB 20" means a 10 or 20% probability of the conditions occurring. "PROB 40" means a 30 to 50% inclusive probability.

d. Cloud and weather terms.

CAVOK— No clouds below 50,000 feet or below the highest minimum sector altitude whichever is greater, and no cumulonibus. Visibility 6 miles or greater. No precipitation, thunderstorms, shallow fog or low drifting snow.

WX NIL— No significant weather (no precipitation, thunderstorms or obstructions to vision).

SKC— Sky clear.

e. Following is a St. Louis MO forecast in TAF code.

EXAMPLE:

KSTL 1212 33025/35 0800 71SN 9//05 INTER 1215 0000 75XXSN 9//000 GRADU 1516 33020 4800 38BLSN 7SC030 PROB 40 85SNSH GRADU 2122 33015 9999 WX NIL 3SC030 RAPID 00 VRB05 9999 SKC GRADU 0304 24015/25 CAVOK Ⓓ

1. The forecast is broken down into the elements lettered "a" to "I" to aid in the discussion. Not included in the example but explained at the end are three optional forecast groups for "m" icing, " n" turbulence and "o" temperature. (See Table 7–29[1]).

(a) *Station identifier.* The TAF code uses ICAO 4–letter station identifiers. In the contiguous 48 states the 3–letter identifier is prefixed with a "K"; i.e., the 3–letter identifier for Seattle is SEA while the ICAO identifier is KSEA. Elsewhere, the first two letters of the ICAO identifier tell what region the station is in. "MB" means Panama/ Canal Zone (MBHO is Howard AFB); "MI" means Virgin Islands (MISX is St. Croix); "MJ" is Puerto Rico (MJSJ is San Juan); "PA" is Alaska (PACD is Cold Bay); PH" is Hawaii (PHTO is Hilo).

(b) *Valid time.* Valid time of the forecast follows station identifier. "1212" means a 24–hour forecast valid from 1200Z until 1200Z the following day.

Table 7–29[1]

KSTL a.	1212 b.	33025/35 c.
0800 d.	71SN e.	9//005 f.

INTER 1215 0000 75XXSN 9//000
g.
GRADU 1516 33020 4800 38BLSN 7SC030
h.
PROB 40 85SNSH
i.
GRADU 2122 33015 9999 WX NIL 3SC030
j.
RAPID 00 VRB05 9999 SKC
k.
GRADU 0304 24015125 CAVOK Ⓓ
l.

(c) *Wind.* Wind is forecast usually by a 5–digit group giving degrees in 3 digits and speed in 2 digits. When wind is expected to be 100 knots or more, the group is 6–digits with speed given is 3 digits. When speed is gusty or variable, peak speed is separated from average speed with a slash. For example, in the KSTL TAF, "33025/ 35" means wind 330 degrees, average speed 25 knots, peak speed 35 knots. A group "160115/ 130" means wind 160 degrees, 115 knots, peak speed 130 knots. "00000" means calm; "VRB" followed by speed indicates direction variable; i.e., "VRBI0" means wind direction variable at 10 knots.

(d) *Visibility.* Visibility is in meters. "0800" means 800 meters converted from table to ½ mile. (See Table 7–29[2] for converting meters to miles and fractions).

Table 7–29[2]

Visibility conversion TAF code to miles

Meters	Miles	Meters	Miles	Meters	Miles
0000	0	1200	¾	3000	1⅞
0100	1/16	1400	⅞	3200	2
0200	⅛	1600	1	3600	2¼
0300	3/16	1800	1	4000	2½
0400	¼	2000	1¼	4800	3
0500	5/16	2200	1⅜	6000	4
0600	⅜	2400	1½	8000	5
0800	½	2600	1⅝	9000	6
1000	⅝	2800	1¾	9999	>6

(e) *Significant weather.* Significant weather is decoded using Table 7–29[3]. Groups in the table are numbered sequentially. Each number is followed by an acronym suggestive of the weather; you can soon learn to read most of the acronyms without reference to the table. Examples: "177TS", thunderstorm; "18SQ", squall; "31SA", sandstorm; "60RA", rain; "85SNSH", snow shower. "XX" freezing rain. In the KSTL forecast, "71SN" means light snow. The TAF encodes only the single most significant type of weather; the U.S. domestic FT permits encoding of multiple weather types. (See Table 7–29[4] to convert weather from FT to TAF).

TAF weather codes

Table 7–29[3]

Code	Simple Definition	Detailed Definition
04FU	Smoke	Visibility reduced by smoke. No visibility restriction.
05HZ	Dust haze	Visibility reduced by haze. No visibility restriction.
06HZ	Dust haze	Visibility reduced by dust suspended in the air but wind not strong enough to be adding more dust. No visibility restriction.*
*While this may seem to be contradictory, it means that while visibility is restricted, the amount of the restriction is not limited.		
07SA	Duststorm, sandatorm, rising dust or sand	Visibility reduced by dust suspended in the air and wind strong enough to be adding sand more dust. No well developed dust devils, duststorm or sand-storm. Visibility 6 miles or less.
08PO	Dust devil	Basically the same as 07SA but with well developed dust devils. Visibility 6 miles or less.
10BR	Mist	Fog, ground fog or ice fog with visibility 5/8 to 6 miles.
11MIFG	Shallow fog	Patchy shallow fog (less than 6 feet deep and coverage less than half) with visibility in the fog less than 5/8 mile.
12MIFG	Shallow fog	Shallow fog (less than 6 feet deep with more or less continuous coverage) with visibility in the fog less than 5/8 miles.
17TS	Thunderstorm	Thunderstorm at the station but with no precipitation.
18SQ	Squall	No precipitation. A sudden increase of at least 15 knots in average wind speed, sustained at 20 knots or more for at least one (1) minute.
19FC	Funnel cloud	Used to forecast a tornado, funnel cloud or waterspout at or near the station. Also not easy to forecast and likely to be overshadowed by some other more violent weather such as thunderstorms.
30SA	Duststorm, sandstorm, rising dust or sand.	Duststorm or sandstorm, visibility 5/16 to less than 5/8 mile, increasing in intensity.
31SA		Basically the same as 30SA but with no change in intensity.
32SA		Basically the same as 30SA but increasing in intensity.
33XXSA	Heavy duststorm or sandstorm.	Severe duststorm or sandstorm, visibility, less than 5/16 mile, decreasing in intensity.
34XXSA		Basically the same as 33XXSA but with no change in intensity.
35XXSA		Basicallv the same as 33XXSA but increasing in intensity.
36DRSN	Low drifting snow.	Low drifting snow (less than 6 feet) with visibility in drifting snow less than 5/16 miles.

TAF weather codes—CONTINUED

Table 7–29[3]

Code	Simple Definition	Detailed Definition
37DRS		Low drifting snow (less than 6 feet) with visibility in drifting snow less than 5/16 miles.
38BLSN	Blowing snow.	Blowing snow (more than 6 feet) with visibility 5/16 to 6 miles.
39BLSN		Blowing snow (more than 6 feet) with visibility 5/16 to 6 miles.
40BCFG	Fog	Distant fog (not at station).
41BCFG	patches	Patchy fog at the station. visibility in the fog patches less than 5/8 of a mile.
42FG		Fog at the station. visibility less than 5/8 mile, sky visible, fog thinning.
43FG		Fog at the station, visibility less than 5/8 mile, sky not visible, fog thinning.
44FG		Fog at the station,less than 5/8 mile, sky visible, no change in intensity. Fog
45FG		Fog at the station, visibility less than 5/8 mile. sky not visible, no change in intensity.
46FG		Fog at the station, visibility less than 5/8 mile, sky visible, fog thickening,
47FG		Fog at the station, visiblity less than 5/8 mile, sky not visible, fog thickening.
NOTE: In code figures 40 through 47,"fog"includes both fog and ice fog. See FMH No.1 (Surface Observations) for definitions of precipitation intensities.		
48FZFG	Freezing	Fog depositing rime ice, visibility fogless than 5/8 mile, sky visible.
49FZFG		Fog depositing rime ice, visibility less than 5/8 mile, sky not visible.
50DZ	Drizzle	Light intermittent drizzle.
51DZ		Light continuous drizzle.
52DZ		Moderate intermittent drizzle.
53DZ		Moderate continuous drizzle.
54XXDZ	Heavy drizzle.	Heavy intermittent drizzle.
55XXDZ		Heavy continuous drizzle.
56XXDZ	Freezing drizzle	Light freezing drizzle.
57XXFZDZ	Heavy freezing drizzle.	Moderate or heavy freezing drizzle.
58RA		Mixed rain and drizzle, light.
59RA		Mixed rain and drizzle, moderate or heavy.
60RA	Rain	Light intermittent rain.
61RA		Light continuous rain.
62RA		Moderate intermittent rain.
63RA		Moderate continuous rain.
64XXRA	Heavy rain.	Heavy intermittent rain.
65XXRA		Heavy continuous rain.
66FZRA	Freezing rain	Freezing rain or mixed freezing rain and freezing drizzle, light.

TAF weather codes—CONTINUED

Table 7–29[3]

Code	Simple Definition	Detailed Definition
67XXFZRA	Heavy freezing rain	Freezing rain or mixed freezing rain and freezig drizzle, moderate or heavy.
68RASN	Rain and snow	Mixed rain and snow or drizzle and snow, light.
69XXRASN	Heavy rain and snow.	Mixed rain and snow or drizzle and and snow.snow, moderate or heavy.
70SN		Light intermittent snow.
71SN		Light continuous snow.
72SN	snow	Moderate intermittent snow.
73SN		Moderate continuous snow.
74XXSN	Heavy snow	Heavy intermittent snow.
75XXSN		Heavy continuous snow.
77SG	Snow grains	Snow grains, any intensity. May be accompanied by fog or ice fog.
79PE	Ice pellets	Ice pellets, any intensity. May be mixed with some other precipitation.
80RASH	Showers	Light rain showers.
81XXSH	Heavy	Moderate or heavy rain showers.
82XXSH	showers	Violent rain showers (more than 1 inch per hour or 0.1 inch in 6 minutes).
83RASN	Showers of rain and	Mixed rain showers and snow showers. Intensity of both showers is light.
84XXRASN	Heavy showers of rain and	Mixed rain showers and snowshowers. Intensity of either shower is moderate or heavy.
85SNSH	Snow showers	Light snow showers.
86XXSNSH	Heavy snow showers	Moderate or heavy snow showers.
87GR		Light ice pellet showers. There may also be rain or mixed rain or snow.
88GR	soft hail	Moderate or heavy ice pellet showers. There may also be rain or mixed rain and snow.
89GR	Hail	Hail, not associated with a Thunder–storm. There may also be rain or mixed rain and snow.
90XXGR	Heavy hail	Moderate or heavy hail, not associated with a thunderstorm. There may also be rain or mixed rain or snow.
91RA	Rain	Light rain or light rain shower at the time of the forecast and thunderstorms during the preceding hour, but not at the time of the forecast.
92XXRA	Heavy rain	Basically the same as 91RA but the intensity of the rain or rain shower is moderate or heavy.
93GR	Hail	Basically the same as 91RA, but the precipitation is light snow or snow showers, or light mixed rain and snow or rain showers and snow showers, or light ice pellets or ice pellet showers.
94XXGR	Heavy hail	Basically the same as 93GR but the intensity of any precipitation is moderate or heavy.
95TS	Thunderstorm	Thunderstorm with rain or snow, or a mixture of rain and snow, but no hail, ice pellets or snow pellets.

TAF weather codes—CONTINUED

Table 7–29[3]

Code	Simple Definition	Detailed Definition
96TSGR	Thunderstorm with hail	Thunderstorm with hail, ice pellets or snow pellets. There may also be rain or snow, or mixed rain and snow.
97XXTS	Heavy thunderstorm	Severe thunderstorm with rain or snow, or a mixture of rain and snow, but no hail, ice pellets or snow pellets.
98TSSA	Thunderstorm with duststorm or sandstorm	Thunderstorm with dust storm or sandstorm. There may also be some form of precipitation with the thunderstorm.
99XXTSGR	Heavy thunderstorm with hail	Basically the same as 97XXTS but in addition to everything else there is hail.

(f) *Clouds.* A cloud group is a 6–character group. The first digit is coverage in octas (eighths) of the individual cloud layer only. The summation of cloud layer to determine total sky cover from a ground observers point of view is NOT used. (See Table 7–29[5]). The two letters identify cloud type as shown in the same table. The last three digits are cloud height in hundreds of feet above ground level (AGL). In the KSTL TAF, "9// 005" means sky obscured (9), clouds not observed (//), vertical visibility 500 feet (005). The TAF may include as many cloud groups as necessary to describe expected sky condition.

(g) Expected variation from prevailing conditions. Variations from prevailing conditions are identified by the contractions INTER and TEMPO as defined earlier. In the KSTL TAF, "INTER 1215 0000 75XXSN 9//000" means intermittently from 1200Z to 1500Z (1215) visibility zero meters (0000) or zero miles, heavy snow (75XXSN), sky obscured, clouds not observed, vertical visibility zero (9//000).

(h), (i), (j), (k), and (l) An expected change in prevailing conditions is indicated by the contraction GRADU and RAPID as defined earlier. In the KSM TAF, "GRADU 1516 33020 4800 38BLSN 7SC030" means a gradual change between 1500Z and 1600Z to wind 330 degrees at 20 knots, visibility 4,800 meters or 3 miles (See Table 7–29[2]), blowing snow, 7/8 stratocumulus (See Table 7–29[5]) at 3000 feet AGL. "PROB 40 85SNSH" means there is a 30 to 50% probability that light snow showers will occur between 1600Z and 2100Z. "GRADU 2122 33015 9999 WX NIL 3SC030" means a gradual change between 2100Z and 2200Z to wind 330 degrees at 15 knots, visibility 10 kilometers or more (more than 6 miles), no significant weather, 3/8 stratocumulus at 3000 feet. "RAPID 00 VRB05 9999 SKC" means a rapid change about 0000Z to wind direction variable at 5 knots, visibility more than 6 miles, sky clear. "GRADU 0304 24015/25 CAVOK φ " means a gradual change between 0300Z and 0400Z to wind 240 degrees at 15 knots, peak gust to 25 knots with CAVOK conditions. φ means end of message.

(m) *Icing.* An icing group may be included. It is a 6–digit group. The first digit is always a 6, identifying it as an icing group. The second digit is the type of ice accretion from Table 7–29[6]. The next three digits are height of the base of the icing layer in hundreds of feet (AGL). The last digit is the thickness of the layer in thousands of feet. For example, let's decode the group "680304". The "6" indicates an icing forecast; the"8" indicates severe icing in cloud; "030" says the base of the icing is at 3,000 feet (AGL); and "4" specifies a layer 4,000 feet thick.

(n) *Turbulence.* A turbulence group also may be included. It also is a 6–digit group coded the same as the icing group except a "5" identifies the group as a turbulence forecast. Type of turbulence is from Table 7–29[7]. For example, decoding the group "590359", the "5" indentifies a turbulence forecast; the "9" specifies frequent severe turbulence in cloud (See Table 7–29[5]); "035" says the base of the turbulent layer is 3,500 feet (AGL); the "9" specifies that the turbulence layer is 9,000 feet thick. When either an icing layer or a turbulent layer is expected to be more than 9,000 feet thick, multiple groups are used. The top specified in one group is incident with the base in the following group. It's assume the forecaster expects frequent turbulence from the surface to 45,000 feet with the most hazardous turbulence at mid–levels. This could be encoded "530005 550509 592309 553209 54104". While you most likely will never see such a complex coding with

Table 7–29[4]

Converting significant weather from U.S. terms to ICAO terms.

Precipitation & Intensity				
US	TAF Code	Light	Moderate	Heavy
a	89GR			
BD or BN (vsby 5/16 to ½ mi)	31SA			
BD or BN (vsby 0 to ¼ mi)	34XXSA			
BS (vsby 6 mi or less)	38BLSN			
D (vsby 6 mi or less)	06HZ			
GF (vsby ½ mi or less)	44FG			
H (vsby 6 mi or less)	05HZ			
F or IF (vsby ½ mi or less)	45FG			
F, GF or IF (vsbv 5/8 to 6 mi)	10BR			
IP	79PE			
IPW	87GR			
K	04FU			
L		51DZ	53DZ	55DZ
R		61RA	63RA	64RA
RS		68RASN	68RASN	69XXRASN
RW	80RASH	80RASH	81XXSH	
RWSW		83RASN	83RASN	84XXRASN
S		71SN	73SN	75XXSN
SG		77SG		
SP	87GR			
SW		85SNSH	85SNSH	86XXSNSH
ZL		56FZDZ	56FZDZ	57XXFZDZ
ZR	66FZRA	67XXFZRA	67XXFZRA	
TRW– or TRW	95TS			INTER 81XXSH
TRW+	95TS			INTER 82XXSH*
TRW–A or TRWA	96TSGR			
T+RW	97XXTS			
T–RW+	97XXTS 97XXTS			INTER 81XXSH INTER 82XXSH*
T+RWA	99XXTSGR			
T+RW+A	99XXTSGR 99XXTSGR			INTER 81XXSH INTER 82XXSH*

*INTER 82XXSH is to be encoded in a TAF only when a violent rainshower (at least 1 inch of rain per hour or 0.10 inch in 6 minutes) is forecast.

NOTE: Conversions from TAF to FT will not be exact in some cases due to a lack of a one to one relationship.

this many groups, the flexible TAF code permits it.

Table 7–29[5]
TAF Cloud Code

Code	Cloud amount	Cloud Type
0	0 (clear)	CI Cirrus
1	1 octa or less but not zero	CC Cirrocumulus
2	2 octas	CS Cirrostratus
3	3 octas	AC Altocumulus
4	4 octas	AS Altostratus
5	5 octas	NS Nimbostratus
6	6 octas	SC Stratocumulus
7	7 octas or more but not 8 octas	ST Stratus CU Cumulus
8	8 octas (overcast)	CB Cumulonimbus
9	Sky obscured or cloud	// cloud not visible amount not estimated due to darkness or obscuring phenomena

Table 7–29[6]
TAF ICING

Figure Code	Amount of ice accretion
0	No icing
I	Light icing
2	Light icing in cloud
3	Light icing in precipitation
4	Moderate icing
5	Moderate icing in cloud
6	Moderate icing in precipitation
7	Severe icing
8	Severe icing in cloud
9	Severe icing in precipitation

Table 7–29[7]
TURBULENCE

Figure Code	Turbulence
0	None
1	Light Turbulence
2	Moderate turbulence in clear air, infrequent
3	Moderate turbulence in clear air, frequent
4	Moderate turbulence in cloud, infrequent
5	Moderate turbulence in cloud, frequent,
6	Severe turbulence in clear air, infrequent
7	Severe turbulence in clear air, frequent
8	Severe turbulence in cloud, infrequent
9	Severe turbulence in cloud, frequent

(o) *Temperature.* A temperature code is seldom included in a terminal forecast. However, it may be included if critical to aviation. It may be used to alert the pilot to high density altitude or possible frost when on the ground. The temperature group is identified by the digit "0". The next two (2) digits are the time to the nearest whole hour (GMT) to which the forecast temperature applies. The last two (2) digits are temperature in degrees Celsius. A minus temperature is preceded by the letter "M".

EXAMPLE:

"02137" means temperature at 2100Z is expected to be 37 degrees Celsius (about 99 degrees F); "012M02" means temperature at 1200Z is expected to be minus 2 degrees Celsius. A forecast may include more than one temperature group.

7–30. RESERVED

Section 2. ALTIMETER SETTING PROCEDURES

7–31. GENERAL

a. The accuracy of aircraft altimeters is subject to the following factors:

1. Nonstandard temperatures of the atmosphere.

2. Nonstandard atmospheric pressure.

3. Aircraft static pressure systems (position error), and

4. Instrument error.

b. EXTREME CAUTION SHOULD BE EXERCISED WHEN FLYING IN PROXIMITY TO OBSTRUCTIONS OR TERRAIN IN LOW TEMPERATURES AND PRESSURES. This is especially true in extremely cold temperatures that cause a large differential between the Standard Day temperature and actual temperature. This circumstance can cause serious errors that result in the aircraft being significantly lower than the indicated altitude.

7–31b NOTE—Standard temperature at sea level is 15 degrees Celsius (59 degrees Fahrenheit). The temperature gradient from sea level is minus 2 degrees Celsius (3.6 degrees Fahrenheit) per 1,000 feet. Pilots should apply corrections for static pressure systems and/or instruments, if appreciable errors exist.

c. The adoption of a standard altimeter setting at the higher altitudes eliminates station barometer errors, some altimeter instrument errors, and errors caused by altimeter settings derived from different geographical sources.

7–32. PROCEDURES

The cruising altitude or flight level of aircraft shall be maintained by reference to an altimeter which shall be set, when operating:

a. Below 18,000 feet MSL:

1. When the barometric pressure is 31.00 inches Hg. or less—to the current reported altimeter setting of a station along the route and within 100 NM of the aircraft, or if there is no station within this area, the current reported altimeter setting of an appropriate available station. When an aircraft is en route on an instrument flight plan, air traffic controllers will furnish this information to the pilot at least once while the aircraft is in the controllers area of jurisdiction. In the case of an aircraft not equipped with a radio, set to the elevation of the departure airport or use an appropriate altimeter setting available prior to departure.

2. When the barometric pressure exceeds 31.00 inches Hg.—the following procedures will be placed in effect by NOTAM defining the geographic area affected:

(a) **For all aircraft**—Set 31.00 inches for en route operations below 18,000 feet MSL. Maintain this setting until beyond the affected area or until reaching final approach segment. At the beginning of the final approach segment, the current altimeter setting will be set, if possible. If not possible, 31.00 inches will remain set throughout the approach. Aircraft on departure or missed approach will set 31.00 inches prior to reaching any mandatory/crossing altitude or 1,500 feet AGL, whichever is lower. (Air traffic control will issue actual altimeter settings and advise pilots to set 31.00 inches in their altimeters for en route operations below 18,000 feet MSL in affected areas.)

(b) During preflight, barometric altimeters shall be checked for normal operation to the extent possible.

(c) For aircraft with the capability of setting the current altimeter setting and operating into airports with the capability of measuring the current altimeter setting, no additional restrictions apply.

(d) For aircraft operating VFR, there are no additional restrictions, however, extra diligence in flight planning and in operating in these conditions is essential.

(e) Airports unable to accurately measure barometric pressures above 31.00 inches of Hg. will report the barometric pressure as "missing" or "in excess of 31.00 inches of Hg." Flight operations to and from those airports are restricted to VFR weather conditions.

(f) For aircraft operating IFR and unable to set the current altimeter setting, the following restrictions apply:

(1) To determine the suitability of departure alternate airports, destination airports, and destination alternate airports, increase ceiling requirements by 100 feet and visibility requirements by ¼ statute mile for each ¹⁄₁₀ of an inch of Hg., or any portion thereof, over 31.00 inches. These adjusted values are then applied in accordance with the requirements of the applicable operating regulations and operations specifications.

EXAMPLE:

Destination altimeter is 31.28 inches, ILS DH 250 feet (200–½). When flight planning, add 300–¾ to the weather requirements which would become 500–1¼.

(2) On approach, 31.00 inches will remain set. Decision Height or minimum descent altitude shall be deemed to have been reached when the published altitude is displayed on the altimeter.

7–32a1f2 NOTE—Although visibility is normally the limiting factor on an approach, pilots should be aware that when reaching DH the aircraft will be higher than indicated. Using the example above the aircraft would be approximately 300 feet higher.

(3) These restrictions do not apply to authorized Category II and III ILS operations nor do they apply to certificate holders using approved QFE altimetry systems.

(g) The FAA Regional Flight Standards Division Manager of the Affected area is authorized to approve temporary waivers to permit emergency resupply or emergency medical service operation.

b. At or above 18,000 feet MSL—to 29.92 inches of mercury (standard setting). The lowest usable flight level is determined by the atmospheric pressure in the area of operation as shown in Table 7–32[1].

Table 7–32[1]

Altimeter Setting (Current Reported)	Lowest Usable Flight Level
29.92 or higher	180
29.91 to 29.42	185
29.41 to 28.92	190
28.91 to 28.42	195
28.41 to 27.92	200

c. Where the minimum altitude, as prescribed in FAR 91.159 and FAR 91.177, is above 18,000 feet MSL, the lowest usable flight level shall be the flight level equivalent of the minimum altitude plus the number of feet specified in Table 7–32[2].

Table 7–32[2]

Altimeter Setting	Correction Factor
29.92 or higher	none
29.91 to 29.42	500 Feet
29.41 to 28.92	1000 Feet
28.91 to 28.42	1500 Feet
28.41 to 27.92	2000 Feet
27.91 to 27.42	2500 Feet

EXAMPLE:

The minimum safe altitude of a route is 19,000 feet MSL and the altimeter setting is reported between 29.92 and 29.42 inches of mercury, the lowest usable flight level will be 195, which is the flight level equivalent of 19,500 feet MSL (minimum altitude plus 500 feet).

7–33. ALTIMETER ERRORS

a. Most pressure altimeters are subject to mechanical, elastic, temperature, and installation errors. (Detailed information regarding the use of pressure altimeters is found in the Instrument Flying Handbook, Chapter IV.) Although manufacturing and installation specifications, as well as the periodic test and inspections required by regulations (Far 43, Appendix E), act to reduce these errors, any scale error may be observed in the following manner:

1. Set the current reported altimeter setting on the altimeter setting scale.

2. Altimeter should now read field elevation if you are located on the same reference level used to establish the altimeter setting.

3. Note the variation between the known field elevation and the altimeter indication. If this variation is in the order of plus or minus 75 feet, the accuracy of the altimeter is questionable and the problem should be referred to an appropriately rated repair station for evaluation and possible correction.

b. Once in flight, it is very important to frequently obtain current altimeter settings en route. If you do not reset your altimeter when flying *from* an area of high pressure or high temperatures *into* an area of low pressures or low temperature, *your aircraft will be closer to the surface than the altimeter indicates.* An inch error on the altimeter equals 1,000 feet of altitude. To quote an old saying: **"GOING FROM A HIGH TO A LOW, LOOK OUT BELOW."**

c. A reverse situation, without resetting the altimeter when going from a low temperature or low pressure area into a high temperature or high pressure area, the aircraft will be higher than the altimeter indicates.

d. The possible results of the above situations is obvious, particularly if operating at the minimum altitude or when conducting an instrument approach. If the altimeter is in error, you may still be on instruments when reaching the minimum altitude (as indicated on the altimeter), whereas you might have been in the clear and able to complete the approach if the altimeter setting was correct.

7–34. HIGH BAROMETRIC PRESSURE

a. Cold, dry air masses may produce barometric pressures in excess of 31.00 inches of Mercury, and many altimeters do not have an accurate means of being adjusted for settings of these levels. When the altimeter cannot be set to the higher pressure setting, the aircraft actual altitude will be higher than the altimeter indicates. (Reference—Altimeter Errors, paragraph 7–33).

b. When the barometric pressure exceeds 31.00 inches, air traffic controllers will issue the actual altimeter setting, and:

1. En Route/Arrivals—Advise pilots to remain set on 31.00 inches until reaching the final approach segment.

2. Departures—Advise pilots to set 31.00 inches prior to reaching any mandatory/crossing altitude or 1,500 feet, whichever is lower.

c. The altimeter error caused by the high pressure will be in the opposite direction to the error caused by the cold temperature.

7–35. LOW BAROMETRIC PRESSURE

When abnormally low barometric pressure conditions occur (below 28.00), flight operations by aircraft unable to set the actual altimeter setting are not recommended.

7–35 Note—The true altitude of the aircraft is **lower** than the indicated altitude if the pilot is unable to set the actual altimeter setting.

7–36 thru 7–40. RESERVED

Section 3. WAKE TURBULENCE

7–41. GENERAL

a. Every aircraft generates a wake while in flight. Initially, when pilots encountered this wake in flight, the disturbance was attributed to "prop wash." It is known, however, that this disturbance is caused by a pair of counter rotating vortices trailing from the wing tips. The vortices from larger aircraft pose problems to encountering aircraft. For instance, the wake of these aircraft can impose rolling moments exceeding the roll control authority of the encountering aircraft. Further, turbulence generated within the vortices can damage aircraft components and equipment if encountered at close range. The pilot must learn to envision the location of the vortex wake generated by larger (transport category) aircraft and adjust the flight path accordingly.

b. During ground operations and during takeoff, jet engine blast (thrust stream turbulence) can cause damage and upsets if encountered at close range. Exhaust velocity versus distance studies at various thrust levels have shown a need for light aircraft to maintain an adequate separation behind large turbojet aircraft. Pilots of larger aircraft should be particularly careful to consider the effects of their "jet blast" on other aircraft, vehicles, and maintenance equipment during ground operations.

7–42. VORTEX GENERATION

Figure 7–42[1]

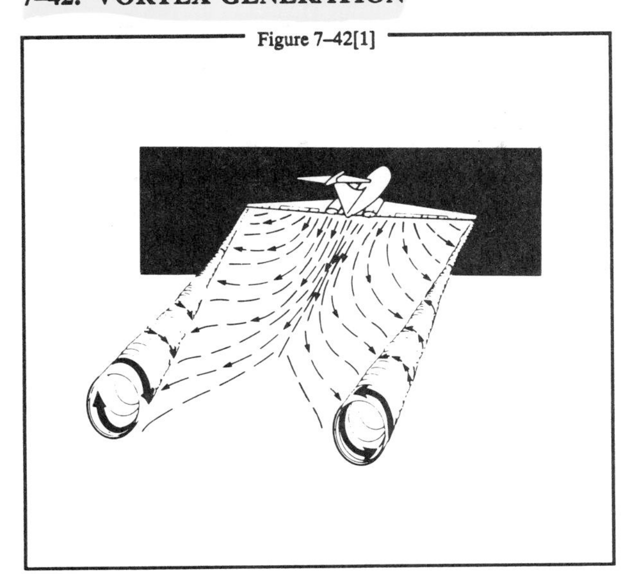

Lift is generated by the creation of a pressure differential over the wing surface. The lowest pressure occurs over the upper wing surface and the highest pressure under the wing. This pressure differential triggers the roll up of the airflow aft of the wing resulting in swirling air masses trailing downstream of the wing tips. After the roll up is completed, the wake consists of two counter rotating cylindrical vortices. (See Figure 7–42[1].) Most of the energy is within a few feet of the center of each vortex, but pilots should avoid a region within about 100 feet of the vortex core.

7–43. VORTEX STRENGTH

a. The strength of the vortex is governed by the weight, speed, and shape of the wing of the generating aircraft. The vortex characteristics of any given aircraft can also be changed by extension of flaps or other wing configuring devices as well as by change in speed. However, as the basic factor is weight, the vortex strength increases proportionately. Peak vortex tangential speeds up to almost 300 feet per second have been recorded. The greatest vortex strength occurs when the generating aircraft is HEAVY, CLEAN, and SLOW.

b. INDUCED ROLL

Figure 7–43[1]

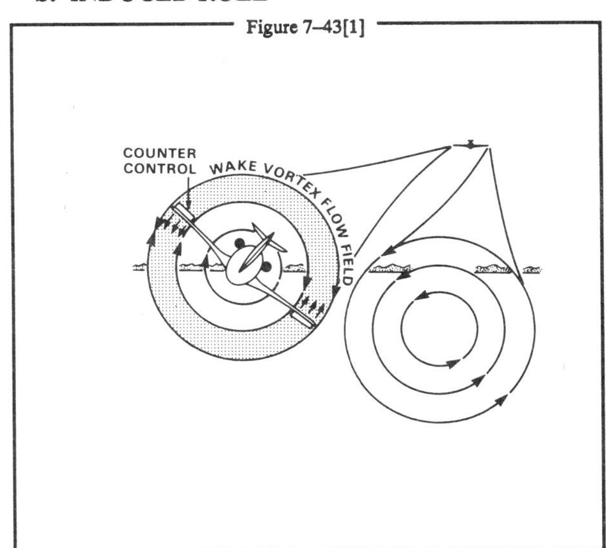

1. In rare instances a wake encounter could cause in-flight structural damage of catastrophic proportions. However, the usual hazard is associated with induced rolling moments which can exceed the roll control authority of the encountering aircraft. In flight experiments, aircraft have been intentionally flown directly up trailing vortex cores of larger aircraft. It was shown that the capability of an aircraft to counteract the roll imposed by the wake vortex primarily depends on the wing span and counter control responsiveness of the encountering aircraft.

2. Counter control is usually effective and induced roll minimal in cases where the wing span and ailerons of the encountering aircraft extend beyond the rotational flow field of the vortex. It is more difficult for aircraft with short wing span (relative to the generating aircraft) to counter the imposed roll induced by vortex flow. Pilots of short span aircraft, even of the high performance type, must be especially alert to vortex encounters. (See Figure 7–43[1].)

3. The wake of larger aircraft requires the respect of all pilots.

7–44. VORTEX BEHAVIOR

a. Trailing vortices have certain behavioral characteristics which can help a pilot visualize the wake location and thereby take avoidance precautions.

Figure 7–44[1]

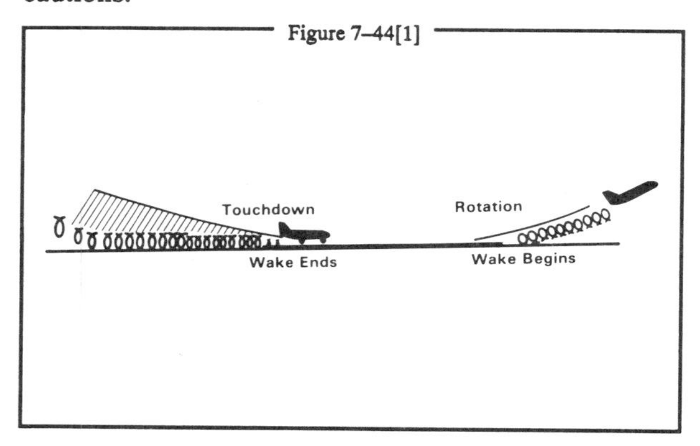

1. Vortices are generated from the moment aircraft leave the ground, since trailing vortices are a by-product of wing lift. Prior to takeoff or touchdown pilots should note the rotation or touchdown point of the preceding aircraft. (See Figure 7–44[1][Wake Begins/Ends].)

2. The vortex circulation is outward, upward and around the wing tips when viewed from either ahead or behind the aircraft. Tests with large aircraft have shown that the vortices remain spaced a bit less than a wing span apart, drifting with the wind, at altitudes greater than a wing span from the ground. In view of this, if persistent vortex turbulence is encountered, a slight change of altitude and lateral position (preferably upwind) will provide a flight path clear of the turbulence.

3. Flight tests have shown that the vortices from larger (transport category) aircraft sink at a rate of several hundered feet per minute, slowing their descent and diminishing in strength with time and distance behind the generating aircraft. Atmospheric turbulence hastens breakup. Pilots should fly at or above the preceding aircraft's

Figure 7–44[2]

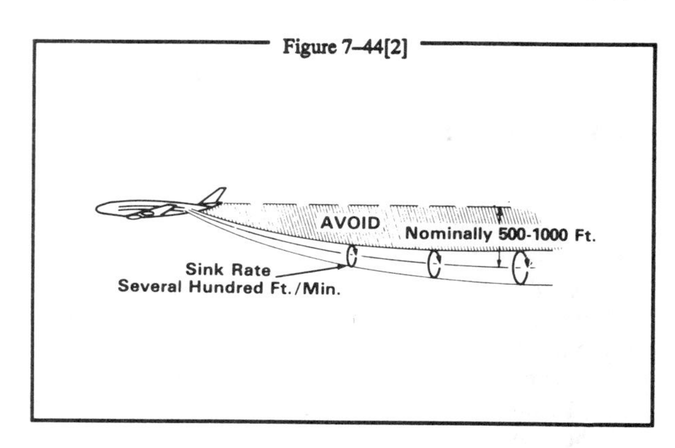

flight path, altering course as necessary to avoid the area behind and below the generating aircraft. (See Figure 7–44[2][Vortex Flow Field].) However vertical separation of 1,000 feet may be considered safe.

Figure 7–44[3]

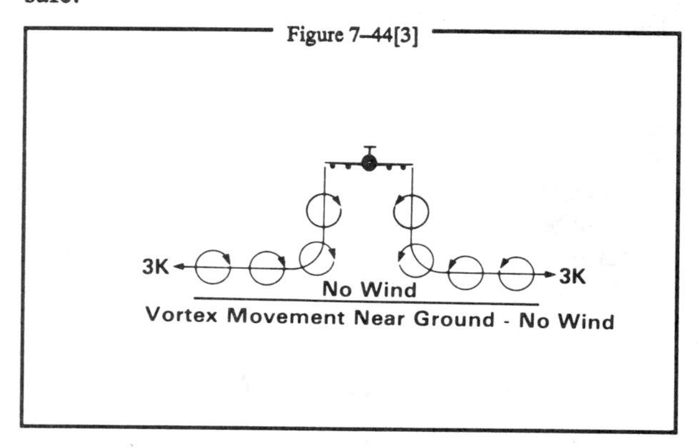

4. When the vortices of larger aircraft sink close to the ground (within 100 to 200 feet), they tend to move laterally over the ground at a speed of 2 or 3 knots. (See Figure 7–44[3][Vortex Sink Rate].)

Figure 7–44[4]

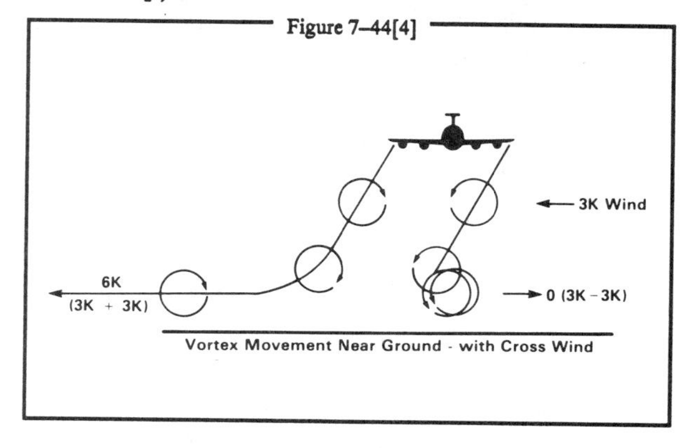

b. A crosswind will decrease the lateral movement of the upwind vortex and increase the movement of the downwind vortex. Thus a light wind with a cross runway component of 1 to 5 knots could result in the upwind vortex remaining in the touchdown zone for a period of time and hasten the drift of the downwind vortex toward another runway. (See Figure 7–44[4][Vortex Movement in Ground Effect (No Wind)].) Similarly, a tailwind condition can move the vortices of the preceding aircraft forward into the touchdown zone. THE LIGHT QUARTERING TAILWIND REQUIRES MAXIMUM CAUTION. Pilots should be alert to large aircraft upwind from their approach and takeoff flight paths. (See Figure 7–44[5][Vortex Movement in Ground Effect (Wind)].)

Figure 7–44[5]

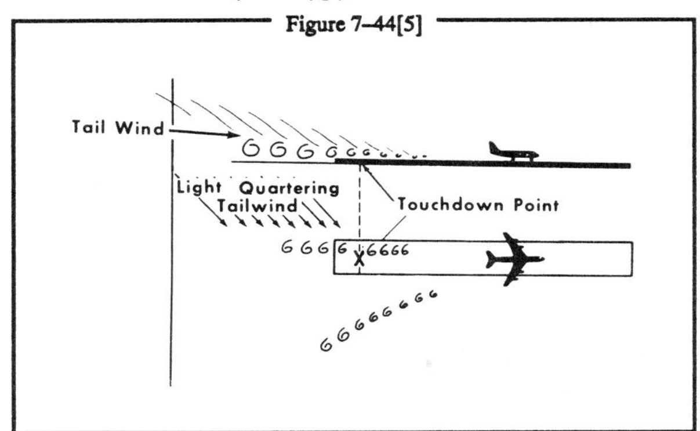

7–45. OPERATIONS PROBLEM AREAS

a. A wake encounter can be catastrophic. In 1972 at Fort Worth a DC–9 got too close to a DC–10 (two miles back), rolled, caught a wingtip, and cartwheeled coming to rest in an inverted position on the runway. All aboard were killed. Serious and even fatal GA accidents induced by wake vortices are not uncommon. However, a wake encounter is not necessarily hazardous. It can be one or more jolts with varying severity depending upon the direction of the encounter, weight of the generating aircraft, size of the encountering aircraft, distance from the generating aircraft, and point of vortex encounter. The probability of induced roll increases when the encountering aircraft's heading is generally aligned with the flight path of the generating aircraft.

b. AVOID THE AREA BELOW AND BEHIND THE GENERATING AIRCRAFT, ESPECIALLY AT LOW ALTITUDE WHERE EVEN A MOMENTARY WAKE ENCOUNTER COULD BE HAZARDOUS.

c. Pilots should be particularly alert in calm wind conditions and situations where the vortices could:

1. Remain in the touchdown area.

2. Drift from aircraft operating on a nearby runway.

3. Sink into the takeoff or landing path from a crossing runway.

4. Sink into the traffic pattern from other airport operations.

5. Sink into the flight path of VFR aircraft operating on the hemispheric altitude 500 feet below.

d. Pilots of all aircraft should visualize the location of the vortex trail behind larger aircraft and use proper vortex avoidance procedures to achieve safe operation. It is equally important that pilots of larger aircraft plan or adjust their flight paths to minimize vortex exposure to other aircraft.

7–46. VORTEX AVOIDANCE PROCEDURES

a. Under certain conditions, airport traffic controllers apply procedures for separating IFR aircraft. The controllers will also provide to VFR aircraft, with whom they are in communication and which in the tower's opinion may be adversely affected by wake turbulence from a larger aircraft, the position, altitude and direction of flight of larger aircraft followed by the phrase "CAUTION—WAKE TURBULENCE." WHETHER OR NOT A WARNING HAS BEEN GIVEN, HOWEVER, THE PILOT IS EXPECTED TO ADJUST HIS OR HER OPERATIONS AND FLIGHT PATH AS NECESSARY TO PRECLUDE SERIOUS WAKE ENCOUNTERS.

b. The following vortex avoidance procedures are recommended for the various situations:

1. Landing behind a larger aircraft—same runway: Stay at or above the larger aircraft's final approach flight path—note its touchdown point—land beyond it.

2. Landing behind a larger aircraft—when parallel runway is closer than 2,500 feet: Consider possible drift to your runway. Stay at or above the larger aircraft's final approach flight path—note its touchdown point.

3. Landing behind a larger aircraft—crossing runway: Cross above the larger aircraft's flight path.

4. Landing behind a departing larger aircraft—same runway: Note the larger aircraft's rotation point—land well prior to rotation point.

5. Landing behind a departing larger aircraft—crossing runway: Note the larger aircraft's rotation point—if past the intersection—continue the approach—land prior to the intersection. If larger aircraft rotates prior to the intersection, avoid flight below the larger aircraft's flight path. Abandon the approach unless a landing is ensured well before reaching the intersection.

6. Departing behind a larger aircraft: Note the larger aircraft's rotation point—rotate prior to larger aircraft's rotation point—continue climb above the larger aircraft's climb path until turning clear of his wake. Avoid subsequent headings which will cross below and behind a larger aircraft. Be alert for any critical takeoff situation which could lead to a vortex encounter.

7. Intersection takeoffs—same runway: Be alert to adjacent larger aircraft operations, particularly upwind of your runway. If intersection takeoff clearance is received, avoid subsequent heading which will cross below a larger aircraft's path.

8. Departing or landing after a larger aircraft executing a low approach, missed approach or touch-and-go landing: Because vortices settle and move laterally near the ground, the vortex hazard may exist along the runway and in your flight path after a larger aircraft has executed a low approach, missed approach or a touch-and-go landing, particular in light quartering wind conditions. You should ensure that an interval of at least 2 minutes has elapsed before your takeoff or landing.

9. En route VFR (thousand-foot altitude plus 500 feet): Avoid flight below and behind a large aircraft's path. If a larger aircraft is observed above on the same track (meeting or overtaking) adjust your position laterally, preferably upwind.

7–47. HELICOPTERS

In a slow hover taxi or stationary hover near the surface, helicopter main rotor(s) generate downwash producing high velocity outwash vortices to a distance approximately three times the diameter of the rotor. When rotor downwash hits the surface, the resulting outwash vortices have behavioral characteristics similar to wing tip vortices produced by fixed wing aircraft. However, the vortex circulation is outward, upward, around, and away from the main rotor(s) in all directions. Pilots of small aircraft should avoid operating within three rotor diameters of any helicopter in a slow hover taxi or stationary hover. In forward flight, departing or landing helicopters produce a pair of strong, high-speed trailing vortices similar to wing tip vortices of larger fixed wing aircraft. Pilots of small aircraft should use caution when operating behind or crossing behind landing and departing helicopters.

7–48. PILOT RESPONSIBILITY

a. Government and industry groups are making concerted efforts to minimize or eliminate the hazards of trailing vortices. However, the flight disciplines necessary to ensure vortex avoidance during VFR operations must be exercised by the pilot. Vortex visualization and avoidance procedures should be exercised by the pilot using the same degree of concern as in collision avoidance.

b. Wake turbulence may be encountered by aircraft in flight as well as when operating on the airport movement area. (Reference—Pilot/Controller Glossary, Wake Turbulence).

c. Pilots are reminded that in operations conducted behind all aircraft, acceptance of instructions from ATC in the following situations is an acknowledgment that the pilot will ensure safe takeoff and landing intervals and accepts the responsibility of providing his own wake turbulence separation.

1. Traffic information,

2. Instructions to follow an aircraft, and

3. The acceptance of a visual approach clearance.

d. For operations conducted behind heavy aircraft, ATC will specify the word "heavy" when this information is known. Pilots of heavy aircraft should always use the word "heavy" in radio communications.

7–49. AIR TRAFFIC WAKE TURBULENCE SEPARATIONS

a. Because of the possible effects of wake turbulence, controllers are required to apply no less than specified minimum separation for aircraft operating behind a heavy jet and, in certain instances, behind large nonheavy aircraft.

1. Separation is applied to aircraft operating directly behind a heavy jet at the same altitude or less than 1,000 feet below:

(a) Heavy jet behind heavy jet—4 miles.

(b) Small/large aircraft behind heavy jet—5 miles.

2. Also, separation, measured at the time the preceding aircraft is over the landing threshold, is provided to small aircraft:

(a) Small aircraft landing behind heavy jet—6 miles.

(b) Small aircraft landing behind large aircraft—4 miles.

7–49a2b NOTE—See Pilot/Controller Glossary, Aircraft Classes.

3. Additionally, appropriate time or distance intervals are provided to departing aircraft:

(a) Two minutes or the appropriate 4 or 5 mile radar separation when takeoff behind a heavy jet will be:

—from the same threshold

—on a crossing runway and projected flight paths will cross

—from the threshold of a parallel runway when staggered ahead of that of the adjacent runway by less than 500 feet and when the runways are separated by less than 2,500 feet.

7–49a3a NOTE—Pilots, after considering possible wake turbulence effects, may specifically request waiver of the 2–minute interval by stating, "request waiver of 2–minute interval" or a similar statement. Controllers may acknowledge this statement as pilot acceptance of responsibility for wake turbulence separation and, if traffic permits, issue takeoff clearance.

b. A 3–minute interval will be provided when a small aircraft will takeoff:

1. From an intersection on the same runway (same or opposite direction) behind a departing large aircraft,

2. In the opposite direction on the same runway behind a large aircraft takeoff or low/missed approach.

7–49b2 NOTE—This 3–minute interval may be waived upon specific pilot request.

c. A 3–minute interval will be provided for all aircraft taking off when the operations are as described in b(1) and (2) above, the preceding aircraft is a heavy jet, and the operations are on either the same runway or parallel runways separated by less than 2,500 feet. Controllers may not reduce or waive this interval.

d. Pilots may request additional separation i.e., 2 minutes instead of 4 or 5 miles for wake turbulence avoidance. This request should be made as soon as practical on ground control and at least before taxiing onto the runway.

7–49d NOTE—FAR 91.3(a) states: "The pilot in command of an aircraft is directly responsible for and is the final authority as to the operation of that aircraft."

7–50 thru 7–60. RESERVED

SECTION 4. BIRD HAZARDS AND FLIGHT OVER NATIONAL REFUGES, PARKS, AND FORESTS

7–61. MIGRATORY BIRD ACTIVITY

a. Bird strike risk increases because of bird migration during the months of March through April, and August through November.

b. The altitudes of migrating birds vary with winds aloft, weather fronts, terrain elevations, cloud conditions, and other environmental variables. While over 90 percent of the reported bird strikes occur at or below 3,000 feet AGL, strikes at higher altitudes are common during migration. Ducks and geese are frequently observed up to 7,000 feet AGL and pilots are cautioned to minimize en route flying at lower altitudes during migration.

c. Considered the greatest potential hazard to aircraft because of their size, abundance, or habit of flying in dense flocks are gulls, waterfowl, vultures, hawks, owls, egrets, blackbirds, and starlings. Four major migratory flyways exist in the United States. The Atlantic flyway parallels the Atlantic Coast. The Mississippi Flyway stretches from Canada through the Great Lakes and follows the Mississippi River. The Central Flyway represents a broad area east of the Rockies, stretching from Canada through Central America. The Pacific Flyway follows the west coast and overflies major parts of Washington, Oregon, and California. There are also numerous smaller flyways which cross these major north-south migratory routes.

7–62. REDUCING BIRD STRIKE RISKS

a. The most serious strikes are those involving ingestion into an engine (turboprops and turbine jet engines) or windshield strikes. These strikes can result in emergency situations requiring prompt action by the pilot.

b. Engine ingestions may result in sudden loss of power or engine failure. Review engine out procedures, especially when operating from airports with known bird hazards or when operating near high bird concentrations.

c. Windshield strikes have resulted in pilots experiencing confusion, disorientation, loss of communications, and aircraft control problems. Pilots are encouraged to review their emergency procedures before flying in these areas.

d. When encountering birds en route, climb to avoid collision, because birds in flocks generally distribute themselves downward, with lead birds being at the highest altitude.

e. Avoid overflight of known areas of bird concentration and flying at low altitudes during bird migration. Charted wildlife refuges and other natural areas contain unusually high local concentration of birds which may create a hazard to aircraft.

7–63. REPORTING BIRD STRIKES

Pilots are urged to report strikes using FAA Form 5200–7, Bird Strike/Incident Report. This form is available at any FSS, General Aviation District Office, Air Carrier District Office, or at an FAA Regional Office. The data derived from these reports is used to develop standards to cope with this extensive hazard to aircraft and for documentation of necessary habitat control on airports.

7–64. REPORTING BIRD AND OTHER WILDLIFE ACTIVITIES

If you observe birds or other animals on or near the runway, request airport management to disperse the wildlife before taking off. Also contact the nearest FAA ARTCC, FSS, or tower (including non-Federal towers) regarding large flocks of birds and report the:

1. geographic location
2. Bird type (geese, ducks, gulls, etc.)
3. Approximate numbers
4. Altitude
5. Direction of bird flight path

7–65. PILOT ADVISORIES ON BIRD AND OTHER WILDLIFE HAZARDS

Many airports advise pilots of other wildlife hazards caused by large animals on the runway through the Airport/Facility Directory and the NOTAM system. Collisions of landing and departing aircraft and animals on the runway are increasing and are not limited to rural airports. These accidents have also occurred at several major airports. Pilots should exercise extreme caution when warned of the presence of wildlife on and in the vicinity of airports. If you observe deer or other large animals in close proximity to movement areas, advise the FSS, tower, or airport management.

7–66. FLIGHTS OVER CHARTED U.S. WILDLIFE REFUGES, PARKS, AND FOREST SERVICE AREAS

a. The landing of aircraft is prohibited on lands or waters administered by the National Park Service, U.S. Fish and Wildlife Service, or U.S. Forest Service without authorization from the respective agency. Exceptions, including

1. when forced to land due to an emergency beyond the control of the operator,

2. at officially designated landing sites, or

3. an approved official business of the Federal Government.

b. All aircraft are requested to maintain a minimum altitude of 2,000 feet above the *surface* of the following: National Parks, Monuments, Seashores, Lakeshores, Recreation Areas and Scenic Riverways administered by the National Park Service, National Wildlife Refuges, Big Game Refuges, Game Ranges and Wildlife Ranges administered by the U.S. Fish and Wildlife Service, and Wilderness and Primitive areas administered by the U.S. Forest Service.

7–66b NOTE—FAA Advisory Circular 91–36, Visual Flight Rules (VFR) Flight Near Noise-Sensitive Areas, defines the *surface* of a National Park Area (including Parks, Forests, Primitive Areas, Wilderness Areas, Recreational Areas, National Seashores, National Monuments, National Lakeshores, and National Wildlife Refuge and Range Areas) as: the highest terrain within 2,000 feet laterally of the route of flight, *or* the upper-most rim of a canyon or valley.

c. Federal statutes prohibit certain types of flight activity and/or provide altitude restrictions over *designated* U.S. Wildlife Refuges, Parks, and Forest Service Areas. These designated areas, for example: Boundary Waters Canoe Wilderness Areas, Minnesota; Haleakala National Park, Hawaii; Yosemite National Park, California; are charted on Sectional Charts.

d. Federal regulations also prohibit airdrops by parachute or other means of persons, cargo, or objects from aircraft on lands administered by the three agencies without authorization from the respective agency. Exceptions include:

1. emergencies involving the safety of human life, or

2. threat of serious property loss.

7–67 thru 7–70. RESERVED.

Section 5. POTENTIAL FLIGHT HAZARDS

7–71. ACCIDENT CAUSE FACTORS

a. The 10 most frequent cause factors for General Aviation Accidents that involve the pilot-in-command are:

1. Inadequate preflight preparation and/or planning.

2. Failure to obtain and/or maintain flying speed.

3. Failure to maintain direction control.

4. Improper level off.

5. Failure to see and avoid objects or obstructions.

6. Mismanagement of fuel.

7. Improper in-flight decisions or planning.

8. Misjudgment of distance and speed.

9. Selection of unsuitable terrain.

10. Improper operation of flight controls.

b. This list remains relatively stable and points out the need for continued refresher training to establish a higher level of flight proficiency for all pilots. A part of the FAA's continuing effort to promote increased aviation safety is the General Aviation Accident Prevention Program. For information on Accident Prevention activities contact your nearest General Aviation or Flight Standards District Office.

c. ALERTNESS—Be alert at all times, especially when the weather is good. Most pilots pay attention to business when they are operating in full IFR weather conditions, but strangely, air collisions almost invariably have occurred under ideal weather conditions. Unlimited visibility appears to encourage a sense of security which is not at all justified. Considerable information of value may be obtained by listening to advisories being issued in the terminal area, even though controller workload may prevent a pilot from obtaining individual service.

d. GIVING WAY—If you think another aircraft is too close to you, give way instead of waiting for the other pilot to respect the right-of-way to which you may be entitled. It is a lot safer to pursue the right-of-way angle after you have completed your flight.

7–72. VFR IN CONGESTED AREAS

A high percentage of near midair collisions occur below 8,000 feet AGL and within 30 miles of an airport. When operating VFR in these highly congested areas, whether you intend to land at an airport within the area or are just flying through, it is recommended that extra vigilance be maintained and that you monitor an appropriate control frequency. Normally the appropriate frequency is an approach control frequency. By such monitoring action you can "get the picture" of the traffic in your area. When the approach controller has radar, radar traffic advisories may be given to VFR pilots upon request. (Reference—Radar Traffic Information Service, paragraph 4–13).

7–73. OBSTRUCTIONS TO FLIGHT

a. GENERAL

Many structures exist that could significantly affect the safety of your flight when operating below 500 feet above ground level (AGL), and particularly below 200 feet AGL. While FAR part 91.119 allows flight below 500 AGL when over sparcely populated areas or open water, such operations are very dangerous. At and below 200 feet AGL there are numerous power lines, antenna towers, etc., that are not marked and lighted as obstructions and therefore may not be seen in time to avoid a collision. Notices to Airmen (NOTAMS) are issued on those lighted structures experiencing temporary light outages. However, some time may pass before the FAA is notified of these outages, and the NOTAM issued, thus pilot vigilance is imperative.

b. Antenna Towers

Extreme caution should be exercised when flying less that 2,000 feet above ground level (AGL) because of numerous skeletal structures, such as radio and television antenna towers, that exceed 1,000 feet AGL with some extending higher than 2,000 feet AGL. Most skeletal structures are supported by guy wires which are very difficult to see in good weather and can be invisible at dusk or during periods of reduced visibility. These wires can extend about 1,500 feet horizontally from a structure; therefore, all skeletal structures should be avoided horizontally by at least 2,000 feet. Additionally, new towers may not be on your current chart because the information was not received prior to the printing of the chart.

c. Overhead Wires

Overhead transmission and utility lines often span approaches to runways, natural flyways such as lakes, rivers, gorges, and canyons, and cross other landmarks pilots frequently follow such as highways, railroad tracks, etc. As with antenna towers, these high voltage/power lines or the supporting structures of these lines may not always be readily visible and the wires may be virtually impossible to see under certain conditions. In some locations, the supporting structures of overhead transmission lines are equipped with unique sequence flashing white strobe light systems to indicate that there are wires between the structures. However, many power lines do not require notice to the FAA and, therefore, are not marked and/or lighted. Many of those that do require notice do not exceed 200 feet AGL or meet the Obstruction Standard of FAR part 77 and, therefore, are not marked and/or lighted. All pilots are cautioned to remain extremely vigilant for these power lines or their supporting structures when following natural flyways or during the approach and landing phase. This is particularly important for seaplane and/or float equipped aircraft when landing on, or departing from, unfamiliar lakes or rivers.

d. Other Objects/Structures

There are other objects or structures that could adversely affect your flight such as construction cranes near an airport, newly constructed buildings, new towers, etc. Many of these structures do not meet charting requirements or may not yet be charted because of the charting cycle. Some structures do not require obstruction marking and/or lighting and some may not be marked and lighted even though the FAA recommended it.

7–74. AVOID FLIGHT BENEATH UNMANNED BALLOONS

a. The majority of unmanned free balloons currently being operated have, extending below them, either a suspension device to which the payload or instrument package is attached, or a trailing wire antenna, or both. In many instances these balloon subsystems may be invisible to the pilot until his aircraft is close to the balloon, thereby creating a potentially dangerous situation. Therefore, good judgment on the part of the pilot dictates that aircraft should remain well clear of all unmanned free balloons and flight below them should be avoided at all times.

b. Pilots are urged to report any unmanned free balloons sighted to the nearest FAA ground facility with which communication is established. Such information will assist FAA ATC facilities to identify and flight follow unmanned free balloons operating in the airspace.

7–75. MOUNTAIN FLYING

a. Your first experience of flying over mountainous terrain (particularly if most of your flight time has been over the flatlands of the midwest) could be a never-to-be-forgotten nightmare if proper planning is not done and if you are not aware of the potential hazards awaiting. Those familiar section lines are not present in the mountains; those flat, level fields for forced landings are practically nonexistent; abrupt changes in wind direction and velocity occur; severe updrafts and downdrafts are common, particularly near or above abrupt changes of terrain such as cliffs or rugged areas; even the clouds look different and can build up with startling rapidity. Mountain flying need not be hazardous if you follow the recommendations below:

b. File a flight plan. Plan your route to avoid topography which would prevent a safe forced landing. The route should be over populated areas and well known mountain passes. Sufficient altitude should be maintained to permit gliding to a safe landing in the event of engine failure.

c. Don't fly a light aircraft when the winds aloft, at your proposed altitude, exceed 35 miles per hour. Expect the winds to be of much greater velocity over mountain passes than reported a few miles from them. Approach mountain passes with as much altitude as possible. Downdrafts of from 1,500 to 2,000 feet per minute are not uncommon on the leeward side.

d. Don't fly near or above abrupt changes in terrain. Severe turbulence can be expected, especially in high wind conditions.

e. Some canyons run into a dead end. Don't fly so far up a canyon that you get trapped. ALWAYS BE ABLE TO MAKE A 180 DEGREE TURN!

f. Plan your trip for the early morning hours. As a rule, the air starts to get bad at about 10 a.m., and grows steadily worse until around 4 p.m., then gradually improves until dark. Mountain flying at night in a single engine light aircraft is asking for trouble.

g. When landing at a high altitude field, the same indicated airspeed should be used as at low elevation fields. *Remember:* that due to the less dense air at altitude, this same indicated airspeed actually results in higher true airspeed, a faster landing speed, and more important, a longer landing distance. During gusty wind conditions which often prevail at high altitude fields, a power approach and power landing is recommended. Additionally, due to the faster groundspeed, your takeoff distance will increase considerably over that required at low altitudes.

h. Effects of Density Altitude. Performance figures in the aircraft owner's handbook for length of takeoff run, horsepower, rate of climb, etc., are generally based on standard atmosphere conditions (59 degrees Fahrenheit (15 degrees Celsius), pressure 29.92 inches of mercury)at sea level. However, inexperienced pilots, as well as experienced pilots, may run into trouble when they encounter an altogether different set of conditions. This is particularly true in hot weather and at higher elevations.

Aircraft operations at altitudes above sea level and at higher than standard temperatures are commonplace in mountainous areas. Such operations quite often result in a drastic reduction of aircraft performance capabilities because of the changing air density. Density altitude is a measure of air density. It is not to be confused with pressure altitude, true altitude or absolute altitude. It is not to be used as a height reference, but as a determining criteria in the performance capability of an aircraft. Air density decreases with altitude. As air density decreases, density altitude increases. The further effects of high temperature and high humidity are cumulative, resulting in an increasing high density altitude condition. High density altitude reduces all aircraft performance parameters. To the pilot, this means that the normal horsepower output is reduced, propeller efficiency is reduced and a higher true airspeed is required to sustain the aircraft throughout its operating parameters. It means an increase in runway length requirements for takeoff and landings, and decreased rate of climb. An average small airplane, for example, requiring 1,000 feet for takeoff at sea level under standard atmospheric conditions will require a takeoff run of approximately 2,000 feet at an operational altitude of 5,000 feet.

7–75h NOTE—A turbo-charged aircraft engine provides some slight advantage in that it provides sea level horsepower up to a specified altitude above sea level.

1. **Density Altitude Advisories**—at airports with elevations of 2,000 feet and higher, control towers and FSSs will broadcast the advisory "Check Density Altitude" when the temperature reaches a predetermined level. These advisories will be broadcast on appropriate tower frequencies or, where available, ATIS. FSSs will broadcast these advisories as a part of Airport Advisory Service, and on TWEB.

2. These advisories are provided by air traffic facilities, as a reminder to pilots that high temperatures and high field elevations will cause significant changes in aircraft characteristics. The pilot retains the responsibility to compute density altitude, when appropriate, as a part of preflight duties.

7–75h2 NOTE—All FSSs will compute the current density altitude upon request.

i. Mountain Wave. Many pilots go all their lives without understanding what a mountain wave is. Quite a few have lost their lives because of this lack of understanding. One need not be a licensed meteorologist to understand the mountain wave phenomenon.

1. Mountain waves occur when air is being blown over a mountain range or even the ridge of a sharp bluff area. As the air hits the upwind side of the range, it starts to climb, thus creating what is generally a smooth updraft which turns into a turbulent downdraft as the air passes the crest of the ridge. From this point, for many miles downwind, there will be a series of downdrafts and updrafts. Satellite photos of the Rockies have shown mountain waves extending as far as 700 miles downwind of the range. Along the east coast area, such photos of the Appalachian chain have picked up the mountain wave phenomenon over a hundred miles eastward. All it takes to form a mountain wave is wind blowing across the range at 15 knots or better at an intersection angle of not less than 30 degrees.

2. Pilots from flatland areas should understand a few things about mountain waves in order to stay out of trouble. When approaching a mountain range from the upwind side (generally the west), there will usually be a smooth updraft; therefore, it is not quite as dangerous an area as the lee of the range. From the leeward side, it is always a good idea to add an extra thousand feet or so of altitude because downdrafts can exceed the climb capability of the aircraft. Never expect an updraft when approaching a mountain chain from the leeward. Always be prepared to cope with a downdraft and turbulence.

3. When approaching a mountain ridge from the downwind side, it is recommended that the ridge be approached at approximately a 45 degrees angle to the horizontal direction of the ridge. This permits a safer retreat from the ridge with less stress on the aircraft should severe turbulence and downdraft be experienced. If severe turbulence is encountered, simultaneously reduce power and adjust pitch until aircraft approaches maneuvering speed, then adjust power and trim to maintain maneuvering speed and fly away from the turbulent area.

7–76. SEAPLANE SAFETY

a. Acquiring a seaplane class rating affords access to many areas not available to landplane pilots. Adding a seaplane class rating to your pilot certificate can be relatively uncomplicated and inexpensive. However, more effort is required to become a safe, efficient, competent "bush" pilot. The natural hazards of the backwoods have given way to modern man-made hazards. Except for the far north, the available bodies of water are no longer the exclusive domain of the airman. Seaplane pilots must be vigilant for hazards such as electric power lines, power, sail and rowboats, rafts, mooring lines, water skiers, swimmers, etc.

b. Seaplane pilots must have a thorough understanding of the right-of-way rules as they apply to aircraft versus boats. Once a seaplane has landed on the water, it is considered a vessel, and nautical rules as well as FAR apply. Seaplane pilots are expected to know and adhere to both Inland Navigation Rules and FAR 91.115 Right of Way Rules; Water Operations which states, in part, that aircraft on the water"...shall, insofar as possible, keep clear of all vessels and avoid impeding their navigation...." In general, while on the surface with engine running, an aircraft must give way to all non-powered vessels. Additionally, good operating procedures apply. Since a seaplane in the water is not as maneuverable as one in the air, the aircraft on the water has right-of-way over one in the air, and one taking off has right-of-way over one landing. Also, as is the case with all vessels, you may be held accountable for any damage caused by your wake while taxiing.

c. Unless they are under Federal jurisdiction, navigable bodies of water are under the jurisdiction of the state, or in a few cases, privately owned. Unless they are specifically restrictied, aircraft have as much right to operate on these bodies of water as other vessels. To avoid problems, check with Federal or local officials in advance of operating on unfamiliar waters. In addition to the agencies listed in Table 7-76[1], the nearest Flight Standards District Office can usually offer some practical suggestions as well as regulatory information. If you land on a restricted body of water because of an inflight emergency, or in ignorance of the restrictions you have violated, report as quickly as practical to the nearest local official having jurisdiction and explain your situation.

d. When operating over or into remote areas, appropriate attention should be given to survival gear. Minimum kits are recommended for summer and winter, and are required by law for flight into sparsely settled areas of Canada and Alaska. Alaska State Department of Transportation and Canadian Ministry of Transport officials can provide specific information on survival gear requirements. The kit should be assembled in one container and be easily reachable and preferably floatable.

e. United States Coast Guard (USCG) regulations require approved personal flotation devices (PFD) on all vessels including seaplanes operating on navigable waters of the United States. All PFDs must be in good and serviceable condition and of an appropriate size for the persons who intend to wear them. Wearable PFDs must be readily accessible and throwable devices must be immediately available for use. Seaplanes must have one USCG Type I, II, III, IV or V PFD on board for each occupant. One additional Type IV (approved device designed to be thrown to a person in the water) is also required. It is imperative that passengers be briefed on the location and proper use of available PFDs prior to leaving the dock. For additional information on approved PFDs contact your local State Boating Office or the USCG, Director of Auxiliary for your district.

Table 7-76[1]

AUTHORITY TO CONSULT FOR USE OF A BODY OF WATER

Location	Authority	Contact
Wilderness Area	U.S. Department of Agriculture, Forest Service	local forest ranger
National Forest	USDA Forest Service	local forest ranger
National Park	U.S. Department of the Interior, National Park Service	local park ranger
Indian Reservation	USDI, Bureau of Indian Affairs	local Bureau office
State Park	State government or state forestry or park service	local state aviation office for further information
Canadian National and Provincial Parks	Supervised and restricted on an individual basis from province to province and by different departments of the Canadian government; consult Canadian Flight Information Manual and/or Water Aerodrome Supplement	Park Superintendent in an emergency

7-77. FLIGHT OPERATIONS IN VOLCANIC DUST

a. Major volcanic eruptions have caused high altitude dust clouds from large amounts of discharged volcanic ash. Volcanic dust clouds create an extreme hazard to aircraft operating near the active volcanoes. Incidents have occurred while flying through volcanic dust clouds at night which resulted in either significant engine thrust loss and/or multiple engine shut downs along with the wing leading edges and windshields were abraded.

b. Pilots should be aware of the following information and instructed to take the appropriate action to preclude flying into volcanic dust clouds. Flight into an area of known volcanic activity that is producing volcanic dust clouds must be avoided. This is particularly important during the hours of darkness or in daytime instrument meteorological conditions when the volcanic dust cloud may not be detected by the flightcrew. If volcanic activity is reported, the planned flight should remain well clear of the area and, if possible, stay on the upwind side of the volcano. Airborne weather radar systems are not designed to detect volcanic dust clouds.

c. The following however, has been reported by flightcrews who have experienced encounters with volcanic dust clouds:

1. Smoke or dust appearing in the cockpit;

2. An acrid odor similar to electrical smoke;

3. Multiple engine malfunctions, such as stalls, increasing egt., torching from tailpipe, and flameouts;

4. At night, St. Elmo's fire or other static discharges accompanied by a bright orange glow in the engine inlets;

5. A fire warning in the forward cargo area.

d. Volcanic dust clouds may extend for several hundred miles. Volcanic dust can cause rapid erosion and damage to the internal components of engines with lost of thrust within 50 seconds. If volcanic dust is encountered, several major engine manufacturers recommend the following procedures:

1. Disengage autothrottle (if engaged). This will prevent the autothrottle from increasing engine thrust;

2. Reduce thrust to idle, altitude permitting. This will provide an additional engine stall margin and lower engine turbine temperatures;

3. Turn on continuous ignition.

4. Turn on all accessory airbleeds including all air conditioning packs, nacelles and wing anti-ice. This will provide an additional engine stall margin by reducing engine pressure.

5. Exit the cloud by the shortest route possible.

e. It may become necessary to shut down and then restart engines to prevent exceeding egt. limits. Volcanic dust may block the pitot system and result in unreliable airspeed indications. Pilots who operate into areas of volcanic activity should be aware of this information.

7–78. EMERGENCY AIRBORNE INSPECTION OF OTHER AIRCRAFT

a. Providing airborne assistance to another aircraft may involve formation flying. Most pilots receive little if any formal training or instruction in formation flying. Formation flying after a face to face planning session is difficult enough. Formation flying without sufficient time to plan (i.e., an emergency situation), coupled with the stress involved in a perceived emergency can be hazardous.

b. The pilot in command of the aircraft experiencing the problem/emergency must take the lead in coordinating the airborne intercept and inspection and take into account the unique flight characteristics and differences of the category(s) of aircraft involved.

c. Some of the safety considerations are:

1. Direction and speed of intercept;

2. Minimum separation distance;

3. Communications requirements, lost communication procedures; and

4. Emergency actions to terminate intercept.

d. Close proximity, in-flight inspection of another aircraft is uniquely hazardous. The pilot in command of the aircraft experiencing the problem/emergency must not relinquish his/her control of the situation and jeopardize the safety of his/her aircraft. The maneuver must be accomplished with minimum risk to both aircraft.

7–79 thru 7–80. RESERVED

Section 6. SAFETY, ACCIDENT, AND HAZARD REPORTS

7–81. AVIATION SAFETY REPORTING PROGRAM

a. The FAA has established a voluntary Aviation Safety Reporting Program designed to stimulate the free and unrestricted flow of information concerning deficiencies and discrepancies in the aviation system. This is a positive program intended to ensure the safest possible system by identifying and correcting unsafe conditions before they lead to accidents. The primary objective of the program is to obtain information to evaluate and enhance the safety and efficiency of the present system.

b. This program applies primarily to that part of the system involving the safety of aircraft operations, including departure, en route, approach and landing operations and procedures, ATC procedures, pilot/controller communications, the aircraft movement area of the airport, and near midair collisions. Pilots, air traffic controllers, and all other members of the aviation community and the general public are asked to file written reports of any discrepancy or deficiency noted in these areas.

c. The report should give the date, time, location, persons and aircraft involved (if applicable), nature of the event, and all pertinent details.

d. To ensure receipt of this information, the program provides for the waiver of certain disciplinary actions against persons, including pilots and air traffic controllers, who file timely written reports concerning potentially unsafe incidents. To be considered timely, reports must be delivered or postmarked within 10 days of the incident unless that period is extended for good cause. Reporting forms are available at FAA facilities.

e. The FAA utilizes the National Aeronautics and Space Administration (NASA) to act as an independent third party to receive and analyze reports submitted under the program. This program is described in ADVISORY CIRCULAR 00–46.

7–82. AIRCRAFT ACCIDENT AND INCIDENT REPORTING

a. Occurrences Requiring Notification—The operator of an aircraft shall immediately, and by the most expeditious means available, notify the nearest National Transportation Safety Board (NTSB) Field Office when:

1. An aircraft accident or any of the following listed incidents occur:

(a) Flight control system malfunction or failure;

(b) Inability of any required flight crew member to perform his normal flight duties as a result of injury or illness;

(c) Failure of structural components of a turbine engine excluding compressor and turbine blades and vanes.

(d) In-flight fire;

(e) Aircraft collide in flight.

(f) Damage to property, other than the aircraft, estimated to exceed $25,000 for repair (including materials and labor) or fair market value in the event of total loss, whichever is less.

(g) For large multi-engine aircraft (more than 12,500 pounds maximum certificated takeoff weight):

(1) In-flight failure of electrical systems which requires the sustained use of an emergency bus powered by a back-up source such as a battery, auxiliary power unit, or air-driven generator to retain flight control or essential instruments;

(2) In-flight failure of hydraulic systems that results in sustained reliance on the sole remaining hydraulic or mechanical system for movement of flight control surfaces;

(3) Sustained loss of the power or thrust produced by two or more engines; and

(4) An evacuation of aircraft in which an emergency egress system is utilized.

2. An aircraft is overdue and is believed to have been involved in an accident.

b. Manner of Notification—

1. The most expeditious method of notification to the NTSB by the operator will be determined by the circumstances existing at that time. The NTSB has advised that any of the following would be considered examples of the type of notification that would be acceptable:

(a) Direct telephone notification.

(b) Telegraphic notification.

(c) Notification to the FAA who would in turn notify the NTSB by direct communication; i.e., dispatch or telephone.

c. Items To Be Notified—The notification required above shall contain the following information, if available:

1. Type, nationality, and registration marks of the aircraft;

2. Name of owner and operator of the aircraft;

3. Name of the pilot-in-command;

4. Date and time of the accident, or incident;

5. Last point of departure, and point of intended landing of the aircraft;

6. Position of the aircraft with reference to some easily defined geographical point;

7. Number of persons aboard, number killed, and number seriously injured;

8. Nature of the accident, or incident, the weather, and the extent of damage to the aircraft so far as is known; and

9. A description of any explosives, radioactive materials, or other dangerous articles carried.

d. Followup Reports—

1. The operator shall file a report on NTSB Form 6120.1 or 6120.2, available from NTSB Field Offices or from the NTSB, Washington, DC, 20594.

(a) Within 10 days after an accident;

(b) When, after 7 days, an overdue aircraft is still missing;

(c) A report on an incident for which notification is required as described in subparagraph a(1) shall be filed only as requested by an authorized representative of the NTSB.

2. Each crewmember, if physically able at the time the report is submitted, shall attach a statement setting forth the facts, conditions and circumstances relating to the accident or incident as they appear to him. If the crewmember is incapacitated, he shall submit the statement as soon as he is physically able.

e. Where To File the Reports—

1. The operator of an aircraft shall file with the NTSB Field Office nearest the accident or incident any report required by this section.

2. The NTSB Field Offices are listed under U.S. Government in the telephone directories in the following cities: Anchorage, AK; Atlanta, GA; Chicago, IL; Denver, CO; Forth Worth, TX; Kansas City, MO; Los Angeles, CA; Miami, FL; New York, NY; Seattle, WA.

7–83. NEAR MIDAIR COLLISION REPORTING

a. Purpose and Data Uses—The primary purpose of the Near Midair Collision (NMAC) Reporting Program is to provide information for use in enhancing the safety and efficiency of the National Airspace System. Data obtained from NMAC reports are used by the FAA to improve the quality of FAA services to users and to develop programs, policies, and procedures aimed at the reduction of NMAC occurrences. All NMAC reports are thoroughly investigated by Flight Standards Facilities in coordination with Air Traffic Facilities. Data from these investigations are transmitted to FAA Headquarters in Washington, DC, where they are compiled and analyzed, and where safety programs and recommendations are developed.

b. Definition—A near midair collision is defined as an incident associated with the operation of an aircraft in which a possibility of collision occurs as a result of proximity of less than 500 feet to another aircraft, or a report is received from a pilot or a flight crew member stating that a collision hazard existed between two or more aircraft.

c. Reporting Responsibility—It is the responsibility of the pilot and/or flight crew to determine whether a near midair collision did actually occur and, if so, to initiate a NMAC report. Be specific, as ATC will not interpret a casual remark to mean that a NMAC is being reported. The pilot should state "I wish to report a near midair collision."

d. Where To File Reports—Pilots and/or flight crew members involved in NMAC occurrences are urged to report each incident immediately:

1. By radio or telephone to the nearest FAA ATC facility or FSS.

2. In writing, in lieu of the above, to the nearest Air Carrier District Office (ACDO), General Aviation District Office (GADO), or Flight Standards District Office (FSDO).

e. Items To Be Reported—

1. Date and Time (UTC) of incident.

2. Location of incident and altitude.

3. Identification and type of reporting aircraft, aircrew destination, name and home base of pilot.

4. Identification and type of other aircraft, aircrew destination, name and home base of pilot.

5. Type of flight plans; station altimeter setting used.

6. Detailed weather conditions at altitude or Flight Level.

7. Approximate courses of both aircraft: indicate if one or both aircraft were climbing or descending.

8. Reported separation in distance at first sighting, proximity at closest point horizontally and vertically, ands length of time in sight prior to evasive action.

9. Degree of evasive action taken, if any (from both aircraft, if possible).

10. Injuries, if any.

f. Investigation—The district office responsible for the investigation and reporting of NMAC's will be:

1. The Air Carrier or Flight Standards District Office in whose area the incident occurred when an air carrier aircraft is involved.

2. The General Aviation or Flight Standards District Office in whose area the incident occurred in all other cases.

g. Existing radar, communication, and weather data will be examined in the conduct of the investigation. When possible, all cockpit crew members will be interviewed regarding factors involving the NMAC incident. Air Traffic controllers will be interviewed in cases where one or more of the involved aircraft was provided ATC service. Both flight and ATC procedures will be evaluated. When the investigation reveals a violation of an FAA regulation, enforcement action will be pursued.

7–84 thru 7–90. RESERVED

Chapter 8. MEDICAL FACTS FOR PILOTS

Section 1. FITNESS FOR FLIGHT

8–1. FITNESS FOR FLIGHT

a. Medical Certification—

1. All pilots except those flying gliders and free air balloons must possess valid medical certificates in order to exercise the privileges of their airman certificates. The periodic medical examinations required for medical certification are conducted by designated Aviation Medical Examiners, who are physicians with a special interest in aviation safety and training in aviation medicine.

2. The standards for medical certification are contained in FAR 67. Pilots who have a history of certain medical conditions described in these standards are mandatorily disqualified from flying. These medical conditions include a personality disorder manifested by overt acts, a psychosis, alcoholism, drug dependence, epilepsy, an unexplained disturbance of consciousness, myocardial infarction, angina pectoris and diabetes requiring medication for its control. Other medical conditions may be temporarily disqualifying, such as acute infections, anemia, and peptic ulcer. Pilots who do not meet medical standards may still be qualified under special issuance provisions or the exemption process. This may require that either additional medical information be provided or practical flight tests be conducted.

3. Student pilots should visit an Aviation Medical Examiner as soon as possible in their flight training in order to avoid unnecessary training expenses should they not meet the medical standards. For the same reason, the student pilot who plans to enter commercial aviation should apply for the highest class of medical certificate that might be necessary in the pilot's career.

Caution: The FAR's prohibit a pilot who possesses a current medical certificate from performing crewmember duties while the pilot has a known medical condition or increase of a known medical condition that would make the pilot unable to meet the standards for the medical certificate.

b. Illness—

1. Even a minor illness suffered in day-to-day living can seriously degrade performance of many piloting tasks vital to safe flight. Illness can produce fever and distracting symptoms that can impair judgment, memory, alertness, and the ability to make calculations. Although symptoms from an illness may be under adequate control with a medication, the medication itself may decrease pilot performance.

2. The safest rule is not to fly while suffering from any illness. If this rule is considered too stringent for a particular illness, the pilot should contact an Aviation Medical Examiner for advice.

c. Medication—

1. Pilot performance can be seriously degraded by both prescribed and over-the-counter medications, as well as by the medical conditions for which they are taken. Many medications, such as tranquilizers, sedatives, strong pain relievers, and cough-suppressant preparations, have primary effects that may impair judgment, memory, alertness, coordination, vision, and the ability to make calculations. Others, such as antihistamines, blood pressure drugs, muscle relaxants, and agents to control diarrhea and motion sickness, have side effects that may impair the same critical functions. Any medication that depresses the nervous system, such as a sedative, tranquilizer or antihistamine, can make a pilot much more susceptible to hypoxia.

2. The FAR's prohibit pilots from performing crewmember duties while using any medication that affects the faculties in any way contrary to safety. The safest rule is not to fly as a crewmember while taking any medication, unless approved to do so by the FAA.

d. Alcohol—

1. Extensive research has provided a number of facts about the hazards of alcohol consumption and flying. As little as one ounce of liquor, one bottle of beer or four ounces of wine can impair flying skills, with the alcohol consumed in these drinks being detectable in the breath and blood for at least 3 hours. Even after the body completely destroys a moderate amount of alcohol, a pilot can still be severely impaired for many hours by hangover. There is simply no way of increasing the destruction of alcohol or alleviating a hangover. Alcohol also renders a pilot much more susceptible to disorientation and hypoxia.

2. A consistently high alcohol related fatal aircraft accident rate serves to emphasize that alcohol and flying are a potentially lethal combination.

The FAR's prohibit pilots from performing crewmember duties within 8 hours after drinking any alcoholic beverage or while under the influence of alcohol. However, due to the slow destruction of alcohol, a pilot may still be under influence 8 hours after drinking a moderate amount of alcohol. Therefore, an excellent rule is to allow at least 12 to 24 hours between "bottle and throttle," depending on the amount of alcoholic beverage consumed.

e. Fatigue—

1. Fatigue continues to be one of the most treacherous hazards to flight safety, as it may not be apparent to a pilot until serious errors are made. Fatigue is best described as either acute (short-term) or chronic (long-term).

2. A normal occurrence of everyday living, acute fatigue is the tiredness felt after long periods of physical and mental strain, including strenuous muscular effort, immobility, heavy mental workload, strong emotional pressure, monotony, and lack of sleep. Consequently, coordination and alertness, so vital to safe pilot performance, can be reduced. Acute fatigue is prevented by adequate rest and sleep, as well as by regular exercise and proper nutrition.

3. Chronic fatigue occurs when there is not enough time for full recovery between episodes of acute fatigue. Performance continues to fall off, and judgment becomes impaired so that unwarranted risks may be taken. Recovery from chronic fatigue requires a prolonged period of rest.

f. Stress—

1. Stress from the pressures of everyday living can impair pilot performance, often in very subtle ways. Difficulties, particularly at work, can occupy thought processes enough to markedly decrease alertness. Distraction can so interfere with judgment that unwarranted risks are taken, such as flying into deteriorating weather conditions to keep on schedule. Stress and fatigue (see above) can be an extremely hazardous combination.

2. Most pilots do not leave stress "on the ground." Therefore, when more than usual difficulties are being experienced, a pilot should consider delaying flight until these difficulties are satisfactorily resolved.

g. Emotion—
Certain emotionally upsetting events, including a serious argument, death of a family member, separation or divorce, loss of job, and financial catastrophe, can render a pilot unable to fly an aircraft safely. The emotions of anger, depression, and anxiety from such events not only decrease alertness but also may lead to taking risks that border on self-destruction. Any pilot who experiences an emotionally upsetting event should not fly until satisfactorily recovered from it.

h. Personal Checklist—Aircraft accident statistics show that pilots should be conducting preflight checklists on themselves as well as their aircraft for pilot impairment contributes to many more accidents than failures of aircraft systems. A personal checklist, which includes all of the categories of pilot impairment as discussed in this section, that can be easily committed to memory is being distributed by the FAA in the form of a wallet-sized card.

PERSONAL CHECKLIST

I'm physically and mentally safe to fly-not being impaired by:

Illness,
Medication,
Stress,
Alcohol,
Fatigue,
Emotion.

8–2. EFFECTS OF ALTITUDE

a. Hypoxia—

1. Hypoxia is a state of oxygen deficiency in the body sufficient to impair functions of the brain and other organs. Hypoxia from exposure to altitude is due only to the reduced barometric pressures encountered at altitude, for the concentration of oxygen in the atmosphere remains about 21 percent from the ground out to space.

2. Although a deterioration in night vision occurs at a cabin pressure altitude as low as 5,000 feet, other significant effects of altitude hypoxia usually do not occur in the normal healthy pilot below 12,000 feet. From 12,000 to 15,000 feet of altitude, judgment, memory, alertness, coordination and ability to make calculations are impaired, and headache, drowsiness, dizziness and either a sense of well-being (euphoria) or belligerence occur. The effects appear following increasingly shorter periods of exposure to increasing altitude. In fact, pilot performance can seriously deteriorate within 15 minutes at 15,000 feet.

3. At cabin pressure altitudes above 15,000 feet, the periphery of the visual field grays out to a point where only central vision remains (tunnel vision). A blue coloration (cyanosis) of the fingernails and lips develops. The ability to take corrective and protective action is lost in 20 to 30 minutes at 18,000 feet and 5 to 12 minutes at 20,000 feet, followed soon thereafter by unconsciousness.

4. The altitude at which significant effects of hypoxia occur can be lowered by a number of factors. Carbon monoxide inhaled in smoking or from exhaust fumes, lowered hemoglobin (anemia), and certain medications can reduce the oxygen-carrying capacity of the blood to the degree that the amount of oxygen provided to body tissues will already be equivalent to the oxygen provided to the tissues when exposed to a cabin pressure altitude of several thousand feet. Small amounts of alcohol and low doses of certain drugs, such as antihistamines, tranquilizers, sedatives and analgesics can, through their depressant action, render the brain much more susceptible to hypoxia. Extreme heat and cold, fever, and anxiety increase the body's demand for oxygen, and hence its susceptibility to hypoxia.

5. The effects of hypoxia are usually quite difficult to recognize, especially when they occur gradually. Since symptoms of hypoxia do not vary in an individual, the ability to recognize hypoxia can be greatly improved by experiencing and witnessing the effects of hypoxia during an altitude chamber "flight." The FAA provides this opportunity through aviation physiology training, which is conducted at the FAA Civil Aeromedical Institute and at many military facilities across the United States, to attend the Physiological Training Program at the Civil Aeromedical Institute, Mike Monroney Aeronautical Center, Oklahoma City, OK, contact by telephone (405) 680–4837, or by writing Airmen Education Branch, AAM–420, CAMI, Mike Monroney Aeronautical Center, P.O. Box 25082, Oklahoma City, OK 73125.

8–2a5 NOTE—To attend the Physiological Training Program at one of the military installations having the training capability, an application form and a fee must be submitted. Full particulars about location, fees, scheduling procedures, course content, individual requirements, etc. are contained in the Physiological Training Application, form number AC–3150–7, which is obtained by contacting the Accident Prevention Specialist or the Office Forms Manager in the nearest FAA office.

6. Hypoxia is prevented by heeding factors that reduce tolerance to altitude, by enriching the inspired air with oxygen from an appropriate oxygen system, and by maintaining a comfortable, safe cabin pressure altitude. For optimum protection, pilots are encouraged to use supplemental oxygen above 10,000 feet during the day, and above 5,000 feet at night. The FAR's require that at the minimum, flight crew be provided with and use supplemental oxygen after 30 minutes of exposure to cabin pressure altitudes between 12,500 and 14,000 feet and immediately on exposure to cabin pressure altitudes above 14,000 feet. Every occupant of the aircraft must be provided with supplemental oxygen at cabin pressure altitudes above 15,000 feet.

b. Ear Block—

1. As the aircraft cabin pressure decreases during ascent, the expanding air in the middle ear pushes the eustachian tube open, and by escaping down it to the nasal passages, equalizes in pressure with the cabin pressure. But during descent, the pilot must periodically open the eustachian tube to equalize pressure. This can be accomplished by swallowing, yawning, tensing muscles in the throat, or if these do not work, by a combination of closing the mouth, pinching the nose closed, and attempting to blow through the nostrils (Valsalva maneuver).

2. Either an upper respiratory infection, such as a cold or sore throat, or a nasal allergic condition can produce enough congestion around the eustachian tube to make equalization difficult. Consequently, the difference in pressure between the middle ear and aircraft cabin can build up to a level that will hold the eustachian tube closed, making equalization difficult if not impossible. The problem is commonly referred to as an "ear block."

3. An ear block produces severe ear pain and loss of hearing that can last from several hours to several days. Rupture of the ear drum can occur in flight or after landing. Fluid can accumulate in the middle ear and become infected.

4. An ear block is prevented by not flying with an upper respiratory infection or nasal allergic condition. Adequate protection is usually not provided by decongestant sprays or drops to reduce congestion around the eustachian tubes. Oral decongestants have side effects that can significantly impair pilot performance.

5. If an ear block does not clear shortly after landing, a physician should be consulted.

c. Sinus Block—

1. During ascent and descent, air pressure in the sinuses equalizes with the aircraft cabin pressure through small openings that connect the sinuses to the nasal passages. Either an upper respiratory infection, such as a cold or sinusitis, or a nasal allergic condition can produce enough congestion around an opening to slow equalization, and as the difference in pressure between the sinus and cabin mounts, eventually plug the opening. This "sinus block" occurs most frequently during descent.

2. A sinus block can occur in the frontal sinuses, located above each eyebrow, or in the maxillary sinuses, located in each upper cheek. It will usually produce excruciating pain over the sinus area. A maxillary sinus block can also make the upper teeth ache. Bloody mucus may discharge from the nasal passages.

3. A sinus block is prevented by not flying with an upper respiratory infection or nasal allergic condition. Adequate protection is usually not provided by decongestant sprays or drops to reduce congestion around the sinus openings. Oral decongestants have side effects that can impair pilot performance.

4. If a sinus block does not clear shortly after landing, a physician should be consulted.

d. Decompression Sickness After Scuba Diving—

1. A pilot or passenger who intends to fly after scuba diving should allow the body sufficient time to rid itself of excess nitrogen absorbed during diving. If not, decompression sickness due to evolved gas can occur during exposure to low altitude and create a serious in-flight emergency.

2. The recommended waiting time before going to flight altitudes of up to 8,000 feet is at least 12 hours after diving which has not required controlled ascent (nondecompression stop diving), and at least 24 hours after diving which has required controlled ascent (decompression stop diving). The waiting time before going to flight altitudes above 8,000 feet should be at least 24 hours after any SCUBA dive. These recommended altitudes are actual flight altitudes above mean sea level (AMSL) and not pressurized cabin altitudes. This takes into consideration the risk of decompression of the aircraft during flight.

8–3. HYPERVENTILATION IN FLIGHT

a. Hyperventilation, or an abnormal increase in the volume of air breathed in and out of the lungs, can occur subconsciously when a stressful situation is encountered in flight. As hyperventilation ''blows off'' excessive carbon dioxide from the body, a pilot can experience symptoms of lightheadedness, suffocation, drowsiness, tingling in the extremities, and coolness and react to them with even greater hyperventilation. Incapacitation can eventually result from incoordination, disorientation, and painful muscle spasms. Finally, unconsciousness can occur.

b. The symptoms of hyperventilation subside within a few minutes after the rate and depth of breathing are consciously brought back under control. The buildup of carbon dioxide in the body can be hastened by controlled breathing in and out of a paper bag held over the nose and mouth.

c. Early symptoms of hyperventilation and hypoxia are similar. Moreover, hyperventilation and hypoxia can occur at the same time. Therefore, if a pilot is using an oxygen system when symptoms are experienced, the oxygen regulator should immediately be set to deliver 100 percent oxygen, and then the system checked to assure that it has been functioning effectively before giving attention to rate and depth of breathing.

8–4. CARBON MONOXIDE POISONING IN FLIGHT

a. Carbon monoxide is a colorless, odorless, and tasteless gas contained in exhaust fumes. When breathed even in minute quantities over a period of time, it can significantly reduce the ability of the blood to carry oxygen. Consequently, effects of hypoxia occur.

b. Most heaters in light aircraft work by air flowing over the manifold. Use of these heaters while exhaust fumes are escaping through manifold cracks and seals is responsible every year for several nonfatal and fatal aircraft accidents from carbon monoxide poisoning.

c. A pilot who detects the odor of exhaust or experiences symptoms of headache, drowsiness, or dizziness while using the heater should suspect carbon monoxide poisoning, and immediately shut off the heater and open air vents. If symptoms are severe or continue after landing, medical treatment should be sought.

8–5. ILLUSIONS IN FLIGHT

a. Introduction—Many different illusions can be experienced in flight. Some can lead to spatial disorientation. Others can lead to landing errors. Illusions rank among the most common factors cited as contributing to fatal aircraft accidents.

b. Illusions Leading to Spatial Disorientation—

1. Various complex motions and forces and certain visual scenes encountered in flight can create illusions of motion and position. Spatial disorientation from these illusions can be prevented only by visual reference to reliable, fixed points on the ground or to flight instruments.

2. The leans—An abrupt correction of a banked attitude, which has been entered too slowly to stimulate the motion sensing system in the inner ear, can create the illusion of banking in the opposite direction. The disoriented pilot will roll the aircraft back into its original dangerous attitude, or if level flight is maintained, will feel compelled to lean in the perceived vertical plane until this illusion subsides.

(a) Coriolis illusion—An abrupt head movement in a prolonged constant-rate turn that has ceased stimulating the motion sensing system can create the illusion of rotation or movement in an entirely different axis. The disoriented pilot will maneuver the aircraft into a dangerous attitude in an attempt to stop rotation. This most overwhelming of all illusions in flight may be prevented by not making sudden, extreme head movements, particularly while making prolonged constant-rate turns under IFR conditions.

(b) Graveyard spin—A proper recovery from a spin that has ceased stimulating the motion sensing system can create the illusion of spinning in the opposite direction. The disoriented pilot will return the aircraft to its original spin.

(c) Graveyard spiral—An observed loss of altitude during a coordinated constant-rate turn that has ceased stimulating the motion sensing system can create the illusion of being in a descent with the wings level. The disoriented pilot will pull back on the controls, tightening the spiral and increasing the loss of altitude.

(d) Somatogravic illusion—A rapid acceleration during takeoff can create the illusion of being in a nose up attitude. The disoriented pilot will push the aircraft into a nose low, or dive attitude. A rapid deceleration by a quick reduction of the throttles can have the opposite effect, with the disoriented pilot pulling the aircraft into a nose up, or stall attitude.

(e) Inversion illusion—An abrupt change from climb to straight and level flight can create the illusion of tumbling backwards. The disoriented pilot will push the aircraft abruptly into a nose low attitude, possibly intensifying this illusion.

(f) Elevator illusion—An abrupt upward vertical acceleration, usually by an updraft, can create the illusion of being in a climb. The disoriented pilot will push the aircraft into a nose low attitude. An abrupt downward vertical acceleration, usually by a downdraft, has the opposite effect, with the disoriented pilot pulling the aircraft into a nose up attitude.

(g) False horizon—Sloping cloud formations, an obscured horizon, a dark scene spread with ground lights and stars, and certain geometric patterns of ground light can create illusions of not being aligned correctly with the actual horizon. The disoriented pilot will place the aircraft in a dangerous attitude.

(h) Autokinesis—In the dark, a static light will appear to move about when stared at for many seconds. The disoriented pilot will lose control of the aircraft in attempting to align it with the light.

3. Illusions Leading to Landing Errors—

(a) Various surface features and atmospheric conditions encountered in landing can create illusions of incorrect height above and distance from the runway threshold. Landing errors from these illusions can be prevented by anticipating them during approaches, aerial visual inspection of unfamiliar airports before landing, using electronic glide slope or VASI systems when available, and maintaining optimum proficiency in landing procedures.

(b) Runway width illusion—A narrower-than-usual runway can create the illusion that the aircraft is at a higher altitude than it actually is. The pilot who does not recognize this illusion will fly a lower approach, with the risk of striking objects along the approach path or landing short. A wider-than-usual runway can have the opposite effect, with the risk of leveling out high and landing hard or overshooting the runway.

(c) Runway and terrain slopes illusion—An upsloping runway, upsloping terrain, or both, can create the illusion that the aircraft is at a higher altitude than it actually is. The pilot who does not recognize this illusion will fly a lower approach. A downsloping runway, downsloping approach terrain, or both, can have the opposite effect.

(d) Featureless terrain illusion—An absence of ground features, as when landing over water, darkened areas, and terrain made featureless by snow, can create the illusion that the aircraft is at a higher altitude than it actually is. The pilot who does not recognize this illusion will fly a lower approach

(e) Atmospheric illusions—Rain on the windscreen can create the illusion of greater height, and atmospheric haze the illusion of being at a greater distance from the runway. The pilot who does not recognize these illusions will fly a lower approach. Penetration of fog can create the illusion of pitching up. The pilot who does not recognize this illusion will steepen the approach, often quite abruptly.

(f) Ground lighting illusions—Lights along a straight path, such as a road, and even lights on moving trains can be mistaken for runway and approach lights. Bright runway and approach lighting systems, especially where few lights illuminate the surrounding terrain, may create the illusion of less distance to the runway. The pilot who does not recognize this illusion will fly a higher approach. Conversely, the pilot overflying terrain which has few lights to provide height cues may make a lower than normal approach.

8–6. VISION IN FLIGHT

a. Introduction—Of the body senses, vision is the most important for safe flight. Major factors that determine how effectively vision can be used are the level of illumination and the technique of scanning the sky for other aircraft.

b. Vision Under Dim and Bright Illumination—

1. Under conditions of dim illumination, small print and colors on aeronautical charts and aircraft instruments become unreadable unless adequate cockpit lighting is available. Moreover, another aircraft must be much closer to be seen unless its navigation lights are on.

2. In darkness, vision becomes more sensitive to light, a process called dark adaptation. Although exposure to total darkness for at least 30 minutes is required for complete dark adaptation, a pilot can achieve a moderate degree of dark adaptation within 20 minutes under dim red cockpit lighting. Since red light severely distorts colors, especially on aeronautical charts, and can cause serious difficulty in focusing the eyes on objects inside the aircraft, its use is advisable only where optimum outside night vision capability is necessary. Even so, white cockpit lighting must be available when needed for map and instrument reading, especially under IFR conditions. Dark adaptation is impaired by exposure to cabin pressure altitudes above 5,000 feet, carbon monoxide inhaled in smoking and from exhaust fumes, deficiency of Vitamin A in the diet, and by prolonged exposure to bright sunlight. Since any degree of dark adaptation is lost within a few seconds of viewing a bright light, a pilot should close one eye when using a light to preserve some degree of night vision.

3. Excessive illumination, especially from light reflected off the canopy, surfaces inside the aircraft, clouds, water, snow, and desert terrain, can produce glare, with uncomfortable squinting, watering of the eyes, and even temporary blindness. Sunglasses for protection from glare should absorb at least 85 percent of visible light (15 percent transmittance) and all colors equally (neutral transmittance), with negligible image distortion from refractive and prismatic errors.

c. Scanning for Other Aircraft—

1. Scanning the sky for other aircraft is a key factor in collision avoidance. It should be used continuously by the pilot and copilot (or right seat passenger) to cover all areas of the sky visible from the cockpit. Although pilots must meet specific visual acuity requirements, the ability to read an eye chart does not ensure that one will be able to efficiently spot other aircraft. Pilots must develop an effective scanning technique which maximizes one's visual capabilities. The probability of spotting a potential collision threat obviously increases with the time spent looking outside the cockpit. Thus, one must use timesharing techniques to efficiently scan the surrounding airspace while monitoring instruments as well.

2. While the eyes can observe an approximate 200 degree arc of the horizon at one glance, only a very small center area called the fovea, in the rear of the eye, has the ability to send clear, sharply focused messages to the brain. All other visual information that is not processed directly through the fovea will be of less detail. An aircraft at a distance of 7 miles which appears in sharp focus within the foveal center of vision would have to be as close as 7/10 of a mile in order to be recognized if it were outside of foveal vision. Because the eyes can focus only on this narrow viewing area, effective scanning is accomplished with a series of short, regularly spaced eye movements that bring successive areas of the sky into the central visual field. Each movement should not exceed 10 degrees, and each area should be observed for at least 1 second to enable detection. Although horizontal back-and-forth eye movements seem preferred by most pilots, each pilot should develop a scanning pattern that is most comfortable and then adhere to it to assure optimum scanning.

3. Studies show that the time a pilot spends on visual tasks inside the cabin should represent no more that 1/4 to 1/3 of the scan time outside, or no more than 4 to 5 seconds on the instrument panel for every 16 seconds outside. Since the brain is already trained to process sight information that is presented from left to right, one may find it easier to start scanning over the left shoulder and proceed across the windshield to the right.

4. Pilots should realize that their eyes may require several seconds to refocus when switching views between items in the cockpit and distant objects. The eyes will also tire more quickly when forced to adjust to distances immediately after close-up focus, as required for scanning the instrument panel. Eye fatigue can be reduced by looking from the instrument panel to the left wing past the wing tip to the center of the first scan quadrant when beginning the exterior scan. After having scanned from left to right, allow the eyes to return to the cabin along the right wing from its tip inward. Once back inside, one should automatically commence the panel scan.

5. Effective scanning also helps avoid "empty-field myopia." This condition usually occurs when flying above the clouds or in a haze layer that provides nothing specific to focus on outside the aircraft. This causes the eyes to relax and seek a comfortable focal distance which may range from 10 to 30 feet. For the pilot, this means looking without seeing, which is dangerous.

8–7. AEROBATIC FLIGHT

a. Airmen planning to engage in aerobatics should be aware of the physiological stresses associated with accelerative forces during aerobatic maneuvers. Many prospective aerobatic trainees enthusiastically enter aerobatic instruction but find their first experiences with G forces to be unanticipated and very uncomfortable. To minimize or avoid potential adverse effects, the aerobatic instructor and trainee must have a basic understanding of the physiology of G force adaptation.

b. Forces experienced with a rapid push-over maneuver result in the blood and body organs being displaced toward the head. Depending on forces involved and individual tolerance, the airman may experience discomfort, headache, "red-out," and even unconsciousness.

c. Forces experienced with a rapid pull-up maneuver result in the blood and body organ displacement toward the lower part of the body away from the head. Since the brain requires continuous blood circulation for an adequate oxygen supply, there is a physiologic limit to the time the pilot can tolerate higher forces before losing consciousness. As the blood circulation to the brain decreases as a result of forces involved, the airman will experience "narrowing" of visual fields, "gray-out," "black-out," and unconsciousness. Even a brief loss of consciousness in a maneuver can lead to improper control movement causing structural failure of the aircraft or collision with another object or terrain.

d. In steep turns, the centrifugal forces tend to push the pilot into the seat, thereby resulting in blood and body organ displacement toward the lower part of the body as in the case of rapid pull-up maneuvers and with the same physiologic effects and symptoms.

e. Physiologically, humans progressively adapt to imposed strains and stress, and with practice, any maneuver will have decreasing effect. Tolerance to G forces is dependent on human physiology and the individual pilot. These factors include the skeletal anatomy, the cardiovascular architecture, the nervous system, the quality of the blood, the general physical state, and experience and recency of exposure. The airman should consult an Aviation Medical Examiner prior to aerobatic training and be aware that poor physical condition can reduce tolerance to accelerative forces.

f. The above information provides the airman a brief summary of the physiologic effects of G forces. It does not address methods of "counteracting" these effects. There are numerous references on the subject of G forces during aerobatics available to the airman. Among these are "G Effects on the Pilot During Aerobatics," FAA-AM-72-28, and "G Incapacitation in Aerobatic Pilots: A Flight Hazard" FAA-AM-82-13. These are available from the National Technical Information Service, Springfield, Virginia 22161. (Reference—FAA Advisory Circular 91-61, "A Hazard in Aerobatics: Effects of G-Forces on Pilots").

8–8. JUDGMENT ASPECTS OF COLLISION AVOIDANCE

a. Introduction—The most important aspects of vision and the techniques to scan for other aircraft are described in Vision in Flight, paragraph 8–6. Pilots should also be familiar with the following information to reduce the possibility of mid-air collisions.

b. Determining Relative Altitude—Use the horizon as a reference point. If the other aircraft is above the horizon, it is probably on a higher flight path. If the aircraft appears to be below the horizon, it is probably flying at a lower altitude.

c. Taking Appropriate Action—Pilots should be familiar with rules on right-of-way, so if an aircraft is on an obvious collision course, one can take immediately evasive action, preferably in compliance with applicable Federal Aviation Regulations.

d. Consider Multiple Threats—The decision to climb, descend, or turn is a matter of personal judgment, but one should anticipate that the other pilot may also be making a quick maneuver. Watch the other aircraft during the maneuver and begin your scanning again immediately since there may be other aircraft in the area.

e. Collision Course Targets—Any aircraft that appears to have no relative motion and stays in one scan quadrant is likely to be on a collision course. Also, if a target shows no lateral or vertical motion, but increases in size, *take evasive action.*

f. Recognize High Hazard Areas—

1. Airways and especially VOR's and Airport Traffic Areas are places where aircraft tend to cluster.

2. Remember, most collisions occur during days when the weather is good. Being in a "radar environment" still requires vigilance to avoid collisions.

g. Cockpit Management—Studying maps, checklists, and manuals before flight, with other proper preflight planning; e.g., noting necessary radio frequencies and organizing cockpit materials, can reduce the amount of time required to look at these items during flight, permitting more scan time.

h. Windshield Conditions—Dirty or bug-smeared windshields can greatly reduce the ability of pilots to see other aircraft. Keep a clean windshield.

i. Visibility Conditions—Smoke, haze, dust, rain, and flying towards the sun can also greatly reduce the ability to detect targets.

j. Visual Obstructions in the Cockpit—

1. Pilots need to move their heads to see around blind spots caused by fixed aircraft structures, such as door posts, wings, etc. It will be necessary at times to maneuver the aircraft; e.g., lift a wing, to facilitate seeing.

2. Pilots must insure curtains and other cockpit objects; e.g., maps on glare shield, are removed and stowed during flight.

k. Lights On—

1. Day or night, use of exterior lights can greatly increase the conspicuity of any aircraft.

2. Keep interior lights low at night.

l. ATC support—ATC facilities often provide radar traffic advisories on a workload-permitting basis. Flight through Airport Radar Service Areas (ARSA's) requires communication with ATC. Use this support whenever possible or when required.

8–9 thru 8–10. RESERVED

Chapter 9. AERONAUTICAL CHARTS AND RELATED PUBLICATIONS

Section 1. TYPES OF CHARTS AVAILABLE

9–1. GENERAL

Aeronautical charts for the U.S., its territories, and possessions are produced by the National Ocean Service (NOS), a part of the Department of Commerce, from information furnished by the FAA.

9–2. OBTAINING CIVIL AERONAUTICAL CHARTS

Enroute Aeronautical Charts, Terminal Procedure Publication Charts, Regional Airport/Facility Directories, and other publications described in this Chapter are available upon subscription and one time sales from:

National Ocean Service
NOAA Distribution Branch (N/CG33)
Riverdale, Maryland 20737
Telephone: (301) 436–6990

9–3. TYPES OF CHARTS AVAILABLE

Sectional and VFR Terminal Area Charts
World Aeronautical Charts (U.S.)
Enroute Low Altitude Charts
Enroute High Altitude Charts
Alaska Enroute Charts (Low and High)
Planning Charts
Terminal Procedures Publication (TPP)
Alaska Terminal Publication
Helicopter Route Charts

9–4. GENERAL DESCRIPTION OF EACH CHART SERIES

a. Sectional and VFR Terminal Area Charts

1. These charts are designed for visual navigation of slow and medium speed aircraft. They are produced to the following scales:

Sectional Charts—
1:500,000 (1 in=6.86 NM)
VFR Terminal Area Charts—
1:250,000 (1 in=3.43 NM)

2. Topographic information features the portrayal of relief and a judicious selection of visual check points for VFR flight. VFR Terminal Area Charts include populated places, drainage, roads, railroads, and other distinctive landmarks. Aeronautical information includes visual and radio aids to navigation, airports, controlled airspace, restricted areas, obstructions, and related data. These charts also depict the airspace designated as "Terminal Control Area" (TCA), which provides for the control or segregation of all aircraft within the terminal control area. The Puerto Rico—Virgin Islands Terminal Area Chart contains basically the same information as that shown on Sectional and Terminal Area Charts. It includes the Gulf of Mexico and Caribbean Planning Chart on the reverse side (See Planning Charts). Charts are revised semiannually, except for several Alaskan Sectionals and the Puerto Rico—Virgin Islands Terminal Area which are revised annually.

b. World Aeronautical Charts—These charts are designed to provide a standard series of aeronautical charts, covering land areas of the world, at a size and scale convenient for navigation by moderate speed aircraft. They are produced at a scale of 1:1,000,000 (1 in=13.7 NM). Topographic information includes cities and towns, principal roads, railroads, distinctive landmarks, drainage, and relief. The latter is shown by spot elevations, contours, and gradient tints. Aeronautical information includes visual and radio aids to navigation, airports, airways, restricted areas, obstructions, and other pertinent data. These charts are revised annually except several Alaskan charts and the Mexican/Caribbean charts which are revised every 2 years.

c. Enroute Low Altitude Charts—These charts are designed to provide aeronautical information for en route navigation under Instrument Flight Rules (IFR) in the low altitude stratum. The series also includes Enroute Area Charts, which furnish terminal data at a large scale in congested areas and are included with the subscription to the series. Information includes the portrayal of L/MF and VHF airways, limits of controlled airspace, position, identification and frequencies of radio aids, selected airports, minimum en route and obstruction clearance altitudes, airway distances, reporting points, special use airspace areas, military training routes, and related information. Charts are printed back to back and are revised every 56 days effective with the date of airspace changes. An Enroute Change Notice may be issued as required.

d. Enroute High Altitude Charts—These charts are designed to provide aeronautical information for en route navigation under IFR in the high altitude stratum. Information includes the portrayal of jet routes, position, identification and frequencies of radio aids, selected airports, distances, time zones, special use airspace areas, and related information. Charts are revised every 56 days effective with the date of airspace changes. An Enroute Change Notice may be issued as required.

e. Alaska Enroute Charts (Low and High)—These charts are produced in a low altitude series and a high altitude series with the purpose and makeup identical to Enroute Low and High Altitude Charts described above. Charts are revised every 56 days effective with the date of airspace changes. An Enroute Change Notice may be issued as required.

f. Charted VFR Flyway Planning Chart—These charts are designed to identify flight paths clear of the major controlled traffic flows. The program is intended to provide charts showing multiple VFR routings through high density traffic areas which may be used as an alternative to flight within TCA's. Ground references are provided as guides for improved visual navigation. These charts are not intended to discourage VFR operations within the TCA's, but are designed for information and planning purposes. They are produced at a scale of 1:250,000 (1 in = 3.43 NM). These charts are revised semi-annually and are published on the back of the existing VFR Terminal Area Charts.

g. Planning Charts

1. VFR/IFR Planning Chart—These charts are designed to fulfill the requirements of preflight planning for flights under VFR/IFR. They are produced at a scale of 1:2,333,232 (1 in=32 NM). The chart is printed in two parts in such a manner that, when assembled, it forms a composite VFR Planning Chart on one side and IFR Planning Chart on the other. Information on the IFR chart depicts low altitude airways and mileages, navigational facilities, special use airspace areas, time zones, airports, isogonic lines, and related data. Information on the VFR chart includes selected populated places, large bodies of water, major drainage, shaded relief, navigational facilities, airports, special use airspace areas, military training routes, and related data.

2. Flight Case Planning Chart—This chart is designed for preflight and en route flight planning for VFR flights. It is produced at a scale of 1:4,374,803 (1 in=60 NM). This chart contains basically the same information as the VFR/IFR Planning Chart with the addition of selected FSS's and Weather Service Offices (WFO's) located at airport sites, parachute jumping areas, a tabulation of special use airspace areas, a mileage table listing distances between 174 major airports, and a city/aerodrome location index.

3. Gulf of Mexico and Caribbean Planning Chart—This chart is designated for preflight planning for VFR flights. It is produced at a scale of 1:6,270,551 (1 in=86 NM). This chart is on the reverse of the Puerto Rico—Virgin Islands Terminal Area Chart. Information includes mileage between Airports of Entry, a selection of special use airspace areas, and a Directory of Airports with their available facilities and servicing.

4. North Atlantic Route Chart—This five-color chart is designed for use by air traffic controllers in monitoring transatlantic flights and by FAA planners. Oceanic control areas, coastal navigation aids, major coastal airports, and oceanic reporting points are depicted. Geographic coordinates for NAVAID's and reporting points are included. The chart may be used for pre- and in-flight planning. This chart is revised each 24 weeks. The chart available in two sizes, full size (58 by 41 inches) scale: 1:5,500,000; half size (29 by 20½ inches) scale: 1:11,000,000.

5. North Pacific Oceanic Route Chart—This chart series, like the North Atlantic Route Chart series, is designed for FAA air traffic controllers' use in monitoring transoceanic air traffic. Charts are available in two scales: one 1:12,000,000 composite small scale planning chart, which covers the entire North Pacific, and four 1:7,000,000 Area Charts. They are revised every 24 weeks. The charts are available unfolded (flat only) and contain established intercontinental air routes including all reporting points with geographic positions.

h. Terminal Procedures Publication (TPP)—This publication contains charts depicting Instrument Approach Procedures (IAP), Standard Terminal Arrivals (STAR), and Standard Instrument Departures (SID).

1. Instrument Approach Procedure (IAP) Charts—IAP charts portray the aeronautical data which is required to execute instrument approaches to airports. Each chart depicts the IAP, all related navigation data, communications information, and an airport sketch. Each procedure is designated for use with a specific electronic navigational aid, such as ILS, VOR, NDB, RNAV, etc. Airport Diagram Charts, where published, are included.

2. Standard Instrument Departure (SID) Charts—These charts are designed to expedite clearance delivery and to facilitate transition between takeoff and en route operations. They furnish pilots departure routing clearance information in graphic and textual form.

3. Standard Terminal Arrival (STAR) Charts—These charts are designed to expedite ATC arrival procedures and to facilitate transition between en route and instrument approach operations. They present to the pilot preplanned IFR ATC arrival procedures in graphic and textual form. Each STAR procedure is presented as a separate chart and may serve a single airport or more than one airport in a given geographic location.

These charts are published in 16 bound volumes covering the conterminous U.S. and the Puerto Rico—Virgin Islands. Each volume is superseded by a new volume each 56 days. Changes to procedures occurring between the 56–day publication cycle is reflected in a Change Notice volume, issued on the 28–day midcycle. These changes are in the form of a new chart. The publication of a new 56–day volume incorporates all the changes and replaces the preceding volume and the change notice. The volumes are 5⅜ x 8¼ inches and are bound on the top edge.

i. Alaska Terminal Publication

1. This publication contains charts depicting all terminal flight procedures in the State of Alaska for civil and military aviation. They are:

(a) Instrument Approach Procedure (IAP) Charts.

(b) Standard Instrument Departure (SID) Charts.

(c) Standard Terminal Arrival (STAR) Charts.

(d) Airport Diagram Charts.

(e) Radar Minimums.

2. All supplementary supporting data; i.e., IFR Takeoff and Departure Procedures, IFR Alternate Minimums, Rate of Descent Table, Inoperative Components Table, etc., is also included.

3. The Alaska Terminal is published in a bound book, 5⅜ inches by 8¼ inches. The publication is issued every 56 days with provisions for an as required "Terminal Change" on the 28–day midpoint.

j. Helicopter Route Charts

1. Prepared under the auspices of the FAA Helicopter Route Chart Program, these charts enhance helicopter operator access into, egress from, and operation within selected high density traffic areas. The scale is 1:125,000; however, some include smaller scale insets. Graphic information includes urban tint, principal roads, pictorial symbols, and spot elevations. Aeronautical information includes routes, operating zones, altitudes or flight ceilings/bases, heliports, helipads, NAVAID's, special use airspace, selected obstacles, ATC and traffic advisory radio communications frequencies, TCA surface area tint, and other important flight aids. These charts are revised when significant aeronautical information changes and/or safety-related events occur. Historically, new editions are published about every 2 years. See the "Dates of Latest Editions" for current editions.

2. Air traffic facility managers are responsible for determining the need for new chart development or existing chart revision. Therefore, requests for new charts or revisions to existing charts should be directed to these managers. Guidance pertinent to mandatory chart features and managerial evaluation of requests is contained in FAA Order 7210.3, Facility Operation and Administration.

9–5. RELATED PUBLICATIONS

a. The Airport Facility Directory—This directory is issued in seven volumes with each volume covering a specific geographic area of the conterminous U.S., including Puerto Rico and the U.S. Virgin Islands. The directory is 5⅜ X 8¼ inches and is bound on the side. Each volume is reissued in its entirety each 56 days. Each volume is indexed alphabetically by state, airport, navigational aid, and ATC facility for the area of coverage. All pertinent information concerning each entry is included.

b. Alaska Supplement—This supplement is a joint Civil and Military Flight Information Publication (FLIP), published and distributed every 56 days by the NOS. It is designed for use with the Flight Information Publication En route Charts, Alaska Terminal, WAC and Sectional Aeronautical Charts. This Supplement contains an Airport/Facility Directory of all airports shown on En route Charts, and those requested by appropriate agencies, communications data, navigational facilities, special notices and procedures applicable to the area of chart coverage.

c. Pacific Supplement—This Chart Supplement is a Civil Flight Information Publication, published and distributed every 56 days by the NOS. It is designed for use with the Flight Information En route Publication Charts and the Sectional Aeronautical Chart covering the State of Hawaii and that area of the Pacific served by U.S. facilities. An Amendment Notice is published 4 weeks after each issue of the Supplement. This chart Supplement contains an Airport/Facility Directory of all airports open to the public, and those requested by appropriate agencies, communications data, navigational facilities, special notices and procedures applicable to the Pacific area.

d. Digital Aeronautical Chart Supplement (DACS)—DACS is a subset of the data NOAA provides to FAA controllers every 56 days. The DACS is designed to assist with flight planning and should not be considered a substitute for a chart. The supplement is divided into nine individual sections. They are:

Section 1–High Altitude Airways, Conterminous U.S.

Section 2–Low Altitude Airways, Conterminous U.S.

Section 3–Selected Instrument Approach Procedure NAVAID and FIX Data

Section 4–Military Training Routes

Section 5–Alaska, Hawaii, Puerto Rico, Bahama and Selected Oceanic Routes

Section 6–STAR's, Standard Terminal Arrivals and Profile Descent Procedures

Section 7–SID's, Standard Instrument Departures

Section 8–Preferred IFR Routes (Low and High Altitudes)

Section 9–Air Route and Airport Surveillance Radar Facilities (updated yearly)

9–5d NOTE—Section 3 has a Change Notice that will be issued at the mid–28 day point and contains changes that occurred after the 56 day publication. Sections 8 and 9 are not digital products, but contain pertinent air route data associated with the other sections.

e. NOAA Aeronautical Chart User's Guide

1. This guide is designed to be used as a teaching aid, reference document, and an introduction to the wealth of information provided on NOAA's aeronautical charts and publications. The guide includes the complete contents of the VFR Chart User's Guide, and a new discussion of IFR chart terms and symbols. This guide also includes a comprehensive index of symbols used on NOAA's VFR, IFR, and Planning charts.

f. Defense Mapping Agency Aerospace Center (DMAAC) Publications

Defense Mapping Agency Aeronautical Charts and Products are available prepaid from

DIRECTOR
DMA Combat Support Center
Attention: PMSR
Washington, DC 20315–0010
Phone: CONUS Toll free telephone number 1–800–826–0342

1. Pilotage Charts (PC/TPC)—Scale 1:500,000 used for detail preflight planning and mission analysis. Emphasis in design is on ground features significant in visual and radar, low-level high speed navigation.

2. Jet Navigation Charts (JNC-A)—Scale 1:3,000,000. Designed to provide greater coverage than the 1:2,000,000 scale Jet Navigation Charts described below. Uses include preflight planning and en route navigation by long-range jet aircraft with dead reckoning, radar, celestial and grid navigation capabilities.

3. LORAN Navigation & Consol LORAN Navigation Charts (LJC/CJC)—Scale 1:2,000,000. Used for preflight planning and in-flight navigation on long-range flights in the Polar areas and adjacent regions utilizing LORAN and CONSOL navigation aids.

4. Continental Entry Chart (CEC)—Scale 1:2,000,000. Used for CONSOLAN and LORAN navigation for entry into the U.S. when a high degree of accuracy is required to comply with Air Defense identification and reporting procedures. Also suitable as a basic dead reckoning sheet and for celestial navigation.

5. Aerospace Planning Chart (ASC)—Scale 1:9,000,000 and 1:18,000,000. Six charts at each scale and with various projections, cover the world. Charts are useful for general planning, briefings, and studies.

6. Air Distance/Geography Chart (GH–2, 2a)—Scales 1:25,000,000 and 1:50,000,000. This chart shows great circle distances between major airports. It also shows major cities, international boundaries, shaded relief and gradient tints.

7. LORAN C Navigation Chart (LCC)—Scale 1:3,000,000. Primarily designed for preflight and in-flight long-range navigation where LORAN-C is used as the basic navigation aid.

8. DOD Weather Plotting Chart (WPC)—Various scales. Designed as non-navigational outline charts which depict locations and identifications of meteorological observing stations. Primarily used to forecast and monitor weather and atmospheric conditions throughout the world.

9. Flight Information Publications (FLIP)—These include En route Low Altitude and High Altitude Charts, En route Supplements, Terminal (Instrument Approach Charts), and other informational publications for various areas of the world.

9–5f9 NOTE—FLIP. Terminal publications do not necessarily include all instrument approach procedures for all airports. They include only those required for military operations.

10. World Aeronautical (WAC) and Operational Navigation Charts (ONC)—The Operational Navigation Charts (ONC) have the same purpose and contain essentially the same information as the WAC series except the terrain is portrayed by shaded relief as well as contours. The ONC series is replacing the WAC series and the WAC's will be available only where the ONC's have not been issued. ONC's are 42 X 57½ inches, WAC's are 22 X 30 inches. These charts are revised on a regular schedule.

11. Jet Navigation Charts—These charts are designed to fulfill the requirements for long-range high altitude, high speed navigation. They are produced at a scale of 1:2,000,000 (1 in=27.4 NM). Topographic features include large cities, roads, railroads, drainage and relief. The latter is indicated by contours, spot elevations and gradient tints. All aeronautical information necessary to conform to the purpose of the chart is shown. This includes restricted areas, L/MF and VOR ranges, radio beacons and a selection of standard broadcasting stations and airports. The runway patterns of the airports are shown to exaggerated scale in order that they may be readily identified as visual landmarks. Universal Jet Navigation Charts are used as plotting charts in the training and practice of celestial and dead reckoning navigation. They may also be used for grid navigational training.

12. Global Navigational Charts—These charts are designed to fulfill the requirements for aeronautical planning, operations over long distances, and en route navigation in long-range, high altitude, high speed aircraft. They are produced at a scale of 1:5,000,000 (1 in=68.58 NM). Global Navigation Charts (GNC) are 42 by 57½ inches. They show principal cities, towns and drainage, primary roads and railroads, prominent culture and shadient relief augmented with tints and spot elevations. Aeronautical data includes radio aids to navigation, aerodrome and restricted areas. Charts 1 and 26 have a polar navigation grid and charts 2 and 6 have sub-polar navigation grids. Global LORAN Navigation Charts (GLC's) are the same size and scale and cover the same area as the GNC charts. They contain major cities only, coast lines, major lakes and rivers, and land tint. No relief or vegetation. Aeronautical data includes radio aids to navigation and LORAN lines of position.

9–6. AUXILIARY CHARTS

a. Airport Obstruction Charts (OC)—The Airport Obstruction Chart is a 1:12,000 scale graphic depicting Federal Aviation Regulations Part 77 (FAR 77) surfaces, a representation of objects that penetrate these surfaces, aircraft movement and apron areas, navigational aids, prominent airport buildings, and a selection of roads and other planimetric detail in the airport vicinity. Also included are tabulations of runway and other operational data.

b. Military Training Routes—

1. Charts and Booklet: The Defense Mapping Agency Aerospace Center (DMAAC) publishes a narrative description in booklet form and charts depicting the IFR and VFR Training Routes.

2. The charts and booklet are published every 56 days. Both the charts and narrative route description booklet are available to the general public as a brochure by single copy or annual subscription.

3. Subscription and single-copy requests should be for the "DOD Area Planning AP/1B, Military Training Routes". (Reference—Military Training Routes (MTR), paragraph 3–63).

9–6b3 NOTE—The DOD provides these booklets and charts to each FSS for use in preflight pilot briefings. Pilots should review this information to acquaint themselves with those routes that are located along their route of flight and in the vicinity of the airports from which they operate.

9–7 thru 9–10. RESERVED

Appendix 1
Aeronautical Charts

INSTRUMENT APPROACH PROCEDURE CHARTS

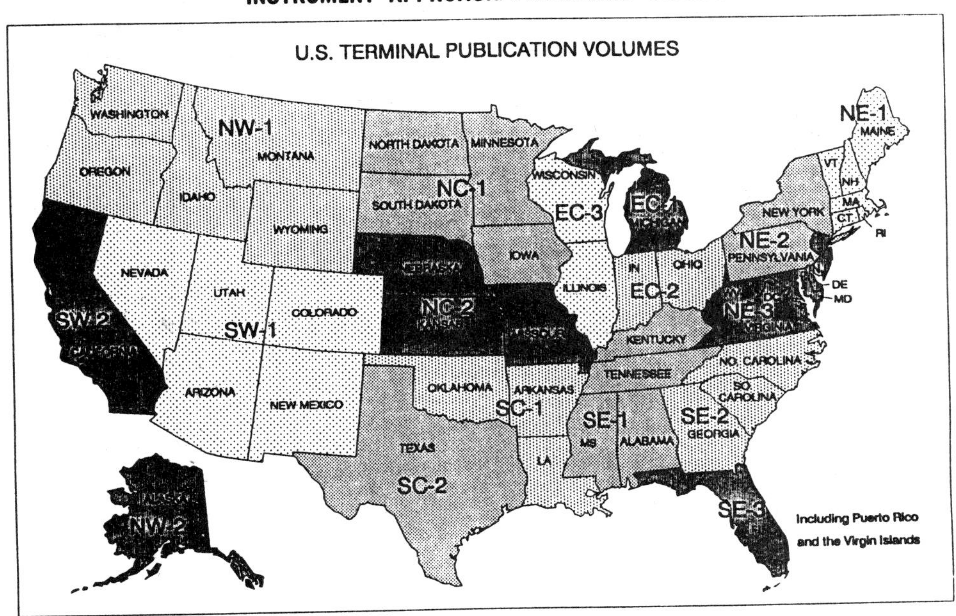

AIRPORT/FACILITY DIRECTORY

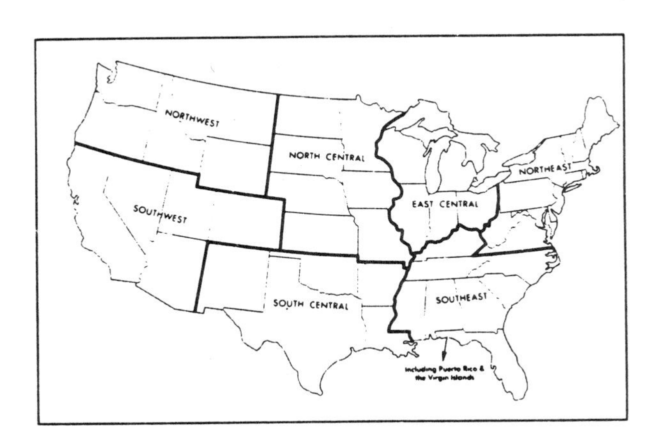

ENROUTE LOW ALTITUDE CHARTS

AREA CHARTS
Atlanta A-1
Chicago-Milwaukee A-1
Dallas-Ft Worth A-2
Denver A-2
Detroit A-1
Jacksonville A-1
Kansas City A-2
Los Angeles A-1
Miami A-1
Minneapolis-St Paul A-2
San Francisco A-2
St Louis A-1
Washington A-1

ENROUTE HIGH ALTITUDE CHARTS

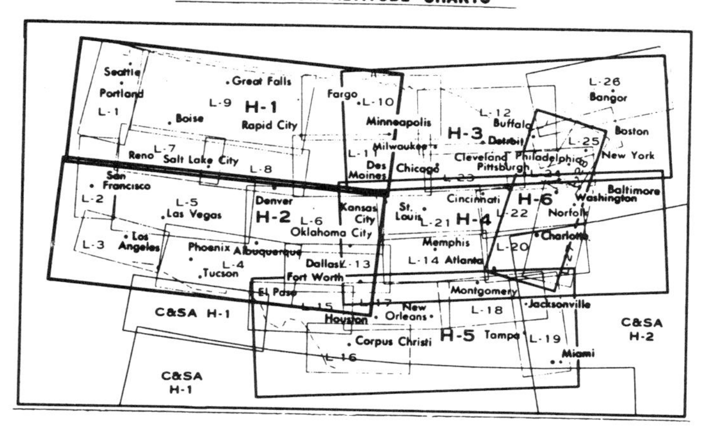

ALASKA ENROUTE CHARTS

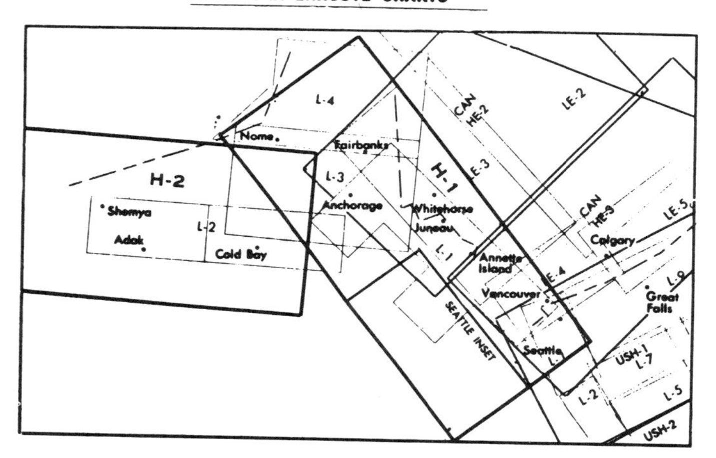

FLIGHT INFORMATION PUBLICATIONS

CARIBBEAN AND SOUTH AMERICA ENROUTE

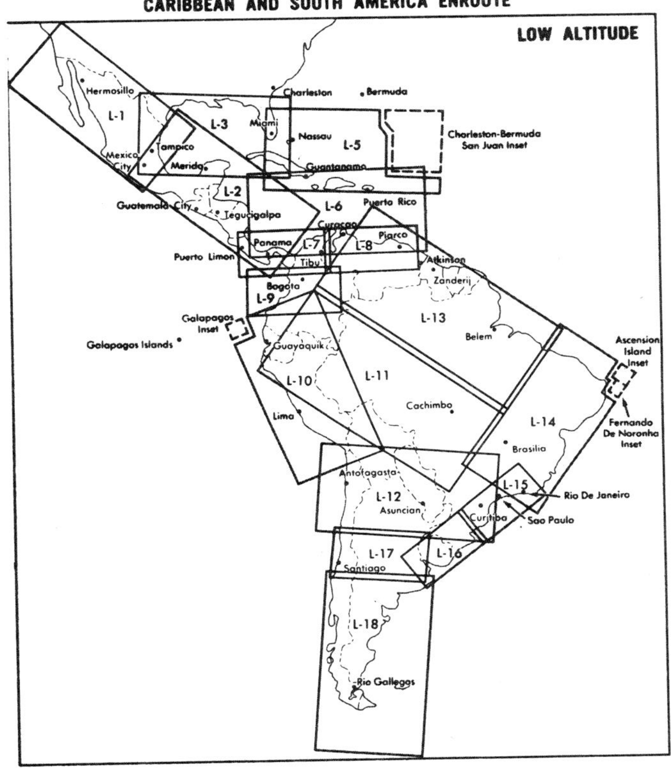

PLANNING CHARTS

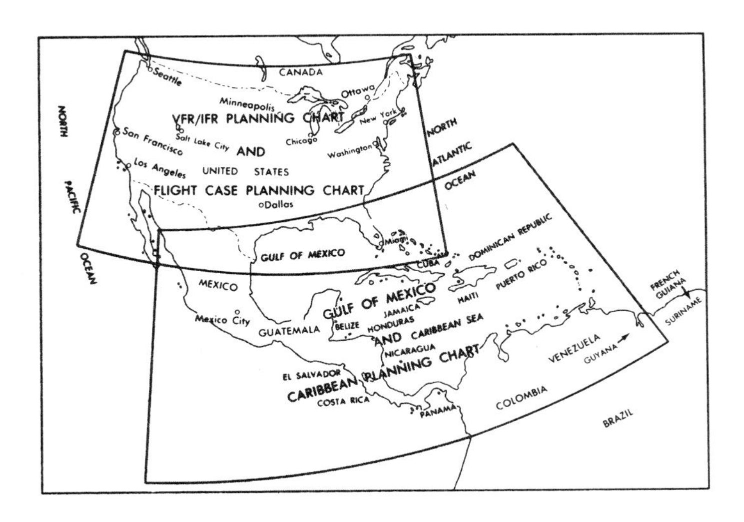

SECTIONAL AND VFR TERMINAL AREA CHARTS

CONTERMINOUS U.S. AND HAWAIIAN ISLANDS

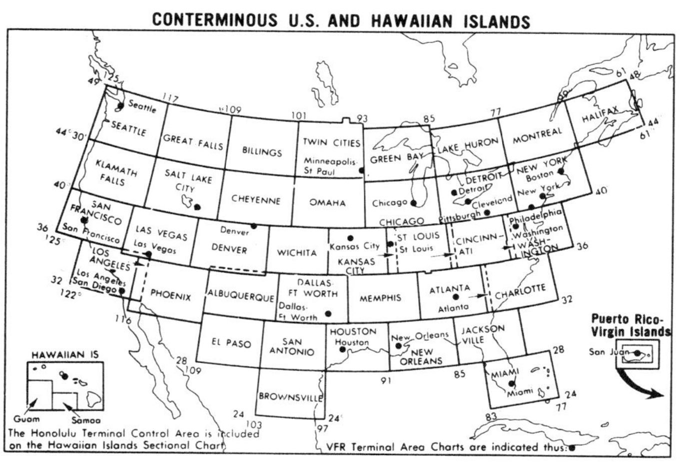

ALASKA

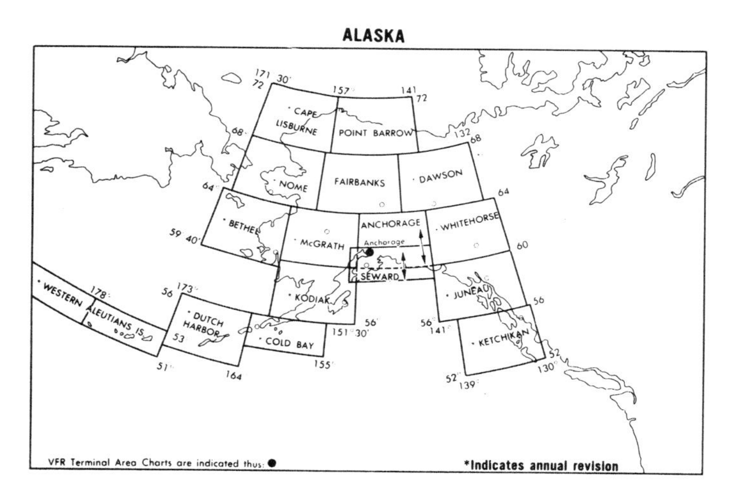

WORLD AERONAUTICAL CHARTS

CONTERMINOUS U.S., MEXICO. AND CARIBBEAN

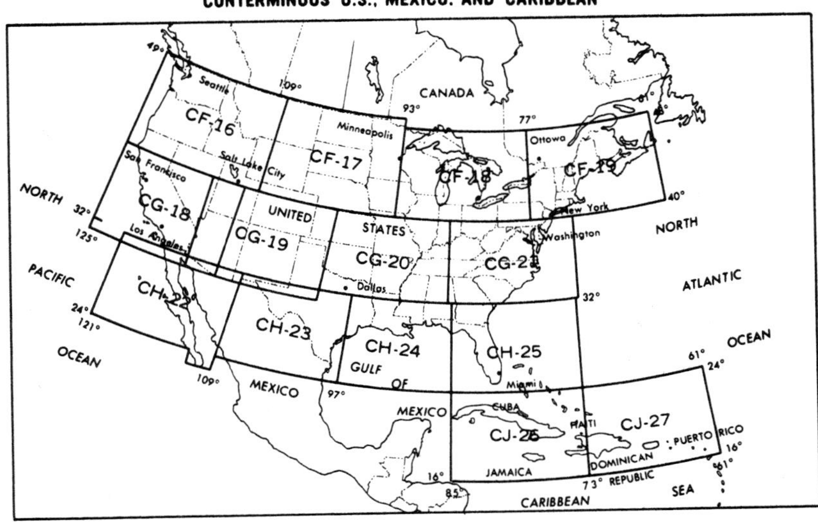

ALASKA

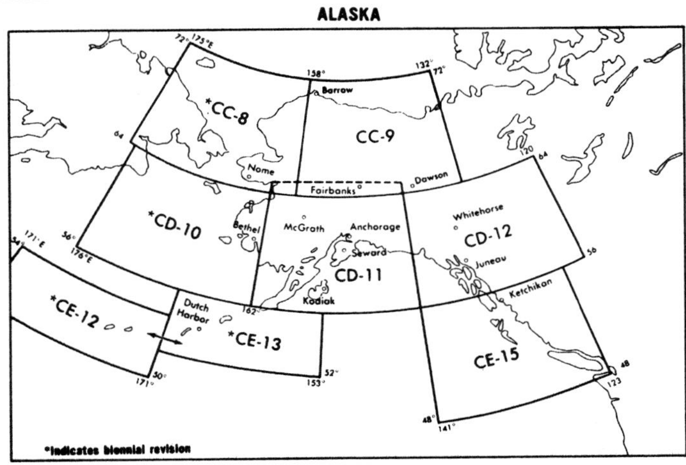

UNITED STATES GOVERNMENT

FLIGHT INFORMATION PUBLICATION

ENROUTE LOW ALTITUDE — U.S.

For use up to but not including 18,000 feet MSL
LEGEND

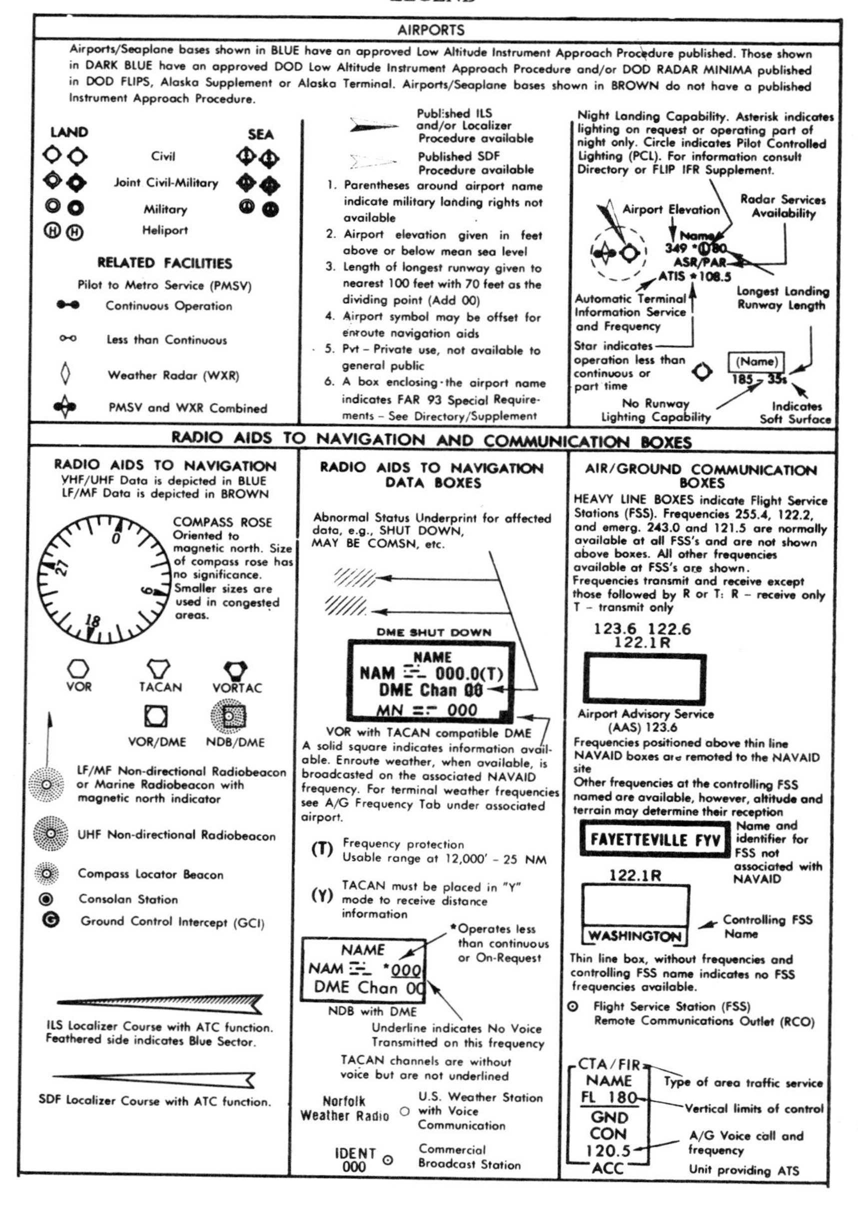

UNITED STATES GOVERNMENT

FLIGHT INFORMATION PUBLICATION

ENROUTE LOW ALTITUDE — U.S.

For use up to but not including 18,000 feet MSL
LEGEND
(Continued)

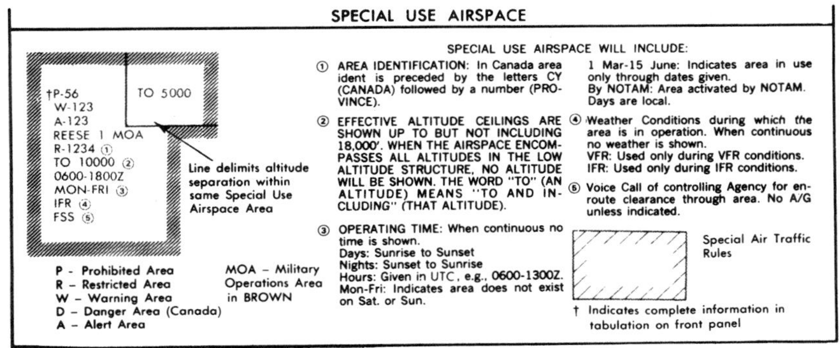

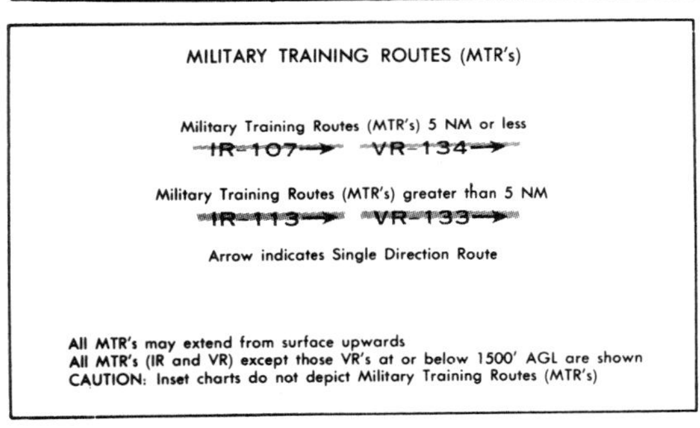

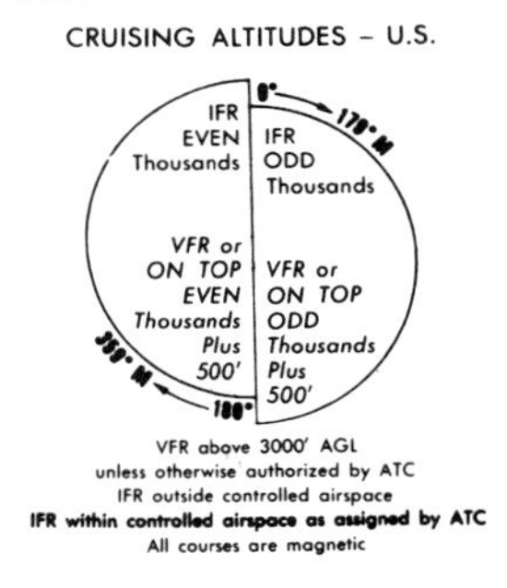

UNITED STATES GOVERNMENT

FLIGHT INFORMATION PUBLICATION

ENROUTE LOW ALTITUDE — U.S.

For use up to but not including 18,000 feet MSL

LEGEND

(Continued)

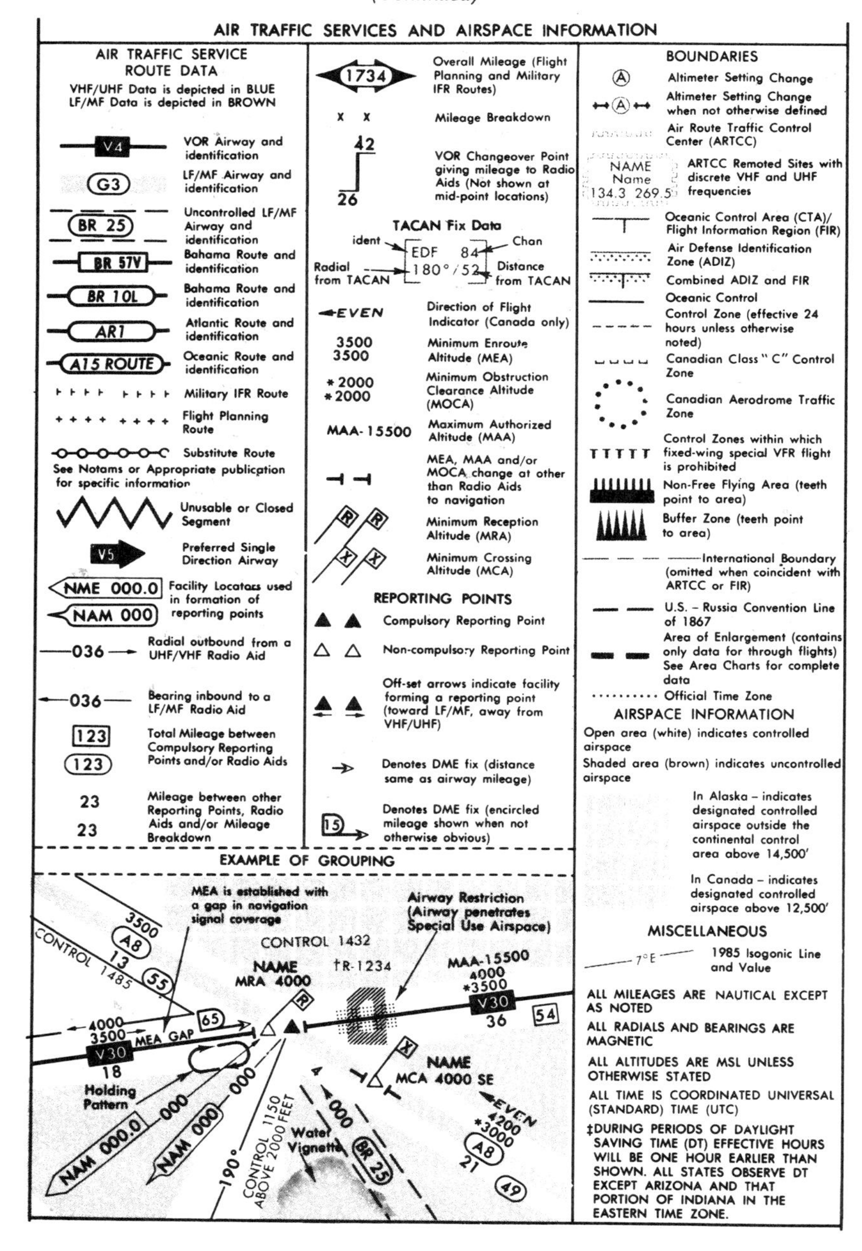

U.S. Department
of Transportation
**Federal Aviation
Administration**

PILOT/CONTROLLER GLOSSARY

October 15, 1992

Airman's Information Manual
Air Traffic Control Handbook
Flight Service Handbook

Airspace–Rules and Aeronautical Information Division (ATP–200)

PILOT/CONTROLLER GLOSSARY

FAA TERMS

This Glossary was compiled to promote a common understanding of the terms used in the Air Traffic Control system. It includes those terms which are intended for pilot/controller communications. Those terms most frequently used in pilot/controller communications are printed in ***bold italics***. The definitions are primarily defined in an operational sense applicable to both users and operators of the National Airspace System. Use of the Glossary will preclude any misunderstandings concerning the system's design, function, and purpose.

Because of the international nature of flying, terms used in the *Lexicon,* published by the International Civil Aviation Organization (ICAO), are included when they differ from FAA definitions. These terms are *italicized.* For the reader's convenience, there are also cross references to related terms in other parts of the Glossary and to other documents, such as the Federal Aviation Regulations (FAR's) and the Airman's Information Manual (AIM).

This Glossary will be revised, as necessary, to maintain a common understanding of the system.-

A

AAI—(See ARRIVAL AIRCRAFT INTERVAL).

AAR—(See AIRPORT ACCEPTANCE RATE).

ABBREVIATED IFR FLIGHT PLANS—An authorization by ATC requiring pilots to submit only that information needed for the purpose of ATC. It includes only a small portion of the usual IFR flight plan information. In certain instances, this may be only aircraft identification, location, and pilot request. Other information may be requested if needed by ATC for separation/control purposes. It is frequently used by aircraft which are airborne and desire an instrument approach or by aircraft which are on the ground and desire a climb to VFR-on-top.
(See VFR-ON-TOP). (Refer to AIM).

ABEAM—An aircraft is "abeam" a fix, point, or object when that fix, point, or object is approximately 90 degrees to the right or left of the aircraft track. Abeam indicates a general position rather than a precise point.

ABORT—To terminate a preplanned aircraft maneuver; e.g., an aborted takeoff.

ACC [ICAO]—(See AREA CONTROL CENTER).

ACCELERATE-STOP DISTANCE AVAILABLE—The runway plus stopway length declared available and suitable for the acceleration and deceleration of an airplane aborting a takeoff

ACCELERATE-STOP DISTANCE AVAILABLE [ICAO]—The length of the take-off run available plus the length of the stopway if provided.

ACDO—(See AIR CARRIER DISTRICT OFFICE).

ACKNOWLEDGE—Let me know that you have received my message.
(See ICAO term ACKNOWLEDGE).

ACKNOWLEDGE [ICAO]—Let me know that you have received and understood this message.

ACLS—(See AUTOMATIC CARRIER LANDING SYSTEM).

ACLT—(See ACTUAL CALCULATED LANDING TIME).

ACROBATIC FLIGHT—An intentional maneuver involving an abrupt change in an aircraft's attitude, an abnormal attitude, or abnormal acceleration not necessary for normal flight.
(Refer to Part 91). (See ICAO term ACROBATIC FLIGHT).

ACROBATIC FLIGHT [ICAO]—Manoeuvres intentionally performed by an aircraft involving an abrupt change in its attitude, an abnormal attitude, or an abnormal variation in speed.

ACTIVE RUNWAY—(See RUNWAY IN USE/ACTIVE RUNWAY/DUTY RUNWAY).

ACTUAL CALCULATED LANDING TIME—ACLT is a flight's frozen calculated landing time. An actual time determined at freeze calculated landing time (FCLT) or meter list display interval (MLDI) for the adapted vertex for each arrival aircraft based upon runway configuration, airport acceptance rate, airport arrival delay period, and other metered arrival aircraft. This time is either the vertex time of arrival (VTA) of the aircraft or the tentative calculated landing time (TCLT)/ACLT of the previous aircraft plus the arrival aircraft interval (AAI), whichever is later. This time will not be updated in response to the aircraft's progress.

ADDITIONAL SERVICES—Advisory information provided by ATC which includes but is not limited to the following:

1. Traffic advisories.
2. Vectors, when requested by the pilot, to assist aircraft receiving traffic advisories to avoid observed traffic.
3. Altitude deviation information of 300 feet or more from an assigned altitude as observed on a verified (reading correctly) automatic altitude readout (Mode C).
4. Advisories that traffic is no longer a factor.
5. Weather and chaff information.
6. Weather assistance.

7. Bird activity information.
8. Holding pattern surveillance. Additional services are provided to the extent possible contingent only upon the controller's capability to fit them into the performance of higher priority duties and on the basis of limitations of the radar, volume of traffic, frequency congestion, and controller workload. The controller has complete discretion for determining if he is able to provide or continue to provide a service in a particular case. The controller's reason not to provide or continue to provide a service in a particular case is not subject to question by the pilot and need not be made known to him.

(See TRAFFIC ADVISORIES). (Refer to AIM).

ADF—(See AUTOMATIC DIRECTION FINDER).

ADIZ—(See AIR DEFENSE IDENTIFICATION ZONE).

ADLY—(See ARRIVAL DELAY).

ADMINISTRATOR—The Federal Aviation Administrator or any person to whom he has delegated his authority in the matter concerned.

ADVISE INTENTIONS—Tell me what you plan to do.

ADVISORY—Advice and information provided to assist pilots in the safe conduct of flight and aircraft movement.
(See ADVISORY SERVICE).

ADVISORY FREQUENCY—The appropriate frequency to be used for Airport Advisory Service.
(See LOCAL AIRPORT ADVISORY). (See UNICOM). (Refer to ADVISORY CIRCULAR NO. 90–42). (Refer to AIM).

ADVISORY SERVICE—Advice and information provided by a facility to assist pilots in the safe conduct of flight and aircraft movement.
(See LOCAL AIRPORT ADVISORY). (See TRAFFIC ADVISORIES). (See SAFETY ALERT). (See ADDITIONAL SERVICES). (See RADAR ADVISORY). (See EN ROUTE FLIGHT ADVISORY SERVICE). (Refer to AIM).

AERIAL REFUELING—A procedure used by the military to transfer fuel from one aircraft to another during flight.
(Refer to VFR / IFR Wall Planning Charts).

AERODROME—A defined area on land or water (including any buildings, installations and equipment) intended to be used either wholly or in part for the arrival, departure, and movement of aircraft.

AERODROME BEACON [ICAO]—Aeronautical beacon used to indicate the location of an aerodrome from the air.

AERODROME CONTROL SERVICE [ICAO]—Air traffic control service for aerodrome traffic.

AERODROME CONTROL TOWER [ICAO]—A unit established to provide air traffic control service to aerodrome traffic.

AERODROME ELEVATION [ICAO]—The elevation of the highest point of the landing area.

AERODROME TRAFFIC CIRCUIT [ICAO]—The specified path to be flown by aircraft operating in the vicinity of an aerodrome.

AERONAUTICAL BEACON—A visual NAVAID displaying flashes of white and/or colored light to indicate the location of an airport, a heliport, a landmark, a certain point of a Federal airway in mountainous terrain, or an obstruction.
(See AIRPORT ROTATING BEACON). (Refer to AIM).

AERONAUTICAL CHART—A map used in air navigation containing all or part of the following: Topographic features, hazards and obstructions, navigation aids, navigation routes, designated airspace, and airports. Commonly used aeronautical charts are:

1. Sectional Charts (1:500,000). Designed for visual navigation of slow or medium speed aircraft. Topographic information on these charts features the portrayal of relief and a judicious selection of visual check points for VFR flight. Aeronautical information includes visual and radio aids to navigation, airports, controlled airspace, restricted areas, obstructions, and related data.

2. VFR Terminal Area Charts (1:250,000). Depict Terminal Control Area (TCA) airspace which provides for the control or segregation of all the aircraft within the TCA. The chart depicts topographic information and aeronautical information which includes visual and radio aids to navigation, airports, controlled airspace, restricted areas, obstructions, and related data.

3. World Aeronautical Charts (WAC) (1:1,000,000). Provide a standard series of aeronautical charts covering land areas of the world at a size and scale convenient for navigation by moderate speed aircraft. Topographic information includes cities and towns, principal roads, railroads, distinctive landmarks, drainage, and relief. Aeronautical information includes visual and radio aids to navigation, airports, airways, restricted areas, obstructions, and other pertinent data.

4. En Route Low Altitude Charts. Provide aeronautical information for en route instrument navigation (IFR) in the low altitude stratum. Information includes the portrayal of airways, limits of controlled airspace, position identification and frequencies of radio aids, selected airports, minimum en route and minimum obstruction clearance altitudes, airway distances, reporting points, restricted areas, and related data. Area charts, which are a part of this series, furnish terminal data at a larger scale in congested areas.

5. En Route High Altitude Charts. Provide aeronautical information for en route instrument navigation (IFR) in the high altitude stratum. Information includes the portrayal of jet routes, identification and frequencies of radio aids, selected airports, distances, time zones, special use airspace, and related information.

6. Instrument Approach Procedures (IAP) Charts. Portray the aeronautical data which is required to execute an instrument approach to an airport. These charts depict the procedures, including all related data, and the airport diagram. Each procedure is designated for use with a specific type of electronic navigation system including NDB, TACAN, VOR, ILS/MLS, and RNAV. These charts are identified by the type of navigational aid(s) which provide final approach guidance.

7. Standard Instrument Departure (SID) Charts. Designed to expedite clearance delivery and to facilitate transition between takeoff and en route operations. Each SID procedure is presented as a separate chart and may serve a single airport or more than one airport in a given geographical location.

8. Standard Terminal Arrival (STAR) Charts. Designed to expedite air traffic control arrival procedures and to facilitate transition between en route and instrument approach operations. Each STAR procedure is presented as a separate chart and may serve a single airport or more than one airport in a given geographical location.

9. Airport Taxi Charts. Designed to expedite the efficient and safe flow of ground traffic at an airport. These charts are identified by the official airport name; e.g., Washington National Airport.

(See ICAO term AERONAUTICAL CHART).

AERONAUTICAL CHART [ICAO]—A representation of a portion of the earth, its culture and relief, specifically designated to meet the requirements of air navigation.

AERONAUTICAL INFORMATION PUBLICATION [AIP] [ICAO]—A publication issued by or with the authority of a State and containing aeronautical information of a lasting character essential to air navigation.

A/FD—(See AIRPORT/FACILITY DIRECTORY).

AFFIRMATIVE—Yes.

AIM—(See AIRMAN'S INFORMATION MANUAL).

AIP [ICAO]—(See AERONAUTICAL INFORMATION PUBLICATION).

AIRBORNE DELAY—Amount of delay to be encountered in airborne holding.

AIR CARRIER DISTRICT OFFICE—An FAA field office serving an assigned geographical area, staffed with Flight Standards personnel serving the aviation industry and the general public on matters related to the certification and operation of scheduled air carriers and other large aircraft operations.

AIRCRAFT—Device(s) that are used or intended to be used for flight in the air, and when used in air traffic control terminology, may include the flight crew.

(See ICAO term AIRCRAFT).

AIRCRAFT [ICAO]—Any machine that can derive support in the atmosphere from the reactions of the air other than the reactions of the air against the earth's surface.

AIRCRAFT APPROACH CATEGORY—A grouping of aircraft based on a speed of 1.3 times the stall speed in the landing configuration at maximum gross landing weight. An aircraft shall fit in only one category. If it is necessary to maneuver at speeds in excess of the upper limit of a speed range for a category, the minimums for the next higher category should be used. For example, an aircraft which falls in Category A, but is circling to land at a speed in excess of 91 knots, should use the approach Category B minimums when circling to land. The categories are as follows:

1. Category A. Speed less than 91 knots.
2. Category B. Speed 91 knots or more but less than 121 knots.
3. Category C. Speed 121 knots or more but less than 141 knots.
4. Category D. Speed 141 knots or more but less than 166 knots.
5. Category E. Speed 166 knots or more.

(Refer to Part 1). (Refer to Part 97).

AIRCRAFT CLASSES—For the purposes of Wake Turbulence Separation Minima, ATC classifies aircraft as Heavy, Large, and Small as follows:

1. Heavy. Aircraft capable of takeoff weights of 300,000 pounds or more whether or not they are operating at this weight during a particular phase of flight.
2. Large. Aircraft of more than 12,500 pounds, maximum certificated takeoff weight, up to 300,000 pounds.
3. Small. Aircraft of 12,500 pounds or less maximum certificated takeoff weight.

(Refer to AIM).

AIRCRAFT SITUATION DISPLAY—ASD is a computer system that receives radar track data from all 20 CONUS ARTCC's, organizes this data into a mosaic display, and presents it on a computer screen. The display allows the traffic management coordinator multiple methods of selection and high-lighting of individual aircraft or groups of aircraft. The user has the option of superimposing these aircraft positions over any number of background displays. These background options include ARTCC boundaries, any stratum of en route sector boundaries, fixes, airways, military and other special use airspace, airports, and geopolitical boundaries. By using the ASD, a coordinator can monitor any number of traffic situations or the entire systemwide traffic flows.

AIR DEFENSE EMERGENCY—A military emergency condition declared by a designated authority. This condition exists when an attack upon the continental U.S., Alaska, Canada, or U.S. installations in Greenland by hostile aircraft or missiles is considered probable, is imminent, or is taking place.

(Refer to AIM).

AIR DEFENSE IDENTIFICATION ZONE—The area of airspace over land or water, extending upward from the surface, within which the ready identification, the location, and the control of aircraft are required in the interest of national security.

1. Domestic Air Defense Identification Zone. An ADIZ within the United States along an international boundary of the United States.
2. Coastal Air Defense Identification Zone. An ADIZ over the coastal waters of the United States.
3. Distant Early Warning Identification Zone (DEWIZ). An ADIZ over the coastal waters of the State of Alaska.

ADIZ locations and operating and flight plan requirements for civil aircraft operations are specified in Part 99.

(Refer to AIM).

AIRMAN'S INFORMATION MANUAL—A primary FAA publication whose purpose is to instruct airmen about operating in the National Airspace System of the U.S. It provides basic flight information, ATC Procedures and general instructional information concerning health, medical facts, factors affecting flight safety, accident and hazard reporting, and types of aeronautical charts and their use.

AIRMAN'S METEOROLOGICAL INFORMATION—(See AIRMET).

AIRMET—In-flight weather advisories issued only to amend the area forecast concerning weather phenomena which are of operational interest to all aircraft and potentially hazardous to aircraft having limited capability because of lack of equipment, instrumentation, or pilot qualifications. AIRMET's concern weather of less severity than that covered by SIGMET's or Convective SIGMET's. AIRMET's cover moderate icing, moderate turbulence, sustained winds of 30 knots or more at the surface, widespread areas of ceilings less than 1,000 feet and/or visibility less than 3 miles, and extensive mountain obscurement.

(See AWW). (See SIGMET). (See CONVECTIVE SIGMET). (See CWA). (Refer to AIM).

AIR NAVIGATION FACILITY—Any facility used in, available for use in, or designed for use in, aid of air navigation, including landing areas, lights, any apparatus or equipment for disseminating weather information, for signaling, for radio-directional finding, or for radio or other electrical communication, and any other structure or mechanism having a similar purpose for guiding or controlling flight in the air or the landing and take-off of aircraft.

(See NAVIGATIONAL AID).

AIRPORT—An area on land or water that is used or intended to be used for the landing and takeoff of aircraft and includes its buildings and facilities, if any.

AIRPORT ACCEPTANCE RATE—A dynamic input parameter specifying the number of arriving aircraft which an airport or airspace can accept from the ARTCC per hour. The AAR is used to calculate the desired interval between successive arrival aircraft.

AIRPORT ADVISORY AREA—The area within ten miles of an airport without a control tower or where the tower is not in operation, and on which a Flight Service Station is located.

(See LOCAL AIRPORT ADVISORY). (Refer to AIM).

AIRPORT ELEVATION—The highest point of an airport's usable runways measured in feet from mean sea level.

(See TOUCHDOWN ZONE ELEVATION). (See ICAO term AERODROME ELEVATION).

AIRPORT/FACILITY DIRECTORY—A publication designed primarily as a pilot's operational manual containing all airports, seaplane bases, and heliports open to the public including communications data, navigational facilities, and certain special notices and procedures. This publication is issued in seven volumes according to geographical area.

AIRPORT INFORMATION AID—(See AIRPORT INFORMATION DESK).

AIRPORT INFORMATION DESK—An airport unmanned facility designed for pilot self-service briefing, flight planning, and filing of flight plans.

(Refer to AIM).

AIRPORT LIGHTING—Various lighting aids that may be installed on an airport. Types of airport lighting include:

1. Approach Light System (ALS). An airport lighting facility which provides visual guidance to landing aircraft by radiating light beams in a directional pattern by which the pilot aligns the aircraft with the extended centerline of the runway on his final approach for landing. Condenser-Discharge Sequential Flashing Lights/ Sequenced Flashing Lights may be installed in conjunction with the ALS at some airports. Types of Approach Light Systems are:

   a. ALSF–1.—Approach Light System with Sequenced Flashing Lights in ILS Cat-I configuration.

   b. ALSF–2.—Approach Light System with Sequenced Flashing Lights in ILS Cat-II configuration. The ALSF–2 may operate as an SSALR when weather conditions permit.

   c. SSALF.—Simplified Short Approach Light System with Sequenced Flashing Lights.

   d. SSALR.—Simplified Short Approach Light System with Runway Alignment Indicator Lights.

   e. MALSF.—Medium Intensity Approach Light System with Sequenced Flashing Lights.

   f. MALSR.—Medium Intensity Approach Light System with Runway Alignment Indicator Lights.

   g. LDIN.—Lead-in-light system: Consists of one or more series of flashing lights installed at or near ground level that provides positive visual guidance along an approach path, either curving or straight, where special problems exist with hazardous terrain, obstructions, or noise abatement procedures.

   h. RAIL.—Runway Alignment Indicator Lights (Sequenced Flashing Lights which are installed only in combination with other light systems).

   i. ODALS.—Omnidirectional Approach Lighting System consists of seven omnidirectional flashing lights located in the approach area of a nonprecision runway. Five lights are located on the runway centerline extended with the first light located 300 feet from the threshold and extending at equal intervals up to 1,500 feet from the threshold. The other two lights are located, one on each side of the runway threshold, at a lateral distance of 40 feet from the runway edge, or 75 feet from the runway edge when installed on a runway equipped with a VASI.

(Refer to Order 6850.2).

2. Runway Lights/Runway Edge Lights. Lights having a prescribed angle of emission used to define the lateral limits of a runway. Runway lights are uniformly spaced at intervals of approximately 200 feet, and the intensity may be controlled or preset.

3. Touchdown Zone Lighting. Two rows of tranverse light bars located symmetrically about the runway centerline normally at 100 foot intervals. The basic system extends 3,000 feet along the runway.

4. Runway Centerline Lighting. Flush centerline lights spaced at 50–foot intervals beginning 75 feet from the landing threshold and extending to within 75 feet of the opposite end of the runway.

5. Threshold Lights. Fixed green lights arranged symmetrically left and right of the runway centerline, identifying the runway threshold.

6. Runway End Identifier Lights (REIL). Two synchronized flashing lights, one on each side of the runway threshold, which provide rapid and positive identification of the approach end of a particular runway.

7. Visual Approach Slope Indicator (VASI). An airport lighting facility providing vertical visual approach slope guidance to aircraft during approach to landing by radiating a directional pattern of high intensity red and white focused light beams which indicate to the pilot that he is "on path" if he sees red/white, "above path' if white/white, and "below path" if red/red. Some airports serving large aircraft have three-bar VASIs which provide two visual glide paths to the same runway.

8. Boundary Lights. Lights defining the perimeter of an airport or landing area.

(Refer to AIM).

AIRPORT MARKING AIDS—Markings used on runway and taxiway surfaces to identify a specific runway, a runway threshold, a centerline, a hold line, etc. A runway should be marked in accordance with its present usage such as:

1. Visual.
2. Nonprecision instrument.
3. Precision instrument.

(Refer to AIM).

AIRPORT RADAR SERVICE AREA—(See CONTROLLED AIRSPACE).

AIRPORT RESERVATION OFFICE—Office responsible for monitoring the operation of the high density rule. Receives and processes requests for IFR operations at high density traffic airports.

AIRPORT ROTATING BEACON—A visual NAVAID operated at many airports. At civil airports, alternating white and green flashes indicate the location of the airport. At military airports, the beacons flash alternately white and green, but are differentiated from civil beacons by dualpeaked (two quick) white flashes between the green flashes.

(See SPECIAL VFR OPERATIONS). (See INSTRUMENT FLIGHT RULES). (Refer to AIM). (See ICAO term AERODROME BEACON).

AIRPORT SURFACE DETECTION EQUIPMENT—Radar equipment specifically designed to detect all principal features on the surface of an airport, including aircraft and vehicular traffic, and to present the entire image on a radar indicator console in the control tower. Used to augment visual observation by tower personnel of aircraft and/or vehicular movements on runways and taxiways.

AIRPORT SURVEILLANCE RADAR—Approach control radar used to detect and display an aircraft's position in the terminal area. ASR provides range and azimuth information but does not provide elevation data. Coverage of the ASR can extend up to 60 miles.

AIRPORT TAXI CHARTS—(See AERONAUTICAL CHART).

AIR TRAFFIC [ICAO]—All aircraft in flight or operating on the manoeuvring area of an aerodrome.

AIRPORT TRAFFIC AREA—Unless otherwise specifically designated in Part 93, that airspace within a horizontal radius of 5 statute miles from the geographical center of any airport at which a control tower is operating, extending from the surface up to, but not including, an altitude of 3,000 feet above the elevation of an airport. Unless otherwise authorized or required by ATC, no person may operate an aircraft within an airport traffic area except for the purpose of landing at or taking off from an airport within that area. ATC authorizations may be given as individual approval of specific operations or may be contained in written agreements between airport users and the tower concerned.

(Refer to Part 1 and Part 91).

AIRPORT TRAFFIC CONTROL SERVICE—A service provided by a control tower for aircraft operating on the movement area and in the vicinity of an airport.

(See MOVEMENT AREA). (See TOWER). (See ICAO term AERODROME CONTROL SERVICE).

AIR TRAFFIC CONTROL CLEARANCE [ICAO]—Authorization for an aircraft to proceed under conditions specified by an air traffic control unit.

Note 1: For convenience, the term air traffic control clearance is frequently abbreviated to clearance when used in appropriate contexts.

Note 2: The abbreviated term clearance may be prefixed by the words taxi, takeoff, departure, en route, approach or landing to indicate the particular portion of flight to which the air traffic control clearance relates.

AIR TRAFFIC CONTROL SERVICE [ICAO]—A service provided for the purpose of:

1. Preventing collisions:
   a. Between aircraft; and
   b. On the maneuvering area between aircraft and obstructions; and
2. Expediting and maintaining an orderly flow of air traffic.

AIR TRAFFIC CONTROL SYSTEM COMMAND CENTER—An Air Traffic Operations Service facility consisting of four operational units.

1. Central Flow Control Function (CFCF). Responsible for coordination and approval of all major intercenter flow control restrictions on a system basis in order to obtain maximum utilization of the airspace.

(See QUOTA FLOW CONTROL).

2. Central Altitude Reservation Function (CARF). Responsible for coordinating, planning, and approving special user requirements under the Altitude Reservation (ALTRV) concept.

(See ALTITUDE RESERVATION).

3. Airport Reservation Office (ARO). Responsible for approving IFR flights at designated high density traffic airports (John F. Kennedy, LaGuardia, O'Hare, and Washington National) during specified hours.

(Refer to Part 93 and AIRPORT/FACILITY DIRECTORY).

4. ATC Contingency Command Post. A facility which enables the FAA to manage the ATC system when significant portions of the system's capabilities have been lost or are threatened.

AIRPORT TRAFFIC CONTROL TOWER—(See TOWER).

AIR ROUTE SURVEILLANCE RADAR—Air route traffic control center (ARTCC) radar used primarily to detect and display an aircraft's position while en route between terminal areas. The ARSR enables controllers to provide radar air traffic control service when aircraft are within the ARSR coverage. In some instances, ARSR may enable an ARTCC to provide terminal radar services similar to but usually more limited than those provided by a radar approach control.

AIR ROUTE TRAFFIC CONTROL CENTER—A facility established to provide air traffic control service to aircraft operating on IFR flight plans within controlled airspace and principally during the en route phase of flight. When equipment capabilities and controller workload permit, certain advisory/assistance services may be provided to VFR aircraft.

(See NAS STAGE A). (See EN ROUTE AIR TRAFFIC CONTROL SERVICES). (Refer to AIM).

AIRSPEED—The speed of an aircraft relative to its surrounding air mass. The unqualified term "airspeed' means one of the following:

1. Indicated Airspeed. The speed shown on the aircraft airspeed indicator. This is the speed used in pilot/controller communications under the general term "airspeed."

(Refer to Part 1).

2. True Airspeed. The airspeed of an aircraft relative to undisturbed air. Used primarily in flight planning and en route portion of flight. When used in pilot/controller communications, it is referred to as "true airspeed" and not shortened to "airspeed."

AIRSTART—The starting of an aircraft engine while the aircraft is airborne, preceded by engine shutdown during training flights or by actual engine failure.

AIR TAXI—Used to describe a helicopter/VTOL aircraft movement conducted above the surface but normally not above 100 feet AGL. The aircraft may proceed either via hover taxi or flight at speeds more than 20 knots. The pilot is solely responsible for selecting a safe airspeed/altitude for the operation being conducted.
(See HOVER TAXI). (Refer to AIM).

AIR TRAFFIC—Aircraft operating in the air or on an airport surface, exclusive of loading ramps and parking areas.
(See ICAO term AIR TRAFFIC).

AIR TRAFFIC CLEARANCE—An authorization by air traffic control, for the purpose of preventing collision between known aircraft, for an aircraft to proceed under specified traffic conditions within controlled airspace.
(See ATC INSTRUCTIONS). (See ICAO term AIR TRAFFIC CONTROL CLEARANCE).

AIR TRAFFIC CONTROL—A service operated by appropriate authority to promote the safe, orderly and expeditious flow of air traffic.
(See ICAO term AIR TRAFFIC CONTROL SERVICE).

AIR TRAFFIC CONTROL SERVICE—(See AIR TRAFFIC CONTROL).

AIR TRAFFIC CONTROL SPECIALIST—A person authorized to provide air traffic control service.
(See AIR TRAFFIC CONTROL). (See FLIGHT SERVICE STATION). (See ICAO term CONTROLLER).

AIRWAY—A control area or portion thereof established in the form of a corridor, the centerline of which is defined by radio navigational aids.
(See FEDERAL AIRWAYS). (Refer to Part 71). (Refer to AIM). (See ICAO term AIRWAY).

AIRWAY [ICAO]—A control area or portion thereof established in the form of corridor equipped with radio navigational aids.

AIRWAY BEACON—Used to mark airway segments in remote mountain areas. The light flashes Morse Code to identify the beacon site.
(Refer to AIM).

AIT—(See AUTOMATED INFORMATION TRANSFER).

ALERFA (Alert Phase) [ICAO]—The code word used to designate an emergency phase wherein apprehension exists as to the safety of an aircraft and its occupants.

ALERT AREA—(See SPECIAL USE AIRSPACE).

ALERT NOTICE—A request originated by a flight service station (FSS) or an air route traffic control center (ARTCC) for an extensive communication search for overdue, unreported, or missing aircraft.

ALERTING SERVICE—A service provided to notify appropriate organizations regarding aircraft in need of search and rescue aid and assist such organizations as required.

ALNOT—(See ALERT NOTICE).

ALPHANUMERIC DISPLAY—Letters and numerals used to show identification, altitude, beacon code, and other information concerning a target on a radar display.
(See AUTOMATED RADAR TERMINAL SYSTEMS). (See NAS STAGE A).

ALTERNATE AERODROME [ICAO]—An aerodrome specified in the flight plan to which a flight may proceed when it becomes inadvisable to land at the aerodrome of intended landing.
Note: An alternate aerodrome may be the aerodrome of departure.

ALTERNATE AIRPORT—An airport at which an aircraft may land if a landing at the intended airport becomes inadvisable.
(See FAA term ICAO term ALTERNATE AERODROME).

ALTIMETER SETTING—The barometric pressure reading used to adjust a pressure altimeter for variations in existing atmospheric pressure or to the standard altimeter setting (29.92).
(Refer to Part 91). (Refer to AIM).

ALTITUDE—The height of a level, point, or object measured in feet Above Ground Level (AGL) or from Mean Sea Level (MSL).

(See FLIGHT LEVEL).

1. MSL Altitude. Altitude expressed in feet measured from mean sea level.
2. AGL Altitude. Altitude expressed in feet measured above ground level.
3. Indicated Altitude. The altitude as shown by an altimeter. On a pressure or barometric altimeter it is altitude as shown uncorrected for instrument error and uncompensated for variation from standard atmospheric conditions.

(See ICAO term ALTITUDE).

ALTITUDE [ICAO]—The vertical distance of a level, a point or an object considered as a point, measured from mean sea level (MSL).

ALTITUDE READOUT—An aircraft's altitude, transmitted via the Mode C transponder feature, that is visually displayed in 100–foot increments on a radar scope having readout capability.

(See AUTOMATED RADAR TERMINAL SYSTEMS). (See NAS STAGE A). (See ALPHANUMERIC DISPLAY). (Refer to AIM).

ALTITUDE RESERVATION—Airspace utilization under prescribed conditions normally employed for the mass movement of aircraft or other special user requirements which cannot otherwise be accomplished. ALTRVs are approved by the appropriate FAA facility.

(See AIR TRAFFIC CONTROL SYSTEM COMMAND CENTER).

ALTITUDE RESTRICTION—An altitude or altitudes, stated in the order flown, which are to be maintained until reaching a specific point or time. Altitude restrictions may be issued by ATC due to traffic, terrain, or other airspace considerations.

ALTITUDE RESTRICTIONS ARE CANCELED—Adherence to previously imposed altitude restrictions is no longer required during a climb or descent.

ALTRV—(See ALTITUDE RESERVATION).

AMVER—(See AUTOMATED MUTUAL–ASSISTANCE VESSEL RESCUE SYSTEM

APPROACH CLEARANCE—Authorization by ATC for a pilot to conduct an instrument approach. The type of instrument approach for which a clearance and other pertinent information is provided in the approach clearance when required.

(See INSTRUMENT APPROACH PROCEDURE). (See CLEARED APPROACH). (Refer to AIM and Part 91).

APPROACH CONTROL FACILITY—A terminal ATC facility that provides approach control service in a terminal area.

(See APPROACH CONTROL SERVICE). (See RADAR APPROACH CONTROL FACILITY).

APPROACH CONTROL SERVICE—Air traffic control service provided by an approach control facility for arriving and departing VFR/IFR aircraft and, on occasion, en route aircraft. At some airports not served by an approach control facility, the ARTCC provides limited approach control service.

(Refer to AIM). (See ICAO term APPROACH CONTROL SERVICE).

APPROACH CONTROL SERVICE [ICAO]—Air traffic control service for arriving or departing controlled flights.

APPROACH GATE—An imaginary point used within ATC as a basis for vectoring aircraft to the final approach course. The gate will be established along the final approach course 1 mile from the outer marker (or the fix used in lieu of the outer marker) on the side away from the airport for precision approaches and 1 mile from the final approach fix on the side away from the airport for nonprecision approaches. In either case when measured along the final approach course, the gate will be no closer than 5 miles from the landing threshold.

APPROACH LIGHT SYSTEM—(See AIRPORT LIGHTING).

APPROACH SEQUENCE—The order in which aircraft are positioned while on approach or awaiting approach clearance.

(See LANDING SEQUENCE). (See ICAO term APPROACH SEQUENCE).

APPROACH SEQUENCE [ICAO]—The order in which two or more aircraft are cleared to approach to land at the aerodrome.

APPROACH SPEED—The recommended speed contained in aircraft manuals used by pilots when making an approach to landing. This speed will vary for different segments of an approach as well as for aircraft weight and configuration.

APPROPRIATE ATS AUTHORITY [ICAO]—The relevant authority designated by the State responsible for providing air traffic services in the airspace concerned. In the United States, the "appropriate ATS authority" is the Director, Office of Air Traffic System Management, ATM-1.

APPROPRIATE AUTHORITY—

1. Regarding flight over the high seas: the relevant authority is the State of Registry.
2. Regarding flight over other than the high seas: the relevant authority is the State having sovereignty over the territory being overflown.

APPROPRIATE OBSTACLE CLEARANCE MINIMUM ALTITUDE—Any of the following:

(See Minimum IFR Altitude MIA). (See Minimum En Route Altitude MEA). (See Minimum Obstruction Clearance Altitude MOCA). (See Minimum Vectoring Altitude MVA).

APPROPRIATE TERRAIN CLEARANCE MINIMUM ALTITUDE—Any of the following:

(See Minimum IFR Altitude MIA). (See Minimum En Route Altitude MEA). (See Minimum Obstruction Clearance Altitude MOCA). (See Minimum Vectoring Altitude MVA).

APRON—A defined area on an airport or heliport intended to accommodate aircraft for purposes of loading or unloading passengers or cargo, refueling, parking, or maintenance. With regard to seaplanes, a ramp is used for access to the apron from the water.

(See ICAO term APRON).

APRON [ICAO]—A defined area, on a land aerodrome, intended to accommodate aircraft for purposes of loading or unloading passengers, mail or cargo, refuelling, parking or maintenance.

ARC—The track over the ground of an aircraft flying at a constant distance from a navigational aid by reference to distance measuring equipment (DME).

AREA CONTROL CENTER [ICAO]—An ICAO term for an air traffic control facility primarily responsible for ATC services being provided IFR aircraft during the en route phase of flight. The U.S. equivalent facility is an air route traffic control center (ARTCC).

AREA NAVIGATION—A method of navigation that permits aircraft operation on any desired course within the coverage of station-referenced navigation signals or within the limits of a self-contained system capability. Random area navigation routes are direct routes, based on area navigation capability, between waypoints defined in terms of latitude/longitude coordinates, degree/distance fixes, or offsets from published or established routes/airways at a specified distance and direction. The major types of equipment are:

1. VORTAC referenced or Course Line Computer (CLC) systems, which account for the greatest number of RNAV units in use. To function, the CLC must be within the service range of a VORTAC.
2. OMEGA/VLF, although two separate systems, can be considered as one operationally. A long-range navigation system based upon Very Low Frequency radio signals transmitted from a total of 17 stations worldwide.
3. Inertial (INS) systems, which are totally self-contained and require no information from external references. They provide aircraft position and navigation information in response to signals resulting from inertial effects on components within the system.
4. MLS Area Navigation (MLS/RNAV), which provides area navigation with reference to an MLS ground facility.
5. LORAN-C is a long-range radio navigation system that uses ground waves transmitted at low frequency to provide user position information at ranges of up to 600 to 1,200 nautical miles at both en route and approach altitudes. The usable signal coverage areas are determined by the signal-to-noise ratio, the envelope-to-cycle difference, and the geometric relationship between the positions of the user and the transmitting stations.

(See ICAO term AREA NAVIGATION).

AREA NAVIGATION [ICAO]—A method of navigation which permits aircraft operation on any desired flight path within the coverage of station-referenced navigation aids or within the limits of the capability of self-contained aids, or a combination of these.

ARINC—An acronym for Aeronautical Radio, Inc., a corporation largely owned by a group of airlines. ARINC is licensed by the FCC as an aeronautical station and contracted by the FAA to provide communications support for air traffic control and meteorological services in portions of international airspace.

ARMY AVIATION FLIGHT INFORMATION BULLETIN—A bulletin that provides air operation data covering Army, National Guard, and Army Reserve aviation activities.

ARO—(See AIRPORT RESERVATION OFFICE).

ARRESTING SYSTEM—A safety device consisting of two major components, namely, engaging or catching devices and energy absorption devices for the purpose of arresting both tailhook and/or nontailhook-equipped aircraft. It is used to prevent aircraft from overrunning runways when the aircraft cannot be stopped after landing or during aborted takeoff. Arresting systems have various names; e.g., arresting gear, hook device, wire barrier cable.
(See ABORT). (Refer to AIM).

ARRIVAL AIRCRAFT INTERVAL—An internally generated program in hundredths of minutes based upon the AAR. AAI is the desired optimum interval between successive arrival aircraft over the vertex.

ARRIVAL CENTER—The ARTCC having jurisdiction for the impacted airport.

ARRIVAL DELAY—A parameter which specifies a period of time in which no aircraft will be metered for arrival at the specified airport.

ARRIVAL SECTOR—An operational control sector containing one or more meter fixes.

ARRIVAL SECTOR ADVISORY LIST—An ordered list of data on arrivals displayed at the PVD of the sector which controls the meter fix.

ARRIVAL SEQUENCING PROGRAM—The automated program designed to assist in sequencing aircraft destined for the same airport.

ARRIVAL TIME—The time an aircraft touches down on arrival.

ARSA—(See AIRPORT RADAR SERVICE AREA).

ARSR—(See AIR ROUTE SURVEILLANCE RADAR).

ARTCC—(See AIR ROUTE TRAFFIC CONTROL CENTER).

ARTS—(See AUTOMATED RADAR TERMINAL SYSTEMS).

ASD—(See AIRCRAFT SITUATION DISPLAY).

ASDA—(See ACCELERATE–STOP DISTANCE AVAILABLE).

ASDA [ICAO]—(See ICAO Term ACCELERATE–STOP DISTANCE AVAILABLE).

ASDE—(See AIRPORT SURFACE DETECTION EQUIPMENT).

ASP—(See ARRIVAL SEQUENCING PROGRAM).

ASR—(See AIRPORT SURVEILLANCE RADAR).

ASR APPROACH—(See SURVEILLANCE APPROACH).

ATC—(See AIR TRAFFIC CONTROL).

ATCAA—(See ATC ASSIGNED AIRSPACE).

ATC ADVISES—Used to prefix a message of noncontrol information when it is relayed to an aircraft by other than an air traffic controller.
(See ADVISORY).

ATC ASSIGNED AIRSPACE—Airspace of defined vertical/lateral limits, assigned by ATC, for the purpose of providing air traffic segregation between the specified activities being conducted within the assigned airspace and other IFR air traffic.

(See SPECIAL USE AIRSPACE).

ATC CLEARANCE—(See AIR TRAFFIC CLEARANCE).

ATC CLEARS—Used to prefix an ATC clearance when it is relayed to an aircraft by other than an air traffic controller.

ATC INSTRUCTIONS—Directives issued by air traffic control for the purpose of requiring a pilot to take specific actions; e.g., "Turn left heading two five zero," "Go around," "Clear the runway."

(Refer to Part 91).

ATCRBS—(See RADAR).

ATC REQUESTS—Used to prefix an ATC request when it is relayed to an aircraft by other than an air traffic controller.

ATCSCC—(See AIR TRAFFIC CONTROL SYSTEM COMMAND CENTER).

ATCSCC DELAY FACTOR—The amount of delay calculated to be assigned prior to departure.

ATCT—(See TOWER).

ATIS—(See AUTOMATIC TERMINAL INFORMATION SERVICE).

ATIS [ICAO]—(See ICAO Term AUTOMATIC TERMINAL INFORMATION SERVICE).

ATS Route [ICAO]—A specified route designed for the flow of traffic as necessary for the provision of air traffic services.

Note: The term "ATS Route" is used to mean airway, advisory route, controlled or uncontrolled route, arrival or departure, etc.

AUTOLAND APPROACH—An autoland approach is a precision instrument approach to touchdown and, in some cases, through the landing rollout. An autoland approach is performed by the aircraft autopilot which is receiving position information and/or steering commands from onboard navigation equipment(See COUPLED APPROACH).

Note: Autoland and coupled approaches are flown in VFR and IFR. It is common for carriers to require their crews to fly coupled approaches and autoland approaches (if certified) when the weather conditions are less than approximately 4,000 RVR.

AUTOMATED INFORMATION TRANSFER—A precoordinated process, specifically defined in facility directives, during which a transfer of altitude control and/or radar identification is accomplished without verbal coordination between controllers using information communicated in a full data block.

AUTOMATED MUTUAL–ASSISTANCE VESSEL RESCUE SYSTEM—A facility which can deliver, in a matter of minutes, a surface picture (SURPIC) of vessels in the area of a potential or actual search and rescue incident, including their predicted positions and their characteristics.

(See Paragraph 10–73, In-Flight Contingencies).

AUTOMATED RADAR TERMINAL SYSTEMS—The generic term for the ultimate in functional capability afforded by several automation systems. Each differs in functional capabilities and equipment. ARTS plus a suffix roman numeral denotes a specific system. A following letter indicates a major modification to that system. In general, an ARTS displays for the terminal controller aircraft identification, flight plan data, other flight associated information; e.g., altitude, speed, and aircraft position symbols in conjunction with his radar presentation. Normal radar co-exists with the alphanumeric display. In addition to enhancing visualization of the air traffic situation, ARTS facilitate intra/inter-facility transfer and coordination of flight information. These capabilities are enabled by specially designed computers and subsystems tailored to the radar and communications equipments and operational requirements of each automated facility. Modular design permits adoption of improvements in computer software and electronic technologies as they become available while retaining the characteristics unique to each system.

1. ARTS II. A programmable nontracking, computer-aided display subsystem capable of modular expansion. ARTS II systems provide a level of automated air traffic control capability at terminals having low to medium activity. Flight identification and altitude may be associated with the display of secondary radar

targets. The system has the capability of communicating with ARTCC's and other ATRS II, IIA, III, and IIIA facilities.

2. ARTS IIA. A programmable radar-tracking computer subsystem capable of modular expansion. The ARTS IIA detects, tracks, and predicts secondary radar targets. The targets are displayed by means of computer-generated symbols, ground speed, and flight plan data. Although it does not track primary radar targets, they are displayed coincident with the secondary radar as well as the symbols and alphanumerics. The system has the capability of communicating with ARTCC's and other ARTS II, IIA, III, and IIIA facilities.

3. ARTS III. The Beacon Tracking Level of the modular programmable automated radar terminal system in use at medium to high activity terminals. ARTS III detects, tracks, and predicts secondary radar-derived aircraft targets. These are displayed by means of computer-generated symbols and alphanumeric characters depicting flight identification, aircraft altitude, ground speed, and flight plan data. Although it does not track primary targets, they are displayed coincident with the secondary radar as well as the symbols and alphanumerics. The system has the capability of communicating with ARTCC's and other ARTS III facilities.

4. ARTS IIIA. The Radar Tracking and Beacon Tracking Level (RT&BTL) of the modular, programmable automated radar terminal system. ARTS IIIA detects, tracks, and predicts primary as well as secondary radar-derived aircraft targets. This more sophisticated computer-driven system upgrades the existing ARTS III system by providing improved tracking, continuous data recording, and fail-soft capabilities.

AUTOMATIC ALTITUDE REPORT—(See ALTITUDE READOUT).

AUTOMATIC ALTITUDE REPORTING—That function of a transponder which responds to Mode C interrogations by transmitting the aircraft's altitude in 100–foot increments.

AUTOMATIC CARRIER LANDING SYSTEM—U.S. Navy final approach equipment consisting of precision tracking radar coupled to a computer data link to provide continuous information to the aircraft, monitoring capability to the pilot, and a backup approach system.

AUTOMATIC DIRECTION FINDER—An aircraft radio navigation system which senses and indicates the direction to a L/MF nondirectional radio beacon (NDB) ground transmitter. Direction is indicated to the pilot as a magnetic bearing or as a relative bearing to the longitudinal axis of the aircraft depending on the type of indicator installed in the aircraft. In certain applications, such as military, ADF operations may be based on airborne and ground transmitters in the VHF/UHF frequency spectrum.

(See BEARING). (See NONDIRECTIONAL BEACON).

AUTOMATIC TERMINAL INFORMATION SERVICE—The continuous broadcast of recorded noncontrol information in selected terminal areas. Its purpose is to improve controller effectiveness and to relieve frequency congestion by automating the repetitive transmission of essential but routine information; e.g., "Los Angeles information Alfa. One three zero zero Coordinated Universal Time. Weather, measured ceiling two thousand overcast, visibility three, haze, smoke, temperature seven one, dew point five seven, wind two five zero at five, altimeter two niner niner six. I-L-S Runway Two Five Left approach in use, Runway Two Five Right closed, advise you have Alfa."

(Refer to AIM). (See ICAO term AUTOMATIC TERMINAL INFORMATION SERVICE).

AUTOMATIC TERMINAL INFORMATION SERVICE [ICAO]—The provision of current, routine information to arriving and departing aircraft by means of continuous and repetitive broadcasts throughout the day or a specified portion of the day.

AUTOROTATION—A rotorcraft flight condition in which the lifting rotor is driven entirely by action of the air when the rotorcraft is in motion.

1. Autorotative Landing/Touchdown Autorotation. Used by a pilot to indicate that he will be landing without applying power to the rotor.

2. Low Level Autorotation. Commences at an altitude well below the traffic pattern, usually below 100 feet AGL and is used primarily for tactical military training.

3. 180 degrees Autorotation. Initiated from a downwind heading and is commenced well inside the normal traffic pattern. "Go around" may not be possible during the latter part of this maneuver.

AVIATION WEATHER SERVICE—A service provided by the National Weather Service (NWS) and FAA which collects and disseminates pertinent weather information for pilots, aircraft operators, and ATC. Available aviation weather reports and forecasts are displayed at each NWS office and FAA FSS.

(See EN ROUTE FLIGHT ADVISORY SERVICE). (See TRANSCRIBED WEATHER BROADCAST). (See WEATHER ADVISORY). (See PILOTS AUTOMATIC TELEPHONE WEATHER ANSWERING SERVICE). (Refer to AIM).

AWW—(See SEVERE WEATHER FORECAST ALERTS).

AZIMUTH (MLS)—A magnetic bearing extending from an MLS navigation facility.

Note: azimuth bearings are described as magnetic and are referred to as "azimuth" in radio telephone communications.-

B

BASE LEG—(See TRAFFIC PATTERN).

BEACON—(See RADAR).

(See NONDIRECTIONAL BEACON). (See MARKER BEACON). (See AIRPORT ROTATING BEACON). (See AERONAUTICAL BEACON). (See AIRWAY BEACON).

BEARING—The horizontal direction to or from any point, usually measured clockwise from true north, magnetic north, or some other reference point through 360 degrees.

(See NONDIRECTIONAL BEACON).

BELOW MINIMUMS—Weather conditions below the minimums prescribed by regulation for the particular action involved; e.g., landing minimums, takeoff minimums.

BLAST FENCE—A barrier that is used to divert or dissipate jet or propeller blast.

BLIND SPEED—The rate of departure or closing of a target relative to the radar antenna at which cancellation of the primary radar target by moving target indicator (MTI) circuits in the radar equipment causes a reduction or complete loss of signal.

(See ICAO term BLIND VELOCITY).

BLIND SPOT—An area from which radio transmissions and/or radar echoes cannot be received. The term is also used to describe portions of the airport not visible from the control tower.

BLIND TRANSMISSION—(See TRANSMITTING IN THE BLIND).

BLIND VELOCITY [ICAO]—The radial velocity of a moving target such that the target is not seen on primary radars fitted with certain forms of fixed echo suppression.

BLIND ZONE—(See BLIND SPOT).

BOUNDARY LIGHTS—(See AIRPORT LIGHTING).

BRAKING ACTION (GOOD, FAIR, POOR, OR NIL)—A report of conditions on the airport movement area providing a pilot with a degree/quality of braking that he might expect. Braking action is reported in terms of good, fair, poor, or nil.

(See RUNWAY CONDITION READING).

BRAKING ACTION ADVISORIES—When tower controllers have received runway braking action reports which include the terms "poor" or "nil," or whenever weather conditions are conducive to deteriorating or rapidly changing runway braking conditions, the tower will include on the ATIS broadcast the statement, "BRAKING ACTION ADVISORIES ARE IN EFFECT." During the time Braking Action Advisories are in effect, ATC will issue the latest braking action report for the runway in use to each arriving and departing aircraft. Pilots should be prepared for deteriorating braking conditions and should request current runway condition information if not volunteered by controllers. Pilots should also be prepared to provide a descriptive runway condition report to controllers after landing.

BROADCAST—Transmission of information for which an acknowledgement is not expected.

(See ICAO term BROADCAST).

BROADCAST [ICAO]—A transmission of information relating to air navigation that is not addressed to a specific station or stations.-

C

CALCULATED LANDING TIME—A term that may be used in place of tentative or actual calculated landing time, whichever applies.

CALL UP—Initial voice contact between a facility and an aircraft, using the identification of the unit being called and the unit initiating the call.

(Refer to AIM).

CALL FOR RELEASE—Wherein the overlying ARTCC requires a terminal facility to initiate verbal coordination to secure ARTCC approval for release of a departure into the en route environment.

CANADIAN MINIMUM NAVIGATION PERFORMANCE SPECIFICATION AIRSPACE—That portion of Canadian domestic airspace within which MNPS separation may be applied.

CARDINAL ALTITUDES—"Odd" or "Even" thousand-foot altitudes or flight levels; e.g., 5,000, 6,000, 7,000, FL 250, FL 260, FL 270.

(See ALTITUDE). (See FLIGHT LEVEL).

CARDINAL FLIGHT LEVELS—(See CARDINAL ALTITUDES).

CAT—(See CLEAR-AIR TURBULENCE).

CDT PROGRAMS—(See CONTROLLED DEPARTURE TIME PROGRAMS).

CEILING—The heights above the earth's surface of the lowest layer of clouds or obscuring phenomena that is reported as "broken," "overcast," or "obscuration," and not classified as "thin" or "partial".

(See ICAO term CEILING).

CEILING [ICAO]—The height above the ground or water of the base of the lowest layer of cloud below 6,000 metres (20,000 feet) covering more than half the sky.

CENRAP—(See CENTER RADAR ARTS PRESENTATION/PROCESSING).

CENRAP–PLUS—(See CENTER RADAR ARTS PRESENTATION/PROCESSING–PLUS).

CENTER—(See AIR ROUTE TRAFFIC CONTROL CENTER).

CENTER'S AREA—The specified airspace within which an air route traffic control center (ARTCC) provides air traffic control and advisory service.

(See AIR ROUTE TRAFFIC CONTROL CENTER). (Refer to AIM).

CENTER RADAR ARTS PRESENTATION/PROCESSING—A computer program developed to provide a back-up system for airport surveillance radar in the event of a failure or malfunction. The program uses air route traffic control center radar for the processing and presentation of data on the ARTS IIA or IIIA displays.

CENTER RADAR ARTS PRESENTATION/PROCESSING-PLUS—A computer program developed to provide a back-up system for airport surveillance radar in the event of a terminal secondary radar system failure. The program uses a combination of Air Route Traffic Control Center Radar and terminal airport surveillance radar primary targets displayed simultaneously for the processing and presentation of data on the ARTS IIA or IIIA displays.

CENTER WEATHER ADVISORY—An unscheduled weather advisory issued by Center Weather Service Unit meteorologists for ATC use to alert pilots of existing or anticipated adverse weather conditions within the next 2 hours. A CWA may modify or redefine a SIGMET.

(See AWW). (See SIGMET). (See CONVECTIVE SIGMET). (See AIRMET). (Refer to AIM).

CENTRAL EAST PACIFIC—An organized route system between the U.S. West Coast and Hawaii.

CEP—(See CENTRAL EAST PACIFIC).

CERAP—(See COMBINED CENTER-RAPCON).

CFR—(See CALL FOR RELEASE).

CHAFF—Thin, narrow metallic reflectors of various lengths and frequency responses, used to reflect radar energy. These reflectors when dropped from aircraft and allowed to drift downward result in large targets on the radar display.

CHARTED VFR FLYWAYS—Charted VFR Flyways are flight paths recommended for use to bypass areas heavily traversed by large turbine-powered aircraft. Pilot compliance with recommended flyways and associated altitudes is strictly voluntary. VFR Flyway Planning charts are published on the back of existing VFR Terminal Area charts.

CHARTED VISUAL FLIGHT PROCEDURE APPROACH—An approach wherein a radar-controlled aircraft on an IFR flight plan, operating in VFR conditions and having an ATC authorization, may proceed to the airport of intended landing via visual landmarks and altitudes depicted on a charted visual flight procedure.

CHASE—An aircraft flown in proximity to another aircraft normally to observe its performance during training or testing.

CHASE AIRCRAFT—(See CHASE).

CIRCLE-TO-LAND MANEUVER—A maneuver initiated by the pilot to align the aircraft with a runway for landing when a straight-in landing from an instrument approach is not possible or is not desirable. This maneuver is made only after ATC authorization has been obtained and the pilot has established required visual reference to the airport

(See CIRCLE TO RUNWAY). (See LANDING MINIMUMS). (Refer to AIM).

CIRCLE TO RUNWAY (RUNWAY NUMBER)—Used by ATC to inform the pilot that he must circle to land because the runway in use is other than the runway aligned with the instrument approach procedure. When the direction of the circling maneuver in relation to the airport/runway is required, the controller will state the direction (eight cardinal compass points) and specify a left or right downwind or base leg as appropriate; e.g., "Cleared VOR Runway Three Six Approach circle to Runway Two Two," or "Circle northwest of the airport for a right downwind to Runway Two Two."

(See CIRCLE-TO-LAND MANEUVER). (See LANDING MINIMUMS). (Refer to AIM).

CIRCLING APPROACH—(See CIRCLE-TO-LAND MANEUVER).

CIRCLING MANEUVER—(See CIRCLE-TO-LAND MANEUVER).

CIRCLING MINIMA—(See LANDING MINIMUMS).

CLEAR-AIR TURBULENCE—Turbulence encountered in air where no clouds are present. This term is commonly applied to high-level turbulence associated with wind shear. CAT is often encountered in the vicinity of the jet stream.

(See WIND SHEAR). (See JET STREAM).

CLEARANCE—(See AIR TRAFFIC CLEARANCE).

CLEARANCE LIMIT—The fix, point, or location to which an aircraft is cleared when issued an air traffic clearance.

(See ICAO term CLEARANCE LIMIT).

CLEARANCE LIMIT [ICAO]—The point of which an aircraft is granted an air traffic control clearance.

CLEARANCE VOID IF NOT OFF BY (TIME)—Used by ATC to advise an aircraft that the departure clearance is automatically canceled if takeoff is not made prior to a specified time. The pilot must obtain a new clearance or cancel his IFR flight plan if not off by the specified time.

(See ICAO term CLEARANCE VOID TIME).

CLEARANCE VOID TIME [ICAO]—A time specified by an air traffic control unit at which a clearance ceases to be valid unless the aircraft concerned has already taken action to comply therewith.

CLEARED AS FILED—Means the aircraft is cleared to proceed in accordance with the route of flight filed in the flight plan. This clearance does not include the altitude, SID, or SID Transition.

(See REQUEST FULL ROUTE CLEARANCE). (Refer to AIM).

CLEARED (TYPE OF) APPROACH—ATC authorization for an aircraft to execute a specific instrument approach procedure to an airport; e.g., "Cleared ILS Runway Three Six Approach."
(See INSTRUMENT APPROACH PROCEDURE). (See APPROACH CLEARANCE). (Refer to AIM). (Refer to Part 91).

CLEARED APPROACH—ATC authorization for an aircraft to execute any standard or special instrument approach procedure for that airport. Normally, an aircraft will be cleared for a specific instrument approach procedure.
(See INSTRUMENT APPROACH PROCEDURE). (See CLEARED (TYPE OF) APPROACH). (Refer to AIM). (Refer to Part 91).

CLEARED FOR TAKEOFF—ATC authorization for an aircraft to depart. It is predicated on known traffic and known physical airport conditions.

CLEARED FOR THE OPTION—ATC authorization for an aircraft to make a touch-and-go, low approach, missed approach, stop and go, or full stop landing at the discretion of the pilot. It is normally used in training so that an instructor can evaluate a student's performance under changing situations.
(See OPTION APPROACH). (Refer to AIM).

CLEARED THROUGH—ATC authorization for an aircraft to make intermediate stops at specified airports without refiling a flight plan while en route to the clearance limit.

CLEARED TO LAND—ATC authorization for an aircraft to land. It is predicated on known traffic and known physical airport conditions.

CLEARWAY—An area beyond the takeoff runway under the control of airport authorities within which terrain or fixed obstacles may not extend above specified limits. These areas may be required for certain turbine-powered operations and the size and upward slope of the clearway will differ depending on when the aircraft was certificated.
(Refer to Part 1).

CLIMBOUT—That portion of flight operation between takeoff and the initial cruising altitude.

CLIMB TO VFR—ATC authorization for an aircraft to climb to VFR conditions within a control zone when the only weather limitation is restricted visibility. The aircraft must remain clear of clouds while climbing to VFR.
(See Special VFR). (Refer to AIM).

CLOSED RUNWAY—A runway that is unusable for aircraft operations. Only the airport management/military operations office can close a runway.

CLOSED TRAFFIC—Successive operations involving takeoffs and landings or low approaches where the aircraft does not exit the traffic pattern.

CLT—(See CALCULATED LANDING TIME).

CLUTTER—In radar operations, clutter refers to the reception and visual display of radar returns caused by precipitation, chaff, terrain, numerous aircraft targets, or other phenomena. Such returns may limit or preclude ATC from providing services based on radar.
(See GROUND CLUTTER). (See CHAFF). (See PRECIPITATION). (See TARGET). (See ICAO term Radar Clutter).

CMNPS—(See CANADIAN MINIMUM NAVIGATION PERFORMANCE SPECIFICATION AIRSPACE).

COASTAL FIX—A navigation aid or intersection where an aircraft transitions between the domestic route structure and the oceanic route structure.

CODES—The number assigned to a particular multiple pulse reply signal transmitted by a transponder.
(See DISCRETE CODE).

COMBINED CENTER-RAPCON—An air traffic facility which combines the functions of an ARTCC and a radar approach control facility.
(See AIR ROUTE TRAFFIC CONTROL CENTER). (See RADAR APPROACH CONTROL FACILITY).

COMMON POINT—A *significant point* over which two or more aircraft will report passing or have reported passing before proceeding on the same or diverging tracks. To establish/maintain longitudinal separation, a controller may determine a common point not originally in the aircraft's flight plan and then clear the aircraft to fly over the point. See significant point.

COMMON PORTION—(See COMMON ROUTE).

COMMON ROUTE—That segment of a North American Route between the inland navigation facility and the coastal fix.

COMMON TRAFFIC ADVISORY FREQUENCY—A frequency designed for the purpose of carrying out airport advisory practices while operating to or from an uncontrolled airport. The CTAF may be a UNICOM, Multicom, FSS, or tower frequency and is identified in appropriate aeronautical publications.

(Refer to AC90–42).

COMPASS LOCATOR—A low power, low or medium frequency (L/MF) radio beacon installed at the site of the outer or middle marker of an instrument landing system (ILS). It can be used for navigation at distances of approximately 15 miles or as authorized in the approach procedure.

1. Outer Compass Locator (LOM). A compass locator installed at the site of the outer marker of an instrument landing system.

(See OUTER MARKER).

2. Middle Compass Locator (LMM). A compass locator installed at the site of the middle marker of an instrument landing system.

(See MIDDLE MARKER). (See ICAO term LOCATOR).

COMPASS ROSE—A circle, graduated in degrees, printed on some charts or marked on the ground at an airport. It is used as a reference to either true or magnetic direction.

COMPOSITE FLIGHT PLAN—A flight plan which specifies VFR operation for one portion of flight and IFR for another portion. It is used primarily in military operations.

(Refer to AIM).

COMPOSITE ROUTE SYSTEM—An organized oceanic route structure, incorporating reduced lateral spacing between routes, in which composite separation is authorized.

COMPOSITE SEPARATION—A method of separating aircraft in a composite route system where, by management of route and altitude assignments, a combination of half the lateral minimum specified for the area concerned and half the vertical minimum is applied.

COMPULSORY REPORTING POINTS—Reporting points which must be reported to ATC. They are designated on aeronautical charts by solid triangles or filed in a flight plan as fixes selected to define direct routes. These points are geographical locations which are defined by navigation aids/fixes. Pilots should discontinue position reporting over compulsory reporting points when informed by ATC that their aircraft is in "radar contact."

CONFLICT ALERT—A function of certain air traffic control automated systems designed to alert radar controllers to existing or pending situations between tracked targets (known IFR or VFR aircraft) that require his immediate attention/action.

(See MODE C INTRUDER ALERT).

CONFLICT RESOLUTION—The resolution of potential conflictions between aircraft that are radar identified and in communication with ATC by ensuring that radar targets do not touch. Pertinent traffic advisories shall be issued when this procedure is applied.

Note: This procedure shall not be provided utilizing mosaic radar systems.

(See CONTROLLED AIRSPACE). (See AIRPORT RADAR SERVICE AREA). (See OUTER AREA).

CONSOLAN—A low frequency, long-distance NAVAID used principally for transoceanic navigations.

CONTACT

1. Establish communication with (followed by the name of the facility and, if appropriate, the frequency to be used).

2. A flight condition wherein the pilot ascertains the attitude of his aircraft and navigates by visual reference to the surface.

(See CONTACT APPROACH). (See RADAR CONTACT).

CONTACT APPROACH—An approach wherein an aircraft on an IFR flight plan, having an air traffic control authorization, operating clear of clouds with at least 1 mile flight visibility and a reasonable expectation of continuing to the destination airport in those conditions, may deviate from the instrument approach procedure and proceed to the destination airport by visual reference to the surface. This approach will only be authorized when requested by the pilot and the reported ground visibility at the destination airport is at least 1 statute mile.

(Refer to AIM).

CONTERMINOUS U.S.—The 48 adjoining States and the District of Columbia.

CONTINENTAL CONTROL AREA—(See CONTROLLED AIRSPACE).

CONTINENTAL UNITED STATES—The 49 States located on the continent of North America and the District of Columbia.

CONTROL AREA—(See CONTROLLED AIRSPACE).

CONTROLLED AIRSPACE—Airspace designated as a control zone, airport radar service area, terminal control area, transition area, control area, continental control area, and positive control area within which some or all aircraft may be subject to air traffic control.

(Refer to AIM). (Refer to Part 71).—Types of U.S. Controlled Airspace:

1. Control Zone. Controlled airspace which extends upward from the surface of the earth and terminates at the base of the continental control area. Control zones that do not underlie the continental control area have no upper limit. A control zone may include one or more airports and is normally a circular area with a radius of 5 statute miles and any extensions necessary to include instrument approach and departure paths.
2. Airport Radar Service Area (ARSA). Regulatory airspace surrounding designated airports wherein ATC provides radar vectoring and sequencing on a full-time basis for all IFR and VFR aircraft. The service provided in an ARSA is called ARSA service which includes: IFR/IFR-standard IFR separation; IFR/VFR-traffic advisories and conflict resolution; and VFR/VFR-traffic advisories and, as appropriate, safety alerts. The AIM contains an explanation of ARSA. The ARSA's are depicted on VFR aeronautical charts.

(See CONFLICT RESOLUTION). (See OUTER AREA). (Refer to AIM). (Refer to A/FD). (Refer to Part 91).

3. Terminal Control Area (TCA). Controlled airspace extending upward from the surface or higher to specified altitudes, within which all aircraft are subject to operating rules and pilot and equipment requirements specified in Part 91. TCA's are depicted on Sectional, World Aeronautical, En Route Low Altitude, DOD FLIP, and TCA charts.

(Refer to Part 91). (Refer to AIM).

4. Transition Area. Controlled airspace extending upward from 700 feet or more above the surface of the earth when designated in conjunction with an airport for which an approved instrument approach procedure has been prescribed; or from 1,200 feet or more above the surface of the earth when designated in conjunction with airway route structures or segments. Unless otherwise specified, transition areas terminate at the base of the overlying controlled airspace. Transition areas are designed to contain IFR operations in controlled airspace during portions of the terminal operation and while transiting between the terminal and en route environment.
5. Control Area. Airspace designated as Colored Federal airways, VOR Federal airways, control areas associated with jet routes outside the continental control area (Part 71.161), additional control areas (Part 71.163), control area extensions (Part 71.165), and area low routes. Control areas do not include the continental control area, but unless otherwise designated, they do include the airspace between a segment of a main VOR Federal airway and its associated alternate segments with the vertical extent of the area corresponding to the vertical extent of the related segment of the main airway. The vertical extent of the various categories of airspace contained in control areas is defined in Part 71.
6. Continental Control Area. The airspace of the 48 contiguous States, the District of Columbia and Alaska, excluding the Alaska peninsula west of Long. 160° 00' 00"W, at and above 14,500 feet MSL, but does not include:
   a. The airspace less than 1,500 feet above the surface of the earth; or
   b. Prohibited and restricted areas, other than the restricted areas listed in Part 71.
7. Positive Control Area (PCA). Airspace designated in Part 71 within which there is positive control of aircraft. Flight in PCA is normally conducted under instrument flight rules. PCA is designated throughout most of the conterminous United States and its vertical extent is from 18,000 feet MSL to and including flight level 600. In Alaska PCA does not include the airspace less than 1,500 feet above the surface

of the earth nor the airspace over the Alaska Peninsula west of longitude 160 degrees West. Rules for operating in PCA are found in Part 91.135 and Part 91.215.

(See ICAO term CONTROLLED AIRSPACE).

CONTROLLED AIRSPACE [ICAO]—Airspace of defined dimensions within which air traffic control service is provided to controlled flights.

Control Zone—A controlled airspace extending upwards from the surface of the earth to a specified upper limit.

Terminal Control Area—A control area normally established at the confluence of ATS routes in the vicinity of one or more major aerodromes.

Control Area—A controlled airspace extending upward from a specified limit above the earth.

CONTROLLED DEPARTURE TIME PROGRAMS—These programs are the flow control process whereby aircraft are held on the ground at the departure airport when delays are projected to occur in either the en route system or the terminal of intended landing. The purpose of these programs is to reduce congestion in the air traffic system or to limit the duration of airborne holding in the arrival center or terminal area. A CDT is a specific departure slot shown on the flight plan as an expected departure clearance time (EDCT).

CONTROLLED TIME OF ARRIVAL—The original estimated time of arrival adjusted by the ATCSCC ground delay factor.

CONTROLLER—(See AIR TRAFFIC CONTROL SPECIALIST).

CONTROLLER [ICAO]—A person authorized to provide air traffic control services.

CONTROL SECTOR—An airspace area of defined horizontal and vertical dimensions for which a controller or group of controllers has air traffic control responsibility, normally within an air route traffic control center or an approach control facility. Sectors are established based on predominant traffic flows, altitude strata, and controller workload. Pilot-communications during operations within a sector are normally maintained on discrete frequencies assigned to the sector.

(See DISCRETE FREQUENCY).

CONTROL SLASH—A radar beacon slash representing the actual position of the associated aircraft. Normally, the control slash is the one closest to the interrogating radar beacon site. When ARTCC radar is operating in narrowband (digitized) mode, the control slash is converted to a target symbol.

CONTROL ZONE—(See CONTROLLED AIRSPACE).

CONVECTIVE SIGMET—A weather advisory concerning convective weather significant to the safety of all aircraft. Convective SIGMET's are issued for tornadoes, lines of thunderstorms, embedded thunderstorms of any intensity level, areas of thunderstorms greater than or equal to VIP level 4 with an area coverage of 4/10 (40%) or more, and hail 3/4 inch or greater.

(See AWW). (See SIGMET). (See CWA). (See AIRMET). (Refer to AIM).

CONVECTIVE SIGNIFICANT METEOROLOGICAL INFORMATION—(See CONVECTIVE SIGMET).

COORDINATES—The intersection of lines of reference, usually expressed in degrees/minutes/seconds of latitude and longitude, used to determine position or location.

COORDINATION FIX—The fix in relation to which facilities will handoff, transfer control of an aircraft, or coordinate flight progress data. For terminal facilities, it may also serve as a clearance for arriving aircraft.

COPTER—(See HELICOPTER).

CORRECTION—An error has been made in the transmission and the correct version follows.

COUPLED APPROACH—A coupled approach is an instrument approach performed by the aircraft autopilot which is receiving position information and/or steering commands from onboard navigation equipment. In general, coupled nonprecision approaches must be discontinued and flown manually at altitudes lower than 50 feet below the minimum descent altitude, and coupled precision approaches must be flown manually below 50 feet ALG

(See AUTOLAND APPROACH).

Note: Coupled and autoland approaches are flown in VFR and IFR. It is common for carriers to require their crews to fly coupled approaches and autoland approaches (if certified) when the weather conditions are less than approximately 4,000 RVR.

COURSE—

1. The intended direction of flight in the horizontal plane measured in degrees from north.
2. The ILS localizer signal pattern usually specified as the front course or the back course.
3. The intended track along a straight, curved, or segmented MLS path.

(See BEARING). (See RADIAL). (See INSTRUMENT LANDING SYSTEM). (See MICROWAVE LANDING SYSTEM).

CPL [ICAO]—(See CURRENT FLIGHT PLAN)

CRITICAL ENGINE—The engine which, upon failure, would most adversely affect the performance or handling qualities of an aircraft.

CROSS (FIX) AT (ALTITUDE)—Used by ATC when a specific altitude restriction at a specified fix is required.

CROSS (FIX) AT OR ABOVE (ALTITUDE)—Used by ATC when an altitude restriction at a specified fix is required. It does not prohibit the aircraft from crossing the fix at a higher altitude than specified; however, the higher altitude may not be one that will violate a succeeding altitude restriction or altitude assignment.
(See ALTITUDE RESTRICTION). (Refer to AIM).

CROSS (FIX) AT OR BELOW (ALTITUDE)—Used by ATC when a maximum crossing altitude at a specific fix is required. It does not prohibit the aircraft from crossing the fix at a lower altitude; however, it must be at or above the minimum IFR altitude.
(See MINIMUM IFR ALTITUDES). (See ALTITUDE RESTRICTION). (Refer to Part 91).

CROSSWIND—

1. When used concerning the traffic pattern, the word means "crosswind leg."

(See TRAFFIC PATTERN).

2. When used concerning wind conditions, the word means a wind not parallel to the runway or the path of an aircraft.

(See CROSSWIND COMPONENT).

CROSSWIND COMPONENT—The wind component measured in knots at 90 degrees to the longitudinal axis of the runway.

CRUISE—Used in an ATC clearance to authorize a pilot to conduct flight at any altitude from the minimum IFR altitude up to and including the altitude specified in the clearance. The pilot may level off at any intermediate altitude within this block of airspace. Climb/descent within the block is to be made at the discretion of the pilot. However, once the pilot starts descent and verbally reports leaving an altitude in the block, he may not return to that altitude without additional ATC clearance. Further, it is approval for the pilot to proceed to and make an approach at destination airport and can be used in conjunction with:

1. An airport clearance limit at locations with a standard/special instrument approach procedure. The FAR's require that if an instrument letdown to an airport is necessary, the pilot shall make the letdown in accordance with a standard/special instrument approach procedure for that airport, or
2. An airport clearance limit at locations that are within/below/outside controlled airspace and without a standard/special instrument approach procedure. Such a clearance is NOT AUTHORIZATION for the pilot to descend under IFR conditions below the applicable minimum IFR altitude nor does it imply that ATC is exercising control over aircraft in uncontrolled airspace; however, it provides a means for the aircraft to proceed to destination airport, descend, and land in accordance with applicable FAR's governing VFR flight operations. Also, this provides search and rescue protection until such time as the IFR flight plan is closed.

(See INSTRUMENT APPROACH PROCEDURE).

CRUISING ALTITUDE—An altitude or flight level maintained during en route level flight. This is a constant altitude and should not be confused with a cruise clearance.
(See ALTITUDE). (See ICAO term CRUISING LEVEL).

CRUISING LEVEL [ICAO]—A level maintained during a significant portion of a flight.

CRUISE CLIMB—A climb technique employed by aircraft, usually at a constant power setting, resulting in an increase of altitude as the aircraft weight decreases.

CRUISING LEVEL—(See CRUISING ALTITUDE).

CT MESSAGE—An EDCT time generated by the ATCSCC to regulate traffic at arrival airports. Normally, a CT message is automatically transferred from the Traffic Management System computer to the NAS en route computer and appears as an EDCT. In the event of a communication failure between the TMS and the NAS, the CT message can be manually entered by the TMC at the en route facility.

CTA—(See CONTROLLED TIME OF ARRIVAL).

CTAF—(See COMMON TRAFFIC ADVISORY FREQUENCY).

CURRENT FLIGHT PLAN [ICAO]—The flight plan, including changes, if any, brought about by subsequent clearances.

CVFP APPROACH—(See CHARTED VISUAL FLIGHT PROCEDURE APPROACH).

CWA—(See CENTER WEATHER ADVISORY).
(See WEATHER ADVISORY).-

D

DA [ICAO]—(See ICAO Term DECISION ALTITUDE/DECISION HEIGHT).

DAIR—(See DIRECT ALTITUDE AND IDENTITY READOUT).

DANGER AREA [ICAO]—An airspace of defined dimensions within which activities dangerous to the flight of aircraft may exist at specified times.

Note: The term "Danger Area" is not used in reference to areas within the United States or any of its possessions or territories.

DATA BLOCK—(See ALPHANUMERIC DISPLAY).

DEAD RECKONING—Dead reckoning, as applied to flying, is the navigation of an airplane solely by means of computations based on airspeed, course, heading, wind direction, and speed, groundspeed, and elapsed time.

DECISION ALTITUDE/DECISION HEIGHT [ICAO]—A specified altitude or height (A/H) in the precision approach at which a missed approach must be initiated if the required visual reference to continue the approach has not been established.

Note 1: Decision altitude [DA] is referenced to mean sea level [MSL] and decision height [DH] is referenced to the threshold elevation.

Note 2: The required visual reference means that section of the visual aids or of the approach area which should have been in view for sufficient time for the pilot to have made an assessment of the aircraft position and rate of change of position, in relation to the desired flight path.

DECISION HEIGHT—With respect to the operation of aircraft, means the height at which a decision must be made during an ILS, MLS, or PAR instrument approach to either continue the approach or to execute a missed approach.

(See ICAO term DECISION ALTITUDE/DECISION HEIGHT).

DECODER—The device used to decipher signals received from ATCRBS transponders to effect their display as select codes.

(See CODES). (See RADAR).

DEFENSE VISUAL FLIGHT RULES—Rules applicable to flights within an ADIZ conducted under the visual flight rules in Part 91.

(See AIR DEFENSE IDENTIFICATION ZONE). (Refer to Part 91). (Refer to Part 99).

DELAY INDEFINITE (REASON IF KNOWN) EXPECT FURTHER CLEARANCE (TIME)—Used by ATC to inform a pilot when an accurate estimate of the delay time and the reason for the delay cannot immediately be determined; e.g., a disabled aircraft on the runway, terminal or center area saturation, weather below landing minimums, etc.

(See EXPECT FURTHER CLEARANCE (TIME)).

DELAY TIME—The amount of time that the arrival must lose to cross the meter fix at the assigned meter fix time. This is the difference between ACLT and VTA.

DEPARTURE CENTER—The ARTCC having jurisdiction for the airspace that generates a flight to the impacted airport.

DEPARTURE CONTROL—A function of an approach control facility providing air traffic control service for departing IFR and, under certain conditions, VFR aircraft.

(See APPROACH CONTROL FACILITY). (Refer to AIM).

DEPARTURE SEQUENCING PROGRAM—A program designed to assist in achieving a specified interval over a common point for departures.

DEPARTURE TIME—The time an aircraft becomes airborne.

DESCENT SPEED ADJUSTMENTS—Speed deceleration calculations made to determine an accurate VTA. These calculations start at the transition point and use arrival speed segments to the vertex.

DETRESFA (DISTRESS PHASE) [ICAO]—The code word used to designate an emergency phase wherein there is reasonable certainty that an aircraft and its occupants are threatened by grave and imminent danger or require immediate assistance.

DEVIATIONS—

1. A departure from a current clearance, such as an off course maneuver to avoid weather or turbulence.
2. Where specifically authorized in the FAR's and requested by the pilot, ATC may permit pilots to deviate from certain regulations.

(Refer to AIM).

DF—(See DIRECTION FINDER).

DF APPROACH PROCEDURE—Used under emergency conditions where another instrument approach procedure cannot be executed. DF guidance for an instrument approach is given by ATC facilities with DF capability.
(See DF GUIDANCE). (See DIRECTION FINDER). (Refer to AIM).

DF FIX—The geographical location of an aircraft obtained by one or more direction finders.
(See DIRECTION FINDER).

DF GUIDANCE—Headings provided to aircraft by facilities equipped with direction finding equipment. These headings, if followed, will lead the aircraft to a predetermined point such as the DF station or an airport. DF guidance is given to aircraft in distress or to other aircraft which request the service. Practice DF guidance is provided when workload permits.
(See DIRECTION FINDER). (See DF FIX). (Refer to AIM).

DF STEER—(See DF GUIDANCE).

DH—(See DECISION HEIGHT).

DH [ICAO]—(See ICAO Term DECISION ALTITUDE/DECISION HEIGHT).

DIRECT—Straight line flight between two navigational aids, fixes, points, or any combination thereof. When used by pilots in describing off-airway routes, points defining direct route segments become compulsory reporting points unless the aircraft is under radar contact.

DIRECT ALTITUDE AND IDENTITY READOUT—The DAIR System is a modification to the AN/TPX-42 Interrogator System. The Navy has two adaptations of the DAIR System-Carrier Air Traffic Control Direct Altitude and Identification Readout System for Aircraft Carriers and Radar Air Traffic Control Facility Direct Altitude and Identity Readout System for land-based terminal operations. The DAIR detects, tracks, and predicts secondary radar aircraft targets. Targets are displayed by means of computer-generated symbols and alphanumeric characters depicting flight identification, altitude, ground speed, and flight plan data. The DAIR System is capable of interfacing with ARTCC's.

DIRECTION FINDER—A radio receiver equipped with a directional sensing antenna used to take bearings on a radio transmitter. Specialized radio direction finders are used in aircraft as air navigation aids. Others are ground-based, primarily to obtain a "fix" on a pilot requesting orientation assistance or to locate downed aircraft. A location "fix" is established by the intersection of two or more bearing lines plotted on a navigational chart using either two separately located Direction Finders to obtain a fix on an aircraft or by a pilot plotting the bearing indications of his DF on two separately located ground-based transmitters, both of which can be identified on his chart. UDF's receive signals in the ultra high frequency radio broadcast band; VDF's in the very high frequency band; and UVDF's in both bands. ATC provides DF service at those air traffic control towers and flight service stations listed in the Airport/Facility Directory and the DOD FLIP IFR En Route Supplement.
(See DF GUIDANCE). (See DF FIX).

DISCRETE BEACON CODE—(See DISCRETE CODE).

DISCRETE CODE—As used in the Air Traffic Control Radar Beacon System (ATCRBS), any one of the 4096 selectable Mode 3/A aircraft transponder codes except those ending in zero zero; e.g., discrete codes: 0010, 1201, 2317, 7777; nondiscrete codes: 0100, 1200, 7700. Nondiscrete codes are normally reserved for radar facilities that are not equipped with discrete decoding capability and for other purposes such as emergencies (7700), VFR aircraft (1200), etc.
(See RADAR). (Refer to AIM).

DISCRETE FREQUENCY—A separate radio frequency for use in direct pilot-controller communications in air traffic control which reduces frequency congestion by controlling the number of aircraft operating on a particular frequency at one time. Discrete frequencies are normally designated for each control sector in en route/terminal ATC facilities. Discrete frequencies are listed in the Airport/Facility Directory and the DOD FLIP IFR En Route Supplement.

(See CONTROL SECTOR).

DISPLACED THRESHOLD—A threshold that is located at a point on the runway other than the designated beginning of the runway.

(See THRESHOLD). (Refer to AIM).

DISTANCE MEASURING EQUIPMENT—Equipment (airborne and ground) used to measure, in nautical miles, the slant range distance of an aircraft from the DME navigational aid.

(See TACAN). (See VORTAC). (See MICROWAVE LANDING SYSTEM).

DISTRESS—A condition of being threatened by serious and/or imminent danger and of requiring immediate assistance.

DIVE BRAKES—(See SPEED BRAKES).

DIVERSE VECTOR AREA—In a radar environment, that area in which a prescribed departure route is not required as the only suitable route to avoid obstacles. The area in which random radar vectors below the MVA/MIA, established in accordance with the TERPS criteria for diverse departures obstacles and terrain avoidance, may be issued to departing aircraft.

DME—(See DISTANCE MEASURING EQUIPMENT).

DME FIX—A geographical position determined by reference to a navigational aid which provides distance and azimuth information. It is defined by a specific distance in nautical miles and a radial, azimuth, or course (i.e., localizer) in degrees magnetic from that aid.

(See DISTANCE MEASURING EQUIPMENT). (See FIX). (See MICROWAVE LANDING SYSTEM).

DME SEPARATION—Spacing of aircraft in terms of distances (nautical miles) determined by reference to distance measuring equipment (DME).

(See DISTANCE MEASURING EQUIPMENT).

DOD FLIP—Department of Defense Flight Information Publications used for flight planning, en route, and terminal operations. FLIP is produced by the Defense Mapping Agency for world-wide use. United States Government Flight Information Publications (en route charts and instrument approach procedure charts) are incorporated in DOD FLIP for use in the National Airspace System (NAS).

DOMESTIC AIRSPACE—Airspace which overlies the continental land mass of the United States plus Hawaii and U.S. possessions. Domestic airspace extends to 12 miles offshore.

DOWNBURST—A strong downdraft which induces an outburst of damaging winds on or near the ground. Damaging winds, either straight or curved, are highly divergent. The sizes of downbursts vary from 1/2 mile or less to more than 10 miles. An intense down burst often causes widespread damage. Damaging winds, lasting 5 to 30 minutes, could reach speeds as high as 120 knots.

DOWNWIND LEG—(See TRAFFIC PATTERN).

DRAG CHUTE—A parachute device installed on certain aircraft which is deployed on landing roll to assist in deceleration of the aircraft.

DSP—(See DEPARTURE SEQUENCING PROGRAM).

DT—(See DELAY TIME).

DUE REGARD—A phase of flight wherein an aircraft commander of a State-operated aircraft assumes responsibility to separate his aircraft from all other aircraft.

(See also Chapter 1, Word Meanings).

DUTY RUNWAY—(See RUNWAY IN USE/ACTIVE RUNWAY/DUTY RUNWAY).

DVA—(See DIVERSE VECTOR AREA).

DVFR—(See DEFENSE VISUAL FLIGHT RULES).

DVFR FLIGHT PLAN—A flight plan filed for a VFR aircraft which intends to operate in airspace within which the ready identification, location, and control of aircraft are required in the interest of national security.

DYNAMIC—Continuous review, evaluation, and change to meet demands.

DYNAMIC RESTRICTIONS—Those restrictions imposed by the local facility on an "as needed" basis to manage unpredictable fluctuations in traffic demands.-

E

EARTS—(See EN ROUTE AUTOMATED RADAR TRACKING SYSTEM).

EDCT—(See EXPECTED DEPARTURE CLEARANCE TIME).

EFC—(See EXPECT FURTHER CLEARANCE (TIME)).

ELT—(See EMERGENCY LOCATOR TRANSMITTER).

EMERGENCY—A *distress* or an *urgency* condition.

EMERGENCY LOCATOR TRANSMITTER—A radio transmitter attached to the aircraft structure which operates from its own power source on 121.5 mHz and 243.0 mHz. It aids in locating downed aircraft by radiating a downward sweeping audio tone, 2–4 times per second. It is designed to function without human action after an accident.

(Refer to Part 91.3). (Refer to AIM).

EMERGENCY SAFE ALTITUDE—(See MINIMUM SAFE ALTITUDE).

E-MSAW—(See EN ROUTE MINIMUM SAFE ALTITUDE WARNING).

ENTRY POINT—The point at which an aircraft transitions from an offshore control area to oceanic airspace.

ENGINEERED PERFORMANCE STANDARDS—A mathematically derived runway capacity standard. EPS's are calculated for each airport on an individual basis and reflect that airport's aircraft mix, operating procedures, runway layout, and specific weather conditions. EPS's do not give consideration to staffing, experience levels, equipment outages, and in-trail restrictions as does the AAR.

EN ROUTE AIR TRAFFIC CONTROL SERVICES—Air traffic control service provided aircraft on IFR flight plans, generally by centers, when these aircraft are operating between departure and destination terminal areas. When equipment, capabilities, and controller workload permit, certain advisory/assistance services may be provided to VFR aircraft.

(See NAS STAGE A). (See AIR ROUTE TRAFFIC CONTROL CENTER). (Refer to AIM).

EN ROUTE AUTOMATED RADAR TRACKING SYSTEM—An automated radar and radar beacon tracking system. Its functional capabilities and design are essentially the same as the terminal ARTS IIIA system except for the EARTS capability of employing both short-range (ASR) and long-range (ARSR) radars, use of full digital radar displays, and fail-safe design.

(See AUTOMATED RADAR TERMINAL SYSTEMS).

EN ROUTE CHARTS—(See AERONAUTICAL CHART).

EN ROUTE DESCENT—Descent from the en route cruising altitude which takes place along the route of flight.

EN ROUTE FLIGHT ADVISORY SERVICE—A service specifically designed to provide, upon pilot request, timely weather information pertinent to his type of flight, intended route of flight, and altitude. The FSS's providing this service are listed in the Airport/Facility Directory.

(See FLIGHT WATCH). (Refer to AIM).

EN ROUTE HIGH ALTITUDE CHARTS—(See AERONAUTICAL CHART).

EN ROUTE LOW ALTITUDE CHARTS—(See AERONAUTICAL CHART).

EN ROUTE MINIMUM SAFE ALTITUDE WARNING—A function of the NAS Stage A en route computer that aids the controller by alerting him when a tracked aircraft is below or predicted by the computer to go below a predetermined minimum IFR altitude (MIA).

EN ROUTE SPACING PROGRAM—A program designed to assist the exit sector in achieving the required in-trail spacing.

EPS—(See ENGINEERED PERFORMANCE STANDARDS).

ESP—(See EN ROUTE SPACING PROGRAM).

ESTIMATED ELAPSED TIME [ICAO]—The estimated time required to proceed from one significant point to another. (See ICAO Term TOTAL ESTIMATED ELAPSED TIME).

ESTIMATED OFF-BLOCK TIME [ICAO]—The estimated time at which the aircraft will commence movement associated with departure.

ESTIMATED TIME OF ARRIVAL—The time the flight is estimated to arrive at the gate (scheduled operators) or the actual runway on times for nonscheduled operators.

ESTIMATED TIME EN ROUTE—The estimated flying time from departure point to destination (lift-off to touchdown).

ETA—(See ESTIMATED TIME OF ARRIVAL).

ETE—(See ESTIMATED TIME EN ROUTE).

EXECUTE MISSED APPROACH—Instructions issued to a pilot making an instrument approach which means continue inbound to the missed approach point and execute the missed approach procedure as described on the Instrument Approach Procedure Chart or as previously assigned by ATC. The pilot may climb immediately to the altitude specified in the missed approach procedure upon making a missed approach. No turns should be initiated prior to reaching the missed approach point. When conducting an ASR or PAR approach, execute the assigned missed approach procedure immediately upon receiving instructions to "execute missed approach."
(Refer to AIM).

EXPECT (ALTITUDE) AT (TIME) OR (FIX)—Used under certain conditions to provide a pilot with an altitude to be used in the event of two-way communications failure. It also provides altitude information to assist the pilot in planning.
(Refer to AIM).

EXPECTED DEPARTURE CLEARANCE TIME—The runway release time assigned to an aircraft in a controlled departure time program and shown on the flight progress strip as an EDCT.

EXPECT FURTHER CLEARANCE (TIME)—The time a pilot can expect to receive clearance beyond a clearance limit.

EXPECT FURTHER CLEARANCE VIA (AIRWAYS, ROUTES OR FIXES)—Used to inform a pilot of the routing he can expect if any part of the route beyond a short range clearance limit differs from that filed.

EXPEDITE—Used by ATC when prompt compliance is required to avoid the development of an imminent situation.-

F

FAF—(See FINAL APPROACH FIX).

FA MESSAGE—The data entered into the ARTCC computers that activates delay processing for an impacted airport. The FA data includes the delay factor for flight plans that have not been assigned delays under CT message processing. The delay factor appears on flight progress strips in the form of an EDCT (e.g., EDCT 1820). FA processing assigns delays in 15-minute time blocks. FA's control numbers of aircraft within a specified time block but do not spread aircraft out evenly throughout the time block.

FAP—(See FINAL APPROACH POINT).

FAST FILE—A system whereby a pilot files a flight plan via telephone that is tape recorded and then transcribed for transmission to the appropriate air traffic facility. Locations having a fast file capability are contained in the Airport/Facility Directory.

(Refer to AIM).

FCLT—(See FREEZE CALCULATED LANDING TIME).

FEATHERED PROPELLER—A propeller whose blades have been rotated so that the leading and trailing edges are nearly parallel with the aircraft flight path to stop or minimize drag and engine rotation. Normally used to indicate shutdown of a reciprocating or turboprop engine due to malfunction.

FEDERAL AIRWAYS—(See LOW ALTITUDE AIRWAY STRUCTURE).

FEEDER FIX—The fix depicted on Instrument Approach Procedure Charts which establishes the starting point of the feeder route.

FEEDER ROUTE—A route depicted on instrument approach procedure charts to designate routes for aircraft to proceed from the en route structure to the initial approach fix (IAF).

(See INSTRUMENT APPROACH PROCEDURE).

FERRY FLIGHT—A flight for the purpose of:

1. Returning an aircraft to base.
2. Delivering an aircraft from one location to another.
3. Moving an aircraft to and from a maintenance base.—Ferry flights, under certain conditions, may be conducted under terms of a special flight permit.

FIELD ELEVATION—(See AIRPORT ELEVATION).

FILED—Normally used in conjunction with flight plans, meaning a flight plan has been submitted to ATC.

FILED EN ROUTE DELAY—Any of the following preplanned delays at points/areas along the route of flight which require special flight plan filing and handling techniques.

1. Terminal Area Delay. A delay within a terminal area for touch-and-go, low approach, or other terminal area activity.
2. Special Use Airspace Delay. A delay within a Military Operating Area, Restricted Area, Warning Area, or ATC Assigned Airspace.
3. Aerial Refueling Delay. A delay within an Aerial Refueling Track or Anchor.

FILED FLIGHT PLAN—The flight plan as filed with an ATS unit by the pilot or his designated representative without any subsequent changes or clearances.

FINAL—Commonly used to mean that an aircraft is on the final approach course or is aligned with a landing area.

(See FINAL APPROACH COURSE). (See FINAL APPROACH-IFR). (See TRAFFIC PATTERN). (See SEGMENTS OF AN INSTRUMENT APPROACH PROCEDURE).

FINAL APPROACH [ICAO]—That part of an instrument approach procedure which commences at the specified final approach fix or point, or where such a fix or point is not specified,

a) At the end of the last procedure turn, base turn or inbound turn of a racetrack procedure, if specified; or *b)* At the point of interception of the last track specified in the approach procedure; and ends at a point in the vicinity of an aerodrome from which:

1) A landing can be made; or

2) A missed approach procedure is initiated.

FINAL APPROACH COURSE—A published MLS course, a straight line extension of a localizer, a final approach radial/bearing, or a runway centerline all without regard to distance.

(See FINAL APPROACH-IFR). (See TRAFFIC PATTERN).

FINAL APPROACH FIX—The fix from which the final approach (IFR) to an airport is executed and which identifies the beginning of the final approach segment. It is designated on Government charts by the Maltese Cross symbol for nonprecision approaches and the lightning bolt symbol for precision approaches; or when ATC directs a lower-than-published Glideslope/path Intercept Altitude, it is the resultant actual point of the glideslope/path intercept.

(See FINAL APPROACH POINT). (See GLIDESLOPE INTERCEPT ALTITUDE). (See SEGMENTS OF AN INSTRUMENT APPROACH PROCEDURE).

FINAL APPROACH-IFR—The flight path of an aircraft which is inbound to an airport on a final instrument approach course, beginning at the final approach fix or point and extending to the airport or the point where a circle-to-land maneuver or a missed approach is executed.

(See SEGMENTS OF AN INSTRUMENT APPROACH PROCEDURE). (See FINAL APPROACH FIX). (See FINAL APPROACH COURSE). (See FINAL APPROACH POINT). (See ICAO term FINAL APPROACH).

FINAL APPROACH POINT—The point, applicable only to a nonprecision approach with no depicted FAF (such as an on-airport VOR), where the aircraft is established inbound on the final approach course from the procedure turn and where the final approach descent may be commenced. The FAP serves as the FAF and identifies the beginning of the final approach segment.

(See FINAL APPROACH FIX). (See SEGMENTS OF AN INSTRUMENT APPROACH PROCEDURE).

FINAL APPROACH SEGMENT—(See SEGMENTS OF AN INSTRUMENT APPROACH PROCEDURE).

FINAL APPROACH SEGMENT [ICAO]—That segment of an instrument approach procedure in which alignment and descent for landing are accomplished.

FINAL APPROACH-VFR—(See TRAFFIC PATTERN).

FINAL CONTROLLER—The controller providing information and final approach guidance during PAR and ASR approaches utilizing radar equipment.

(See RADAR APPROACH).

FINAL MONITOR AID—A high resolution color display that is equipped with the controller alert system hardware/software which is used in the precision runway monitor (PRM) system. The display includes alert algorithms providing the target predictors, a color change alert when a target penetrates or is predicted to penetrate the no transgression zone (NTZ), a color change alert if the aircraft transponder becomes inoperative, synthesized voice alerts, digital mapping, and like features contained in the PRM system.

(See RADAR APPROACH).

FIR—(See FLIGHT INFORMATION REGION).

FIRST TIER CENTER—The ARTCC immediately adjacent to the impacted center.

FIX—A geographical position determined by visual reference to the surface, by reference to one or more radio NAVAIDs, by celestial plotting, or by another navigational device.

FIX BALANCING—A process whereby aircraft are evenly distributed over several available arrival fixes reducing delays and controller workload.

FLAG—A warning device incorporated in certain airborne navigation and flight instruments indicating that:

1. Instruments are inoperative or otherwise not operating satisfactorily, or
2. Signal strength or quality of the received signal falls below acceptable values.

FLAG ALARM—(See FLAG).

FLAMEOUT—Unintended loss of combustion in turbine engines resulting in the loss of engine power.

FLIGHT CHECK—A call-sign prefix used by FAA aircraft engaged in flight inspection/certification of navigational aids and flight procedures. The word "recorded" may be added as a suffix; e.g., "Flight Check 320 recorded" to indicate that an automated flight inspection is in progress in terminal areas.

(See FLIGHT INSPECTION). (Refer to AIM).

FLIGHT FOLLOWING—(See TRAFFIC ADVISORIES).

FLIGHT INFORMATION REGION—An airspace of defined dimensions within which Flight Information Service and Alerting Service are provided.

1. Flight Information Service. A service provided for the purpose of giving advice and information useful for the safe and efficient conduct of flights.
2. Alerting Service. A service provided to notify appropriate organizations regarding aircraft in need of search and rescue aid and to assist such organizations as required.

FLIGHT INFORMATION SERVICE—A service provided for the purpose of giving advice and information useful for the safe and efficient conduct of flights.

FLIGHT INSPECTION—Inflight investigation and evaluation of a navigational aid to determine whether it meets established tolerances.

(See NAVIGATIONAL AID). (See FLIGHT CHECK).

FLIGHT LEVEL—A level of constant atmospheric pressure related to a reference datum of 29.92 inches of mercury. Each is stated in three digits that represent hundreds of feet. For example, flight level 250 represents a barometric altimeter indication of 25,000 feet; flight level 255, an indication of 25,500 feet.

(See ICAO term FLIGHT LEVEL).

FLIGHT LEVEL [ICAO]—A surface of constant atmospheric pressure which is related to a specific pressure datum, 1013.2 hPa (1013.2 mb), and is separated from other such surfaces by specific pressure intervals.

Note 1: A pressure type altimeter calibrated in accordance with the standard atmosphere:

a) When set to a QNH altimeter setting, will indicate altitude;

b) When set to a QFE altimeter setting, will indicate height above the QFE reference datum; and

c) When set to a pressure of 1013.2 hPa (1013.2 mb), may be used to indicate flight levels.

Note 2: The terms 'height' and 'altitude', used in Note 1 above, indicate altimetric rather than geometric heights and altitudes.

FLIGHT LINE—A term used to describe the precise movement of a civil photogrammetric aircraft along a predetermined course(s) at a predetermined altitude during the actual photographic run.

FLIGHT MANAGEMENT SYSTEMS—A computer system that uses a large data base to allow routes to be preprogrammed and fed into the system by means of a data loader. The system is constantly updated with respect to position accuracy by reference to conventional navigation aids. The sophisticated program and its associated data base insures that the most appropriate aids are automatically selected during the information update cycle.

FLIGHT PATH—A line, course, or track along which an aircraft is flying or intended to be flown.

(See TRACK). (See COURSE).

FLIGHT PLAN—Specified information relating to the intended flight of an aircraft that is filed orally or in writing with an FSS or an ATC facility.

(See FAST FILE). (See FILED). (Refer to AIM).

FLIGHT PLAN AREA—The geographical area assigned by regional air traffic divisions to a flight service station for the purpose of search and rescue for VFR aircraft, issuance of notams, pilot briefing, in-flight services, broadcast, emergency services, flight data processing, international operations, and aviation weather services. Three letter identifiers are assigned to every flight service station and are annotated in AFD's and Order 7350.6 as tie-in-facilities.

(See FAST FILE). (See FILED). (Refer to AIM).

FLIGHT RECORDER—A general term applied to any instrument or device that records information about the performance of an aircraft in flight or about conditions encountered in flight. Flight recorders may make records of airspeed, outside air temperature, vertical acceleration, engine RPM, manifold pressure, and other pertinent variables for a given flight.

(See ICAO term FLIGHT RECORDER).

FLIGHT RECORDER [ICAO]—Any type of recorder installed in the aircraft for the purpose of complementing accident/incident investigation.

Note: See Annex 6 Part I, for specifications relating to flight recorders.

FLIGHT SERVICE STATION—Air traffic facilities which provide pilot briefing, en route communications and VFR search and rescue services, assist lost aircraft and aircraft in emergency situations, relay ATC clearances, originate Notices to Airmen, broadcast aviation weather and NAS information, receive and process IFR flight plans, and monitor NAVAID's. In addition, at selected locations, FSS's provide Enroute Flight Advisory Service (Flight Watch), take weather observations, issue airport advisories, and advise Customs and Immigration of transborder flights.

(Refer to AIM).

FLIGHT STANDARDS DISTRICT OFFICE—An FAA field office serving an assigned geographical area and staffed with Flight Standards personnel who serve the aviation industry and the general public on matters relating to the certification and operation of air carrier and general aviation aircraft. Activities include general surveillance of operational safety, certification of airmen and aircraft, accident prevention, investigation, enforcement, etc.

FLIGHT TEST—A flight for the purpose of:

1. Investigating the operation/flight characteristics of an aircraft or aircraft component.
2. Evaluating an applicant for a pilot certificate or rating.

FLIGHT VISIBILITY—(See VISIBILITY).

FLIGHT WATCH—A shortened term for use in air-ground contacts to identify the flight service station providing En Route Flight Advisory Service; e.g., "Oakland Flight Watch."

(See EN ROUTE FLIGHT ADVISORY SERVICE).

FLIP—(See DOD FLIP).

FLOW CONTROL—Measures designed to adjust the flow of traffic into a given airspace, along a given route, or bound for a given aerodrome (airport) so as to ensure the most effective utilization of the airspace.

(See QUOTA FLOW CONTROL). (Refer to AIRPORT / FACILITY DIRECTORY).

FLY HEADING (DEGREES)—Informs the pilot of the heading he should fly. The pilot may have to turn to, or continue on, a specific compass direction in order to comply with the instructions. The pilot is expected to turn in the shorter direction to the heading unless otherwise instructed by ATC.

FMA—(See FINAL MONITOR AID).

FMS— (See FLIGHT MANAGEMENT SYSTEM)

FORMATION FLIGHT—More than one aircraft which, by prior arrangement between the pilots, operate as a single aircraft with regard to navigation and position reporting. Separation between aircraft within the formation is the responsibility of the flight leader and the pilots of the other aircraft in the flight. This includes transition periods when aircraft within the formation are maneuvering to attain separation from each other to effect individual control and during join-up and breakaway.

1. A standard formation is one in which a proximity of no more than 1 mile laterally or longitudinally and within 100 feet vertically from the flight leader is maintained by each wingman.
2. Nonstandard formations are those operating under any of the following conditions:
   a. When the flight leader has requested and ATC has approved other than standard formation dimensions.
   b. When operating within an authorized altitude reservation (ALTRV) or under the provisions of a letter of agreement.
   c. When the operations are conducted in airspace specifically designed for a special activity.

(See ALTITUDE RESERVATION). (Refer to Part 91).

FRC—(See REQUEST FULL ROUTE CLEARANCE).

FREEZE/FROZEN—Terms used in referring to arrivals which have been assigned ACLT's and to the lists in which they are displayed.

FREEZE CALCULATED LANDING TIME—A dynamic parameter number of minutes prior to the meter fix calculated time of arrival for each aircraft when the TCLT is frozen and becomes an ACLT (i.e., the VTA is updated and consequently the TCLT is modified as appropriate until FCLT minutes prior to meter fix calculated time of arrival, at which time updating is suspended and an ACLT and a frozen meter fix crossing time (MFT) is assigned).

FREEZE SPEED PARAMETER—A speed adapted for each aircraft to determine fast and slow aircraft. Fast aircraft freeze on parameter FCLT and slow aircraft freeze on parameter MLDI.

FSDO—(See FLIGHT STANDARDS DISTRICT OFFICE).

FSPD—(See FREEZE SPEED PARAMETER).

FSS—(See FLIGHT SERVICE STATION).

FUEL DUMPING—Airborne release of usable fuel. This does not include the dropping of fuel tanks.
(See JETTISONING OF EXTERNAL STORES).

FUEL REMAINING—A phrase used by either pilots or controllers when relating to the fuel remaining on board until actual fuel exhaustion. When transmitting such information in response to either a controller question or pilot initiated cautionary advisory to air traffic control, pilots will state the APPROXIMATE NUMBER OF MINUTES the flight can continue with the fuel remaining. All reserve fuel SHOULD BE INCLUDED in the time stated, as should an allowance for established fuel gauge system error.

FUEL SIPHONING—Unintentional release of fuel caused by overflow, puncture, loose cap, etc.

FUEL VENTING—(See FUEL SIPHONING).-

G

GADO—(See GENERAL AVIATION DISTRICT OFFICE).

GATE HOLD PROCEDURES—Procedures at selected airports to hold aircraft at the gate or other ground location whenever departure delays exceed or are anticipated to exceed 15 minutes. The sequence for departure will be maintained in accordance with initial call-up unless modified by flow control restrictions. Pilots should monitor the ground control/clearance delivery frequency for engine start/taxi advisories or new proposed start/taxi time if the delay changes.

(See FLOW CONTROL).

GCA—(See GROUND CONTROLLED APPROACH).

GENERAL AVIATION—That portion of civil aviation which encompasses all facets of aviation except air carriers holding a certificate of public convenience and necessity from the Civil Aeronautics Board and large aircraft commercial operators.

(See ICAO term GENERAL AVIATION).

GENERAL AVIATION [ICAO]—All civil aviation operations other than scheduled air services and nonscheduled air transport operations for remuneration or hire.

GENERAL AVIATION DISTRICT OFFICE—An FAA field office serving a designated geographical area and staffed with Flight Standards personnel who have the responsibility for serving the aviation industry and the general public on all matters relating to the certification and operation of general aviation aircraft.

GLIDEPATH—(See GLIDESLOPE).

GLIDEPATH INTERCEPT ALTITUDE—(See GLIDESLOPE INTERCEPT ALTITUDE).

GLIDESLOPE—Provides vertical guidance for aircraft during approach and landing. The glideslope/glidepath is based on the following:

1. Electronic components emitting signals which provide vertical guidance by reference to airborne instruments during instrument approaches such as ILS/MLS, or
2. Visual ground aids, such as VASI, which provide vertical guidance for a VFR approach or for the visual portion of an instrument approach and landing.
3. PAR. Used by ATC to inform an aircraft making a PAR approach of its vertical position (elevation) relative to the descent profile.

(See ICAO term GLIDEPATH).

GLIDEPATH [ICAO]—A descent profile determined for vertical guidance during a final approach.

GLIDESLOPE INTERCEPT ALTITUDE—The minimum altitude to intercept the glideslope/path on a precision approach. The intersection of the published intercept altitude with the glideslope/path, designated on Government charts by the lightning bolt symbol, is the precision FAF; however, when ATC directs a lower altitude, the resultant lower intercept position is then the FAF.

(See FINAL APPROACH FIX). (See SEGMENTS OF AN INSTRUMENT APPROACH PROCEDURE).

GLOBAL POSITIONING SYSTEM—A space-base radio positioning, navigation, and time-transfer system being developed by Department of Defense. When fully deployed, the system is intended to provide highly accurate position and velocity information, and precise time, on a continuous global basis, to an unlimited number of properly equipped users. The system will be unaffected by weather, and will provide a worldwide common grid reference system. The GPS concept is predicated upon accurate and continuous knowledge of the spatial position of each satellite in the system with respect to time and distance from a transmitting satellite to the user. The GPS receiver automatically selects appropriate signals from the satellites in view and translates these into a three-dimensional position, velocity, and time. Predictable system accuracy for civil users is projected to be 100 meters horizontally. Performance standards and certification criteria have not yet been established.

GO AHEAD—Proceed with your message. Not to be used for any other purpose.

GO AROUND—Instructions for a pilot to abandon his approach to landing. Additional instructions may follow. Unless otherwise advised by ATC, a VFR aircraft or an aircraft conducting visual approach should overfly the runway while climbing to traffic pattern altitude and enter the traffic pattern via the crosswind leg. A pilot on an IFR flight plan making an instrument approach should execute the published missed approach procedure or proceed as instructed by ATC; e.g., "Go around" (additional instructions if required).

(See LOW APPROACH). (See MISSED APPROACH).

GPS—(See Global Positioning System)

GROUND CLUTTER—A pattern produced on the radar scope by ground returns which may degrade other radar returns in the affected area. The effect of ground clutter is minimized by the use of moving target indicator (MTI) circuits in the radar equipment resulting in a radar presentation which displays only targets which are in motion.

(See CLUTTER).

GROUND CONTROLLED APPROACH—A radar approach system operated from the ground by air traffic control personnel transmitting instructions to the pilot by radio. The approach may be conducted with surveillance radar (ASR) only or with both surveillance and precision approach radar (PAR). Usage of the term "GCA" by pilots is discouraged except when referring to a GCA facility. Pilots should specifically request a "PAR" approach when a precision radar approach is desired or request an "ASR" or "surveillance" approach when a nonprecision radar approach is desired.

(See RADAR APPROACH).

GROUND DELAY—The amount of delay attributed to ATC, encountered prior to departure, usually associated with a CDT program.

GROUND SPEED—The speed of an aircraft relative to the surface of the earth.

GROUND STOP—Normally, the last initiative to be utilized; this method mandates that the terminal facility will not allow any departures to enter the ARTCC airspace until further notified.

GROUND VISIBILITY—(See VISIBILITY).-

H

HAA—(See HEIGHT ABOVE AIRPORT).

HAL—(See HEIGHT ABOVE LANDING).

HANDOFF—An action taken to transfer the radar identification of an aircraft from one controller to another if the aircraft will enter the receiving controller's airspace and radio communications with the aircraft will be transferred.

HAT—(See HEIGHT ABOVE TOUCHDOWN).

HAVE NUMBERS—Used by pilots to inform ATC that they have received runway, wind, and altimeter information only.

HAZARDOUS INFLIGHT WEATHER ADVISORY SERVICE—Continuous recorded hazardous inflight weather forecasts broadcasted to airborne pilots over selected VOR outlets defined as an HIWAS BROADCAST AREA.

HAZARDOUS WEATHER INFORMATION—Summary of significant meteorological information (SIGMET/WS), convective significant meteorological information (convective SIGMET/WST), urgent pilot weather reports (urgent PIREP/UUA), center weather advisories (CWA), airmen's meteorological information (AIRMET/WA) and any other weather such as isolated thunderstorms that are rapidly developing and increasing in intensity, or low ceilings and visibilities that are becoming widespread which is considered significant and are not included in a current hazardous weather advisory.

HEAVY (AIRCRAFT)—(See AIRCRAFT CLASSES).

HEIGHT ABOVE AIRPORT—The height of the Minimum Descent Altitude above the published airport elevation. This is published in conjunction with circling minimums.
(See MINIMUM DESCENT ALTITUDE).

HEIGHT ABOVE LANDING—The height above a designated helicopter landing area used for helicopter instrument approach procedures.
(Refer to Part 97).

HEIGHT ABOVE TOUCHDOWN—The height of the Decision Height or Minimum Descent Altitude above the highest runway elevation in the touchdown zone (first 3,000 feet of the runway). HAT is published on instrument approach charts in conjunction with all straight-in minimums.
(See DECISION HEIGHT). (See MINIMUM DESCENT ALTITUDE).

HELICOPTER—Rotorcraft that, for its horizontal motion, depends principally on its engine-driven rotors.
(See ICAO term HELICOPTER).

HELICOPTER [ICAO]—A heavier-than-air aircraft supported in flight chiefly by the reactions of the air on one or more power-driven rotors on substantially vertical axes.

HELIPAD—A small, designated area, usually with a prepared surface, on a heliport, airport, landing/takeoff area, apron/ramp, or movement area used for takeoff, landing, or parking of helicopters.

HELIPORT—An area of land, water, or structure used or intended to be used for the landing and takeoff of helicopters and includes its buildings and facilities if any.

HERTZ—The standard radio equivalent of frequency in cycles per second of an electromagnetic wave. Kilohertz (kHz) is a frequency of one thousand cycles per second. Megahertz (mHz) is a frequency of one million cycles per second.

HF—(See HIGH FREQUENCY).

HF COMMUNICATIONS—(See HIGH FREQUENCY COMMUNICATIONS).

HIGH FREQUENCY—The frequency band between 3 and 30 mHz.
(See HIGH FREQUENCY COMMUNICATIONS).

HIGH FREQUENCY COMMUNICATIONS—High radio frequencies (HF) between 3 and 30 mHz used for air-to-ground voice communication in overseas operations.

HIGH SPEED EXIT—(See HIGH SPEED TAXIWAY).

HIGH SPEED TAXIWAY—A long radius taxiway designed and provided with lighting or marking to define the path of aircraft, traveling at high speed (up to 60 knots), from the runway center to a point on the center of a taxiway. Also referred to as long radius exit or turn-off taxiway. The high speed taxiway is designed to expedite aircraft turning off the runway after landing, thus reducing runway occupancy time.

HIGH SPEED TURNOFF—(See HIGH SPEED TAXIWAY).

HIWAS—(See HAZARDOUS INFLIGHT WEATHER ADVISORY SERVICE).

HIWAS AREA—(See HAZARDOUS INFLIGHT WEATHER ADVISORY SERVICE).

HIWAS BROADCAST AREA—A geographical area of responsibility including one or more HIWAS outlet areas assigned to an AFSS/FSS for hazardous weather advisory broadcasting.

HIWAS OUTLET AREA—An area defined as a 150 NM radius of a HIWAS outlet, expanded as necessary to provide coverage.

HOLDING PROCEDURE—(See HOLD PROCEDURE).

HOLD PROCEDURE—A predetermined maneuver which keeps aircraft within a specified airspace while awaiting further clearance from air traffic control. Also used during ground operations to keep aircraft within a specified area or at a specified point while awaiting further clearance from air traffic control.

(See HOLDING FIX). (Refer to AIM).

HOLDING FIX—A specified fix identifiable to a pilot by NAVAID's or visual reference to the ground used as a reference point in establishing and maintaining the position of an aircraft while holding.

(See FIX). (See VISUAL HOLDING). (Refer to AIM).

HOLDING POINT [ICAO]—A specified location, identified by visual or other means, in the vicinity of which the position of an aircraft in flight is maintained in accordance with air traffic control clearances.

HOLD FOR RELEASE—Used by ATC to delay an aircraft for traffic management reasons; i.e., weather, traffic volume, etc. Hold for release instructions (including departure delay information) are used to inform a pilot or a controller (either directly or through an authorized relay) that a departure clearance is not valid until a release time or additional instructions have been received.

(See ICAO term HOLDING POINT).

HOMING—Flight toward a NAVAID, without correcting for wind, by adjusting the aircraft heading to maintain a relative bearing of zero degrees.

(See BEARING). (See ICAO term HOMING).

HOMING [ICAO]—The procedure of using the direction-finding equipment of one radio station with the emission of another radio station, where at least one of the stations is mobile, and whereby the mobile station proceeds continuously towards the other station.

HOVER CHECK—Used to describe when a helicopter/VTOL aircraft requires a stabilized hover to conduct a performance/power check prior to hover taxi, air taxi, or takeoff. Altitude of the hover will vary based on the purpose of the check.

HOVER TAXI—Used to describe a helicopter/VTOL aircraft movement conducted above the surface and in ground effect at airspeeds less than approximately 20 knots. The actual height may vary, and some helicopters may require hover taxi above 25 feet AGL to reduce ground effect turbulence or provide clearance for cargo slingloads.

(See AIR TAXI). (See HOVER CHECK). (Refer to AIM).

HOW DO YOU HEAR ME?—A question relating to the quality of the transmission or to determine how well the transmission is being received.

HZ—(See HERTZ).-

I

IAF—(See INITIAL APPROACH FIX).

IAP—(See INSTRUMENT APPROACH PROCEDURE).

ICAO—(See INTERNATIONAL CIVIL AVIATION ORGANIZATION).

ICAO [ICAO]—(See ICAO Term INTERNATIONAL CIVIL AVIATION ORGANIZATION).

IDENT—A request for a pilot to activate the aircraft transponder identification feature. This will help the controller to confirm an aircraft identity or to identify an aircraft.
(Refer to AIM).

IDENT FEATURE—The special feature in the Air Traffic Control Radar Beacon System (ATCRBS) equipment. It is used to immediately distinguish one displayed beacon target from other beacon targets.
(See IDENT).

IF—(See INTERMEDIATE FIX).

IFIM—(See INTERNATIONAL FLIGHT INFORMATION MANUAL).

IF NO TRANSMISSION RECEIVED FOR (TIME)—Used by ATC in radar approaches to prefix procedures which should be followed by the pilot in event of lost communications.
(See LOST COMMUNICATIONS).

IFR—(See INSTRUMENT FLIGHT RULES).

IFR AIRCRAFT—An aircraft conducting flight in accordance with instrument flight rules.

IFR CONDITIONS—Weather conditions below the minimum for flight under visual flight rules.
(See INSTRUMENT METEOROLOGICAL CONDITIONS).

IFR DEPARTURE PROCEDURE—(See IFR TAKEOFF MINIMUMS AND DEPARTURE PROCEDURES).
(Refer to AIM).

IFR FLIGHT—(See IFR AIRCRAFT).

IFR LANDING MINIMUMS—(See LANDING MINIMUMS).

IFR MILITARY TRAINING ROUTES (IR)—Routes used by the Department of Defense and associated Reserve and Air Guard units for the purpose of conducting low-altitude navigation and tactical training in both IFR and VFR weather conditions below 10,000 feet MSL at airspeeds in excess of 250 knots IAS.

IFR TAKEOFF MINIMUMS AND DEPARTURE PROCEDURES—Federal Aviation Regulations, Part 91, prescribes standard takeoff rules for certain civil users. At some airports, obstructions or other factors require the establishment of nonstandard takeoff minimums, departure procedures, or both to assist pilots in avoiding obstacles during climb to the minimum en route altitude. Those airports are listed in NOS/DOD Instrument Approach Charts (IAP's) under a section entitled "IFR Takeoff Minimums and Departure Procedures." The NOS/DOD IAP chart legend illustrates the symbol used to alert the pilot to nonstandard takeoff minimums and departure procedures. When departing IFR from such airports or from any airports where there are no departure procedures, SID's, or ATC facilities available, pilots should advise ATC of any departure limitations. Controllers may query a pilot to determine acceptable departure directions, turns, or headings after takeoff. Pilots should be familiar with the departure procedures and must assure that their aircraft can meet or exceed any specified climb gradients.

ILS—(See INSTRUMENT LANDING SYSTEM).

ILS CATEGORIES—1. ILS Category I. An ILS approach procedure which provides for approach to a height above touchdown of not less than 200 feet and with runway visual range of not less than 1,800 feet.—2. ILS Category II. An ILS approach procedure which provides for approach to a height above touchdown of not less than 100 feet and with runway visual range of not less than 1,200 feet.—3. ILS Category III:

a. IIIA.—An ILS approach procedure which provides for approach without a decision height minimum and with runway visual range of not less than 700 feet.

b. IIIB.—An ILS approach procedure which provides for approach without a decision height minimum and with runway visual range of not less than 150 feet.

c. IIIC.—An ILS approach procedure which provides for approach without a decision height minimum and without runway visual range minimum.

IM—(See INNER MARKER).

IMC—(See INSTRUMENT METEOROLOGICAL CONDITIONS).

IMMEDIATELY—Used by ATC when such action compliance is required to avoid an imminent situation.

INCERFA (Uncertainty Phase) [ICAO]—The code word used to designate an emergency phase wherein there is concern about the safety of an aircraft or its occupants. In most cases this phase involves an aircraft which is overdue or unreported.

INCREASE SPEED TO (SPEED)—(See SPEED ADJUSTMENT).

INERTIAL NAVIGATION SYSTEM—An RNAV system which is a form of self-contained navigation.

(See Area Navigation / RNAV.).

INFLIGHT REFUELING—(See AERIAL REFUELING).

INFLIGHT WEATHER ADVISORY—(See WEATHER ADVISORY).

INFORMATION REQUEST—A request originated by an FSS for information concerning an overdue VFR aircraft.

INITIAL APPROACH FIX—The fixes depicted on instrument approach procedure charts that identify the beginning of the initial approach segment(s).

(See FIX). (See SEGMENTS OF AN INSTRUMENT APPROACH PROCEDURE).

INITIAL APPROACH SEGMENT—(See SEGMENTS OF AN INSTRUMENT APPROACH PROCEDURE).

INITIAL APPROACH SEGMENT [ICAO]—That segment of an instrument approach procedure between the initial approach fix and the intermediate approach fix or, where applicable, the final approach fix or point.

INLAND NAVIGATION FACILITY—A navigation aid on a North American Route at which the common route and/or the noncommon route begins or ends.

INNER MARKER—A marker beacon used with an ILS (CAT II) precision approach located between the middle marker and the end of the ILS runway, transmitting a radiation pattern keyed at six dots per second and indicating to the pilot, both aurally and visually, that he is at the designated decision height (DH), normally 100 feet above the touchdown zone elevation, on the ILS CAT II approach. It also marks progress during a CAT III approach.

(See INSTRUMENT LANDING SYSTEM). (Refer to AIM).

INNER MARKER BEACON—(See INNER MARKER).

INREQ—(See INFORMATION REQUEST).

INS—(See INERTIAL NAVIGATION SYSTEM).

INSTRUMENT APPROACH—(See INSTRUMENT APPROACH PROCEDURE).

INSTRUMENT APPROACH PROCEDURE—A series of predetermined maneuvers for the orderly transfer of an aircraft under instrument flight conditions from the beginning of the initial approach to a landing or to a point from which a landing may be made visually. It is prescribed and approved for a specific airport by competent authority.

(See SEGMENTS OF AN INSTRUMENT APPROACH PROCEDURE). (Refer to Part 91). (See AIM).

1. U.S. civil standard instrument approach procedures are approved by the FAA as prescribed under Part 97 and are available for public use.
2. U.S. military standard instrument approach procedures are approved and published by the Department of Defense.

3. Special instrument approach procedures are approved by the FAA for individual operators but are not published in Part 97 for public use.

(See ICAO term INSTRUMENT APPROACH PROCEDURE).

INSTRUMENT APPROACH PROCEDURE [ICAO]—A series of predetermined manoeuvres by reference to flight instruments with specified protection from obstacles from the initial approach fix, or where applicable, from the beginning of a defined arrival route to a point from which a landing can be completed and thereafter, if a landing is not completed, to a position at which holding or en route obstacle clearance criteria apply.

INSTRUMENT APPROACH PROCEDURES CHARTS—(See AERONAUTICAL CHART).

INSTRUMENT FLIGHT RULES—Rules governing the procedures for conducting instrument flight. Also a term used by pilots and controllers to indicate type of flight plan.

(See VISUAL FLIGHT RULES). (See INSTRUMENT METEOROLOGICAL CONDITIONS). (See VISUAL METEOROLOGICAL CONDITIONS). (Refer to AIM). (See ICAO term INSTRUMENT FLIGHT RULES).

INSTRUMENT FLIGHT RULES [ICAO]—A set of rules governing the conduct of flight under instrument meteorological conditions.

INSTRUMENT LANDING SYSTEM—A precision instrument approach system which normally consists of the following electronic components and visual aids:

1. Localizer.

(See LOCALIZER).

2. Glideslope.

(See GLIDESLOPE).

3. Outer Marker.

(See OUTER MARKER).

4. Middle Marker.

(See MIDDLE MARKER).

5. Approach Lights.

(See AIRPORT LIGHTING).

(Refer to Part 91). (See AIM).

INSTRUMENT METEOROLOGICAL CONDITIONS—Meteorological conditions expressed in terms of visibility, distance from cloud, and ceiling less than the minima specified for visual meteorological conditions.

(See VISUAL METEOROLOGICAL CONDITIONS). (See INSTRUMENT FLIGHT RULES). (See VISUAL FLIGHT RULES).

INSTRUMENT RUNWAY—A runway equipped with electronic and visual navigation aids for which a precision or nonprecision approach procedure having straight-in landing minimums has been approved.

(See ICAO term INSTRUMENT RUNWAY).

INSTRUMENT RUNWAY [ICAO]—One of the following types of runways intended for the operation of aircraft using instrument approach procedures:

a) Nonprecision Approach Runway—An instrument runway served by visual aids and a nonvisual aid providing at least directional guidance adequate for a straight-in approach.

b) Precision Approach Runway, Category I—An instrument runway served by ILS and visual aids intended for operations down to 60 m (200 feet) decision height and down to an RVR of the order of 800 m.

c) Precision Approach Runway, Category II—An instrument runway served by ILS and visual aids intended for operations down to 30 m (100 feet) decision height and down to an RVR of the order of 400 m.

d) Precision Approach Runway, Category III—An instrument runway served by ILS to and along the surface of the runway and:

A. Intended for operations down to an RVR of the order of 200 m (no decision height being applicable) using visual aids during the final phase of landing;

B. Intended for operations down to an RVR of the order of 50 m (no decision height being applicable) using visual aids for taxiing;

C. Intended for operations without reliance on visual reference for landing or taxiing.

Note 1: See Annex 10 Volume I, Part I Chapter 3, for related ILS specifications.

Note 2: Visual aids need not necessarily be matched to the scale of nonvisual aids provided. The criterion for the selection of visual aids is the conditions in which operations are intended to be conducted.

INTERMEDIATE APPROACH SEGMENT—(See SEGMENTS OF AN INSTRUMENT APPROACH PROCEDURE).

INTERMEDIATE APPROACH SEGMENT [ICAO]—That segment of an instrument approach procedure between either the intermediate approach fix and the final approach fix or point, or between the end of a reversal, race track or dead reckoning track procedure and the final approach fix or point, as appropriate.

INTERMEDIATE FIX—The fix that identifies the beginning of the intermediate approach segment of an instrument approach procedure. The fix is not normally identified on the instrument approach chart as an intermediate fix (IF).
(See SEGMENTS OF AN INSTRUMENT APPROACH PROCEDURE).

INTERMEDIATE LANDING—On the rare occasion that this option is requested, it should be approved. The departure center, however, must advise the ATCSCC so that the appropriate delay is carried over and assigned at the intermediate airport. An intermediate landing airport within the arrival center will not be accepted without coordination with and the approval of the ATCSCC.

INTERNATIONAL AIRPORT—Relating to international flight, it means:

1. An airport of entry which has been designated by the Secretary of Treasury or Commissioner of Customs as an international airport for customs service.
2. A landing rights airport at which specific permission to land must be obtained from customs authorities in advance of contemplated use.
3. Airports designated under the Convention on International Civil Aviation as an airport for use by international commercial air transport and/or international general aviation.

(Refer to AIRPORT / FACILITY DIRECTORY). (Refer to IFIM). (See ICAO term INTERNATIONAL AIRPORT).

INTERNATIONAL AIRPORT [ICAO]—Any airport designated by the Contracting State in whose territory it is situated as an airport of entry and departure for international air traffic, where the formalities incident to customs, immigration, public health, animal and plant quarantine and similar procedures are carried out.

INTERNATIONAL CIVIL AVIATION ORGANIZATION [ICAO]—A specialized agency of the United Nations whose objective is to develop the principles and techniques of international air navigation and to foster planning and development of international civil air transport.

ICAO Regions include:
AFI African-Indian Ocean Region
CAR Caribbean Region
EUR European Region
MID/ASIA Middle East/Asia Region
NAM North American Region
NAT North Atlantic Region
PAC Pacific Region
SAM South American Region

INTERNATIONAL FLIGHT INFORMATION MANUAL—A publication designed primarily as a pilot's preflight planning guide for flights into foreign airspace and for flights returning to the U.S. from foreign locations.

INTERROGATOR—The ground-based surveillance radar beacon transmitter-receiver, which normally scans in synchronism with a primary radar, transmitting discrete radio signals which repetitiously request all transponders on the mode being used to reply. The replies received are mixed with the primary radar returns and displayed on the same plan position indicator (radar scope). Also, applied to the airborne element of the TACAN/DME system.
(See TRANSPONDER). (Refer to AIM).

INTERSECTING RUNWAYS—Two or more runways which cross or meet within their lengths.
(See INTERSECTION).

INTERSECTION—

1. A point defined by any combination of courses, radials, or bearings of two or more navigational aids.
2. Used to describe the point where two runways, a runway and a taxiway, or two taxiways cross or meet.

INTERSECTION DEPARTURE—A departure from any runway intersection except the end of the runway. *(See INTERSECTION).*

INTERSECTION TAKEOFF—(See INTERSECTION DEPARTURE).

IR—(See IFR MILITARY TRAINING ROUTES).

I SAY AGAIN—The message will be repeated.-

J

JAMMING—Electronic or mechanical interference which may disrupt the display of aircraft on radar or the transmission/reception of radio communications/navigation.

JET BLAST—Jet engine exhaust (thrust stream turbulence).
(See WAKE TURBULENCE).

JET ROUTE—A route designed to serve aircraft operations from 18,000 feet MSL up to and including flight level 450. The routes are referred to as "J" routes with numbering to identify the designated route; e.g., J105.
(See ROUTE). (Refer to Part 71).

JET STREAM—A migrating stream of high-speed winds present at high altitudes.

JETTISONING OF EXTERNAL STORES—Airborne release of external stores; e.g., tiptanks, ordnance.
(See FUEL DUMPING). (Refer to Part 91).

JOINT USE RESTRICTED AREA—(See RESTRICTED AREA).-

K

KNOWN TRAFFIC—With respect to ATC clearances, means aircraft whose altitude, position, and intentions are known to ATC.-

L

LAA—(See LOCAL AIRPORT ADVISORY).

LAAS—(See LOW ALTITUDE ALERT SYSTEM).

LANDING AREA—Any locality either on land, water, or structures, including airports/heliports and intermediate landing fields, which is used, or intended to be used, for the landing and takeoff of aircraft whether or not facilities are provided for the shelter, servicing, or for receiving or discharging passengers or cargo.

(See ICAO term LANDING AREA).

LANDING AREA [ICAO]—That part of a movement area intended for the landing or takeoff of aircraft.

LANDING DIRECTION INDICATOR—A device which visually indicates the direction in which landings and takeoffs should be made.

(See TETRAHEDRON). (Refer to AIM).

LANDING DISTANCE AVAILABLE [ICAO]—The length of runway which is declared available and suitable for the ground run of an aeroplane landing.

LANDING MINIMUMS—The minimum visibility prescribed for landing a civil aircraft while using an instrument approach procedure. The minimum applies with other limitations set forth in Part 91 with respect to the Minimum Descent Altitude (MDA) or Decision Height (DH) prescribed in the instrument approach procedures as follows:

1. Straight-in landing minimums. A statement of MDA and visibility, or DH and visibility, required for a straight-in landing on a specified runway, or
2. Circling minimums. A statement of MDA and visibility required for the circle-to-land maneuver.

Descent below the established MDA or DH is not authorized during an approach unless the aircraft is in a position from which a normal approach to the runway of intended landing can be made and adequate visual reference to required visual cues is maintained.

(See STRAIGHT-IN LANDING). (See CIRCLE-TO-LAND MANEUVER). (See DECISION HEIGHT). (See MINIMUM DESCENT ALTITUDE). (See VISIBILITY). (See INSTRUMENT APPROACH PROCEDURE). (Refer to Part 91).

LANDING ROLL—The distance from the point of touchdown to the point where the aircraft can be brought to a stop or exit the runway.

LANDING SEQUENCE—The order in which aircraft are positioned for landing.

(See APPROACH SEQUENCE).

LAST ASSIGNED ALTITUDE—The last altitude/flight level assigned by ATC and acknowledged by the pilot.

(See MAINTAIN). (Refer to Part 91).

LATERAL SEPARATION—The lateral spacing of aircraft at the same altitude by requiring operation on different routes or in different geographical locations.

(See SEPARATION).

LDA—(See LOCALIZER TYPE DIRECTIONAL AID).

LDA [ICAO]—(See ICAO Term LANDING DISTANCE AVAILABLE).

LF—(See LOW FREQUENCY).

LIGHTED AIRPORT—An airport where runway and obstruction lighting is available.

(See AIRPORT LIGHTING). (Refer to AIM).

LIGHT GUN—A handheld directional light signaling device which emits a brilliant narrow beam of white, green, or red light as selected by the tower controller. The color and type of light transmitted can be used to approve or disapprove anticipated pilot actions where radio communication is not available. The light gun is used for controlling traffic operating in the vicinity of the airport and on the airport movement area.

(Refer to AIM).

LOCALIZER—The component of an ILS which provides course guidance to the runway.

(See INSTRUMENT LANDING SYSTEM). (Refer to AIM). (See ICAO term LOCALIZER COURSE).

LOCALIZER COURSE [ICAO]—The locus of points, in any given horizontal plane, at which the DDM (difference in depth of modulation) is zero.

LOCALIZER TYPE DIRECTIONAL AID—A NAVAID used for nonprecision instrument approaches with utility and accuracy comparable to a localizer but which is not a part of a complete ILS and is not aligned with the runway.

(Refer to AIM).

LOCALIZER USABLE DISTANCE—The maximum distance from the localizer transmitter at a specified altitude, as verified by flight inspection, at which reliable course information is continuously received.

(Refer to AIM).

LOCAL AIRPORT ADVISORY [LAA]—A service provided by flight service stations or the military at airports not serviced by an operating control tower. This service consists of providing information to arriving and departing aircraft concerning wind direction and speed, favored runway, altimeter setting, pertinent known traffic, pertinent known field conditions, airport taxi routes and traffic patterns, and authorized instrument approach procedures. This information is advisory in nature and does not constitute an ATC clearance.

(See AIRPORT ADVISORY AREA).

LOCAL TRAFFIC—Aircraft operating in the traffic pattern or within sight of the tower, or aircraft known to be departing or arriving from flight in local practice areas, or aircraft executing practice instrument approaches at the airport.

(See TRAFFIC PATTERN).

LOCATOR [ICAO]—An LM/MF NDB used as an aid to final approach.

Note: A locator usually has an average radius of rated coverage of between 18.5 and 46.3 km (10 and 25 NM).

LONGITUDINAL SEPARATION—The longitudinal spacing of aircraft at the same altitude by a minimum distance expressed in units of time or miles.

(See SEPARATION). (Refer to AIM).

LONG RANGE NAVIGATION—(See LORAN).

LORAN—An electronic navigational system by which hyperbolic lines of position are determined by measuring the difference in the time of reception of synchronized pulse signals from two fixed transmitters. Loran A operates in the 1750–1950 kHz frequency band. Loran C and D operate in the 100–110 kHz frequency band.

(Refer to AIM).

LOST COMMUNICATIONS—Loss of the ability to communicate by radio. Aircraft are sometimes referred to as NORDO (No Radio). Standard pilot procedures are specified in Part 91. Radar controllers issue procedures for pilots to follow in the event of lost communications during a radar approach when weather reports indicate that an aircraft will likely encounter IFR weather conditions during the approach.

(Refer to Part 91). (See AIM).

LOW ALTITUDE AIRWAY STRUCTURE—The network of airways serving aircraft operations up to but not including 18,000 feet MSL.

(See AIRWAY). (Refer to AIM).

LOW ALTITUDE ALERT, CHECK YOUR ALTITUDE IMMEDIATELY—(See SAFETY ALERT).

LOW ALTITUDE ALERT SYSTEM—An automated function of the TPX–42 that alerts the controller when a Mode C transponder-equipped aircraft on an IFR flight plan is below a predetermined minimum safe altitude. If requested by the pilot, LAAS monitoring is also available to VFR Mode C transponder-equipped aircraft.

LOW APPROACH—An approach over an airport or runway following an instrument approach or a VFR approach including the go-around maneuver where the pilot intentionally does not make contact with the runway.

(Refer to AIM).

LOW FREQUENCY—The frequency band between 30 and 300 kHz.

(Refer to AIM).-

M

MAA—(See MAXIMUM AUTHORIZED ALTITUDE).

MACH NUMBER—The ratio of true airspeed to the speed of sound; e.g., MACH .82, MACH 1.6.
(See AIRSPEED).

MACH TECHNIQUE [ICAO]—Describes a control technique used by air traffic control whereby turbojet aircraft operating successively along suitable routes are cleared to maintain appropriate MACH numbers for a relevant portion of the en route phase of flight. The principle objective is to achieve improved utilization of the airspace and to ensure that separation between successive aircraft does not decrease below the established minima.

MAINTAIN—

1. Concerning altitude/flight level, the term means to remain at the altitude/flight level specified. The phrase "climb and" or "descend and" normally precedes "maintain" and the altitude assignment; e.g., "descend and maintain 5,000."
2. Concerning other ATC instructions, the term is used in its literal sense; e.g., maintain VFR.

MAKE SHORT APPROACH—Used by ATC to inform a pilot to alter his traffic pattern so as to make a short final approach.
(See TRAFFIC PATTERN).

MANDATORY ALTITUDE—An altitude depicted on an instrument Approach Procedure Chart requiring the aircraft to maintain altitude at the depicted value.

MAP—(See MISSED APPROACH POINT).

MARKER BEACON—An electronic navigation facility transmitting a 75 mHz vertical fan or boneshaped radiation pattern. Marker beacons are identified by their modulation frequency and keying code, and when received by compatible airborne equipment, indicate to the pilot, both aurally and visually, that he is passing over the facility.
(See OUTER MARKER). (See MIDDLE MARKER). (See INNER MARKER). (Refer to AIM).

MARSA—(See MILITARY AUTHORITY ASSUMES RESPONSIBILITY FOR SEPARATION OF AIRCRAFT).

MAXIMUM AUTHORIZED ALTITUDE—A published altitude representing the maximum usable altitude or flight level for an airspace structure or route segment. It is the highest altitude on a Federal airway, jet route, area navigation low or high route, or other direct route for which an MEA is designated in Part 95 at which adequate reception of navigation aid signals is assured.

MAYDAY—The international radiotelephony distress signal. When repeated three times, it indicates imminent and grave danger and that immediate assistance is requested.
(See PAN-PAN). (Refer to AIM).

MCA—(See MINIMUM CROSSING ALTITUDE).

MDA—(See MINIMUM DESCENT ALTITUDE).

MEA—(See MINIMUM EN ROUTE IFR ALTITUDE).

METEOROLOGICAL IMPACT STATEMENT—An unscheduled planning forecast describing conditions expected to begin within 4 to 12 hours which may impact the flow of air traffic in a specific center's (ARTCC) area.

METER FIX TIME/SLOT TIME—A calculated time to depart the meter fix in order to cross the vertex at the ACLT. This time reflects descent speed adjustment and any applicable time that must be absorbed prior to crossing the meter fix.

METER LIST DISPLAY INTERVAL—A dynamic parameter which controls the number of minutes prior to the flight plan calculated time of arrival at the meter fix for each aircraft, at which time the TCLT is frozen and becomes an ACLT; i.e., the VTA is updated and consequently the TCLT modified as appropriate until frozen at which time updating is suspended and an ACLT is assigned. When frozen, the flight entry is inserted into the arrival sector's meter list for display on the sector PVD. MLDI is used if filed true airspeed is less than or equal to freeze speed parameters (FSPD).

METERING—A method of time-regulating arrival traffic flow into a terminal area so as not to exceed a predetermined terminal acceptance rate.

METERING AIRPORTS—Airports adapted for metering and for which optimum flight paths are defined. A maximum of 15 airports may be adapted.

METERING FIX—A fix along an established route from over which aircraft will be metered prior to entering terminal airspace. Normally, this fix should be established at a distance from the airport which will facilitate a profile descent 10,000 feet above airport elevation [AAE] or above.

METERING POSITION(S)—Adapted PVD's and associated "D" positions eligible for display of a metering position list. A maximum of four PVD's may be adapted.

METERING POSITION LIST—An ordered list of data on arrivals for a selected metering airport displayed on a metering position PVD.

MFT—(See METER FIX TIME/SLOT TIME).

MHA—(See MINIMUM HOLDING ALTITUDE).

MIA—(See MINIMUM IFR ALTITUDES).

MICROBURST—A small downburst with outbursts of damaging winds extending 2.5 miles or less. In spite of its small horizontal scale, an intense microburst could induce wind speeds as high as 150 knots

(Refer to AIM).

MICROWAVE LANDING SYSTEM—A precision instrument approach system operating in the microwave spectrum which normally consists of the following components:

1. Azimuth Station.
2. Elevation Station.
3. Precision Distance Measuring Equipment.

(See MLS CATEGORIES).

MIDDLE COMPASS LOCATOR—(See COMPASS LOCATOR).

MIDDLE MARKER—A marker beacon that defines a point along the glideslope of an ILS normally located at or near the point of decision height (ILS Category I). It is keyed to transmit alternate dots and dashes, with the alternate dots and dashes keyed at the rate of 95 dot/dash combinations per minute on a 1300 Hz tone, which is received aurally and visually by compatible airborne equipment.

(See MARKER BEACON). (See INSTRUMENT LANDING SYSTEM). (Refer to AIM).

MID RVR—(See VISIBILITY).

MILES-IN-TRAIL—A specified distance between aircraft, normally, in the same stratum associated with the same destination or route of flight.

MILITARY AUTHORITY ASSUMES RESPONSIBILITY FOR SEPARATION OF AIRCRAFT—A condition whereby the military services involved assume responsibility for separation between participating military aircraft in the ATC system. It is used only for required IFR operations which are specified in letters of agreement or other appropriate FAA or military documents.

MILITARY OPERATIONS AREA—(See SPECIAL USE AIRSPACE).

MILITARY TRAINING ROUTES—Airspace of defined vertical and lateral dimensions established for the conduct of military flight training at airspeeds in excess of 250 knots IAS.

(See IFR MILITARY TRAINING ROUTES). (See VFR MILITARY TRAINING ROUTES).

MINIMA—(See MINIMUMS).

MINIMUM CROSSING ALTITUDE—The lowest altitude at certain fixes at which an aircraft must cross when proceeding in the direction of a higher minimum en route IFR altitude (MEA).

(See MINIMUM EN ROUTE IFR ALTITUDE).

MINIMUM DESCENT ALTITUDE—The lowest altitude, expressed in feet above mean sea level, to which descent is authorized on final approach or during circle-to-land maneuvering in execution of a standard instrument approach procedure where no electronic glideslope is provided.

(See NONPRECISION APPROACH PROCEDURE).

MINIMUM EN ROUTE IFR ALTITUDE—The lowest published altitude between radio fixes which assures acceptable navigational signal coverage and meets obstacle clearance requirements between those fixes. The MEA prescribed for a Federal airway or segment thereof, area navigation low or high route, or other direct route applies to the entire width of the airway, segment, or route between the radio fixes defining the airway, segment, or route.

(Refer to Part 91). (Refer to Part 95). (Refer to AIM).

MINIMUM FUEL—Indicates that an aircraft's fuel supply has reached a state where, upon reaching the destination, it can accept little or no delay. This is not an emergency situation but merely indicates an emergency situation is possible should any undue delay occur.

(Refer to AIM).

MINIMUM HOLDING ALTITUDE—The lowest altitude prescribed for a holding pattern which assures navigational signal coverage, communications, and meets obstacle clearance requirements.

MINIMUM IFR ALTITUDES—Minimum altitudes for IFR operations as prescribed in Part 91. These altitudes are published on aeronautical charts and prescribed in Part 95 for airways and routes, and in Part 97 for standard instrument approach procedures. If no applicable minimum altitude is prescribed in FAR 95 or FAR 97, the following minimum IFR altitude applies:

1. In designated mountainous areas, 2,000 feet above the highest obstacle within a horizontal distance of 4 nautical miles from the course to be flown; or
2. Other than mountainous areas, 1,000 feet above the highest obstacle within a horizontal distance of 4 nautical miles from the course to be flown; or
3. As otherwise authorized by the Administrator or assigned by ATC.

(See MINIMUM EN ROUTE IFR ALTITUDE). (See MINIMUM OBSTRUCTION CLEARANCE ALTITUDE). (See MINIMUM CROSSING ALTITUDE). (See MINIMUM SAFE ALTITUDE). (See MINIMUM VECTORING ALTITUDE). (Refer to Part 91).

MINIMUM OBSTRUCTION CLEARANCE ALTITUDE—The lowest published altitude in effect between radio fixes on VOR airways, off-airway routes, or route segments which meets obstacle clearance requirements for the entire route segment and which assures acceptable navigational signal coverage only within 25 statute (22 nautical) miles of a VOR.

(Refer to Part 91). (Refer to Part 95).

MINIMUM NAVIGATION PERFORMANCE SPECIFICATION—A set of standards which require aircraft to have a minimum navigation performance capability in order to operate in MNPS designated airspace. In addition, aircraft must be certified by their State of Registry for MNPS operation.

MINIMUM NAVIGATION PERFORMANCE SPECIFICATIONS AIRSPACE—Designated airspace in which MNPS procedures are applied between MNPS certified and equipped aircraft. Under certain conditions, non-MNPS aircraft can operate in MNPSA. However, standard oceanic separation minima is provided between the non-MNPS aircraft and other traffic. Currently, the only designated MNPSA is described as follows:

1. Between FL 275 and FL 400;
2. Between latitudes 27– N and the North Pole;
3. In the east, the eastern boundaries of the CTA's Santa Maria Oceanic, Shanwick Oceanic, and Reykjavik;
4. In the west, the western boundaries of CTA's Reykjavik and Gander Oceanic and New York Oceanic excluding the area west of 60– W and south of 38– 30' N.

MINIMUM RECEPTION ALTITUDE—The lowest altitude at which an intersection can be determined.

(Refer to Part 95).

MINIMUM SAFE ALTITUDE—

1. The minimum altitude specified in Part 91 for various aircraft operations.
2. Altitudes depicted on approach charts which provide at least 1,000 feet of obstacle clearance for emergency use within a specified distance from the navigation facility upon which a procedure is predicated. These altitudes will be identified as Minimum Sector Altitudes or Emergency Safe Altitudes and are established as follows:

a. Minimum Sector Altitudes. Altitudes depicted on approach charts which provide at least 1,000 feet of obstacle clearance within a 25-mile radius of the navigation facility upon which the procedure is predicated. Sectors depicted on approach charts must be at least 90 degrees in scope. These altitudes are for emergency use only and do not necessarily assure acceptable navigational signal coverage.

(See ICAO term Minimum Sector Altitude).

b. Emergency Safe Altitudes. Altitudes depicted on approach charts which provide at least 1,000 feet of obstacle clearance in nonmountainous areas and 2,000 feet of obstacle clearance in designated mountainous areas within a 100-mile radius of the navigation facility upon which the procedure is predicated and normally used only in military procedures. These altitudes are identified on published procedures as "Emergency Safe Altitudes."

MINIMUM SAFE ALTITUDE WARNING—A function of the ARTS III computer that aids the controller by alerting him when a tracked Mode C—equipped aircraft is below or is predicted by the computer to go below a predetermined minimum safe altitude.

(Refer to AIM).

MINIMUM SECTOR ALTITUDE [ICAO]—The lowest altitude which may be used under emergency conditions which will provide a minimum clearance of 300 m (1,000 feet) above all obstacles located in an area contained within a sector of a circle of 46 km (25 NM) radius centered on a radio aid to navigation.

MINIMUMS—Weather condition requirements established for a particular operation or type of operation; e.g., IFR takeoff or landing, alternate airport for IFR flight plans, VFR flight, etc.

(See LANDING MINIMUMS). (See IFR TAKEOFF MINIMUMS AND DEPARTURE PROCEDURES). (See VFR CONDITIONS). (See IFR CONDITIONS). (Refer to Part 91). (Refer to AIM).

MINIMUM VECTORING ALTITUDE—The lowest MSL altitude at which an IFR aircraft will be vectored by a radar controller, except as otherwise authorized for radar approaches, departures, and missed approaches. The altitude meets IFR obstacle clearance criteria. It may be lower than the published MEA along an airway or J-route segment. It may be utilized for radar vectoring only upon the controller's determination that an adequate radar return is being received from the aircraft being controlled. Charts depicting minimum vectoring altitudes are normally available only to the controllers and not to pilots.

(Refer to AIM).

MINUTES-IN-TRAIL—A specified interval between aircraft expressed in time. This method would more likely be utilized regardless of altitude.

MIS—(See METEOROLOGICAL IMPACT STATEMENT).

MISSED APPROACH—

1. A maneuver conducted by a pilot when an instrument approach cannot be completed to a landing. The route of flight and altitude are shown on instrument approach procedure charts. A pilot executing a missed approach prior to the Missed Approach Point (MAP) must continue along the final approach to the MAP. The pilot may climb immediately to the altitude specified in the missed approach procedure.
2. A term used by the pilot to inform ATC that he is executing the missed approach.
3. At locations where ATC radar service is provided, the pilot should conform to radar vectors when provided by ATC in lieu of the published missed approach procedure.

(See Missed Approach Point) (Refer to AIM).

MISSED APPROACH POINT—A point prescribed in each instrument approach procedure at which a missed approach procedure shall be executed if the required visual reference does not exist.

(See MISSED APPROACH). (See SEGMENTS OF AN INSTRUMENT APPROACH PROCEDURE).

MISSED APPROACH PROCEDURE [ICAO]—The procedure to be followed if the approach cannot be continued.

MISSED APPROACH SEGMENT—(See SEGMENTS OF AN INSTRUMENT APPROACH PROCEDURE).

MLDI—(See METER LIST DISPLAY INTERVAL).

MLS—(See MICROWAVE LANDING SYSTEM).

MLS CATEGORIES—

1. MLS Category I. An MLS approach procedure which provides for an approach to a height above touchdown of not less than 200 feet and a runway visual range of not less than 1,800 feet.
2. MLS Category II. Undefined until data gathering/analysis completion.
3. MLS Category III. Undefined until data gathering/analysis completion.

MM—(See MIDDLE MARKER).

MNPS—(See MINIMUM PERFORMANCE SPECIFICATION).

MNPSA—(See MINIMUM PERFORMANCE SPECIFICATIONS AIRSPACE).

MOA—(See MILITARY OPERATIONS AREA).

MOCA—(See MINIMUM OBSTRUCTION CLEARANCE ALTITUDE).

MODE—The letter or number assigned to a specific pulse spacing of radio signals transmitted or received by ground interrogator or airborne transponder components of the Air Traffic Control Radar Beacon System (ATCRBS). Mode A (military Mode 3) and Mode C (altitude reporting) are used in air traffic control.

(See TRANSPONDER). (See INTERROGATOR). (See RADAR). (Refer to AIM). (See ICAO term MODE).

MODE (SSR MODE) [ICAO]—The letter or number assigned to a specific pulse spacing of the interrogation signals transmitted by an interrogator. There are 4 modes, A, B, C and D specified in Annex 10, corresponding to four different interrogation pulse spacings.

MODE C INTRUDER ALERT—A function of certain air traffic control automated systems designed to alert radar controllers to existing or pending situations between a tracked target (known IFR or VFR aircraft) and an untracked target (unknown IFR or VFR aircraft) that requires immediate attention/action.

(See CONFLICT ALERT).

MONITOR—(When used with communication transfer) listen on a specific frequency and stand by for instructions. Under normal circumstances do not establish communications.

MOVEMENT AREA—The runways, taxiways, and other areas of an airport/heliport which are utilized for taxiing/hover taxiing, air taxiing, takeoff, and landing of aircraft, exclusive of loading ramps and parking areas. At those airports/heliports with a tower, specific approval for entry onto the movement area must be obtained from ATC.

(See ICAO term MOVEMENT AREA).

MOVEMENT AREA [ICAO]—That part of an aerodrome to be used for the takeoff, landing and taxiing of aircraft, consisting of the manoeuvring area and the apron(s).

MOVING TARGET INDICATOR—An electronic device which will permit radar scope presentation only from targets which are in motion. A partial remedy for ground clutter.

MRA—(See MINIMUM RECEPTION ALTITUDE).

MSA—(See MINIMUM SAFE ALTITUDE).

MSAW—(See MINIMUM SAFE ALTITUDE WARNING).

MTI—(See MOVING TARGET INDICATOR).

MTR—(See MILITARY TRAINING ROUTES).

MULTICOM—A mobile service not open to public correspondence used to provide communications essential to conduct the activities being performed by or directed from private aircraft.

MULTIPLE RUNWAYS—The utilization of a dedicated arrival runway(s) for departures and a dedicated departure runway(s) for arrivals when feasible to reduce delays and enhance capacity.

MVA—(See MINIMUM VECTORING ALTITUDE).-

N

NAS—(See NATIONAL AIRSPACE SYSTEM).

NAS STAGE A—The en route ATC system's radar, computers and computer programs, controller plan view displays (PVDs/Radar Scopes), input/output devices, and the related communications equipment which are integrated to form the heart of the automated IFR air traffic control system. This equipment performs Flight Data Processing (FDP) and Radar Data Processing (RDP). It interfaces with automated terminal systems and is used in the control of en route IFR aircraft.

(Refer to AIM).

NATIONAL AIRSPACE SYSTEM—The common network of U.S. airspace; air navigation facilities, equipment and services, airports or landing areas; aeronautical charts, information and services; rules, regulations and procedures, technical information, and manpower and material. Included are system components shared jointly with the military.

NATIONAL BEACON CODE ALLOCATION PLAN AIRSPACE—Airspace over United States territory located within the North American continent between Canada and Mexico, including adjacent territorial waters outward to about boundaries of oceanic control areas (CTA)/Flight Information Regions (FIR).

(See FLIGHT INFORMATION REGION).

NATIONAL FLIGHT DATA CENTER—A facility in Washington D.C., established by FAA to operate a central aeronautical information service for the collection, validation, and dissemination of aeronautical data in support of the activities of government, industry, and the aviation community. The information is published in the National Flight Data Digest.

(See NATIONAL FLIGHT DATA DIGEST).

NATIONAL FLIGHT DATA DIGEST—A daily (except weekends and Federal holidays) publication of flight information appropriate to aeronautical charts, aeronautical publications, Notices to Airmen, or other media serving the purpose of providing operational flight data essential to safe and efficient aircraft operations.

NATIONAL SEARCH AND RESCUE PLAN—An interagency agreement which provides for the effective utilization of all available facilities in all types of search and rescue missions.

NAVAID—(See NAVIGATIONAL AID).

NAVAID CLASSES—VOR, VORTAC, and TACAN aids are classed according to their operational use. The three classes of NAVAID's are:

T-Terminal.

L-Low altitude.

H-High altitude.

The normal service range for T, L, and H class aids is found in the AIM. Certain operational requirements make it necessary to use some of these aids at greater service ranges than specified. Extended range is made possible through flight inspection determinations. Some aids also have lesser service range due to location, terrain, frequency protection, etc. Restrictions to service range are listed in Airport / Facility Directory.

NAVIGABLE AIRSPACE—Airspace at and above the minimum flight altitudes prescribed in the FAR's including airspace needed for safe takeoff and landing.

(Refer to Part 91).

NAVIGATIONAL AID—Any visual or electronic device airborne or on the surface which provides point-to-point guidance information or position data to aircraft in flight.

(See AIR NAVIGATION FACILITY).

NBCAP AIRSPACE—(See NATIONAL BEACON CODE ALLOCATION PLAN AIRSPACE).

NDB—(See NONDIRECTIONAL BEACON).

NEGATIVE—"No," or "permission not granted," or "that is not correct."

NEGATIVE CONTACT—Used by pilots to inform ATC that:

1. Previously issued traffic is not in sight. It may be followed by the pilot's request for the controller to provide assistance in avoiding the traffic.
2. They were unable to contact ATC on a particular frequency.

NFDC—(See NATIONAL FLIGHT DATA CENTER).

NFDD—(See NATIONAL FLIGHT DATA DIGEST).

NIGHT—The time between the end of evening civil twilight and the beginning of morning civil twilight, as published in the American Air Almanac, converted to local time.
(See ICAO term NIGHT).

NIGHT [ICAO]—The hours between the end of evening civil twilight and the beginning of morning civil twilight or such other period between sunset and sunrise as may be specified by the appropriate authority.

Note. Civil twilight ends in the evening when the centre of the sun's disk is 6 degrees below the horizon and begins in the morning when the centre of the sun's disk is 6 degrees below the horizon.

NO GYRO APPROACH—A radar approach/vector provided in case of a malfunctioning gyro-compass or directional gyro. Instead of providing the pilot with headings to be flown, the controller observes the radar track and issues control instructions "turn right/left" or "stop turn" as appropriate.
(Refer to AIM).

NO GYRO VECTOR—(See NO GYRO APPROACH).

NONAPPROACH CONTROL TOWER—Authorizes aircraft to land or takeoff at the airport controlled by the tower or to transit the airport traffic area. The primary function of a nonapproach control tower is the sequencing of aircraft in the traffic pattern and on the landing area. Nonapproach control towers also separate aircraft operating under instrument flight rules clearances from approach controls and centers. They provide ground control services to aircraft, vehicles, personnel, and equipment on the airport movement area.

NONCOMMON ROUTE/PORTION—That segment of a North American Route between the inland navigation facility and a designated North American terminal.

NONCOMPOSITE SEPARATION—Separation in accordance with minima other than the composite separation minimum specified for the area concerned.

NONDIRECTIONAL BEACON—An L/MF or UHF radio beacon transmitting nondirectional signals whereby the pilot of an aircraft equipped with direction finding equipment can determine his bearing to or from the radio beacon and "home" on or track to or from the station. When the radio beacon is installed in conjunction with the Instrument Landing System marker, it is normally called a Compass Locator.
(See COMPASS LOCATOR). (See AUTOMATIC DIRECTION FINDER).

NONMOVEMENT AREAS—Taxiways and apron (ramp) areas not under the control of air traffic.

NONPRECISION APPROACH—(See NONPRECISION APPROACH PROCEDURE).

NONPRECISION APPROACH PROCEDURE—A standard instrument approach procedure in which no electronic glideslope is provided; e.g., VOR, TACAN, NDB, LOC, ASR, LDA, or SDF approaches.

NONRADAR—Precedes other terms and generally means without the use of radar, such as:

1. Nonradar Approach. Used to describe instrument approaches for which course guidance on final approach is not provided by ground-based precision or surveillance radar. Radar vectors to the final approach course may or may not be provided by ATC. Examples of nonradar approaches are VOR, NDB, TACAN, and ILS/MLS approaches.

(See FINAL APPROACH-IFR). (See FINAL APPROACH COURSE). (See RADAR APPROACH). (See INSTRUMENT APPROACH PROCEDURE).

2. Nonradar Approach Control. An ATC facility providing approach control service without the use of radar.
(See APPROACH CONTROL FACILITY). (See APPROACH CONTROL SERVICE).

3. Nonradar Arrival. An aircraft arriving at an airport without radar service or at an airport served by a radar facility and radar contact has not been established or has been terminated due to a lack of radar service to the airport. *(See RADAR ARRIVAL). (See RADAR SERVICE).*

4. Nonradar Route. A flight path or route over which the pilot is performing his own navigation. The pilot may be receiving radar separation, radar monitoring, or other ATC services while on a nonradar route.

(See RADAR ROUTE).

5. Nonradar Separation. The spacing of aircraft in accordance with established minima without the use of radar; e.g., vertical, lateral, or longitudinal separation.

(See RADAR SEPARATION). (See ICAO term NONRADAR SEPARATION).

NONRADAR SEPARATION [ICAO]—The separation used when aircraft position information is derived from sources other than radar.

NOPAC—(See NORTH PACIFIC).

NORDO—(See LOST COMMUNICATIONS).

NORTH AMERICAN ROUTE—A numerically coded route preplanned over existing airway and route systems to and from specific coastal fixes serving the North Atlantic. North American Routes consist of the following:

1. Common Route/Portion. That segment of a North American Route between the inland navigation facility and the coastal fix.
2. NonCommon Route/Portion. That segment of a North American Route between the inland navigation facility and a designated North American terminal.
3. Inland Navigation Facility. A navigation aid on a North American Route at which the common route and/or the noncommon route begins or ends.
4. Coastal Fix. A navigation aid or intersection where an aircraft transitions between the domestic route structure and the oceanic route structure.

NORTH MARK—A beacon data block sent by the host computer to be displayed by the ARTS on a 360 degree bearing at a locally selected radar azimuth and distance. The North Mark is used to ensure correct range/azimuth orientation during periods of CENRAP.

NORTH PACIFIC—An organized route system between the Alaskan west coast and Japan.

NOTAM—(See NOTICE TO AIRMEN).

NOTICE TO AIRMEN—A notice containing information (not known sufficiently in advance to publicize by other means) concerning the establishment, condition, or change in any component (facility, service, or procedure of, or hazard in the National Airspace System) the timely knowledge of which is essential to personnel concerned with flight operations.

1. NOTAM(D). A NOTAM given (in addition to local dissemination) distant dissemination beyond the area of responsibility of the Flight Service Station. These NOTAM's will be stored and available until canceled.
2. NOTAM(L). A NOTAM given local dissemination by voice and other means, such as telautograph and telephone, to satisfy local user requirements.
3. FDC NOTAM. A NOTAM regulatory in nature, transmitted by USNOF and given system wide dissemination.

(See ICAO term NOTAM).

NOTAM [ICAO]—A notice containing information concerning the establishment, condition or change in any aeronautical facility, service, procedure or hazard, the timely knowledge of which is essential to personnel concerned with flight operations.

Class I Distribution—Distribution by means of telecommunication.

Class II Distribution—Distribution by means other than telecommunications.

NOTICES TO AIRMEN PUBLICATION—A publication issued every 14 days, designed primarily for the pilot, which contains current NOTAM information considered essential to the safety of flight as well as supplemental data to other aeronautical publications. The contraction NTAP is used in NOTAM text.

(See NOTICE TO AIRMEN).

NTAP—(See NOTICES TO AIRMEN PUBLICATION).

NUMEROUS TARGETS VICINITY (LOCATION)—A traffic advisory issued by ATC to advise pilots that targets on the radar scope are too numerous to issue individually.

(See TRAFFIC ADVISORIES).-

O

OALT—(See OPERATIONAL ACCEPTABLE LEVEL OF TRAFFIC).

OBSTACLE—An existing object, object of natural growth, or terrain at a fixed geographical location or which may be expected at a fixed location within a prescribed area with reference to which vertical clearance is or must be provided during flight operation.

STRUCTION—Any object/obstacle exceeding the obstruction standards specified by Part 77, Subpart C.

OBSTRUCTION LIGHT—A light or one of a group of lights, usually red or white, frequently mounted on a surface structure or natural terrain to warn pilots of the presence of an obstruction.

OCEANIC AIRSPACE—Airspace over the oceans of the world, considered international airspace, where oceanic separation and procedures per the International Civil Aviation Organization are applied. Responsibility for the provisions of air traffic control service in this airspace is delegated to various countries, based generally upon geographic proximity and the availability of the required resources.

OCEANIC DISPLAY AND PLANNING SYSTEM—An automated digital display system which provides flight data processing, conflict probe, and situation display for oceanic air traffic control.

OCEANIC NAVIGATIONAL ERROR REPORT—A report filed when an aircraft exiting oceanic airspace has been observed by radar to be off course. ONER reporting parameters and procedures are contained in Order 7110.82, Monitoring of Navigational Performance In Oceanic Areas.

OCEANIC PUBLISHED ROUTE—A route established in international airspace and charted or described in flight information publications, such as Route Charts, DOD Enroute Charts, Chart Supplements, NOTAM's, and Track Messages.

OCEANIC TRANSITION ROUTE—An ATS route established for the purpose of transitioning aircraft to/from an organized track system.

ODAPS—(See OCEANIC DISPLAY AND PLANNING SYSTEM).

OFF COURSE—A term used to describe a situation where an aircraft has reported a position fix or is observed on radar at a point not on the ATC-approved route of flight.

OFFSHORE CONTROL AREA—That portion of airspace between the U.S 12–mile limit and the oceanic CTA/FIR boundary within which air traffic control is exercised. These areas are established to permit the application of domestic procedures in the provision of air traffic control services. Offshore control area is generally synonymous with Federal Aviation Regulations, Part 71, Subpart E, "Control Areas and Control Area Extensions."

OFF-ROUTE VECTOR—A vector by ATC which takes an aircraft off a previously assigned route. Altitudes assigned by ATC during such vectors provide required obstacle clearance.

OFFSET PARALLEL RUNWAYS—Staggered runways having centerlines which are parallel.

OFT—(See OUTER FIX TIME).

OM—(See OUTER MARKER).

OMEGA—An RNAV system designed for long-range navigation based upon ground-based electronic navigational aid signals.

ONER—(See OCEANIC NAVIGATIONAL ERROR REPORT).

OPERATIONAL—(See DUE REGARD).

ON COURSE—

1. Used to indicate that an aircraft is established on the route centerline.
2. Used by ATC to advise a pilot making a radar approach that his aircraft is lined up on the final approach course.

(See ON-COURSE INDICATION).

ON-COURSE INDICATION—An indication on an instrument, which provides the pilot a visual means of determining that the aircraft is located on the centerline of a given navigational track, or an indication on a radar scope that an aircraft is on a given track.

OPERATIONAL ACCEPTABLE LEVEL OF TRAFFIC—An air traffic activity level associated with the designed capacity for a sector or airport. The OALT considers dynamic changes in staffing, personnel experience levels, equipment outages, operational configurations, weather, traffic complexity, aircraft performance mixtures, transitioning flights, adjacent airspace, handoff/point-out responsibilities, and other factors that may affect an air traffic operational position or system element. The OALT is normally considered to be the total number of aircraft that any air traffic functional position can accommodate for a defined period of time under a given set of circumstances.

OPPOSITE DIRECTION AIRCRAFT—Aircraft are operating in opposite directions when:

1. They are following the same track in reciprocal directions; or
2. Their tracks are parallel and the aircraft are flying in reciprocal directions; or
3. Their tracks intersect at an angle of more than 135°.

OPTION APPROACH—An approach requested and conducted by a pilot which will result in either a touch-and-go, missed approach, low approach, stop-and-go, or full stop landing.

(See CLEARED FOR THE OPTION). (Refer to AIM).

ORGANIZED TRACK SYSTEM—A movable system of oceanic tracks that traverses the North Atlantic between Europe and North America the physical position of which is determined twice daily taking the best advantage of the winds aloft.

ORGANIZED TRACK SYSTEM—A series of ATS routes which are fixed and charted; i.e., CEP, NOPAC, or flexible and described by NOTAM; i.e., NAT TRACK MESSAGE.

OTR—(See OCEANIC TRANSITION ROUTE).

OTS—(See ORGANIZED TRACK SYSTEM).

OUT—The conversation is ended and no response is expected.

OUTER AREA (associated with ARSA)—Nonregulatory airspace surrounding designated ARSA airports wherein ATC provides radar vectoring and sequencing on a full-time basis for all IFR and participating VFR aircraft. The service provided in the outer area is called ARSA service which includes: IFR/IFR-standard IFR separation; IFR/VFR-traffic advisories and conflict resolution; and VFR/VFR-traffic advisories and, as appropriate, safety alerts. The normal radius will be 20 nautical miles with some variations based on site-specific requirements. The outer area extends outward from the primary ARSA airport and extends from the lower limits of radar/radio coverage up to the ceiling of the approach control's delegated airspace excluding the ARSA and other airspace as appropriate.

(See CONTROLLED AIRSPACE). (See CONFLICT RESOLUTION).

OUTER COMPASS LOCATOR—(See COMPASS LOCATOR).

OUTER FIX—A general term used within ATC to describe fixes in the terminal area, other than the final approach fix. Aircraft are normally cleared to these fixes by an Air Route Traffic Control Center or an Approach Control Facility. Aircraft are normally cleared from these fixes to the final approach fix or final approach course.

OUTER FIX—An adapted fix along the converted route of flight, prior to the meter fix, for which crossing times are calculated and displayed in the metering position list.

OUTER FIX TIME—A calculated time to depart the outer fix in order to cross the vertex at the ACLT. The time reflects descent speed adjustments and any applicable delay time that must be absorbed prior to crossing the meter fix.

OUTER MARKER—A marker beacon at or near the glideslope intercept altitude of an ILS approach. It is keyed to transmit two dashes per second on a 400 Hz tone, which is received aurally and visually by compatible airborne equipment. The OM is normally located four to seven miles from the runway threshold on the extended centerline of the runway.

(See MARKER BEACON). (See INSTRUMENT LANDING SYSTEM). (Refer to AIM).

OVER—My transmission is ended; I expect a response.

OVERHEAD APPROACH—A series of predetermined maneuvers prescribed for VFR arrival of military aircraft (often in formation) for entry into the VFR traffic pattern and to proceed to a landing. The pattern usually specifies the following:

1. The radio contact required of the pilot.
2. The speed to be maintained.
3. An initial approach 3 to 5 miles in length.
4. An elliptical pattern consisting of two 180 degree turns.
5. A break point at which the first 180 degree turn is started.
6. The direction of turns.
7. Altitude (at least 500 feet above the conventional pattern).
8. A "Roll-out" on final approach not less than 1/4 mile from the landing threshold and not less than 300 feet above the ground.

OVERLYING CENTER—The ARTCC facility that is responsible for arrival/departure operations at a specific terminal.

OVERRUN—An area beyond the takeoff runway no less wide than the runway and centered upon the extended centerline of the runway, able to support the airplane during an aborted takeoff, without causing structural damage to the airplane, and designated by the airport authorities for use in decelerating the airplane during an aborted takeoff. (STOPWAY term used for civilian airports).-

P

P TIME—(See PROPOSED DEPARTURE TIME).

PAN-PAN—The international radio-telephony urgency signal. When repeated three times, indicates uncertainty or alert followed by the nature of the urgency.
(See MAYDAY). (Refer to AIM).

PAR—(See PRECISION APPROACH RADAR).

PAR [ICAO]—(See ICAO Term PRECISION APPROACH RADAR).

PARALLEL ILS APPROACHES—Approaches to parallel runways by IFR aircraft which, when established inbound toward the airport on the adjacent final approach courses, are radar-separated by at least 2 miles.
(See FINAL APPROACH COURSE). (See SIMULTANEOUS ILS APPROACHES).

PARALLEL MLS APPROACHES—(See PARALLEL ILS APPROACHES).

PARALLEL OFFSET ROUTE—A parallel track to the left or right of the designated or established airway/ route. Normally associated with Area Navigation (RNAV) operations.
(See AREA NAVIGATION).

PARALLEL RUNWAYS—Two or more runways at the same airport whose centerlines are parallel. In addition to runway number, parallel runways are designated as L (left) and R (right) or, if three parallel runways exist, L (left), C (center), and R (right).

PATWAS—(See PILOTS AUTOMATIC TELEPHONE WEATHER ANSWERING SERVICE).

PBCT—(See PROPOSED BOUNDARY CROSSING TIME).

PCA—(See POSITIVE CONTROL AREA).

PERMANENT ECHO—Radar signals reflected from fixed objects on the earth's surface; e.g., buildings, towers, terrain. Permanent echoes are distinguished from "ground clutter" by being definable locations rather than large areas. Under certain conditions they may be used to check radar alignment.

PHOTO RECONNAISSANCE—Military activity that requires locating individual photo targets and navigating to the targets at a preplanned angle and altitude. The activity normally requires a lateral route width of 16 NM and altitude range of 1,500 feet to 10,000 feet AGL.

PIDP—(See PROGRAMMABLE INDICATOR DATA PROCESSOR).

PILOT BRIEFING—A service provided by the FSS to assist pilots in flight planning. Briefing items may include weather information, NOTAMS, military activities, flow control information, and other items as requested.
(Refer to AIM).

PILOT IN COMMAND—The pilot responsible for the operation and safety of an aircraft during flight time.
(Refer to Part 91).

PILOTS AUTOMATIC TELEPHONE WEATHER ANSWERING SERVICE—A continuous telephone recording containing current and forecast weather information for pilots.
(See FLIGHT SERVICE STATION). (Refer to AIM).

PILOT'S DISCRETION—When used in conjunction with altitude assignments, means that ATC has offered the pilot the option of starting climb or descent whenever he wishes and conducting the climb or descent at any rate he wishes. He may temporarily level off at any intermediate altitude. However, once he has vacated an altitude, he may not return to that altitude.

PILOT WEATHER REPORT—A report of meteorological phenomena encountered by aircraft in flight.
(Refer to AIM).

PIREP—(See PILOT WEATHER REPORT).

POINT OUT—(See RADAR POINT OUT).

POLAR TRACK STRUCTURE—A system of organized routes between Iceland and Alaska which overlie Canadian MNPS Airspace.

POSITION REPORT—A report over a known location as transmitted by an aircraft to ATC.
(Refer to AIM).

POSITION SYMBOL—A computer-generated indication shown on a radar display to indicate the mode of tracking.

POSITIVE CONTROL—The separation of all air traffic within designated airspace by air traffic control.
(See POSITIVE CONTROL AREA).

POSITIVE CONTROL AREA—(See CONTROLLED AIRSPACE).

PRACTICE INSTRUMENT APPROACH—An instrument approach procedure conducted by a VFR or an IFR aircraft for the purpose of pilot training or proficiency demonstrations.

PREARRANGED COORDINATION—A standardized procedure which permits an air traffic controller to enter the airspace assigned to another air traffic controller without verbal coordination. The procedures are defined in a facility directive which ensures standard separation between aircraft.

PRECIPITATION—Any or all forms of water particles (rain, sleet, hail, or snow) that fall from the atmosphere and reach the surface.

PRECISION APPROACH—(See PRECISION APPROACH PROCEDURE).

PRECISION APPROACH PROCEDURE—A standard instrument approach procedure in which an electronic glideslope/glidepath is provided; e.g., ILS/MLS and PAR.
(See INSTRUMENT LANDING SYSTEM). (See MICROWAVE LANDING SYSTEM). (See PRECISION APPROACH RADAR).

PRECISION APPROACH RADAR—Radar equipment in some ATC facilities operated by the FAA and/or the military services at joint-use civil/military locations and separate military installations to detect and display azimuth, elevation, and range of aircraft on the final approach course to a runway. This equipment may be used to monitor certain nonradar approaches, but is primarily used to conduct a precision instrument approach (PAR) wherein the controller issues guidance instructions to the pilot based on the aircraft's position in relation to the final approach course (azimuth), the glidepath (elevation), and the distance (range) from the touchdown point on the runway as displayed on the radar scope.
(See GLIDEPATH). (See PAR). (Refer to AIM).—The abbreviation "PAR" is also used to denote preferential arrival routes in ARTCC computers.
(See PREFERENTIAL ROUTES). (See ICAO term PRECISION APPROACH RADAR).

PRECISION APPROACH RADAR [ICAO]—Primary radar equipment used to determine the position of an aircraft during final approach, in terms of lateral and vertical deviations relative to a nominal approach path, and in range relative to touchdown.

Note: Precision approach radars are designed to enable pilots of aircraft to be given guidance by radio communication during the final stages of the approach to land.

PRECISION RUNWAY MONITOR—Provides air traffic controllers with high precision secondary surveillance data for aircraft on final approach to closely spaced parallel runways. High resolution color monitoring displays (FMA) are required to present surveillance track data to controllers along with detailed maps depicting approaches and no transgression zone.

PREFERENTIAL ROUTES—Preferential routes (PDR's, PAR's, and PDAR's) are adapted in ARTCC computers to accomplish inter/intrafacility controller coordination and to assure that flight data is posted at the proper control positions. Locations having a need for these specific inbound and outbound routes normally publish such routes in local facility bulletins, and their use by pilots minimizes flight plan route amendments. When the workload or traffic situation permits, controllers normally provide radar vectors or assign requested routes to minimize circuitous routing. Preferential routes are usually confined to one ARTCC's area and are referred to by the following names or acronyms:

1. Preferential Departure Route (PDR). A specific departure route from an airport or terminal area to an en route point where there is no further need for flow control. It may be included in a Standard Instrument Departure (SID) or a Preferred IFR Route.
2. Preferential Arrival Route (PAR). A specific arrival route from an appropriate en route point to an airport or terminal area. It may be included in a Standard Terminal Arrival (STAR) or a Preferred IFR Route.

The abbreviation "PAR" is used primarily within the ARTCC and should not be confused with the abbreviation for Precision Approach Radar.

3. Preferential Departure and Arrival Route (PDAR). A route between two terminals which are within or immediately adjacent to one ARTCC's area. PDAR's are not synonymous with Preferred IFR Routes but may be listed as such as they do accomplish essentially the same purpose.

(See PREFERRED IFR ROUTES). (See NAS STAGE A).

PREFERRED IFR ROUTES—Routes established between busier airports to increase system efficiency and capacity. They normally extend through one or more ARTCC areas and are designed to achieve balanced traffic flows among high density terminals. IFR clearances are issued on the basis of these routes except when severe weather avoidance procedures or other factors dictate otherwise. Preferred IFR Routes are listed in the Airport/ Facility Directory. If a flight is planned to or from an area having such routes but the departure or arrival point is not listed in the Airport/Facility Directory, pilots may use that part of a Preferred IFR Route which is appropriate for the departure or arrival point that is listed. Preferred IFR Routes are correlated with SID's and STAR's and may be defined by airways, jet routes, direct routes between NAVAID's, Waypoints, NAVAID radials/DME, or any combinations thereof.

(See STANDARD INSTRUMENT DEPARTURE). (See STANDARD TERMINAL ARRIVAL). (See PREFERENTIAL ROUTES). (See CENTER'S AREA). (Refer to AIRPORT/FACILITY DIRECTORY). (Refer to NOTICES TO AIRMEN PUBLICATION).

PRE-FLIGHT PILOT BRIEFING—(See PILOT BRIEFING).

PREVAILING VISIBILITY—(See VISIBILITY).

PRM—(See PRECISION RUNWAY MONITOR).

PROCEDURE TURN—The maneuver prescribed when it is necessary to reverse direction to establish an aircraft on the intermediate approach segment or final approach course. The outbound course, direction of turn, distance within which the turn must be completed, and minimum altitude are specified in the procedure. However, unless otherwise restricted, the point at which the turn may be commenced and the type and rate of turn are left to the discretion of the pilot.

(See ICAO term PROCEDURE TURN).

PROCEDURE TURN [ICAO]—A manoeuvre in which a turn is made away from a designated track followed by a turn in the opposite direction to permit the aircraft to intercept and proceed along the reciprocal of the designated track.

Note 1: Procedure turns are designated "left" or "right" according to the direction of the initial turn.

Note 2: Procedure turns may be designated as being made either in level flight or while descending, according to the circumstances of each individual approach procedure.

PROCEDURE TURN INBOUND—That point of a procedure turn maneuver where course reversal has been completed and an aircraft is established inbound on the intermediate approach segment or final approach course. A report of "procedure turn inbound" is normally used by ATC as a position report for separation purposes.

(See FINAL APPROACH COURSE). (See PROCEDURE TURN). (See SEGMENTS OF AN INSTRUMENT APPROACH PROCEDURE).

PROFILE DESCENT—An uninterrupted descent (except where level flight is required for speed adjustment; e.g., 250 knots at 10,000 feet MSL) from cruising altitude/level to interception of a glideslope or to a minimum altitude specified for the initial or intermediate approach segment of a nonprecision instrument approach. The profile descent normally terminates at the approach gate or where the glideslope or other appropriate minimum altitude is intercepted.

PROGRAMMABLE INDICATOR DATA PROCESSOR—The PIDP is a modification to the AN/TPX–42 interrogator system currently installed in fixed RAPCON's. The PIDP detects, tracks, and predicts secondary radar aircraft targets. These are displayed by means of computer-generated symbols and alphanumeric characters depicting flight identification, aircraft altitude, ground speed, and flight plan data. Although primary radar targets are not tracked, they are displayed coincident with the secondary radar targets as well as with the other symbols and alphanumerics. The system has the capability of interfacing with ARTCC's.

PROGRESS REPORT—(See POSITION REPORT).

PROGRESSIVE TAXI—Precise taxi instructions given to a pilot unfamiliar with the airport or issued in stages as the aircraft proceeds along the taxi route.

PROHIBITED AREA—(See SPECIAL USE AIRSPACE).
(See ICAO term PROHIBITED AREA).

PROHIBITED AREA [ICAO]—An airspace of defined dimensions, above the land areas or territorial waters of a State, within which the flight of aircraft is prohibited.

PROPOSED BOUNDARY CROSSING TIME—Each center has a PBCT parameter for each internal airport. Proposed internal flight plans are transmitted to the adjacent center if the flight time along the proposed route from the departure airport to the center boundary is less than or equal to the value of PBCT or if airport adaptation specifies transmission regardless of PBCT.

PROPOSED DEPARTURE TIME—The time a scheduled flight will depart the gate (scheduled operators) or the actual runway off time for nonscheduled operators. For EDCT purposes, the ATCSCC adjusts the "P" time for scheduled operators to reflect the runway off times.

PROTECTED AIRSPACE—The airspace on either side of an oceanic route/track that is equal to one-half the lateral separation minimum except where reduction of protected airspace has been authorized.

PT—(See PROCEDURE TURN).

PTS—(See POLAR TRACK STRUCTURE).

PUBLISHED ROUTE—A route for which an IFR altitude has been established and published; e.g., Federal Airways, Jet Routes, Area Navigation Routes, Specified Direct Routes.-

Q

QUEUING—(See STAGING/QUEUING).

LOW—(See QUOTA FLOW CONTROL).

QNE—The barometric pressure used for the standard altimeter setting (29.92 inches Hg.).

QNH—The barometric pressure as reported by a particular station.

QUADRANT—A quarter part of a circle, centered on a NAVAID, oriented clockwise from magnetic north as follows: NE quadrant 000–089, SE quadrant 090–179, SW quadrant 180–269, NW quadrant 270–359.

QUICK LOOK—A feature of NAS Stage A and ARTS which provides the controller the capability to display full data blocks of tracked aircraft from other control positions.

QUOTA FLOW CONTROL—A flow control procedure by which the Central Flow Control Function (CFCF) restricts traffic to the ARTC Center area having an impacted airport, thereby avoiding sector/area saturation.
(See AIR TRAFFIC CONTROL SYSTEM COMMAND CENTER). (Refer to AIRPORT/FACILITY DIRECTORY).-

R

RADAR—A device which, by measuring the time interval between transmission and reception of radio pulses and correlating the angular orientation of the radiated antenna beam or beams in azimuth and/or elevation, provides information on range, azimuth, and/or elevation of objects in the path of the transmitted pulses.

1. Primary Radar. A radar system in which a minute portion of a radio pulse transmitted from a site is reflected by an object and then received back at that site for processing and display at an air traffic control facility.

2. Secondary Radar/Radar Beacon (ATCRBS). A radar system in which the object to be detected is fitted with cooperative equipment in the form of a radio receiver/transmitter (transponder). Radar pulses transmitted from the searching transmitter/receiver (interrogator) site are received in the cooperative equipment and used to trigger a distinctive transmission from the transponder. This reply transmission, rather than a reflected signal, is then received back at the transmitter/receiver site for processing and display at an air traffic control facility.

(See TRANSPONDER). (See INTERROGATOR). (Refer to AIM). (See ICAO term RADAR). (See ICAO term PRIMARY RADAR). (See ICAO term SECONDARY RADAR).

RADAR [ICAO]—A radio detection device which provides information on range, azimuth and/or elevation of objects.

Primary Radar.—Radar system which uses reflected radio signals.

Secondary Radar.—Radar system wherein a radio signal transmitted from a radar station initiates the transmission of a radio signal from another station.

RADAR ADVISORY—The provision of advice and information based on radar observations.

(See ADVISORY SERVICE).

RADAR ALTIMETER—(See RADIO ALTIMETER).

RADAR APPROACH—An instrument approach procedure which utilizes Precision Approach Radar (PAR) or Airport Surveillance Radar (ASR).

(See SURVEILLANCE APPROACH). (See AIRPORT SURVEILLANCE RADAR). (See PRECISION APPROACH RADAR). (See INSTRUMENT APPROACH PROCEDURE). (Refer to AIM). (See ICAO term RADAR APPROACH).

RADAR APPROACH [ICAO]—An approach, executed by an aircraft, under the direction of a radar controller.

RADAR APPROACH CONTROL FACILITY—A terminal ATC facility that uses radar and nonradar capabilities to provide approach control services to aircraft arriving, departing, or transiting airspace controlled by the facility

(See APPROACH CONTROL SERVICE).

Provides radar ATC services to aircraft operating in the vicinity of one or more civil and/or military airports in a terminal area. The facility may provide services of a ground controlled approach (GCA); i.e., ASR and PAR approaches. A radar approach control facility may be operated by FAA, USAF, US Army, USN, USMC, or jointly by FAA and a military service. Specific facility nomenclatures are used for administrative purposes only and are related to the physical location of the facility and the operating service generally as follows:

Army Radar Approach Control (ARAC) (Army).
Radar Air Traffic Control Facility (RATCF) (Navy/FAA).
Radar Approach Control (RAPCON) (Air Force/FAA).
Terminal Radar Approach Control (TRACON) (FAA).
Tower/Airport Traffic Control Tower (ATCT) (FAA). (Only those towers delegated approach control authority.).

RADAR ARRIVAL—An aircraft arriving at an airport served by a radar facility and in radar contact with the facility.

(See NONRADAR).

RADAR BEACON—(See RADAR).

RADAR CONTACT—

1. Used by ATC to inform an aircraft that it is identified on the radar display and radar flight following will be provided until radar identification is terminated. Radar service may also be provided within the limits of necessity and capability. When a pilot is informed of "radar contact," he automatically discontinues reporting over compulsory reporting points.

(See RADAR FLIGHT FOLLOWING). (See RADAR CONTACT LOST). (See RADAR SERVICE). (See RADAR SERVICE TERMINATED). (Refer to AIM).

2. The term used to inform the controller that the aircraft is identified and approval is granted for the aircraft to enter the receiving controllers airspace.

(See ICAO term RADAR CONTACT).

RADAR CONTACT LOST—Used by ATC to inform a pilot that radar identification of his aircraft has been lost. The loss may be attributed to several things including the aircraft's merging with weather or ground clutter, the aircraft's flying below radar line of sight, the aircraft's entering an area of poor radar return, or a failure of the aircraft transponder or the ground radar equipment.

(See CLUTTER). (See RADAR CONTACT).

RADAR CLUTTER [ICAO]—The visual indication on a radar display of unwanted signals.

RADAR CONTACT [ICAO]—The situation which exists when the radar blip or radar position symbol of a particular aircraft is seen and identified on a radar display.

RADAR ENVIRONMENT—An area in which radar service may be provided.

(See RADAR CONTACT). (See RADAR SERVICE). (See ADDITIONAL SERVICES). (See TRAFFIC ADVISORIES).

RADAR FLIGHT FOLLOWING—The observation of the progress of radar identified aircraft, whose primary navigation is being provided by the pilot, wherein the controller retains and correlates the aircraft identity with the appropriate target or target symbol displayed on the radar scope.

(See RADAR CONTACT). (See RADAR SERVICE). (Refer to AIM).

RADAR IDENTIFICATION—The process of ascertaining that an observed radar target is the radar return from a particular aircraft.

(See RADAR CONTACT). (See RADAR SERVICE). (See ICAO term RADAR IDENTIFICATION).

RADAR IDENTIFICATION [ICAO]—The process of correlating a particular radar blip or radar position symbol with a specific aircraft.

RADAR IDENTIFIED AIRCRAFT—An aircraft, the position of which has been correlated with an observed target or symbol on the radar display.

(See RADAR CONTACT). (See RADAR CONTACT LOST).

RADAR MONITORING—(See RADAR SERVICE).

RADAR NAVIGATIONAL GUIDANCE—(See RADAR SERVICE).

RADAR POINT OUT—An action taken by a controller to transfer the radar identification of an aircraft to another controller if the aircraft will or may enter the airspace or protected airspace of another controller and radio communications will not be transferred.

RADAR REQUIRED—A term displayed on charts and approach plates and included in FDC Notams to alert pilots that segments of either an instrument approach procedure or a route are not navigable because of either the absence or unusability of a NAVAID. The pilot can expect to be provided radar navigational guidance while transiting segments labeled with this term.

(See RADAR ROUTE). (See RADAR SERVICE).

RADAR ROUTE—A flight path or route over which an aircraft is vectored. Navigational guidance and altitude assignments are provided by ATC.

(See FLIGHT PATH). (See ROUTE).

RADAR SEPARATION—(See RADAR SERVICE).

RADAR SERVICE—A term which encompasses one or more of the following services based on the use of radar which can be provided by a controller to a pilot of a radar identified aircraft.

1. Radar Monitoring. The radar flight-following of aircraft, whose primary navigation is being performed by the pilot, to observe and note deviations from its authorized flight path, airway, or route. When being applied specifically to radar monitoring of instrument approaches; i.e., with precision approach radar (PAR) or radar monitoring of simultaneous ILS/MLS approaches, it includes advice and instructions whenever an aircraft nears or exceeds the prescribed PAR safety limit or simultaneous ILS/MLS no transgression zone.

(See ADDITIONAL SERVICES). (See TRAFFIC ADVISORIES).

2. Radar Navigational Guidance. Vectoring aircraft to provide course guidance.
3. Radar Separation. Radar spacing of aircraft in accordance with established minima.

(See ICAO term RADAR SERVICE).

RADAR SERVICE [ICAO]—Term used to indicate a service provided directly by means of radar.

Radar Monitoring.—The use of radar for the purpose of providing aircraft with information and advice relative to significant deviations from nominal flight path.

Radar Separation.—The separation used when aircraft position information is derived from radar sources.

RADAR SERVICE TERMINATED—Used by ATC to inform a pilot that he will no longer be provided any of the services that could be received while in radar contact. Radar service is automatically terminated, and the pilot is not advised in the following cases:

1. An aircraft cancels its IFR flight plan, except within a TCA, TRSA, ARSA, or where Stage II service is provided.
2. An aircraft conducting an instrument, visual, or contact approach has landed or has been instructed to change to advisory frequency.
3. An arriving VFR aircraft, receiving radar service to a tower-controlled airport within a TCA, TRSA, ARSA, or where Stage II service is provided, has landed; or to all other airports, is instructed to change to tower or advisory frequency.
4. An aircraft completes a radar approach.

RADAR SURVEILLANCE—The radar observation of a given geographical area for the purpose of performing some radar function.

RADAR TRAFFIC ADVISORIES—Advisories issued to alert pilots to known or observed radar traffic which may affect the intended route of flight of their aircraft.

(See TRAFFIC ADVISORIES).

RADAR TRAFFIC INFORMATION SERVICE—(See TRAFFIC ADVISORIES).

RADAR VECTORING [ICAO]—Provision of navigational guidance to aircraft in the form of specific headings, based on the use of radar.

RADAR WEATHER ECHO INTENSITY LEVELS—Existing radar systems cannot detect turbulence. However, there is a direct correlation between the degree of turbulence and other weather features associated with thunderstorms and the radar weather echo intensity. The National Weather Service has categorized radar weather echo intensity for precipitation into six levels. These levels are sometimes expressed during communications as "VIP LEVEL" 1 through 6 (derived from the component of the radar that produces the information-Video Integrator and Processor). The following list gives the "VIP LEVELS" in relation to the precipitation intensity within a thunderstorm:

Level 1. WEAK

Level 2. MODERATE

Level 3. STRONG

Level 4. VERY STRONG

Level 5. INTENSE

Level 6. EXTREME

(See AC00–45).

RADIAL—A magnetic bearing extending from a VOR/VORTAC/TACAN navigation facility.

RADIO—

1. A device used for communication.
2. Used to refer to a flight service station; e.g., "Seattle Radio" is used to call Seattle FSS.

RADIO ALTIMETER—Aircraft equipment which makes use of the reflection of radio waves from the ground to determine the height of the aircraft above the surface.

RADIO BEACON—(See NONDIRECTIONAL BEACON).

RADIO DETECTION AND RANGING—(See RADAR).

RADIO MAGNETIC INDICATOR—An aircraft navigational instrument coupled with a gyro compass or similar compass that indicates the direction of a selected NAVAID and indicates bearing with respect to the heading of the aircraft.

RAMP—(See APRON).

RANDOM ALTITUDE—An altitude inappropriate for direction of flight and/or not in accordance with paragraph 4–40.

RANDOM ROUTE—Any route not established or charted/published or not otherwise available to all users.

RC—(See ROAD RECONNAISSANCE).

RCAG—(See REMOTE COMMUNICATIONS AIR/GROUND FACILITY).

RCC—(See RESCUE COORDINATION CENTER).

RCO—(See REMOTE COMMUNICATIONS OUTLET).

RCR—(See RUNWAY CONDITION READING).

READ BACK—Repeat my message back to me.

RECEIVING CONTROLLER—A controller/facility receiving control of an aircraft from another controller/facility.

RECEIVING FACILITY—(See RECEIVING CONTROLLER).

REDUCE SPEED TO (SPEED)—(See SPEED ADJUSTMENT).

REIL—(See RUNWAY END IDENTIFIER LIGHTS).

RELEASE TIME—A departure time restriction issued to a pilot by ATC (either directly or through an authorized relay) when necessary to separate a departing aircraft from other traffic.
(See ICAO term RELEASE TIME).

RELEASE TIME [ICAO]—Time prior to which an aircraft should be given further clearance or prior to which it should not proceed in case of radio failure.

REMOTE COMMUNICATIONS AIR/GROUND FACILITY—An unmanned VHF/UHF transmitter/receiver facility which is used to expand ARTCC air/ground communications coverage and to facilitate direct contact between pilots and controllers. RCAG facilities are sometimes not equipped with emergency frequencies 121.5 mHz and 243.0 mHz.
(Refer to AIM).

REMOTE COMMUNICATIONS OUTLET—An unmanned communications facility remotely controlled by air traffic personnel. RCO's serve FSS's. RTR's serve terminal ATC facilities. An RCO or RTR may be UHF or VHF and will extend the communication range of the air traffic facility. There are several classes of RCO's and RTR's. The class is determined by the number of transmitters or receivers. Classes A through G are used primarily for air/ground purposes. RCO and RTR class O facilities are nonprotected outlets subject to undetected and prolonged outages. RCO (O's) and RTR (O's) were established for the express purpose of providing ground-to-ground communications between air traffic control specialists and pilots located at a satellite airport for delivering en route clearances, issuing departure authorizations, and acknowledging instrument flight rules cancellations or departure/landing times. As a secondary function, they may be used for advisory purposes whenever the aircraft is below the coverage of the primary air/ground frequency.

REMOTE TRANSMITTER/RECEIVER—(See REMOTE COMMUNICATIONS OUTLET).

REPORT—Used to instruct pilots to advise ATC of specified information; e.g., "Report passing Hamilton VOR."

REPORTING POINT—A geographical location in relation to which the position of an aircraft is reported.
(See COMPULSORY REPORTING POINTS). (Refer to AIM). (See ICAO term REPORTING POINT).

REPORTING POINT [ICAO]—A specified geographical location in relation to which the position of an aircraft can be reported.

REQUEST FULL ROUTE CLEARANCE—Used by pilots to request that the entire route of flight be read verbatim in an ATC clearance. Such request should be made to preclude receiving an ATC clearance based on the original filed flight plan when a filed IFR flight plan has been revised by the pilot, company, or operations prior to departure.

RESCUE COORDINATION CENTER—A search and rescue (SAR) facility equipped and manned to coordinate and control SAR operations in an area designated by the SAR plan. The U.S. Coast Guard and the U.S. Air Force have responsibility for the operation of RCC's.
(See ICAO term RESCUE CO-ORDINATION CENTRE).

RESCUE CO-ORDINATION CENTRE [ICAO]—A unit responsible for promoting efficient organization of search and rescue service and for co-ordinating the conduct of search and rescue operations within a search and rescue region.

RESTRICTED AREA—(See SPECIAL USE AIRSPACE).
(See ICAO term RESTRICTED AREA).

RESTRICTED AREA [ICAO]—An airspace of defined dimensions, above the land areas or territorial waters of a State, within which the flight of aircraft is restricted in accordance with certain specified conditions.

RESUME OWN NAVIGATION—Used by ATC to advise a pilot to resume his own navigational responsibility. It is issued after completion of a radar vector or when radar contact is lost while the aircraft is being radar vectored.
(See RADAR CONTACT LOST). (See RADAR SERVICE TERMINATED).

RMI—(See RADIO MAGNETIC INDICATOR).

RNAV—(See AREA NAVIGATION).

RNAV [ICAO]—(See ICAO Term AREA NAVIGATION).

RNAV APPROACH—An instrument approach procedure which relies on aircraft area navigation equipment for navigational guidance.
(See INSTRUMENT APPROACH PROCEDURE). (See AREA NAVIGATION).

ROAD RECONNAISSANCE—Military activity requiring navigation along roads, railroads, and rivers. Reconnaissance route/route segments are seldom along a straight line and normally require a lateral route width of 10 NM to 30 NM and an altitude range of 500 feet to 10,000 feet AGL.

ROGER—I have received all of your last transmission. It should not be used to answer a question requiring a yes or a no answer.
(See AFFIRMATIVE). (See NEGATIVE).

ROLLOUT RVR—(See VISIBILITY).

ROUTE—A defined path, consisting of one or more courses in a horizontal plane, which aircraft traverse over the surface of the earth.
(See AIRWAY). (See JET ROUTE). (See PUBLISHED ROUTE). (See UNPUBLISHED ROUTE).

ROUTE SEGMENT—As used in Air Traffic Control, a part of a route that can be defined by two navigational fixes, two NAVAID's, or a fix and a NAVAID.
(See FIX). (See ROUTE). (See ICAO term ROUTE SEGMENT).

ROUTE SEGMENT [ICAO]—A portion of a route to be flown, as defined by two consecutive significant points specified in a flight plan.

RTR—(See REMOTE TRANSMITTER/RECEIVER).

RUNWAY—A defined rectangular area on a land airport prepared for the landing and takeoff run of aircraft along its length. Runways are normally numbered in relation to their magnetic direction rounded off to the nearest 10 degrees; e.g., Runway 01, Runway 25.

(See PARALLEL RUNWAYS). (See ICAO term RUNWAY).

RUNWAY [ICAO]—A defined rectangular area on a land aerodrome prepared for the landing and takeoff of aircraft.

RUNWAY CENTERLINE LIGHTING—(See AIRPORT LIGHTING).

RUNWAY CONDITION READING—Numerical decelerometer readings relayed by air traffic controllers at USAF and certain civil bases for use by the pilot in determining runway braking action. These readings are routinely relayed only to USAF and Air National Guard Aircraft.

(See BRAKING ACTION).

RUNWAY END IDENTIFIER LIGHTS—(See AIRPORT LIGHTING).

RUNWAY GRADIENT—The average slope, measured in percent, between two ends or points on a runway. Runway gradient is depicted on Government aerodrome sketches when total runway gradient exceeds 0.3%.

RUNWAY HEADING—The magnetic direction that corresponds with the runway centerline extended, not the painted runway number. When cleared to "fly or maintain runway heading," pilots are expected to fly or maintain the heading that corresponds with the extended centerline of the departure runway. Drift correction shall not be applied; e.g., Runway 4, actual magnetic heading of the runway centerline 044, fly 044.

RUNWAY IN USE/ACTIVE RUNWAY/DUTY RUNWAY—Any runway or runways currently being used for takeoff or landing. When multiple runways are used, they are all considered active runways. In the metering sense, a selectable adapted item which specifies the landing runway configuration or direction of traffic flow. The adapted optimum flight plan from each transition fix to the vertex is determined by the runway configuration for arrival metering processing purposes.

RUNWAY LIGHTS—(See AIRPORT LIGHTING).

RUNWAY MARKINGS—(See AIRPORT MARKING AIDS).

RUNWAY PROFILE DESCENT—An instrument flight rules (IFR) air traffic control arrival procedure to a runway published for pilot use in graphic and/or textual form and may be associated with a STAR. Runway Profile Descents provide routing and may depict crossing altitudes, speed restrictions, and headings to be flown from the en route structure to the point where the pilot will receive clearance for and execute an instrument approach procedure. A Runway Profile Descent may apply to more than one runway if so stated on the chart.

(Refer to AIM).

RUNWAY USE PROGRAM—A noise abatement runway selection plan designed to enhance noise abatement efforts with regard to airport communities for arriving and departing aircraft. These plans are developed into runway use programs and apply to all turbojet aircraft 12,500 pounds or heavier; turbojet aircraft less than 12,500 pounds are included only if the airport proprietor determines that the aircraft creates a noise problem. Runway use programs are coordinated with FAA offices, and safety criteria used in these programs are developed by the Office of Flight Operations. Runway use programs are administered by the Air Traffic Service as "Formal" or "Informal" programs.

1. Formal Runway Use Program. An approved noise abatement program which is defined and acknowledged in a Letter of Understanding between Flight Operations, Air Traffic Service, the airport proprietor, and the users. Once established, participation in the program is mandatory for aircraft operators and pilots as provided for in Part 91.129.
2. Informal Runway Use Program. An approved noise abatement program which does not require a Letter of Understanding, and participation in the program is voluntary for aircraft operators/pilots.

RUNWAY VISIBILITY VALUE—(See VISIBILITY).

RUNWAY VISUAL RANGE—(See VISIBILITY).-

S

SAFETY ALERT—A safety alert issued by ATC to aircraft under their control if ATC is aware the aircraft is at an altitude which, in the controller's judgment, places the aircraft in unsafe proximity to terrain, obstructions, or other aircraft. The controller may discontinue the issuance of further alerts if the pilot advises he is taking action to correct the situation or has the other aircraft in sight.

1. Terrain/Obstruction Alert. A safety alert issued by ATC to aircraft under their control if ATC is aware the aircraft is at an altitude which, in the controller's judgment, places the aircraft in unsafe proximity to terrain/obstructions; e.g., "Low Altitude Alert, check your altitude immediately."
2. Aircraft Conflict Alert. A safety alert issued by ATC to aircraft under their control if ATC is aware of an aircraft that is not under their control at an altitude which, in the controller's judgment, places both aircraft in unsafe proximity to each other. With the alert, ATC will offer the pilot an alternate course of action when feasible; e.g., "Traffic Alert, advise you turn right heading zero niner zero or climb to eight thousand immediately."

The issuance of a safety alert is contingent upon the capability of the controller to have an awareness of an unsafe condition. The course of action provided will be predicated on other traffic under ATC control. Once the alert is issued, it is solely the pilot's prerogative to determine what course of action, if any, he will take.

SAIL BACK—A maneuver during high wind conditions (usually with power off) where float plane movement is controlled by water rudders/opening and closing cabin doors.

SAME DIRECTION AIRCRAFT—Aircraft are operating in the same direction when:

1. They are following the same track in the same direction; or
2. Their tracks are parallel and the aircraft are flying in the same direction; or
3. Their tracks intersect at an angle of less than 45 degrees.

SAR—(See SEARCH AND RESCUE).

SAY AGAIN—Used to request a repeat of the last transmission. Usually specifies transmission or portion thereof not understood or received; e.g., "Say again all after ABRAM VOR."

SAY ALTITUDE—Used by ATC to ascertain an aircraft's specific altitude/flight level. When the aircraft is climbing or descending, the pilot should state the indicated altitude rounded to the nearest 100 feet.

SAY HEADING—Used by ATC to request an aircraft heading. The pilot should state the actual heading of the aircraft.

SDF—(See SIMPLIFIED DIRECTIONAL FACILITY).

SEA LANE—A designated portion of water outlined by visual surface markers for and intended to be used by aircraft designed to operate on water.

SEARCH AND RESCUE—A service which seeks missing aircraft and assists those found to be in need of assistance. It is a cooperative effort using the facilities and services of available Federal, state and local agencies. The U.S. Coast Guard is responsible for coordination of search and rescue for the Maritime Region, and the U.S. Air Force is responsible for search and rescue for the Inland Region. Information pertinent to search and rescue should be passed through any air traffic facility or be transmitted directly to the Rescue Coordination Center by telephone.

(See FLIGHT SERVICE STATION). (See RESCUE COORDINATION CENTER). (Refer to AIM).

SEARCH AND RESCUE FACILITY—A facility responsible for maintaining and operating a search and rescue (SAR) service to render aid to persons and property in distress. It is any SAR unit, station, NET, or other operational activity which can be usefully employed during an SAR Mission; e.g., a Civil Air Patrol Wing, or a Coast Guard Station.

(See SEARCH AND RESCUE).

SECTIONAL CHARTS—(See AERONAUTICAL CHART).

SECTOR LIST DROP INTERVAL—A parameter number of minutes after the meter fix time when arrival aircraft will be deleted from the arrival sector list.

SEE AND AVOID—A visual procedure wherein pilots of aircraft flying in visual meteorological conditions (VMC), regardless of type of flight plan, are charged with the responsibility to observe the presence of other aircraft and to maneuver their aircraft as required to avoid the other aircraft. Right-of-way rules are contained in Part 91.

(See INSTRUMENT FLIGHT RULES). (See VISUAL FLIGHT RULES). (See VISUAL METEOROLOGICAL CONDITIONS). (See INSTRUMENT METEOROLOGICAL CONDITIONS).

SEGMENTED CIRCLE—A system of visual indicators designed to provide traffic pattern information at airports without operating control towers.

(Refer to AIM).

SEGMENTS OF AN INSTRUMENT APPROACH PROCEDURE—An instrument approach procedure may have as many as four separate segments depending on how the approach procedure is structured.

1. Initial Approach. The segment between the initial approach fix and the intermediate fix or the point where the aircraft is established on the intermediate course or final approach course.

(See ICAO term INITIAL APPROACH SEGMENT).

2. Intermediate Approach. The segment between the intermediate fix or point and the final approach fix.

(See ICAO term INTERMEDIATE APPROACH SEGMENT).

3. Final Approach. The segment between the final approach fix or point and the runway, airport, or missed approach point.

(See ICAO term FINAL APPROACH SEGMENT).

4. Missed Approach. The segment between the missed approach point or the point of arrival at decision height and the missed approach fix at the prescribed altitude.

(Refer to Part 97). (See ICAO term MISSED APPROACH PROCEDURE).

SELECTED GROUND DELAYS—A traffic management procedure whereby selected flights are issued ground delays to better regulate traffic flows over a particular fix or area.

SEPARATION—In air traffic control, the spacing of aircraft to achieve their safe and orderly movement in flight and while landing and taking off.

(See SEPARATION MINIMA). (See ICAO term SEPARATION).

SEPARATION [ICAO]—Spacing between aircraft, levels or tracks.

SEPARATION MINIMA—The minimum longitudinal, lateral, or vertical distances by which aircraft are spaced through the application of air traffic control procedures.

(See SEPARATION).

SEVERE WEATHER AVOIDANCE PLAN—An approved plan to minimize the affect of severe weather on traffic flows in impacted terminal and/or ARTCC areas. SWAP is normally implemented to provide the least disruption to the ATC system when flight through portions of airspace is difficult or impossible due to severe weather.

SEVERE WEATHER FORECAST ALERTS—Preliminary messages issued in order to alert users that a Severe Weather Watch Bulletin (WW) is being issued. These messages define areas of possible severe thunderstorms or tornado activity. The messages are unscheduled and issued as required by the National Severe Storm Forecast Center at Kansas City, Missouri.

(See SIGMET). (See CONVECTIVE SIGMET). (See CWA). (See AIRMET).

SFA—(See SINGLE FREQUENCY APPROACH).

SFO—(See SIMULATED FLAMEOUT).

SHF—(See SUPER HIGH FREQUENCY).

SHORT RANGE CLEARANCE—A clearance issued to a departing IFR flight which authorizes IFR flight to a specific fix short of the destination while air traffic control facilities are coordinating and obtaining the complete clearance.

SHORT TAKEOFF AND LANDING AIRCRAFT AIRCRAFT—An aircraft which, at some weight within its approved operating weight, is capable of operating from a STOL runway in compliance with the applicable STOL characteristics, airworthiness, operations, noise, and pollution standards.

(See VERTICAL TAKEOFF AND LANDING AIRCRAFT).

SIAP—(See STANDARD INSTRUMENT APPROACH PROCEDURE).

SIDESTEP MANEUVER—A visual maneuver accomplished by a pilot at the completion of an instrument approach to permit a straight-in landing on a parallel runway not more than 1,200 feet to either side of the runway to which the instrument approach was conducted.
(Refer to AIM).

SIGMET—A weather advisory issued concerning weather significant to the safety of all aircraft. SIGMET advisories cover severe and extreme turbulence, severe icing, and widespread dust or sandstorms that reduce visibility to less than 3 miles.
(See AWW). (See CONVECTIVE SIGMET). (See CWA). (See AIRMET). (Refer to AIM). (See ICAO term SIGMET INFORMATION).

SIGMET INFORMATION [ICAO]—Information issued by a meteorological watch office concerning the occurrence or expected occurrence of specified en-route weather phenomena which may affect the safety of aircraft operations.

SIGNIFICANT METEOROLOGICAL INFORMATION—(See SIGMET).

SIGNIFICANT POINT—A point, whether a named intersection, a NAVAID, a fix derived from a NAVAID(s), or geographical coordinate expressed in degrees of latitude and longitude, which is established for the purpose of providing separation, as a reporting point, or to delineate a route of flight.

SIMPLIFIED DIRECTIONAL FACILITY—A NAVAID used for nonprecision instrument approaches. The final approach course is similar to that of an ILS localizer except that the SDF course may be offset from the runway, generally not more than 3 degrees, and the course may be wider than the localizer, resulting in a lower degree of accuracy.
(Refer to AIM).

SIMULATED FLAMEOUT—A practice approach by a jet aircraft (normally military) at idle thrust to a runway. The approach may start at a relatively high altitude over a runway (high key) and may continue on a relatively high and wide downwind leg with a high rate of descent and a continuous turn to final. It terminates in a landing or low approach. The purpose of this approach is to simulate a flameout.
(See FLAMEOUT).

SIMULTANEOUS ILS APPROACHES—An approach system permitting simultaneous ILS/MLS approaches to airports having parallel runways separated by at least 4,300 feet between centerlines. Integral parts of a total system are ILS/MLS, radar, communications, ATC procedures, and appropriate airborne equipment.
(See PARALLEL RUNWAYS). (Refer to AIM).

SIMULTANEOUS MLS APPROACHES—(See SIMULTANEOUS ILS APPROACHES).

SINGLE DIRECTION ROUTES—Preferred IFR Routes which are sometimes depicted on high altitude en route charts and which are normally flown in one direction only.
(See PREFERRED IFR ROUTES). (Refer to AIRPORT/FACILITY DIRECTORY).

SINGLE FREQUENCY APPROACH—A service provided under a letter of agreement to military single-piloted turbojet aircraft which permits use of a single UHF frequency during approach for landing. Pilots will not normally be required to change frequency from the beginning of the approach to touchdown except that pilots conducting an en route descent are required to change frequency when control is transferred from the air route traffic control center to the terminal facility. The abbreviation "SFA" in the DOD FLIP IFR Supplement under "Communications" indicates this service is available at an aerodrome.

SINGLE-PILOTED AIRCRAFT—A military turbojet aircraft possessing one set of flight controls, tandem cockpits, or two sets of flight controls but operated by one pilot is considered single-piloted by ATC when determining the appropriate air traffic service to be applied.
(See SINGLE FREQUENCY APPROACH).

SLASH—A radar beacon reply displayed as an elongated target.

SLDI—(See SECTOR LIST DROP INTERVAL).

SLOT TIME—(See METER FIX TIME/SLOT TIME).

SLOW TAXI—To taxi a float plane at low power or low RPM.

SN—(See SYSTEM STRATEGIC NAVIGATION).

SPEAK SLOWER—Used in verbal communications as a request to reduce speech rate.

SPECIAL EMERGENCY—A condition of air piracy or other hostile act by a person(s) aboard an aircraft which threatens the safety of the aircraft or its passengers.

SPECIAL INSTRUMENT APPROACH PROCEDURE—(See INSTRUMENT APPROACH PROCEDURE).

SPECIAL USE AIRSPACE—Airspace of defined dimensions identified by an area on the surface of the earth wherein activities must be confined because of their nature and/or wherein limitations may be imposed upon aircraft operations that are not a part of those activities. Types of special use airspace are:

1. Alert Area. Airspace which may contain a high volume of pilot training activities or an unusual type of aerial activity, neither of which is hazardous to aircraft. Alert Areas are depicted on aeronautical charts for the information of nonparticipating pilots. All activities within an Alert Area are conducted in accordance with Federal Aviation Regulations, and pilots of participating aircraft as well as pilots transiting the area are equally responsible for collision avoidance.
2. Controlled Firing Area. Airspace wherein activities are conducted under conditions so controlled as to eliminate hazards to nonparticipating aircraft and to ensure the safety of persons and property on the ground.
3. Military Operations Area (MOA). An MOA is an airspace assignment of defined vertical and lateral dimensions established outside positive control areas to separate/segregate certain military activities from IFR traffic and to identify for VFR traffic where these activities are conducted.

(Refer to AIM).

4. Prohibited Area. Designated airspace within which the flight of aircraft is prohibited.

(Refer to En Route Charts, AIM).

5. Restricted Area. Airspace designated under Part 73, within which the flight of aircraft, while not wholly prohibited, is subject to restriction. Most restricted areas are designated joint use and IFR/VFR operations in the area may be authorized by the controlling ATC facility when it is not being utilized by the using agency. Restricted areas are depicted on en route charts. Where joint use is authorized, the name of the ATC controlling facility is also shown.

(Refer to Part 73). (Refer to AIM).

6. Warning Area. Airspace which may contain hazards to nonparticipating aircraft in international airspace.

SPECIAL VFR CONDITIONS—Weather conditions in a control zone which are less than basic VFR and in which some aircraft are permitted flight under Visual Flight Rules.

(See SPECIAL VFR OPERATIONS). (Refer to Part 91).

SPECIAL VFR FLIGHT [ICAO]—A controlled VFR flight authorized by air traffic control to operate within a control zone under meteorological conditions below the visual meteorological conditions.

SPECIAL VFR OPERATIONS—Aircraft operating in accordance with clearances within control zones in weather conditions less than the basic VFR weather minima. Such operations must be requested by the pilot and approved by ATC.

(See SPECIAL VFR CONDITIONS). (See ICAO term SPECIAL VFR FLIGHT).

SPEED—(See AIRSPEED).

(See GROUND SPEED).

SPEED ADJUSTMENT—An ATC procedure used to request pilots to adjust aircraft speed to a specific value for the purpose of providing desired spacing. Pilots are expected to maintain a speed of plus or minus 10 knots or 0.02 mach number of the specified speed.

Examples of speed adjustments are:

1. "Increase/reduce speed to mach point (number)."
2. "Increase/reduce speed to (speed in knots)" or "Increase/reduce speed (number of knots) knots."

SPEED BRAKES—Moveable aerodynamic devices on aircraft that reduce airspeed during descent and landing.

SPEED SEGMENTS—Portions of the arrival route between the transition point and the vertex along the optimum flight path for which speeds and altitudes are specified. There is one set of arrival speed segments adapted from each transition point to each vertex. Each set may contain up to six segments.

SQUAWK (MODE, CODE, FUNCTION)—Activate specific modes/codes/functions on the aircraft transponder; e.g., "Squawk three/alpha, two one zero five, low."
(See TRANSPONDER).

STAGE I SERVICE—(See TERMINAL RADAR PROGRAM).

STAGE II SERVICE—(See TERMINAL RADAR PROGRAM).

STAGE III SERVICE—(See TERMINAL RADAR PROGRAM).

STAGING/QUEUING—The placement, integration, and segregation of departure aircraft in designated movement areas of an airport by departure fix, EDCT, and/or restriction.

STANDARD INSTRUMENT APPROACH PROCEDURE—(See INSTRUMENT APPROACH PROCEDURE).

STANDARD INSTRUMENT DEPARTURE—A preplanned instrument flight rule (IFR) air traffic control departure procedure printed for pilot use in graphic and/or textual form. SID's provide transition from the terminal to the appropriate en route structure.
(See IFR TAKEOFF MINIMUMS AND DEPARTURE PROCEDURES). (Refer to AIM).

STANDARD INSTRUMENT DEPARTURE CHARTS—(See AERONAUTICAL CHART).

STANDARD RATE TURN—A turn of three degrees per second.

STANDARD TERMINAL ARRIVAL—A preplanned instrument flight rule (IFR) air traffic control arrival procedure published for pilot use in graphic and/or textual form. STAR's provide transition from the en route structure to an outer fix or an instrument approach fix/arrival waypoint in the terminal area.

STANDARD TERMINAL ARRIVAL CHARTS—(See AERONAUTICAL CHART).

STAND BY—Means the controller or pilot must pause for a few seconds, usually to attend to other duties of a higher priority. Also means to wait as in "stand by for clearance." The caller should reestablish contact if a delay is lengthy. "Stand by" is not an approval or denial.

STAR—(See STANDARD TERMINAL ARRIVAL).

STATE AIRCRAFT—Aircraft used in military, customs and police service, in the exclusive service of any government, or of any political subdivision, thereof including the government of any state, territory, or possession of the United States or the District of Columbia, but not including any government-owned aircraft engaged in carrying persons or property for commercial purposes.

STATIC RESTRICTIONS—Those restrictions that are usually not subject to change, fixed, in place, and/or published.

STATIONARY RESERVATIONS—Altitude reservations which encompass activities in a fixed area. Stationary reservations may include activities, such as special tests of weapons systems or equipment, certain U.S. Navy carrier, fleet, and anti-submarine operations, rocket, missile and drone operations, and certain aerial refueling or similar operations.

STEPDOWN FIX—A fix permitting additional descent within a segment of an instrument approach procedure by identifying a point at which a controlling obstacle has been safely overflown.

STEP TAXI—To taxi a float plane at full power or high RPM.

STEP TURN—A maneuver used to put a float plane in a planing configuration prior to entering an active sea lane for takeoff. The STEP TURN maneuver should only be used upon pilot request.

STEREO ROUTE—A routinely used route of flight established by users and ARTCC's identified by a coded name; e.g., ALPHA 2. These routes minimize flight plan handling and communications.

STOL AIRCRAFT—(See SHORT TAKEOFF AND LANDING AIRCRAFT).

STOP ALTITUDE SQUAWK—Used by ATC to inform an aircraft to turn-off the automatic altitude reporting feature of its transponder. It is issued when the verbally reported altitude varies 300 feet or more from the automatic altitude report.
(See ALTITUDE READOUT). (See TRANSPONDER).

STOP AND GO—A procedure wherein an aircraft will land, make a complete stop on the runway, and then commence a takeoff from that point.

(See LOW APPROACH). (See OPTION APPROACH).

STOP BURST—(See STOP STREAM).

STOP BUZZER—(See STOP STREAM).

STOPOVER FLIGHT PLAN—A flight plan format which permits in a single submission the filing of a sequence of flight plans through interim full-stop destinations to a final destination.

STOP SQUAWK (MODE OR CODE)—Used by ATC to tell the pilot to turn specified functions of the aircraft transponder off.

(See STOP ALTITUDE SQUAWK). (See TRANSPONDER).

STOP STREAM—Used by ATC to request a pilot to suspend electronic countermeasure activity.

(See JAMMING).

STOPWAY—An area beyond the takeoff runway no less wide than the runway and centered upon the extended centerline of the runway, able to support the airplane during an aborted takeoff, without causing structural damage to the airplane, and designated by the airport authorities for use in decelerating the airplane during an aborted takeoff. (OVERRUN term used for USAF airfields).

STRAIGHT-IN APPROACH-IFR—An instrument approach wherein final approach is begun without first having executed a procedure turn, not necessarily completed with a straight-in landing or made to straight-in landing minimums.

(See STRAIGHT-IN LANDING). (See LANDING MINIMUMS). (See STRAIGHT-IN APPROACH-VFR).

STRAIGHT-IN APPROACH-VFR—Entry into the traffic pattern by interception of the extended runway centerline (final approach course) without executing any other portion of the traffic pattern.

(See TRAFFIC PATTERN).

STRAIGHT-IN LANDING—A landing made on a runway aligned within 30° of the final approach course following completion of an instrument approach.

(See STRAIGHT-IN APPROACH-IFR).

STRAIGHT-IN LANDING MINIMUMS—(See LANDING MINIMUMS).

STRAIGHT-IN MINIMUMS—(See STRAIGHT-IN LANDING MINIMUMS).

SUBSTITUTIONS—Users are permitted to exchange CTA's. Normally, the airline dispatcher will contact the ATCSCC with this request. The ATCSCC shall forward approved substitutions to the TMU's who will notify the appropriate terminals. Permissible swapping must not change the traffic load for any given hour of an EQF program.

SUBSTITUTE ROUTE—A route assigned to pilots when any part of an airway or route is unusable because of NAVAID status. These routes consist of:

1. Substitute routes which are shown on U.S. Government charts.
2. Routes defined by ATC as specific NAVAID radials or courses.
3. Routes defined by ATC as direct to or between NAVAID's.

SUNSET AND SUNRISE—The mean solar times of sunset and sunrise as published in the Nautical Almanac, converted to local standard time for the locality concerned. Within Alaska, the end of evening civil twilight and the beginning of morning civil twilight, as defined for each locality.

SUPER HIGH FREQUENCY—The frequency band between 3 and 30 gigahertz (gHz). The elevation and azimuth stations of the microwave landing system operate from 5031 mHz to 5091 mHz in this spectrum.

SUPPS—Refers to ICAO Document 7030 Regional Supplementary Procedures. SUPPS contain procedures for each ICAO Region which are unique to that Region and are not covered in the worldwide provisions identified in the ICAO Air Navigation Plan. Procedures contained in chapter 8 are based in part on those published in SUPPS.

SURPIC—A description of surface vessels in the area of a Search and Rescue incident including their predicted positions and their characteristics.

(See Paragraph 10–83, Emergency Assistance).

SURVEILLANCE APPROACH—An instrument approach wherein the air traffic controller issues instructions, for pilot compliance, based on aircraft position in relation to the final approach course (azimuth), and the distance (range) from the end of the runway as displayed on the controller's radar scope. The controller will provide recommended altitudes on final approach if requested by the pilot.

(Refer to AIM).

SWAP—(See SEVERE WEATHER AVOIDANCE PLAN).

SYSTEM STRATEGIC NAVIGATION—Military activity accomplished by navigating along a preplanned route using internal aircraft systems to maintain a desired track. This activity normally requires a lateral route width of 10 NM and altitude range of 1,000 feet to 6,000 feet AGL with some route segments that permit terrain following.-

T

TACAN—(See TACTICAL AIR NAVIGATION).

TACAN-ONLY AIRCRAFT—An aircraft, normally military, possessing TACAN with DME but no VOR navigational system capability. Clearances must specify TACAN or VORTAC fixes and approaches.

TACTICAL AIR NAVIGATION—An ultra-high frequency electronic rho-theta air navigation aid which provides suitably equipped aircraft a continuous indication of bearing and distance to the TACAN station.

(See VORTAC). (Refer to AIM).

TAKEOFF AREA—(See LANDING AREA).

TAKE–OFF DISTANCE AVAILABLE [ICAO]—The length of the take–off run available plus the length of the clearway, if provided.

TAKE–OFF RUN AVAILABLE [ICAO]—The length of runway declared available and suitable for the ground run of an aeroplane take–off.

TARGET—The indication shown on a radar display resulting from a primary radar return or a radar beacon reply.

(See RADAR). (See TARGET SYMBOL). (See ICAO term TARGET).

TARGET [ICAO]—In radar:

1. Generally, any discrete object which reflects or retransmits energy back to the radar equipment.

2. Specifically, an object of radar search or surveillance.

TARGET RESOLUTION—A process to ensure that correlated radar targets do not touch. Target resolution shall be applied as follows:

1. Between the edges of two primary targets or the edges of the ASR–9 primary target symbol.
2. Between the end of the beacon control slash and the edge of a primary target.
3. Between the ends of two beacon control slashes.—MANDATORY TRAFFIC ADVISORIES AND SAFETY ALERTS SHALL BE ISSUED WHEN THIS PROCEDURE IS USED.

Note: This procedure shall not be provided utilizing mosaic radar systems.

TARGET SYMBOL—A computer-generated indication shown on a radar display resulting from a primary radar return or a radar beacon reply.

TAXI—The movement of an airplane under its own power on the surface of an airport (Part 135.100–Note). Also, it describes the surface movement of helicopters equipped with wheels.

(See AIR TAXI). (See HOVER TAXI). (Refer to AIM). (Refer to Part 135.100).

TAXI INTO POSITION AND HOLD—Used by ATC to inform a pilot to taxi onto the departure runway in takeoff position and hold. It is not authorization for takeoff. It is used when takeoff clearance cannot immediately be issued because of traffic or other reasons.

(See CLEARED FOR TAKEOFF).

TAXI PATTERNS—Patterns established to illustrate the desired flow of ground traffic for the different runways or airport areas available for use.

TCAS—(See TRAFFIC ALERT AND COLLISION AVOIDANCE SYSTEM).

TCH—(See THRESHOLD CROSSING HEIGHT).

TCLT—(See TENTATIVE CALCULATED LANDING TIME).

TDZE—(See TOUCHDOWN ZONE ELEVATION).

TELEPHONE INFORMATION BRIEFING SERVICE—A continuous telephone recording of meteorological and/or aeronautical information.

(Refer to AIM).

TENTATIVE CALCULATED LANDING TIME—A projected time calculated for adapted vertex for each arrival aircraft based upon runway configuration, airport acceptance rate, airport arrival delay period, and other metered arrival aircraft. This time is either the VTA of the aircraft or the TCLT/ACLT of the previous aircraft plus the AAI, whichever is later. This time will be updated in response to an aircraft's progress and its current relationship to other arrivals.

TERMINAL AREA—A general term used to describe airspace in which approach control service or airport traffic control service is provided.

TERMINAL AREA FACILITY—A facility providing air traffic control service for arriving and departing IFR, VFR, Special VFR, and on occasion en route aircraft.

(See APPROACH CONTROL FACILITY). (See TOWER).

TERMINAL CONTROL AREA—(See CONTROLLED AIRSPACE).

TERMINAL RADAR PROGRAM—A national program instituted to extend the terminal radar services provided IFR aircraft to VFR aircraft. Pilot participation in the program is urged but is not mandatory. The program is divided into two parts and referred to as Stage II and Stage III. The Stage service provided at a particular location is contained in the Airport/Facility Directory.

1. Stage I originally comprised two basic radar services (traffic advisories and limited vectoring to VFR aircraft). These services are provided by all commissioned terminal radar facilities, but the term "Stage I" has been deleted from use.
2. Stage II/Radar Advisory and Sequencing for VFR Aircraft. Provides, in addition to the basic radar services, vectoring and sequencing on a full-time basis to arriving VFR aircraft. The purpose is to adjust the flow of arriving IFR and VFR aircraft into the traffic pattern in a safe and orderly manner and to provide traffic advisories to departing VFR aircraft.
3. Stage III/Radar Sequencing and Separation Service for VFR Aircraft. Provides, in addition to the basic radar services and Stage II, separation between all participating VFR aircraft. The purpose is to provide separation between all participating VFR aircraft and all IFR aircraft operating within the airspace defined as a Terminal Radar Service Area (TRSA) or Terminal Control Area (TCA).

(See CONTROLLED AIRSPACE). (See TERMINAL RADAR SERVICE AREA). (Refer to AIM). (Refer to AIRPORT/FACILITY DIRECTORY).

TERMINAL RADAR SERVICE AREA—Airspace surrounding designated airports wherein ATC provides radar vectoring, sequencing, and separation on a full-time basis for all IFR and participating VFR aircraft. Service provided in a TRSA is called Stage III Service. The AIM contains an explanation of TRSA. TRSA's are depicted on VFR aeronautical charts. Pilot participation is urged but is not mandatory.

(See TERMINAL RADAR PROGRAM). (Refer to AIM). (Refer to AIRPORT/FACILITY DIRECTORY).

TERMINAL-VERY HIGH FREQUENCY OMNIDIRECTIONAL RANGE STATION—A very high frequency terminal omnirange station located on or near an airport and used as an approach aid.

(See NAVIGATIONAL AID). (See VOR).

TERRAIN FOLLOWING—The flight of a military aircraft maintaining a constant AGL altitude above the terrain or the highest obstruction. The altitude of the aircraft will constantly change with the varying terrain and/or obstruction.

TETRAHEDRON—A device normally located on uncontrolled airports and used as a landing direction indicator. The small end of a tetrahedron points in the direction of landing. At controlled airports, the tetrahedron, if installed, should be disregarded because tower instructions supersede the indicator.

(See SEGMENTED CIRCLE). (Refer to AIM).

TF—(See TERRAIN FOLLOWING).

THAT IS CORRECT—The understanding you have is right.

360 OVERHEAD—(See OVERHEAD APPROACH).

THRESHOLD—The beginning of that portion of the runway usable for landing.

(See AIRPORT LIGHTING). (See DISPLACED THRESHOLD).

THRESHOLD CROSSING HEIGHT—The theoretical height above the runway threshold at which the aircraft's glideslope antenna would be if the aircraft maintains the trajectory established by the mean ILS glideslope or MLS glidepath.

(See GLIDESLOPE). (See THRESHOLD).

THRESHOLD LIGHTS—(See AIRPORT LIGHTING).

TIBS—(See TELEPHONE INFORMATION BRIEFING SERVICE).

TIME GROUP—Four digits representing the hour and minutes from the 24–hour clock. Time groups without time zone indicators are understood to be UTC (Coordinated Universal Time); e.g., "0205." The term "Zulu" is used when ATC procedures require a reference to UTC. A time zone designator is used to indicate local time; e.g., "0205M." The end and the beginning of the day are shown by "2400" and "0000," respectively.

TMPA—(See TRAFFIC MANAGEMENT PROGRAM ALERT).

TMU—(See TRAFFIC MANAGEMENT UNIT).

TODA [ICAO]—(See ICAO Term TAKE–OFF DISTANCE AVAILABLE).

TORA [ICAO]—(See ICAO Term TAKE–OFF RUN AVAILABLE).

TORCHING—The burning of fuel at the end of an exhaust pipe or stack of a reciprocating aircraft engine, the result of an excessive richness in the fuel air mixture.

TOTAL ESTIMATED ELAPSED TIME [ICAO]—For IFR flights, the estimated time required from take-off to arrive over that designated point, defined by reference to navigation aids, from which it is intended that an instrument approach procedure will be commenced, or, if no navigation aid is associated with the destination aerodrome, to arrive over the destination aerodrome. For VFR flights, the estimated time required from takeoff to arrive over the destination aerodrome.

(See ESTIMATED ELAPSED TIME).

TOUCH-AND-GO—An operation by an aircraft that lands and departs on a runway without stopping or exiting the runway.

TOUCH-AND-GO LANDING—(See TOUCH-AND-GO).

TOUCHDOWN—

1. The point at which an aircraft first makes contact with the landing surface.
2. Concerning a precision radar approach (PAR), it is the point where the glide path intercepts the landing surface.

(See ICAO term TOUCHDOWN).

TOUCHDOWN [ICAO]—The point where the nominal glide path intercepts the runway.

Note: Touchdown as defined above is only a datum and is not necessarily the actual point at which the aircraft will touch the runway.

TOUCHDOWN RVR—(See VISIBILITY).

TOUCHDOWN ZONE—The first 3,000 feet of the runway beginning at the threshold. The area is used for determination of Touchdown Zone Elevation in the development of straight-in landing minimums for instrument approaches.

(See ICAO term TOUCHDOWN ZONE).

TOUCHDOWN ZONE [ICAO]—The portion of a runway, beyond the threshold, where it is intended landing aircraft first contact the runway.

TOUCHDOWN ZONE ELEVATION—The highest elevation in the first 3,000 feet of the landing surface. TDZE is indicated on the instrument approach procedure chart when straight-in landing minimums are authorized.

(See TOUCHDOWN ZONE).

TOUCHDOWN ZONE LIGHTING—(See AIRPORT LIGHTING).

TOWER—A terminal facility that uses air/ground communications, visual signaling, and other devices to provide ATC services to aircraft operating in the vicinity of an airport or on the movement area. Authorizes aircraft to land or takeoff at the airport controlled by the tower or to transit the airport traffic area regardless of flight plan or weather conditions (IFR or VFR). A tower may also provide approach control services (radar or nonradar).

(See AIRPORT TRAFFIC AREA). (See AIRPORT TRAFFIC CONTROL SERVICE). (See APPROACH CONTROL FACILITY). (See APPROACH CONTROL SERVICE). (See MOVEMENT AREA). (See TOWER EN ROUTE CONTROL SERVICE). (Refer to AIM). (See ICAO term AERODROME CONTROL TOWER).

TOWER EN ROUTE CONTROL SERVICE—The control of IFR en route traffic within delegated airspace between two or more adjacent approach control facilities. This service is designed to expedite traffic and reduce control and pilot communication requirements.

TOWER TO TOWER—(See TOWER EN ROUTE CONTROL SERVICE).

TPX–42—A numeric beacon decoder equipment/system. It is designed to be added to terminal radar systems for beacon decoding. It provides rapid target identification, reinforcement of the primary radar target, and altitude information from Mode C.

(See AUTOMATED RADAR TERMINAL SYSTEMS). (See TRANSPONDER).

TRACK—The actual flight path of an aircraft over the surface of the earth.

(See COURSE). (See ROUTE). (See FLIGHT PATH). (See ICAO term TRACK).

TRACK [ICAO]—The projection on the earth's surface of the path of an aircraft, the direction of which path at any point is usually expressed in degrees from North (True, Magnetic, or Grid).

TRAFFIC—

1. A term used by a controller to transfer radar identification of an aircraft to another controller for the purpose of coordinating separation action. Traffic is normally issued:
   (a) in response to a handoff or point out,
   (b) in anticipation of a handoff or point out, or
   (c) in conjunction with a request for control of an aircraft.
2. A term used by ATC to refer to one or more aircraft.

TRAFFIC ADVISORIES—Advisories issued to alert pilots to other known or observed air traffic which may be in such proximity to the position or intended route of flight of their aircraft to warrant their attention. Such advisories may be based on:

1. Visual observation.
2. Observation of radar identified and nonidentified aircraft targets on an ATC radar display, or
3. Verbal reports from pilots or other facilities.

The word "traffic" followed by additional information, if known, is used to provide such advisories; e.g., "Traffic, 2 o'clock, one zero miles, southbound, eight thousand."

Traffic advisory service will be provided to the extent possible depending on higher priority duties of the controller or other limitations; e.g., radar limitations, volume of traffic, frequency congestion, or controller workload. Radar/nonradar traffic advisories do not relieve the pilot of his responsibility to see and avoid other aircraft. Pilots are cautioned that there are many times when the controller is not able to give traffic advisories concerning all traffic in the aircraft's proximity; in other words, when a pilot requests or is receiving traffic advisories, he should not assume that all traffic will be issued.

(Refer to AIM).

(IDENTIFICATION), TRAFFIC ALERT. ADVISE YOU TURN LEFT/RIGHT (SPECIFIC HEADING IF APPROPRIATE), AND/OR CLIMB/DESCEND (SPECIFIC ALTITUDE IF APPROPRIATE) IMMEDIATELY.—(See SAFETY ALERT).

TRAFFIC ALERT AND COLLISION AVOIDANCE SYSTEM—An airborne collision avoidance system based on radar beacon signals which operates independent of ground-based equipment. TCAS-I generates traffic advisories only. TCAS-II generates traffic advisories, and resolution (collision avoidance) advisories in the vertical plane.

TRAFFIC INFORMATION—(See TRAFFIC ADVISORIES).

TRAFFIC IN SIGHT—Used by pilots to inform a controller that previously issued traffic is in sight.

(See NEGATIVE CONTACT). (See TRAFFIC ADVISORIES).

TRAFFIC MANAGEMENT PROGRAM ALERT—A term used in a Notice to Airmen (NOTAM) issued in conjunction with a special traffic management program to alert pilots to the existence of the program and to refer them to either the Notices to Airmen publication or a special traffic management program advisory message for program details. The contraction TMPA is used in NOTAM text.

TRAFFIC MANAGEMENT UNIT—The entity in ARTCC's and designated terminals responsible for direct involvement in the active management of facility traffic. Usually under the direct supervision of an assistant manager for traffic management.

TRAFFIC NO FACTOR—Indicates that the traffic described in a previously issued traffic advisory is no factor.

TRAFFIC PATTERN—The traffic flow that is prescribed for aircraft landing at, taxiing on, or taking off from an airport. The components of a typical traffic pattern are upwind leg, crosswind leg, downwind leg, base leg, and final approach.

1. Upwind Leg. A flight path parallel to the landing runway in the direction of landing.
2. Crosswind Leg. A flight path at right angles to the landing runway off its upwind end.
3. Downwind Leg. A flight path parallel to the landing runway in the direction opposite to landing. The downwind leg normally extends between the crosswind leg and the base leg.
4. Base Leg. A flight path at right angles to the landing runway off its approach end. The base leg normally extends from the downwind leg to the intersection of the extended runway centerline.
5. Final Approach. A flight path in the direction of landing along the extended runway centerline. The final approach normally extends from the base leg to the runway. An aircraft making a straight-in approach VFR is also considered to be on final approach.

(See STRAIGHT-IN APPROACH-VFR). (See TAXI PATTERNS). (Refer to AIM). (Refer to Part 91). (See ICAO term AERODROME TRAFFIC CIRCUIT).

TRANSCRIBED WEATHER BROADCAST—A continuous recording of meteorological and aeronautical information that is broadcast on L/MF and VOR facilities for pilots.

(Refer to AIM).

TRANSFER OF CONTROL—That action whereby the responsibility for the separation of an aircraft is transferred from one controller to another.

(See ICAO term TRANSFER OF CONTROL).

TRANSFER OF CONTROL [ICAO]—Transfer of responsibility for providing air traffic control service.

TRANSFERRING CONTROLLER—A controller/facility transferring control of an aircraft to another controller/facility.

(See ICAO term TRANSFERRING UNIT / CONTROLLER).

TRANSFERRING FACILITY—(See TRANSFERRING CONTROLLER).

TRANSFERRING UNIT/CONTROLLER [ICAO]—Air traffic control unit/air traffic controller in the process of transferring the responsibility for providing air traffic control service to an aircraft to the next air traffic control unit/air traffic controller along the route of flight.

Note: See definition of *accepting unit / controller.*

TRANSITION—

1. The general term that describes the change from one phase of flight or flight condition to another; e.g., transition from en route flight to the approach or transition from instrument flight to visual flight.
2. A published procedure (SID Transition) used to connect the basic SID to one of several en route airways/jet routes, or a published procedure (STAR Transition) used to connect one of several en route airways/jet routes to the basic STAR.

(Refer to SID / STAR Charts).

TRANSITION AREA—(See CONTROLLED AIRSPACE).

TRANSITIONAL AIRSPACE—That portion of controlled airspace wherein aircraft change from one phase of flight or flight condition to another.

TRANSITION POINT—A point at an adapted number of miles from the vertex at which an arrival aircraft would normally commence descent from its en route altitude. This is the first fix adapted on the arrival speed segments.

TRANSMISSOMETER—An apparatus used to determine visibility by measuring the transmission of light through the atmosphere. It is the measurement source for determining runway visual range (RVR) and runway visibility value (RVV).

(See VISIBILITY).

TRANSMITTING IN THE BLIND—A transmission from one station to other stations in circumstances where two-way communication cannot be established, but where it is believed that the called stations may be able to receive the transmission.

TRANSPONDER—The airborne radar beacon receiver/transmitter portion of the Air Traffic Control Radar Beacon System (ATCRBS) which automatically receives radio signals from interrogators on the ground, and selectively replies with a specific reply pulse or pulse group only to those interrogations being received on the mode to which it is set to respond.

(See INTERROGATOR). (Refer to AIM). (See ICAO term TRANSPONDER).

TRANSPONDER [ICAO]—A receiver/transmitter which will generate a reply signal upon proper interrogation; the interrogation and reply being on different frequencies.

TRANSPONDER CODES—(See CODES).

TRSA—(See TERMINAL RADAR SERVICE AREA).

TURBOJET AIRCRAFT—An aircraft having a jet engine in which the energy of the jet operates a turbine which in turn operates the air compressor.

TURBOPROP AIRCRAFT—An aircraft having a jet engine in which the energy of the jet operates a turbine which drives the propeller.

TWEB—(See TRANSCRIBED WEATHER BROADCAST).

TVOR—(See TERMINAL-VERY HIGH FREQUENCY OMNIDIRECTIONAL RANGE STATION).

TWO-WAY RADIO COMMUNICATIONS FAILURE—(See LOST COMMUNICATIONS).-

U

UDF—(See DIRECTION FINDER).

UHF—(See ULTRAHIGH FREQUENCY).

ULTRAHIGH FREQUENCY—The frequency band between 300 and 3,000 mHz. The bank of radio frequencies used for military air/ground voice communications. In some instances this may go as low as 225 mHz and still be referred to as UHF.

ULTRALIGHT VEHICLE—An aeronautical vehicle operated for sport or recreational purposes which does not require FAA registration, an airworthiness certificate, nor pilot certification. They are primarily single occupant vehicles, although some two-place vehicles are authorized for training purposes. Operation of an ultralight vehicle in certain airspace requires authorization from ATC.

(See Part 103).

UNABLE—Indicates inability to comply with a specific instruction, request, or clearance.

UNCONTROLLED AIRSPACE—Uncontrolled airspace is that portion of the airspace that has not been designated as continental control area, control area, control zone, terminal control area, or transition area and within which ATC has neither the authority nor the responsibility for exercising control over air traffic.

(See CONTROLLED AIRSPACE).

UNDER THE HOOD—Indicates that the pilot is using a hood to restrict visibility outside the cockpit while simulating instrument flight. An appropriately rated pilot is required in the other control seat while this operation is being conducted.

(Refer to Part 91).

UNICOM—A nongovernment communication facility which may provide airport information at certain airports. Locations and frequencies of UNICOMs are shown on aeronautical charts and publications.

(See AIRPORT/FACILITY DIRECTORY). (Refer to AIM).

UNPUBLISHED ROUTE—A route for which no minimum altitude is published or charted for pilot use. It may include a direct route between NAVAIDS, a radial, a radar vector, or a final approach course beyond the segments of an instrument approach procedure.

(See PUBLISHED ROUTE). (See ROUTE).

UPWIND LEG—(See TRAFFIC PATTERN).

URGENCY—A condition of being concerned about safety and of requiring timely but not immediate assistance; a potential *distress* condition.

(See ICAO term URGENCY).

URGENCY [ICAO]—A condition concerning the safety of an aircraft or other vehicle, or of person on board or in sight, but which does not require immediate assistance.

USAFIB—(See ARMY AVIATION FLIGHT INFORMATION BULLETIN).

UVDF—(See DIRECTION FINDER).-

V

VASI—(See VISUAL APPROACH SLOPE INDICATOR).

VDF—(See DIRECTION FINDER).

VDP—(See VISUAL DESCENT POINT).

VECTOR—A heading issued to an aircraft to provide navigational guidance by radar.
(See ICAO term RADAR VECTORING).

VERIFY—Request confirmation of information; e.g., "verify assigned altitude."

VERIFY SPECIFIC DIRECTION OF TAKEOFF (OR TURNS AFTER TAKEOFF)—Used by ATC to ascertain an aircraft's direction of takeoff and/or direction of turn after takeoff. It is normally used for IFR departures from an airport not having a control tower. When direct communication with the pilot is not possible, the request and information may be relayed through an FSS, dispatcher, or by other means.
(See IFR TAKEOFF MINIMUMS AND DEPARTURE PROCEDURES).

VERTEX—The last fix adapted on the arrival speed segments. Normally, it will be the outer marker of the runway in use. However, it may be the actual threshold or other suitable common point on the approach path for the particular runway configuration.

VERTEX TIME OF ARRIVAL—A calculated time of aircraft arrival over the adapted vertex for the runway configuration in use. The time is calculated via the optimum flight path using adapted speed segments.

VERTICAL SEPARATION—Separation established by assignment of different altitudes or flight levels.
(See SEPARATION). (See ICAO term VERTICAL SEPARATION).

VERTICAL SEPARATION [ICAO]—Separation between aircraft expressed in units of vertical distance.

VERTICAL TAKEOFF AND LANDING AIRCRAFT—Aircraft capable of vertical climbs and/or descents and of using very short runways or small areas for takeoff and landings. These aircraft include, but are not limited to, helicopters.
(See SHORT TAKEOFF AND LANDING AIRCRAFT).

VERY HIGH FREQUENCY—The frequency band between 30 and 300 mHz. Portions of this band, 108 to 118 mHz, are used for certain NAVAIDS; 118 to 136 mHz are used for civil air/ground voice communications. Other frequencies in this band are used for purposes not related to air traffic control.

VERY HIGH FREQUENCY OMNIDIRECTIONAL RANGE STATION—(See VOR).

VERY LOW FREQUENCY—The frequency band between 3 and 30 kHz.

VFR—(See VISUAL FLIGHT RULES).

VFR AIRCRAFT—An aircraft conducting flight in accordance with visual flight rules.
(See VISUAL FLIGHT RULES).

VFR CONDITIONS—Weather conditions equal to or better than the minimum for flight under visual flight rules. The term may be used as an ATC clearance/instruction only when:

1. An IFR aircraft requests a climb/descent in VFR conditions.
2. The clearance will result in noise abatement benefits where part of the IFR departure route does not conform to an FAA approved noise abatement route or altitude.
3. A pilot has requested a practice instrument approach and is not on an IFR flight plan.—All pilots receiving this authorization must comply with the VFR visibility and distance from cloud criteria in Part 91. Use of the term does not relieve controllers of their responsibility to separate aircraft in TCAs/TRSAs as required by FAA Order 7110.65. When used as an ATC clearance/instruction, the term may be abbreviated "VFR;" e.g., "MAINTAIN VFR," "CLIMB/DESCEND VFR," etc.

VFR FLIGHT—(See VFR AIRCRAFT).

VFR MILITARY TRAINING ROUTES—Routes used by the Department of Defense and associated Reserve and Air Guard units for the purpose of conducting low-altitude navigation and tactical training under VFR below 10,000 feet MSL at airspeeds in excess of 250 knots IAS.

VFR NOT RECOMMENDED—An advisory provided by a flight service station to a pilot during a preflight or inflight weather briefing that flight under visual flight rules is not recommended. To be given when the current and/or forecast weather conditions are at or below VFR minimums. It does not abrogate the pilot's authority to make his own decision.

VFR-ON-TOP—ATC authorization for an IFR aircraft to operate in VFR conditions at any appropriate VFR altitude (as specified in FAR and as restricted by ATC). A pilot receiving this authorization must comply with the VFR visibility, distance from cloud criteria, and the minimum IFR altitudes specified in Part 91. The use of this term does not relieve controllers of their responsibility to separate aircraft in TCA's/TRSA's as required by FAA Order 7110.65.

VFR TERMINAL AREA CHARTS—(See AERONAUTICAL CHART).

VHF—(See VERY HIGH FREQUENCY).

VHF OMNIDIRECTIONAL RANGE/TACTICAL AIR NAVIGATION—(See VORTAC).

VIDEO MAP—An electronically displayed map on the radar display that may depict data such as airports, heliports, runway centerline extensions, hospital emergency landing areas, NAVAID's and fixes, reporting points, airway/route centerlines, boundaries, handoff points, special use tracks, obstructions, prominent geographic features, map alignment indicators, range accuracy marks, minimum vectoring altitudes.

VISIBILITY—The ability, as determined by atmospheric conditions and expressed in units of distance, to see and identify prominent unlighted objects by day and prominent lighted objects by night. Visibility is reported as statute miles, hundreds of feet or meters.

(Refer to Part 91). (See AIM).

1. Flight Visibility. The average forward horizontal distance, from the cockpit of an aircraft in flight, at which prominent unlighted objects may be seen and identified by day and prominent lighted objects may be seen and identified by night.
2. Ground Visibility. Prevailing horizontal visibility near the earth's surface as reported by the United States National Weather Service or an accredited observer.
3. Prevailing Visibility. The greatest horizontal visibility equaled or exceeded throughout at least half the horizon circle which need not necessarily be continuous.
4. Runway Visibility Value (RVV). The visibility determined for a particular runway by a transmissometer. A meter provides a continuous indication of the visibility (reported in miles or fractions of miles) for the runway. RVV is used in lieu of prevailing visibility in determining minimums for a particular runway.
5. Runway Visual Range (RVR). An instrumentally derived value, based on standard calibrations, that represents the horizontal distance a pilot will see down the runway from the approach end. It is based on the sighting of either high intensity runway lights or on the visual contrast of other targets whichever yields the greater visual range. RVR, in contrast to prevailing or runway visibility, is based on what a pilot in a moving aircraft should see looking down the runway. RVR is horizontal visual range, not slant visual range. It is based on the measurement of a transmissometer made near the touchdown point of the instrument runway and is reported in hundreds of feet. RVR is used in lieu of RVV and/or prevailing visibility in determining minimums for a particular runway.
   a. Touchdown RVR. The RVR visibility readout values obtained from RVR equipment serving the runway touchdown zone.
   b. Mid-RVR. The RVR readout values obtained from RVR equipment located midfield of the runway.
   c. Rollout RVR. The RVR readout values obtained from RVR equipment located nearest the rollout end of the runway.

(See ICAO term VISIBILITY). (See ICAO term FLIGHT VISIBILITY). (See ICAO term GROUND VISIBILITY). (See ICAO term RUNWAY VISUAL RANGE).

VISIBILITY [ICAO]—The ability, as determined by atmospheric conditions and expressed in units of distance, to see and identify prominent unlighted objects by day and prominent lighted objects by night.

Flight Visibility.—The visibility forward from the cockpit of an aircraft in flight.

Ground Visibility.—The visibility at an aerodrome as reported by an accredited observer.

Runway Visual Range [RVR].—The range over which the pilot of an aircraft on the centre line of a runway can see the runway surface markings or the lights delineating the runway or identifying its centre line.

VISUAL APPROACH—An approach wherein an aircraft on an IFR flight plan, operating in VFR conditions under the control of an air traffic control facility and having an air traffic control authorization, may proceed to the airport of destination in VFR conditions.

(See ICAO term VISUAL APPROACH).

VISUAL APPROACH [ICAO]—An approach by an IFR flight when either part or all of an instrument approach procedure is not completed and the approach is executed in visual reference to terrain.

VISUAL APPROACH SLOPE INDICATOR—(See AIRPORT LIGHTING).

VISUAL DESCENT POINT—A defined point on the final approach course of a nonprecision straight-in approach procedure from which normal descent from the MDA to the runway touchdown point may be commenced, provided the approach threshold of that runway, or approach lights, or other markings identifiable with the approach end of that runway are clearly visible to the pilot.

VISUAL FLIGHT RULES—Rules that govern the procedures for conducting flight under visual conditions. The term "VFR" is also used in the United States to indicate weather conditions that are equal to or greater than minimum VFR requirements. In addition, it is used by pilots and controllers to indicate type of flight plan.

(See INSTRUMENT FLIGHT RULES). (See INSTRUMENT METEOROLOGICAL CONDITIONS). (See VISUAL METEOROLOGICAL CONDITIONS). (Refer to Part 91). (Refer to AIM).

VISUAL HOLDING—The holding of aircraft at selected, prominent geographical fixes which can be easily recognized from the air.

(See HOLDING FIX).

VISUAL METEOROLOGICAL CONDITIONS—Meteorological conditions expressed in terms of visibility, distance from cloud, and ceiling equal to or better than specified minima.

(See INSTRUMENT FLIGHT RULES). (See INSTRUMENT METEOROLOGICAL CONDITIONS). (See VISUAL FLIGHT RULES).

VISUAL SEPARATION—A means employed by ATC to separate aircraft in terminal areas. There are two ways to effect this separation:

1. The tower controller sees the aircraft involved and issues instructions, as necessary, to ensure that the aircraft avoid each other.
2. A pilot sees the other aircraft involved and upon instructions from the controller provides his own separation by maneuvering his aircraft as necessary to avoid it. This may involve following another aircraft or keeping it in sight until it is no longer a factor.

(See and Avoid). (Refer to Part 91).

VLF—(See VERY LOW FREQUENCY).

VMC—(See VISUAL METEOROLOGICAL CONDITIONS).

VOR—A ground-based electronic navigation aid transmitting very high frequency navigation signals, 360 degrees in azimuth, oriented from magnetic north. Used as the basis for navigation in the National Airspace System. The VOR periodically identifies itself by Morse Code and may have an additional voice identification feature. Voice features may be used by ATC or FSS for transmitting instructions/information to pilots.

(See NAVIGATIONAL AID). (Refer to AIM).

VORTAC—A navigation aid providing VOR azimuth, TACAN azimuth, and TACAN distance measuring equipment (DME) at one site.

(See DISTANCE MEASURING EQUIPMENT). (See NAVIGATIONAL AID). (See TACAN). (See VOR). (Refer to AIM).

VORTICES—Circular patterns of air created by the movement of an airfoil through the air when generating lift. As an airfoil moves through the atmosphere in sustained flight, an area of area of low pressure is created above it. The air flowing from the high pressure area to the low pressure area around and about the tips of the airfoil tends to roll up into two rapidly rotating vortices, cylindrical in shape. These vortices are the most predominant parts of aircraft wake turbulence and their rotational force is dependent upon the wing loading, gross weight, and speed of the generating aircraft. The vortices from medium to heavy aircraft can be of extremely high velocity and hazardous to smaller aircraft.

(See AIRCRAFT CLASSES). (See WAKE TURBULENCE). (Refer to AIM).

VOR TEST SIGNAL—(See VOT).

VOT—A ground facility which emits a test signal to check VOR receiver accuracy. Some VOT's are available to the user while airborne, and others are limited to ground use only.
(Refer to Part 91). (See AIM). (See AIRPORT/FACILITY DIRECTORY).

VR—(See VFR MILITARY TRAINING ROUTES).

VTA—(See VERTEX TIME OF ARRIVAL).

VTOL AIRCRAFT—(See VERTICAL TAKEOFF AND LANDING AIRCRAFT).

W

WA—(See AIRMET).
(See WEATHER ADVISORY).

WAKE TURBULENCE—Phenomena resulting from the passage of an aircraft through the atmosphere. The term includes vortices, thrust stream turbulence, jet blast, jet wash, propeller wash, and rotor wash both on the ground and in the air.
(See AIRCRAFT CLASSES). (See JET BLAST). (See VORTICES). (Refer to AIM).

WARNING AREA—(See SPECIAL USE AIRSPACE).

WAYPOINT—A predetermined geographical position used for route/instrument approach definition, or progress reporting purposes, that is defined relative to a VORTAC station or in terms of latitude/longitude coordinates.

WEATHER ADVISORY—In aviation weather forecast practice, an expression of hazardous weather conditions not predicted in the area forecast, as they affect the operation of air traffic and as prepared by the NWS.
(See SIGMET). (See AIRMET).

WHEN ABLE—When used in conjunction with ATC instructions, gives the pilot the latitude to delay compliance until a condition or event has been reconciled. Unlike "pilot discretion," when instructions are prefaced "when able," the pilot is expected to seek the first opportunity to comply. Once a maneuver has been initiated, the pilot is expected to continue until the specifications of the instructions have been met. "When able," should not be used when expeditious compliance is required.

WILCO—I have received your message, understand it, and will comply with it.

WIND SHEAR—A change in wind speed and/or wind direction in a short distance resulting in a tearing or shearing effect. It can exist in a horizontal or vertical direction and occasionally in both.

WING TIP VORTICES—(See VORTICES).

WORDS TWICE—

1. As a request: "Communication is difficult. Please say every phrase twice."
2. As information: "Since communications are difficult, every phrase in this message will be spoken twice."

WORLD AERONAUTICAL CHARTS—(See AERONAUTICAL CHART).

WS—(See SIGMET).
(See WEATHER ADVISORY).

WST—(See CONVECTIVE SIGMET).
(See WEATHER ADVISORY).

INDEX

References are to Paragraph numbers

References are to Paragraph numbers

References are to Paragraph numbers

References are to Paragraph numbers

References are to Paragraph numbers

Free Update Service
for Airman's Information Manual (AIM)

Fill out this coupon and mail it to ASA in an envelope. Send it in right away to be sure you receive the 1993 FAR-AIM Midyear Update. The update will include information current through June 30, 1993, and will be mailed to you in mid August.

This will be your mailing label, so please print your name and address clearly.

Please send me ASA's 1993 FAR-AIM Midyear Update:

Name ______________________________

Address ______________________________

City ________________ **State** ________ **Zip** ________

Mail this coupon to:

Aviation Supplies & Academics, Inc.
7005 132nd Place SE
Renton, WA 98059-3153

AIM '93